Arterial Chemoreception

Advances in Experimental Medicine and Biology

For further volumes:
http://www.springer.com/series/5584

Colin A. Nurse • Constancio Gonzalez • Chris Peers
Nanduri R. Prabhakar
Editors

Arterial Chemoreception

From Molecules to Systems

Springer

Editors
Colin A. Nurse
Department of Biology
McMaster University
Hamilton, ON, Canada

Chris Peers
Division of Cardiovascular and Neuronal Remodelling
Faculty of Medicine and Health
University of Leeds
Leeds, Yorkshire, UK

Constancio Gonzalez
Departamento de Bioquímica y Biología Molecular y Fisiología
Facultad de Medicina
Universidad de Valladolid
Valladolid, Spain

Nanduri R. Prabhakar
Department of Medicine
Biological Sciences Division
University of Chicago
Chicago, IL, USA

ISSN 0065-2598
ISBN 978-94-007-4583-4 ISBN 978-94-007-4584-1 (eBook)
DOI 10.1007/978-94-007-4584-1
Springer Dordrecht Heidelberg New York London

Library of Congress Control Number: 2012947999

Printed on acid-free paper

Springer is part of Springer Science+Business Media (www.springer.com)

Preface

This volume is based on the Proceedings of the XVIII ISAC Meeting which was held at McMaster University, Hamilton, Ontario, Canada, from July 10–15, 2011. It contains the combined contributions of many of the attendees who participated in the scientific sessions. The meeting was launched in the evening of July 10 with a plenary lecture given by one of Canada's 2010 Gairdner Award Recipients, Prof. Gregg Semenza, in the beautiful confines of the Art Gallery of Hamilton. For most of the following 5 days, the meeting shifted to the Michael De Groote Centre for Learning and Discovery on the McMaster Campus, where attendees actively participated in a series of five thematic symposia comprising a series of ~20 min oral presentations. Each symposium was preceded by a keynote lecture given by one of our invited speakers who kindly accepted our offer of participation. Among the keynote speakers were Profs. Naweed Syed, Nino Ramirez, Patrice Guyenet, Ellis Cooper, and Jerome Dempsey. Poster presentations were available for viewing throughout the meeting and the lead author on each poster had the opportunity to summarize their key findings in a brief oral format. We wish to thank our many long-standing ISAC contributors for their continued support of these meetings, collegiality, and suggestions for future improvements. We also welcome the many first-time ISAC participants, and hope that the experience was sufficiently rewarding so as to entice them to attend future meetings.

The success of the XVIII ISAC Meeting owes much to the collaborative efforts of many individuals. The Local Organizing Committee was especially appreciative of the advice received during the planning stages from Prof. Nanduri Prabhakar from Chicago, USA, Prof. Constancio Gonzalez from Valladolid, Spain, Prof. Prem Kumar from Birmingham UK, Prof. Chris Peers, Leeds, UK, and Prof. Ernest Cutz, Toronto, Canada. The assistance of Prof. Prabhakar in organizing and planning the scientific sessions, in selecting the keynote speakers, and in the important task of fund raising, was greatly appreciated and contributed significantly to the success of the meeting.

The smooth planning and organization of the meeting, including the social events, would not have been possible without the help and professionalism of the local organizing committee. We sincerely thank Nikol Piskuric, Cathy Vollmer, Karen Haines, Simon Livermore, Shaima Salman, Min Zhang, Dianne Carment, Wendy Reid, and Paula Nurse for their contributions in various capacities. A special "thank you" is extended to Nikol Piskuric for her exceptional organizational skills and energy that was in evidence throughout the planning stages, and over the course of the entire meeting.

The Heymans-De Castro-Neil Awards for young investigators were selected by Profs. P. Kumar, P. Zapata, and M. Pokorski, and the awardees were Ana Rita Nunes (Portugal), Erica Wehrwein (Rochester, USA), and Noah Marcus (Omaha, USA); in addition, Shaima Salman (Hamilton, Ontario) and Edward O'Connor (Dublin, Ireland) received honorable mention. We extend our congratulations to the recipients and wish them continued success in their future careers.

The XVIII ISAC Meeting was generously supported from several sources including McMaster University, Canadian Institutes of Health Research, Galleon Pharmaceuticals, Inc., Sova Pharmaceuticals, Inc., and Cedarlane Labs. We are extremely grateful to all of them. In addition, we would like to thank Marlies Vlot and Max Haring of Springer for their expert management in ensuring timely production of this volume.

At the business meeting, it was decided that the next ISAC Symposium will be held in Leeds, UK, in 2014, with Prof. Chris Peers as ISAC and Meeting President.

Hamilton, Canada — Colin Nurse
Valladolid, Spain — Constancio Gonzalez
Leeds, UK — Chris Peers
Chicago, USA — Nanduri R. Prabhakar

A Tribute to Professor Sukhamay Lahiri: Past President of International Society for Arterial Chemoreceptors (ISAC)

Sukhamay Lahiri

Professor Sukhamay Lahiri, past president of ISAC, and a leading authority on physiology of high altitude and carotid body oxygen sensing died May 2, 2009 of prostate cancer in Philadelphia. Born in Natore now in Bangladesh, he earned his bachelor's with physiology honors from Presidency College in Calcutta in 1951, master's in 1953, and a doctorate in physiology in 1956 as a Jubilee scholar. Later, with a scholarship from the state of Bengal, he went to the University of Oxford where he obtained a second doctorate in physiology in 1959. He came to USA to join the Downstate Medical Center in New York City in 1965 and worked there till 1967. He then moved to the cardiovascular department at Michael Reese Hospital in Chicago. In 1969, he went to the University of Pennsylvania as a faculty member in the department of physiology, where he remained as Professor of Physiology until his death.

Professor Lahiri made several seminal contributions to the areas of high altitude as well as carotid body physiology. Dr. Lahiri's initial studies were focused on physiological responses to high altitude on Mount Everest, which he climbed for the first time in the 1960 expedition led by Sir Edmund Hillary and made his final trip to Mount Everest in 1981. His studies on Andes showed that birth at high altitude leads to blunting of breathing responses to hypoxia, which he later confirmed in children born with cyanotic congenital heart disease. Because of his interest in hypoxia and high altitude, in early 1970s he began work on O_2 sensing by arterial chemoreceptors, especially the carotid body. He was one of the first to demonstrate using "single" fiber approach interactions of hypoxia with hypercapnia at the carotid body. His work further demonstrated fundamental differences between the behavior of aortic and carotid body responses to hypoxia leading to the proposal that aortic chemoreceptors are more sensitive to oxygen content and carotid bodies to partial pressure of oxygen. Using inhibitors of oxidative phosphorylation he demonstrated that hypoxia and carbon dioxide are independently sensed by the carotid bodies. He also examined various theories of oxygen sensing and proposed a role for cytochrome a3 in oxygen sensing. Despite his ill health, he continued his scientific work until just before he died. I remember vividly Dr. Lahiri presenting his poster at the Experimental Biology meeting in April 2009, shortly before his death in May 2009. He wrote over 150 scientific papers and edited many books. He was the recipient of the prestigious NIH MERIT award as well as the Humboldt award from Germany to a senior scientist. Late Prof. Neil Cherniack described Dr. Lahiri ..."for decades arguably the most prominent scientist in area of physiology of oxygen homeostasis."

Professor Lahiri was a driving force in promoting the field of hypoxia and carotid body physiology. Nearly four decades ago, Dr. Lahiri was responsible in establishing the "Hypoxia Interest Group" at the American Physiological Society (APS), whose activities I had the privilege of overlooking for over a decade. It continues to be one of the important cross-disciplinary interest groups in the APS and has over 1,000 memberships. Since its inception, Dr. Lahiri was active and an important member of ISAC. He was the president of ISAC from 1998 to 2000 and organized the ISAC meeting held at Philadelphia in 1999. I was the co-organizer of this meeting and had the privilege of working with him closely in this endeavor. Philadelphia meeting was a great success and attended by nearly 300 delegates from various areas of oxygen biology resulting in ~850 pages of book volume entitled "*Oxygen Sensing: Molecule to Man*." I recall it was at this meeting a proposal was made that Prof. Colin Nurse will host ISAC meeting in 2010 in Hamilton. Dr. Lahiri attended his last ISAC meeting held in Valladolid (Spain) on July 2008. Despite his ill health, he still had admirable optimism, and even proposed Philadelphia again as a venue for one of the forth coming ISAC meetings.

Dr. Lahiri was a very modest and generous human being. Besides physiology, he had a great passion for art and photography. In fact he was a serious and excellent photographer. One should see his collection of Himalayan photographs... truly breath taking. He was an active member of committees for Indian Art for Philadelphia Museum. In all his endeavors, he was supported by his wife, Mrs. Krishna Lahiri, an amazing lady with compassion, who is his only survivor. He left many scientific heirs, hundreds of students, collaborators, and admirers who will miss him greatly. I take this opportunity to express my sincere gratitude to Prof. Camillo Di Giulio, a longstanding collaborator and a close friend of Prof. Lahiri for providing me with some of the information included in this book.

Institute for integrative Physiology and
The Centre for Systems Biology of O_2 Sensing
University of Chicago
Chicago

Nanduri R. Prabhakar

Contents

Contributors

Sebastián Abarzúa Departamento de Ciencias Biologicas, Facultad de Ciencias Biologicas, Universidad Andres Bello, Santiago, Chile

Stephen B.G. Abbott Department of Pharmacology, University of Virginia, Charlottesville, VA, USA

Sara J. Abdallah Department of Biology, University of Ottawa, Ottawa, ON, Canada

Dirk Adriaensen Department of Veterinary Sciences, Laboratory of Cell Biology and Histology, University of Antwerp, Antwerp, Belgium

Teresa Agapito Department of Biochemistry and Molecular Biology and Physiology, IBGM Universidad de Valladolid and CSIC, Valladolid, Spain

Ciber de Enfermedades Respiratorias, Valladolid, Spain

Amit Agarwal University of Arkansas for Medical Sciences, Department of Pediatrics, Division of Pediatric Pulmonary Medicine, Little Rock, AR, USA

Julio Alcayaga Laboratorio de Fisiología Celular, Facultad de Ciencias Biológicas, P Universidad Católica de Chile, Santiago, Chile

Laura Almaraz Instituto de Neurociencias, Universidad Miguel Hernández-CSIC, San Juan de Alicante, Spain

Justyna Antosiewicz Laboratory of Molecular Pharmacology, Drug Research and Development Center (CIDEM), Havana, Cuba

Reyes Ávalos Facultad de Medicina, Universidad de Colima, Colima, Mexico

Aida Bairam Department of pediatrics, Unité de Recherche en Périnatologie, Centre de Recherche, D0-717, Hôpital Saint-François d'Assise, Laval University, Québec, Qc, Canada

Klaus Ballanyi Department of Physiology, Faculty of Medicine and Dentistry, University of Alberta, Edmonton, AB, Canada

Joan Albert Barbera Department of Pulmonary Medicine Hospital Clínic-Institut d'Investigacions Biomèdiques August Pi i Sunyer (IDIBAPS), Universitat de Barcelona, Valladolid, Spain

Ciber de Enfermedades Respiratorias, Valladolid, Spain

Barbara L. Barr Department of Neuroscience, Cell Biology and Physiology, Wright State University, Dayton, OH, USA

Ananda Basu Endocrine Research Unit, Division of Endocrinology, Diabetes, Metabolism and Nutrition, Mayo Clinic College of Medicine, Rochester, MN, USA

Rita Basu Endocrine Research Unit, Division of Endocrinology, Diabetes, Metabolism and Nutrition, Mayo Clinic College of Medicine, Rochester, MN, USA

Tammie Bishop The Henry Wellcome Building for Molecular Physiology, University of Oxford, Oxford, UK

Gregory M. Blain Departments of Population Health Sciences, University of Wisconsin – Madison, Madison, WI, USA

Aidan Bradford Department of Physiology and Medical Physics, Royal College of Surgeons in Ireland, Dublin 2, Ireland

Inge Brouns Department of Veterinary Sciences, Laboratory of Cell Biology and Histology, University of Antwerp, Antwerp, Belgium

Leslie T. Buck Department of Cell and Systems Biology, University of Toronto, Toronto, ON, Canada

Department of Ecology and Evolutionary Biology, University of Toronto, Toronto, ON, Canada

Keith Buckler Department of Physiology Anatomy & Genetics, Oxford, United Kingdom

A. Cataldi Department of Medicine and Aging Sciences, University of Chieti, Chieti, Italy

Céline Caravagna Department of Pediatrics, Centre de Recherche de l'Hôpital St-François d'Assise (CR-SFA), Centre Hospitalier Universitaire de Québec (CHUQ), Faculty of Medicine, Laval University, Québec, QC, Canada

John L. Carroll Pediatric Pulmonary Medicine, Arkansas Children's Hospital, Little Rock, AR, USA

University of Arkansas for Medical Sciences, Department of Pediatrics, Division of Pediatric Pulmonary Medicine, Little Rock, AR, USA

Andy Chang Department of Biochemistry, Stanford University, Stanford, CA, USA

Irene Chang Department of Environmental Health Sciences, Bloomberg School of Public Health, The Johns Hopkins University, Baltimore, MD, USA

Silvia V. Conde CEDOC Departamento de Farmacologia, Faculdade de Ciências Médicas Universidade Nova de Lisboa, Lisbon, Portugal

Ellis Cooper Department of Physiology, McGill University, Montreal, Quebec, Canada

Paula P. Cortés Departamento de Ciencias Biologicas, Facultad de Ciencias Biologicas, Universidad Andres Bello, Santiago, Chile

Ricardo Cuéllar Centro Universitario de Investigaciones Biomédicas y, Universidad de Colima, Colima, Col, Mexico

Timothy B. Curry Departement of Anesthesiology, Mayo Clinic, Rochester, MN, USA

Ernest Cutz Division of Pathology, Department of Pediatric Laboratory Medicine, The Reserch Institute, The Hospital for Sick Children, Toronto, ON, Canada

Raffaele De Caro Department of Human Anatomy and Physiology, University of Padova, Padova, Italy

Rodrigo Del Rio Laboratorio de Neurobiología, Facultad de Ciencias Biológicas, Pontificia Universidad Católica de Chile, Santiago, Region Metropolitana, Chile

Rene Delgado Laboratory of Molecular Pharmacology, Drug Research and Development Center (CIDEM), Havana, Cuba

Jerome A. Dempsey Departments of Population Health Sciences, University of Wisconsin – Madison, Madison, WI, USA

Seth D. Depuy Department of Pharmacology, University of Virginia, Charlottesville, VA, USA

Elena Roces de Álvarez-Buylla Centro Universitario de Investigaciones Biomédicas y, Universidad de Colima, Colima, Col, Mexico

J.M. Di Fiore Department of Pediatrics, Division of Neonatology, Rainbow Babies & Children's Hospital Case Western Reserve University, Cleveland, OH, USA

Camillo Di Giulio Department of Neuroscience and Imaging, University of Chieti-Pescara, 'G. d'Annunzio', Chieti, CH, Italy

Van Diep Doan Department of Pediatrics, Laval University, Centre de Recherche (D0-711), Hôpital St-François d'Assise, Quebec, QC, Canada

Nicolle Domnik Department of Physiology, Queen's University, Kingston, ON, Canada

David F. Donnelly Department of Pediatrics, Yale University, School of Medicine, New Haven, CT, USA

Deirdre Edge UCD School of Medicine and Medical Science, University College Dublin, Dublin 4, Ireland

A. Mark Evans Centre for Integrative Physiology, College of Medicine and Veterinary Medicine, University of Edinburgh, Midlothian, UK

Lidia Faff Department of Respiratory Research, Medical Research Center, Polish Academy of Sciences, Warsaw, Mazovian, Poland

Ricardo Fernández Departamento de Ciencias Biologicas, Facultad de Ciencias Biologicas, Universidad Andres Bello, Santiago, Chile

Facultad de Ciencias Biologicas y Facultad de Medicina, Escuela de Medicina, Universidad Andres Bello, Lo Barrenechea, Santiago, Metropolitana, Chile

Elisabet Ferrer Department of Pulmonary Medicine Hospital Clínic-Institut d'Investigacions Biomèdiques August Pi i Sunyer (IDIBAPS), Universitat de Barcelona, Valladolid, Spain

John T. Fisher Department of Biomedical and Molecular Sciences, Paediatrics, Medicine, Queen's University, Kingston, ON, Canada

Robert S. Fitzgerald Departments of Environmental Health Sciences, of Physiology, and of Medicine, The Johns Hopkins Medical Institutions

Brian Foo Department of Biomedical and Molecular Sciences, Queen's University, Kingston, ON, Canada

Stéphanie Fournier Department of Pediatrics, Laval University, Centre de Recherche (D0-711), Hôpital St-François d'Assise, Quebec, QC, Canada

Matías Freire Laboratorio de Fisiología Celular, Facultad de Ciencias, Universidad de Chile, Santiago, Chile

Man Lung Fung Departments of Physiology and Anatomy, University of Hong Kong, Hong Kong, China

Teresa Gallego-Martin Department of Biochemistry Molecular Biology and Physiology, IBGM School of Medicine, University of Valladolid, Valladolid, Spain

Estelle Gauda Department of Pediatrics, Johns Hopkins University, Baltimore, MD, USA

Susana P. Gaytan Department of Physiology and Zoology, University of Seville, Seville, Spain

M. Angela Gomez-Niño CIBER Enfermedades Respiratorias, Valladolid, Spain

Department of Biochemistry and Molecular Biology and Physiology, and IBGM, Universidad de Valladolid and CSIC, Valladolid, Spain

Yuansheng Gong Departments of Population Health Sciences, University of Wisconsin – Madison, Madison, WI, USA

Constancio Gonzalez Departamento de Bioquímica y Biología Molecular y Fisiología, Facultad de Medicina, School of Medicine, Universidad de Valladolid, Valladolid, Spain

Carmen Gonzalez-Martin Department of Biochemistry and Molecular Biology and Physiology, IBGM Universidad de Valladolid and CSIC, Valladolid, Spain

Ciber de Enfermedades Respiratorias, Valladolid, Spain

Elvira Gonzalez-Obeso Department of Biochemistry and Molecular Biology and Physiology, IBGM Universidad de Valladolid and CSIC, Valladolid, Spain

Department of Pulmonary Medicine Hospital Clínic-Institut d'Investigacions Biomèdiques August Pi i Sunyer (IDIBAPS), Universitat de Barcelona, Valladolid, Spain

Patrice G. Guyenet Department of Pharmacology, University of Virginia, Charlottesville, VA, USA

Yasushi Hara Division of Cardiology and Pulmonary Circulation, Department of Medicine, National Cardiovascular Center, Suita, Osaka, Japan

D. Grahame Hardie College of Life Sciences, University of Dundee, Dundee, UK

David Hauton School of Clinical and Experimental Medicine, The Medical School, University of Birmingham, Birmingham, West Midlands, UK

Oubeidallah Hichri Unité de Recherche en Périnatologie, Département de Pédiatrie, Centre Hospitalier Universitaire de Québec, Hôpital Saint-François d'Assise, Université Laval, Québec, QC, Canada

D.W.R Hogg Department of Cell and Systems Biology, University of Toronto, Toronto, ON, Canada

Andrew P.S. Holmes School of Clinical and Experimental Medicine, The Medical School, University of Birmingham, Birmingham, West Midlands, UK

Rodrigo Iturriaga Laboratorio de Neurobiología, Facultad de Ciencias Biológicas, Pontificia Universidad Católica de Chile, Santiago, Region Metropolitana, 22, Chile

Tara A. Janes Department of Cell Biology and Anatomy and the Hotchkiss Brain Institute, Faculty of Medicine, University of Calgary, Calgary, AB, Canada

James F.X. Jones School of Medicine and Medical Science, University College Dublin, Belfield, Dublin 4, Ireland

Michael G. Jonz Department of Biology, University of Ottawa, Ottawa, ON, Canada

Heidi Jordan Wright State University, Dayton, OH, USA

Vincent Joseph Département de Pédiatrie, Unité de Recherche en Périnatologie, Centre Hospitalier Universitaire de Québec (D0-711), Hôpital Saint-François d'Assise, Université Laval, Québec, QC, Canada

CRCHUQ/Hop Laval, Québec, Canada

Michael J. Joyner Department of Anesthesiology, Mayo Clinic, Rochester, MN, USA

Cécile A. Julien Department of pediatrics, Unité de Recherche en PérinatologieCentre de Recherche St-François d'Assise, Hôpital Saint-François d'Assise, Laval University, Québec, QC, Canada

Roy Kanbar Department of Pharmacology, University of Virginia, Charlottesville, VA, USA

Hanan Khemiri Department of Pediatrics, Centre de Recherche de l'Hôpital St-François d'Assise (CR-SFA), Centre Hospitalier Universitaire de Québec (CHUQ), Faculty of Medicine, Laval University, Québec, QC, Canada

Samara Kiihl Department of Biostatistics, Bloomberg School of Public Health, The Johns Hopkins University, Baltimore, MD, USA

Insook Kim University of Arkansas for Medical Sciences, Department of Pediatrics, Division of Pediatric Pulmonary Medicine, Little Rock, AR, USA

Richard Kinkead Department of pediatrics, Unité de Recherche en PérinatologieCentre de Recherche St-François d'Assise, Centre de Recherche St-François d'Assise, Hôpital Saint-François d'Assise, Laval University, Québec, QC, Canada

Eric W. Kostuk Department of Environmental Health Sciences, Bloomberg School of Public Health, The Johns Hopkins University, Baltimore, MD, USA

Ganesh K. Kumar Department of Medicine, Institute for Integrative Physiology, University of Chicago, Chicago, IL, USA

Prem Kumar School of Clinical and Experimental Medicine, University of Birmingham, Birmingham, West Midlands, UK

Shingo Kyotani Division of Cardiology and Pulmonary Circulation, Department of Medicine, National Cardiovascular Center, Suita, Osaka, Japan

S.Y. Lam Departments of Physiology and Anatomy, University of Hong Kong, Hong Kong SAR, China

Carolina Larraín Facultad de Medicina, Clínica Alemana, Universidad del Desarrollo, Lo Barnechea, Santiago, Region Metropolitana, Chile

Jean-C Laurin Département de Pédiatrie, Unité de Recherche en Périnatologie, Centre Hospitalier Universitaire de Québec, Hôpital Saint-François d'Assise, Université Laval, Québec, QC, Canada

Raphael Lavoie Université Laval

Robrecht Lembrechts Department of Veterinary Sciences, Laboratory of Cell Biology and Histology, University of Antwerp, Antwerp, Belgium

Mónica Lemus Centro Universitario de Investigaciones Biomédicas y, Universidad de Colima, Colima, Col, Mexico

E.C. Liong Departments of Physiology and Anatomy, University of Hong Kong, Hong Kong SAR, China

Y. Liu Departments of Physiology and Anatomy, University of Hong Kong, Hong Kong SAR, China

Simon Livermore Department of Biology, McMaster University, Hamilton, ON, Canada

Veronica Macchi Department of Human Anatomy and Physiology, University of Padua, Padova, Italy

P.W. MacFarlane Department of Pediatrics, Division of Neonatology, Rainbow Babies & Children's Hospital Case Western Reserve University, Cleveland, OH, USA

Amira Mahmoud Centre of Integrative Physiology, University of Edinburgh, Edinburgh, Midlothian, United Kingdom

Ludwik K. Malendowicz Department of Histology and Embryology, Poznan University of Medical Sciences, Poznan, Poland

Noah J. Marcus Department of Cellular and Integrative Physiology, University of Nebraska Medical Center, Omaha, NE, USA

Aldo Martin Departamento de Ciencias Biologicas, Facultad de Ciencias Biologicas, Universidad Andres Bello, Santiago, Chile

R.J. Martin Department of Pediatrics Division of Neonatology, Rainbow Babies & Children's Hospital Case Western Reserve University, Cleveland, OH, USA

Sergio Montero Centro Universitario de Investigaciones Biomédicas y, Universidad de Colima, Colima, Col, MexicoFacultad de Medicina, Universidad de Colima, Colima, Mexico

Emília C. Monteiro CEDOC, Departamento de Farmacologia, Faculdade de Ciências Médicas Universidade Nova de Lisboa, Lisbon, Portugal

Sergio Montero Centro Universitario de Investigaciones Biomédicas y, Universidad de Colima, Colima, Col, MexicoFacultad de Medicina, Universidad de Colima, Colima, Mexico

Jose M. Montserrat CIBER Enfermedades Respiratorias, Valladolid, SpainLaboratori de la Son, Pneumologia, Hospital Clínic-IDIBAPS, Barcelona, Spain

Esteban A. Moya Laboratorio de Neurobiología, Facultad de Ciencias Biológicas, P. Universidad Católica de Chile, Santiago, Region Metropolitana, Chile

Noritoshi Nagaya Division of Cardiology and Pulmonary Circulation, Department of Medicine, National Cardiovascular Center, Suita, Osaka, Japan

Norifumi Nakanishi Division of Cardiology and Pulmonary Circulation, Department of Medicine, National Cardiovascular Center, Suita, Osaka, Japan

Lalah M. Niane Pediatrics Department, Centre de Recherche CHUQ-HSFA, Laval University, Québec, QC, Canada

Ana Rita Nunes Department of Pediatrics, Johns Hopkins University, Baltimore, MD, USA

CEDOC, Departamento de Farmacologia, Faculdade de Ciências Médicas, Universidade NOVA de Lisboa, Lisbon, Portugal

Colin A. Nurse Department of Biology, McMaster University, Hamilton, ON, Canada

Ana Obeso Universidad de Valladolid, Valladolid, Spain

Department of Biochemistry and Molecular Biology and Physiology and IBGM, CIBERES-Institute Carlos III, University of Valladolid School of Medicine and CSIC, Valladolid, Spain

Edward T. O'Connor School of Medicine and Medical Science, University College Dublin, Belfield, Dublin 4, Ireland

Ken D. O'Halloran School of Medicine and Medical Science, University College Dublin, Belfield, Dublin 4, Ireland

Yasumasa Okada Department of Medicine, Keio University Tsukigase Rehabilitation Center, Amagiyugashima-cho, Japan

Elena Olea Department of Biochemistry and Molecular Biology and Physiology and IBGM, CIBERES-Institute Carlos III, University of Valladolid School of Medicine and CSIC, Valladolid, Spain

Fernando C. Ortiz Laboratorio de Neurobiología, Facultad de Ciencias Biológicas, Pontificia Universidad Católica de Chile, Santiago, Chile

M.E. Pamenter Department of Pediatrics (Division of Respiratory Medicine), University of California San Diego, La Jolla, CA, USA

Jie Pan Division of Pathology, Department of The Paediatric Laboratory Medicine Research Institute, The Hospital for Sick Children, and Department of Laboratory Medicine and Pathobiology, University of Toronto, Toronto, ON, Canada

Rosario Pasaro Department of Physiology and Zoology, University of Seville, Seville, Spain

Pejmon Pashai Division of Physiology, Department of Environmental Health Sciences, Bloomberg School of Public Health, Johns Hopkins University, Baltimore, MD, USA

Chris Peers Faculty of Medicine and Health, University of Leeds, Leeds, Yorkshire, UK

Victor Peinado Department of Pulmonary Medicine Hospital Clínic-Institut d'Investigacions Biomèdiques August Pi i Sunyer (IDIBAPS), Universitat de Barcelona, Valladolid, Spain

Ciber de Enfermedades Respiratorias, Valladolid, Spain

Steve F. Perry Department of Biology, University of Ottawa, Ottawa, ON, Canada

Luis E. Pilchard Division of Physiology, Department of Environmental Health Sciences, Bloomberg School of Public Health, Johns Hopkins University, Baltimore, MD, USA

Isabel Pintelon Department of Veterinary Sciences, Laboratory of Cell Biology and Histology, University of Antwerp, Antwerp, Belgium

Nikol A. Piskuric Department of Biology, McMaster University, Hamilton, ON, Canada

Mieczyslaw Pokorski Department of Respiratory Research, Medical Research Center, Polish Academy of Sciences, Warsaw, Mazovian, Poland

Department of Psychology, Institute of Psychology, Opole University, Opole, Poland

Andrea Porzionato Department of Human Anatomy and Physiology, University of Padua, Padova, Italy

Section of Anatomy, Department of Human Anatomy and Physiology, University of Padua, Padova, Italy

Nanduri R. Prabhakar Department of Medicine and Biological Sciences Division, Institute for Integrative Physiology and Center for Systems Biology of O_2 Sensing, Biological Sciences Division, University of Chicago, Chicago, IL, USA

Jesus Prieto-Lloret Department of Biochemistry and Molecular Biology and Physiology, IBGM Universidad de Valladolid and CSIC, Valladolid, Spain

Ciber de Enfermedades Respiratorias, Valladolid, Spain

Robert W. Putnam Department of Neuroscience, Cell Biology and Physiology, Wright State University, Dayton, OH, USA

Gayatri Raghuraman Department of Medicine, Institute for Integrative Physiology, University of Chicago, Chicago, IL, USA

Jan Ramirez Seattle Children's Research Institute, Seattle, WA, USA

Gladys Ramírez Facultad de Medicina, Universidad de Colima, Colima, Mexico

Maria Ramirez Departamento de Bioquímica y Biología Molecular y Fisiología. IBGM. Facultad de Medicina, CIBERES, Instituto de Salud Carlos III, Universidad de Valladolid-CSIC, Valladolid, Spain

Peter Ratcliffe The Henry Wellcome Building for Molecular Physiology, University of Oxford, Oxford, UK

Agnieszka Rekawek Department of Respiratory Research, Medical Research Center, Polish Academy of Sciences, Warsaw, Poland

Edison-Pablo Reyes Facultad de Medicina, Clínica Alemana, Universidad del Desarrollo, Lo Barnechea, Santiago, Region Metropolitana, Chile

Ricardo Rigual CIBER Enfermedades Respiratorias, Valladolid, Spain

Department of Biochemistry and Molecular Biology and Physiology, and IBGM, Universidad de Valladolid and CSIC, Valladolid, Spain

Robert A. Rizza Endocrine Research Unit, Division of Endocrinology, Diabetes, Metabolism & Nutrition, Mayo Clinic College of Medicine, Rochester, MN, USA

Asuncion Rocher Departamento de Bioquímica y Biología Molecular y Fisiología. IBGM. Facultad de Medicina, CIBERES, Instituto de Salud Carlos III, School of Medicine, Universidad de Valladolid-CSIC, Valladolid, Spain

C. Rodgers-Garlick Department of Cell and Systems Biology, University of Toronto, Toronto, ON, Canada

Jorge Rodríguez Facultad de Medicina, Escuela de Medicina, Universidad Andres Bello, Santiago, Chile

Araya Ruangkittisakul Department of Physiology, Faculty of Medicine and Dentistry, University of Alberta, Edmonton, AB, Canada

Marcin Rucinski Department of Histology and Embryology, Poznan University of Medical Sciences, Poznan, Poland

Fumio Sakamaki Division of Cardiology and Pulmonary Circulation, Department of Medicine, National Cardiovascular Center, Suita, Osaka, Japan

Shaima Salman Department of Biology, McMaster University, Hamilton, ON, Canada

Vedangi Sample Department of Pharmacology and Molecular Sciences, Johns Hopkins University, Baltimore, MD, USA

Gloria Sarasin Department of Human Anatomy and Physiology, University of Padua, Padova, Italy

Toru Satoh Cardiology Division, Department of Medicine, Kyorin University School of Medicine, Mitakashi, Tokyo, Japan

Kathy Schnorbusch Department of Veterinary Sciences, Laboratory of Cell Biology and Histology, University of Antwerp, Antwerp, Belgium

Harold D. Schultz Department of Cellular and Integrative Physiology, College of Medicine, University of Nebraska Medical Center, Omaha, NE, USA

Tommy Seaborn Department of Pediatrics, Centre de Recherche de l'Hôpital St-François d'Assise (CR-SFA), Centre Hospitalier Universitaire de Québec (CHUQ), Faculty of Medicine, Laval University, Québec, QC, Canada

Gregg L. Semenza Vascular Program, Institute for Cell Engineering; Departments of Pediatrics, Medicine, Oncology, and Radiation Oncology; and McKusick-Nathans Institute of Genetic Medicine, The Johns Hopkins University School of Medicine, Baltimore, MD, USA

Maria M. Sfriso Department of Human Anatomy and Physiology, University of Padua, Padova, Italy

Ryan L. Shapiro Department of Neuroscience, Cell Biology and Physiology, Wright State University, Dayton, OH, USA

Machiko Shirahata Departments of Environmental Health Sciences, and of Anesthesiology/Critical Care Medicine, The Johns Hopkins Medical Institutions

Curtis A. Smith Departments of Population Health Sciences, University of Wisconsin – Madison, Madison, WI, USA

Jorge Soliz Department of Pediatrics, Centre de Recherche de l'Hôpital St-François d'Assise (CR-SFA), Centre Hospitalier Universitaire de Québec (CHUQ), Faculty of Medicine, Laval University, Québec, QC, Canada

Carla Stecco Department of Human Anatomy and Physiology, University of Padua, Padova, Italy

Ruth L. Stornetta Department of Pharmacology, University of Virginia, Charlottesville, VA, USA

Kingman Strohl Case Western Reserve University, Cleveland, OH, USA

Naweed I. Syed Department of Cell Biology and Anatomy and the Hotchkiss Brain Institute, Faculty of Medicine, University of Calgary, Calgary, AB, Canada

Mihaela Teodorescu Departments of Medicine, University of Wisconsin – Madison, Madison, WI, USA

Luc Teppema Leiden University Medical Center, Leiden, RC, Netherlands

Jean-Pierre Timmermans Department of Veterinary Sciences, Laboratory of Cell Biology and Histology, University of Antwerp, Antwerp, Belgium

G.L. Tipoe Departments of Physiology and Anatomy, University of Hong Kong, Hong Kong SAR, China

Takeshi Tomita Division of Cardiology and Pulmonary Circulation, Department of Medicine, National Cardiovascular Center, Suita, Osaka, Japan

Araya Ungpakawa University of Alberta, 232D Heritage Medical Research Centre, Edmonton, Alberta, Canada

Rodrigo Varas Laboratorio de Neurobiología, Facultad de Ciencias Biológicas, Pontificia Universidad Católica de Chile, Santiago, Chile

Victoria Vega-Agapito Department of Biochemistry and Molecular Biology and Physiology, IBGM Universidad de Valladolid and CSIC, Valladolid, Spain

Ciber de Enfermedades Respiratorias, Valladolid, Spain

Sandra G. Vincent Department of Biomedical and Molecular Sciences, Queen's University, Kingston, ON, Canada

Erica A. Wehrwein Department of Anesthesiology, Mayo Clinic, Rochester, MN, USA

C.G. Wilson Department of Pediatrics, Division of Neonatology, Rainbow Babies & Children's Hospital Case Western Reserve University, Cleveland, OH, USA

Richard Wilson Departmentof Physiology and Pharmacology, University of Calgary, Calgary, Alberta, Canada

Christopher N. Wyatt Department of Neuroscience, Cell Biology and Physiology, Wright State University, Boonshoft School of Medicine, Dayton, OH, USA

Yang K. Xiang Department of Molecular and Integrative Physiology, University of Illinois, Urbana, IL, USA

Ailiang Xie Departments of Population Health Sciences, University of Wisconsin – Madison, Madison, WI, USA

Herman Yeger Division of Pathology, Department of the Paediatric Laboratory MedicineResearch Institute, The Hospital for Sick Children, and Department of Laboratory Medicine and Pathobiology, University of Toronto, Toronto, ON, Canada

Sara Yubero CIBER Enfermedades Respiratorias, Valladolid, Spain

Department of Biochemistry and Molecular Biology and Physiology, and IBGM, Universidad de Valladolid and CSIC, Valladolid, Spain

Nathaniel Y.W. Yuen Department of Biomedical and Molecular Sciences, Queen's University, Kingston, ON, Canada

Peter C. Zachar Department of Biology, University of Ottawa, Ottawa, ON, Canada

Patricio Zapata Facultad de Medicina, Clínica Alemana, Universidad del Desarrollo, Lo Barnechea, Santiago, Region Metropolitana, Chile

S. Zara Department of Drug Sciences, University of Chieti, Chieti, Italy

Izabela Zasada Department of Respiratory Research, Medical Research Center, Polish Academy of Sciences, Warsaw, Poland

Jin Zhang Department of Pharmacology and Molecular Sciences, Johns Hopkins University, Baltimore, MD, USADepartment of Neuroscience and Oncology, School of Medicine, Johns Hopkins University, Baltimore, MD, USA

Li Zhang Department of Biology, McMaster University, Hamilton, ON, Canada

Min Zhang Department of Biology, McMaster University, Hamilton, ON, Canada

Chapter 1
The Role of Hypoxia-Inducible Factors in Oxygen Sensing by the Carotid Body

Gregg L. Semenza and Nanduri R. Prabhakar

Abstract Chronic intermittent hypoxia (IH) associated with sleep-disordered breathing is an important cause of hypertension, which results from carotid body-mediated activation of the sympathetic nervous system. IH triggers increased levels of reactive oxygen species (ROS) in the carotid body, which induce increased synthesis and stability of hypoxia-inducible factor 1α (HIF-1α) and calpain-dependent degradation of HIF-2α. HIF-1 activates transcription of the *Nox2* gene, encoding NADPH oxidase 2, which generates superoxide. Loss of HIF-2 activity leads to decreased transcription of the *Sod2* gene, encoding manganese superoxide dismutase, which converts superoxide to hydrogen peroxide. Thus, IH disrupts the balance between HIF-1-dependent pro-oxidant and HIF-2-dependent anti-oxidant activities, and this loss of redox homeostasis underlies the pathogenesis of autonomic morbidities associated with IH.

Keywords Cardiorespiratory homeostasis • Obstructive sleep apnea • Oxidative stress • Oxygen homeostasis

Hypoxia-inducible factor 1 (HIF-1) is a transcriptional activator that functions as a master regulator of oxygen homeostasis in all metazoan species (Semenza 2009, 2010). HIF-1 is a heterodimeric protein that is composed of an O_2-regulated HIF-1α subunit and a constitutively expressed HIF-1β subunit (Wang and Semenza 1995; Wang et al. 1995). Proline and asparagine residues in HIF-1α are hydroxylated under aerobic conditions (Kaelin and Ratcliffe 2008), which provides a mechanism for the decreased HIF-1α protein stability and transactivation function that are observed in well-oxygenated, as compared to hypoxic cells (Jiang et al. 1997). Database searches for homologs of HIF-1α led to the

G.L. Semenza (✉)
Vascular Program, Institute for Cell Engineering; Departments of Pediatrics, Medicine, Oncology, and Radiation Oncology; and McKusick-Nathans Institute of Genetic Medicine, The Johns Hopkins University School of Medicine, Broadway Research Building, Suite 671, 733 N. Broadway, Baltimore, MD 21205, USA
e-mail: gsemenza@jhmi.edu

N.R. Prabhakar
Institute for Integrative Physiology and Center for Systems Biology of O2 Sensing, Biological Sciences Division, University of Chicago, 5841 S Maryland Ave, MC 5068 Room N-711, Chicago, IL 60637, USA
e-mail: nprabhak@medicine.bsd.uchicago.edu

C.A. Nurse et al. (eds.), *Arterial Chemoreception: From Molecules to Systems*, Advances in Experimental Medicine and Biology 758, DOI 10.1007/978-94-007-4584-1_1,

discovery of HIF-2α (Ema et al. 1997; Flamme et al. 1997; Hogenesch et al. 1997; Tian et al. 1997), which is also subject to O_2-dependent regulation and heterodimerization with HIF-1β (Wiesener et al. 1998). Whereas HIF-1α is expressed by most nucleated cells of all metazoan species, HIF-2α is present only in the vertebrate lineage and is only expressed in a restricted number of cell types (Loenarz et al. 2011).

Homozygosity for a null (knockout) allele at the *Hif1a* locus encoding HIF-1α results in embryonic lethality at midgestation with defects in erythrocytosis, vascularization, and cardiogenesis (Iyer et al. 1998; Ryan et al. 1998; Yoon et al. 2006), demonstrating that all three components of the circulatory system are dependent upon HIF-1 for normal development. Mice that are heterozygous for the knockout allele develop normally but have impaired responses to hypoxic and ischemic stimuli (Yu et al. 1999; Shimoda et al. 2001, 2006; Cai et al. 2003, 2008; Li et al. 2006; Bosch-Marce et al. 2007; Whitman et al. 2008; Feinman et al. 2010; Zhang et al. 2010; Kannan et al. 2011; Keswani et al. 2011).

Perhaps the most remarkable phenotype of *Hif1a*$^{+/-}$ mice was the demonstration that the carotid bodies (CBs) of these mice do not sense/respond to hypoxia, despite the fact that they are histologically normal, including glomus cell morphometry, and respond appropriately to other stimuli such as hypercarbia and cyanide (Kline et al. 2002; Peng et al. 2006). In contrast to the failure of the CBs from these mice to sense hypoxia, *Hif1a*$^{+/-}$ mice manifest normal ventilatory responses to a hypoxic challenge, indicating that other chemosensors compensate for the loss of CB function *in vivo*. However, hypoxic ventilatory acclimatization responses are defective in *Hif1a*$^{+/-}$ mice, suggesting that this is a CB-specific function (Kline et al. 2002; Peng et al. 2006).

The CB is also known to sense and respond to intermittent hypoxia (IH) (Prabhakar et al. 2005). Patients with sleep-disordered breathing (also known as sleep apnea) are subjected to chronic IH, which leads to increased sympathetic nerve activity, hypertension, and its sequelae (Somers et al. 1995; Nieto et al. 2000; Peppard et al. 2000; Shahar et al. 2001). Exposure of rodents to IH for several weeks results in activation of the sympathetic nervous system, leading to elevated plasma catecholamine levels and systemic hypertension (Fletcher et al. 1992; Bao et al. 1997; Lesske et al. 1997; Kumar et al. 2006; Dick et al. 2007; Prabhakar et al. 2007a). Remarkably, *Hif1a*$^{+/-}$ mice are completely protected from the elevation of blood pressure and plasma catecholamines that are observed in wild type mice littermates after 10 days of IH (consisting of 15 s of hypoxia, followed by 5 min of normoxia, nine episodes per hour for 8 h per day) (Peng et al. 2006).

IH results in the generation of reactive oxygen species (ROS) in the CB, as reflected in the levels of thiobarbituric acid-reactive substances, and treatment of wild-type mice with the free radical scavenger MnTMPyP blocks the development of IH-induced hypertension (Prabhakar et al. 2007b). The induction of HIF-1α that is observed in the brains of mice exposed to IH is also blocked by radical scavengers, indicating that increased ROS levels are required for HIF-1 activation in response to IH (Peng et al. 2006). When *Hif1a*$^{+/-}$ mice are subjected to IH, increased levels of ROS are not detected in the brain, indicating that HIF-1 activation is required for increased ROS in response to IH. These results suggested a feed-forward mechanism, in which the increased generation of ROS induces HIF-1 activity, which in turn leads to increased ROS generation. Studies in PC12 rat pheochromocytoma cells revealed a complex signal transduction pathway by which IH stimulates HIF-1α protein synthesis, stabilization, and transactivation involving: NADPH oxidase; Ca^{2+} signaling via phospholipase Cγ, protein kinase C, and CaM kinase; prolyl hydroxylases; and mTOR (Yuan et al. 2005, 2008). A critical role for NADPH oxidase in mediating IH-induced oxidative damage has also been demonstrated in mice (Zhan et al. 2005; Peng et al. 2009; Khan et al. 2011).

Recent studies indicate that the transactivation of the *Nox2* gene, which encodes NADPH oxidase 2, is critical for the increase in ROS associated with chronic IH (Yuan et al. 2011). Induction of Nox2 mRNA expression was demonstrated in the CB, cerebral cortex, and brainstem, but not in the cerebellum, of wild-type mice, whereas Nox2 mRNA expression was not induced in any of these tissues in *Hif1a*$^{+/-}$ mice (Yuan et al. 2011). Analysis of the cerebellar tissues from wild-type mice exposed to IH revealed no activation (by phosphorylation) of phospholipase Cγ or mTOR and no induction of

HIF-1α protein, indicating that the IH-induced feed-forward mechanism, which is active in CB, cortex, and brainstem, is not active in the cerebellum (Yuan et al. 2011). Further studies are required to determine whether the response to IH is localized to specific regions of the cortex and brainstem.

Whereas HIF-1α is expressed at low levels in the CB under normoxic conditions and is dramatically induced by IH, HIF-2α is expressed at high levels in the CB under normoxic condition and its levels are dramatically decreased in response to IH, due to calpain-dependent protein degradation (Nanduri et al. 2009). The decreased HIF-2α levels were associated with decreased expression of the *Sod2* gene, which encodes superoxide dismutase 2, the enzyme that converts mitochondrial superoxide to hydrogen peroxide. Remarkably, systemic treatment of IH-exposed rats with the calpain inhibitor ALLM blocked HIF-2α degradation, restored SOD2 activity, and prevented oxidative stress and hypertension (Nanduri et al. 2009).

Based on the striking results obtained from *Hif1a*$^{+/-}$ mice, CBs from *Hif2a*$^{+/-}$ mice were analyzed. Whereas CBs from *Hif1a*$^{+/-}$ mice did not respond to hypoxia (Kline et al. 2002), CBs from *Hif2a*$^{+/-}$ mice manifested augmented sinus nerve activity in response to hypoxia, whereas responses to hypercarbia or cyanide were normal, as was CB histology including glomus cell morphometry (Peng et al. 2011). *Hif2a*$^{+/-}$ mice manifested an augmented hypoxic ventilatory response and instability of breathing. *Hif2a*$^{+/-}$ mice were also found to be hypertensive with elevated plasma norepinephrine levels even under normoxic conditions (Peng et al. 2011). Expression of Sod2 mRNA was significantly reduced in the CB of *Hif2a*$^{+/-}$ mice. Remarkably, treatment of *Hif2a*$^{+/-}$ mice with the superoxide scavenger MnTMPyP normalized the CB response to hypoxia, normalized the hypoxic ventilatory response and blood pressure, and corrected the abnormalities of breathing (Peng et al. 2011). Taken together, the increased oxidative stress associated with IH (as well as all of its downstream sequelae) results both from increased HIF-1 → Nox2 signaling and decreased HIF-2 → Sod2 signaling in response to IH. Thus, an imbalance between pro-oxidant and anti-oxidant activities underlies the pathogenesis of autonomic morbidities associated with IH. More broadly, these results suggest that the functional antagonism between HIF-1 and HIF-2 may play a fundamental role in redox regulation and the maintenance of cardiorespiratory homeostasis.

Acknowledgements Research from authors' laboratories was supported by contracts/grants HHS-N268201000032C, PO1-HL65608, P20-GM78494, RO1-HL55338, U54-CA143868 (G.L.S.) and HL-76537, HL-90554, and HL-86493 (N.R.P) from the National Institutes of Health.

References

Bao G, Metreveli N, Li R, Taylor A, Fletcher EC (1997) Blood pressure response to chronic episodic hypoxia: role of the sympathetic nervous system. J Appl Physiol 83:95–101

Bosch-Marcé M, Okuyama H, Wesley JB, Sarkar K, Kimura H, Liu YV, Zhang H, Strazza M, Rey S, Savino L, Zhou YF, McDonald KR, Na Y, Vandiver S, Rabi A, Shaked Y, Kerbel R, LaVallee T, Semenza GL (2007) Effects of aging and HIF-1 activity on angiogenic cell mobilization and recovery of perfusion following limb ischemia. Circ Res 101:1310–1318

Cai Z, Manalo DJ, Wei G, Rodriguez ER, Fox-Talbot K, Lu H, Zweier JL, Semenza GL (2003) Hearts from rodents exposed to intermittent hypoxia or erythropoietin are protected against ischemia-reperfusion injury. Circulation 108:79–85

Cai Z, Zhong H, Bosch-Marce M, Fox-Talbot K, Wang L, Wei C, Trush MA, Semenza GL (2008) Complete loss of ischaemic preconditioning-induced cardioprotection in mice with partial deficiency of HIF-1α. Cardiovasc Res 77:463–470

Dick TE, Hsieh YH, Wang N, Prabhakar NR (2007) Acute intermittent hypoxia increases both phrenic and sympathetic nerve activities in the rat. Exp Physiol 92:87–97

Ema M, Taya S, Yokotani N, Sogawa K, Matsuda Y, Fujii-Kuriyama Y (1997) A novel bHLH-PAS factor with close sequence similarity to hypoxia-inducible factor 1α regulates the VEGF expression and is potentially involved in lung and vascular development. Proc Natl Acad Sci U S A 94:4273–4278

Feinman R, Deitch EA, Watkins AC, Abungu B, Colorado I, Kannan KB, Sheth S, Caputo FJ, Lu Q, Ramanathan M, Attan S, Badami CJ, Doucet D, Barlos D, Bosch-Marcé M, Semenza GL, Xu DZ (2010) HIF-1 mediates pathogenic inflammatory responses to intestinal ischemia reperfusion injury. Am J Physiol Gastrointest Liver Physiol 299:G833–G843

Flamme I, Fröhlich T, von Reutern M, Kappel A, Damert A, Risau W (1997) HRF, a putative basic helix-loop-helix-PAS-domain transcription factor is closely related to hypoxia-inducible factor-1α and developmentally expressed in blood vessels. Mech Dev 63:51–60

Fletcher EC, Lesske J, Qian W, Miller CC 3rd, Unger T (1992) Repetitive, episodic hypoxia causes diurnal elevation of blood pressure in rats. Hypertension 19:555–561

Hogenesch JB, Chan WK, Jackiw VH, Brown RC, Gu YZ, Pray-Grant M, Perdew GH, Bradfield CA (1997) Characterization of a subset of the basic-helix-loop-helix-PAS superfamily that interacts with components of the dioxin signaling pathway. J Biol Chem 272:8581–8593

Iyer NV, Kotch LE, Agani F, Leung SW, Laughner E, Wenger RH, Gassmann M, Gearhart JD, Lawler AM, Yu AY, Semenza GL (1998) Cellular and developmental control of O_2 homeostasis by hypoxia-inducible factor 1α. Genes Dev 12:149–162

Jiang BH, Zheng JZ, Leung SW, Roe R, Semenza GL (1997) Transactivation and inhibitory domains of hypoxia-inducible factor 1α. Modulation of transcriptional activity by oxygen tension. J Biol Chem 272:19253–19260

Kaelin WG Jr, Ratcliffe PJ (2008) Oxygen sensing by metazoans: the central role of the HIF hydroxylase pathway. Mol Cell 30:393–402

Kannan KB, Colorado I, Reino D, Palange D, Lu Q, Qin X, Watkins A, Caputo FJ, Semenza GL, Deitch EA, Feinman R (2011) Hypoxia-inducible factor plays a gut-injurious role in intestinal ischemia reperfusion injury. Am J Physiol Gastrointest Liver Physiol 300:G853–G861

Keswani SC, Bosch-Marcé M, Reed N, Fischer A, Semenza GL, Hoke A (2011) Nitric oxide prevents axonal degeneration by inducing HIF-1-dependent expression of erythropoietin. Proc Natl Acad Sci U S A 108:4986–4990

Khan SA, Nanduri J, Yuan G, Kinsman B, Kumar GK, Joseph J, Kalyanaraman B, Prabhakar NR (2011) NADPH oxidase 2 mediates intermittent hypoxia-induced mitochondrial complex I inhibition: relevance to blood pressure changes in rats. Antioxid Redox Signal 14:533–542

Kline DD, Peng Y, Manalo DJ, Semenza GL, Prabhakar NR (2002) Defective carotid body function and impaired ventilatory responses to chronic hypoxia in mice partially deficient for hypoxia-inducible factor 1α. Proc Natl Acad Sci U S A 99:821–826

Kumar GK, Rai V, Sharma SD, Ramakrishnan DP, Peng YJ, Souvannakitti D, Prabhakar NR (2006) Chronic intermittent hypoxia induces hypoxia-evoked catecholamine efflux in adult rat adrenal medulla via oxidative stress. J Physiol 575:229–239

Lesske J, Fletcher EC, Bao G, Unger T (1997) Hypertension caused by chronic intermittent hypoxia–influence of chemoreceptors and sympathetic nervous system. J Hypertens 12:1593–1603

Li J, Bosch-Marcé M, Nanayakkara A, Savransky V, Fried SK, Semenza GL, Polotsky VY (2006) Altered metabolic responses to intermittent hypoxia in mice with partial deficiency of hypoxia-inducible factor 1α. Physiol Genomics 25:450–457

Loenarz C, Coleman ML, Boleininger A, Schierwater B, Holland PW, Ratcliffe PJ, Schofield CJ (2011) The hypoxia-inducible transcription factor pathway regulates oxygen sensing in the simplest animal, Trichoplax adhaerens. EMBO Rep 12:63–70

Nanduri J, Wang N, Yuan G, Khan SA, Souvannakitti D, Peng YJ, Kumar GK, Garcia JA, Prabhakar NR (2009) Intermittent hypoxia degrades HIF-2α via calpains resulting in oxidative stress: implications for recurrent apnea-induced morbidities. Proc Natl Acad Sci U S A 106:1199–1204

Nieto FJ, Young TB, Lind BK, Shahar E, Samet JM, Redline S, D'Agostino RB, Newman AB, Lebowitz MD, Pickering TG, Nieto FJ, Young TB, Lind BK, Shahar E, Samet JM, Redline S, D'Agostino RB, Newman AB, Lebowitz MD, Pickering TG (2000) Association of sleep-disordered breathing, sleep apnea, and hypertension in a large community-based study, Sleep Heart Health Study. JAMA 283:1829–1836

Peng Y, Yuan G, Ramakrishnan D, Sharma SD, Bosch-Marcé M, Kumar GK, Semenza GL, Prabhakar NR (2006) Heterozygous HIF-1α deficiency impairs carotid body-mediated systemic responses and reactive oxygen species generation in mice exposed to intermittent hypoxia. J Physiol 577:705–716

Peng YJ, Nanduri J, Yuan G, Wang N, Deneris E, Pendyala S, Natarajan V, Kumar GK, Prabhakar NR (2009) NADPH oxidase is required for the sensory plasticity of the carotid body by chronic intermittent hypoxia. J Neurosci 29:4903–4910

Peng Y, Nanduri J, Khan SA, Yuan G, Wang N, Kinsman B, Vaddi DR, Kumar GK, Garcia JA, Semenza GL, Prabhakar NR (2011) Hypoxia-inducible factor 2α (HIF-2α) heterozygous-null mice exhibit exaggerated carotid body sensitivity to hypoxia, breathing instability, and hypertension. Proc Natl Acad Sci U S A 108:3065–3070

Peppard PE, Young T, Palta M, Skatrud J (2000) Prospective study of the association between sleep disordered breathing and hypertension. N Engl J Med 342:1378–1384

Prabhakar NR, Peng YJ, Jacono FJ, Kumar GK, Dick TE (2005) Cardiovascular alterations by chronic intermittent hypoxia: importance of carotid body chemoreflexes. Clin Exp Pharmacol Physiol 32:447–449

Prabhakar NR, Dick TE, Nanduri J, Kumar GK (2007a) Systemic, cellular and molecular analysis of chemoreflex-mediated sympathoexcitation by chronic intermittent hypoxia. Exp Physiol 92:39–44

Prabhakar NR, Kumar GK, Nanduri J, Semenza GL (2007b) ROS signaling in systemic and cellular responses to chronic intermittent hypoxia. Antioxid Redox Signal 9:1397–1403

Ryan HE, Lo J, Johnson RS (1998) HIF-1α is required for solid tumor formation and embryonic vascularization. EMBO J 17:3005–3015

Semenza GL (2009) Regulation of oxygen homeostasis by hypoxia-inducible factor 1. Physiology (Bethesda) 24:97–106

Semenza GL (2010) Oxygen homeostasis. Wiley Interdiscip RevSyst Bio Med 2:336–361

Shahar E, Whitney CW, Redline S, Lee ET, Newman AB, Nieto J, O'Connor GT, Boland LL, Schwartz JE, Samet JM (2001) Sleep-disordered breathing and cardiovascular disease. Am J Respir Crit Care Med 163:19–25

Shimoda LA, Manalo DJ, Sham JSK, Semenza GL, Sylvester JT (2001) Partial HIF-1α deficiency impairs pulmonary arterial myocyte electrophysiological responses to chronic hypoxia. Am J Physiol 281:L202–L208

Shimoda LA, Fallon M, Pisarcik S, Wang J, Semenza GL (2006) HIF-1 regulates hypoxic induction of NHE1 expression and alkalinization of intracellular pH in pulmonary arterial myocytes. Am J Physiol 291:941–949

Somers VK, Dyken ME, Clary MP, Abboud FM (1995) Sympathetic neural mechanisms in obstructive sleep apnea. J Clin Invest 96:1897–1904

Tian H, McKnight SL, Russell DW (1997) Endothelial PAS domain protein 1 (EPAS1), a transcription factor selectively expressed in endothelial cells. Genes Dev 11:72–82

Wang GL, Semenza GL (1995) Purification and characterization of hypoxia-inducible factor 1. J Biol Chem 270:1230–1237

Wang GL, Jiang B-H, Rue EA, Semenza GL (1995) Hypoxia-inducible factor 1 is a basic-helix-loop-helix-PAS heterodimer regulated by cellular O_2 tension. Proc Natl Acad Sci U S A 92:5510–5514

Whitman EM, Pisarcik S, Luke T, Fallon M, Wang J, Sylvester JT, Semenza GL, Shimoda LA (2008) Endothelin-1 mediates hypoxia-induced inhibition of voltage-gated K^+ channel expression in pulmonary arterial myocytes. Am J Physiol 294:309–310

Wiesener MS, Turley H, Allen WE, Willam C, Eckardt KU, Talks KL, Wood SM, Gatter KC, Harris AL, Pugh CW, Ratcliffe PJ, Maxwell PH (1998) Induction of endothelial PAS domain protein-1 by hypoxia: characterization and comparison with hypoxia-inducible factor 1α. Blood 92:2260–2268

Yoon D, Pastore YD, Divoky V, Liu E, Mlodnicka AE, Rainey K, Ponka P, Semenza GL, Schumacher A, Prchal JT (2006) HIF-1α deficiency results in dysregulated EPO signaling and iron homeostasis in mouse development. J Biol Chem 281:25703–25711

Yu AY, Shimoda LA, Iyer NV, Huso DL, Sun X, McWilliams R, Beaty T, Sham JSK, Wiener CM, Sylvester JT, Semenza GL (1999) Impaired physiological responses to chronic hypoxia in mice partially deficient for hypoxia-inducible factor 1α. J Clin Invest 103:691–696

Yuan G, Nanduri J, Bhasker RC, Semenza GL, Prabhakar NR (2005) Ca^{2+}/calmodulin kinase-dependent activation of hypoxia-inducible factor 1 transcriptional activity in cells subjected to intermittent hypoxia. J Biol Chem 280:4321–4328

Yuan G, Nanduri J, Khan S, Semenza GL, Prabhakar NR (2008) Induction of HIF-1α expression by intermittent hypoxia: involvement of NADPH oxidase, Ca^{2+} signaling, prolyl hydroxylases, and mTOR. J Cell Physiol 217:674–685

Yuan G, Khan SA, Luo W, Nanduri J, Semenza GL, Prabhakar NR (2011) Hypoxia-inducible factor 1 mediates increased expression of NADPH oxidase-2 in response to intermittent hypoxia. J Cell Physiol. doi:10.1002/jcp.22640, 2011 Feb 1 [Epub ahead of print]

Zhan G, Serrano F, Fenik P, Hsu R, Kong L, Pratico D, Klann E, Veasey SC (2005) NADPH oxidase mediates hypersomnolence and brain oxidative injury in a murine model of sleep apnea. Am J Respir Crit Care Med 172:921–929

Zhang X, Liu L, Wei X, Tan YS, Tong L, Chang R, Ghanamah MS, Reinblatt M, Marti GP, Harmon JW, Semenza GL (2010) Impaired angiogenesis and mobilization of circulating angiogenic cells in HIF-1α heterozygous-null mice after burn wounding. Wound Repair Regen 18:193–201

Chapter 2
Neuronal Mechanisms of Oxygen Chemoreception: An Invertebrate Perspective

Tara A. Janes and Naweed I. Syed

Abstract Since the evolution of aerobic metabolism, cellular requirements for molecular oxygen have been the major driver for the development of sophisticated mechanisms underlying both invertebrate and vertebrate respiratory behaviour. Among the most important characteristics of respiration is its adaptability, which allows animals to maintain oxygen homeostasis over a wide range of environmental and metabolic conditions. In all animals, the respiratory behaviour is controlled by neural networks often termed respiratory central pattern generators (rCPG). While rCPG neurons are intrinsically capable of generating rhythmical outputs, the respiratory needs are generally "sensed" by either central or peripheral chemoreceptive neurons. The mechanisms by which chemoreceptors respond to changes in oxygen and modulate central respiratory control centers have been the focus of decades of research. However, our understanding of these mechanisms has been limited due to an inability to precisely locate oxygen chemoreceptor populations, combined with the overwhelming complexity of vertebrate neural circuits. Although mammalian models remain the gold standard for research in general, invertebrates do nevertheless offer greatly simplified neural networks that share fundamental similarities with vertebrates. The following review will provide evidence for the existence of oxygen chemoreceptors in many invertebrate groups and reveal the mechanisms by which these neurons may "perceive" environmental oxygen and drive central rCPG activity. For this, we will specifically highlight an invertebrate model, the pond snail *Lymnaea stagnalis* whose episodic respiratory behaviour resembles that of diving mammals. The rCPG neurons have been identified and fully characterized in this model both *in vivo* and *in vitro*. The *Lymnaea* respiratory network has also been reconstructed *in vitro* and the contributions of individual rCPG neurons towards rhythm generation characterized through direct intracellular recordings. We now provide evidence for the presence of genuine peripheral oxygen chemoreceptors in *Lymnaea,* and demonstrate that these neurons respond to hypoxia in a manner analogous to that of mammalian carotid bodies. These chemoreceptor cells not only drive the activity of the rCPG neurons but their synaptic connections also exhibit hypoxia-induced plasticity. The lessons learned from this model will likely reveal fundamental principles underlying both peripheral and central respiratory control mechanisms, which may be conserved in both invertebrate and vertebrate species.

T.A. Janes • N.I. Syed (✉)
Department of Cell Biology and Anatomy and the Hotchkiss Brain Institute, Faculty of Medicine, University of Calgary, 3280 Hospital Drive N.W, Calgary AB T2N 4Z6, Canada
e-mail: nisyed@ucalgary.ca

C.A. Nurse et al. (eds.), *Arterial Chemoreception: From Molecules to Systems*,
Advances in Experimental Medicine and Biology 758, DOI 10.1007/978-94-007-4584-1_2,

Keywords *Lymnaea* • Respiration • Invertebrate • Oxygen chemoreceptor • Central pattern generator • Hypoxia • Neuron-chip interface

2.1 Introduction: Oxygen Sensing

The emergence of the first aerobic organisms during the Proterozoic, approximately 2.3 billion years ago, was arguably the most important event in metazoan evolution. During this period, rapid atmospheric oxidation facilitated the use of molecular oxygen in mitochondrial respiration (Hedges et al. 2004). This change from inefficient substrates (carbon dioxide, sulphate) to the thermodynamically superior oxygen allowed more efficient energy production and the rapid evolution of complex, multicellular life (Thannickal 2009). Indeed, molecular oxygen has since become so integrated into metazoan biology that without its adequate supply the survival of individual cells, tissues and even the entire organism is impossible. However, a caveat to aerobic metabolism is the production of reactive oxygen species (ROS) and oxidative stress leading to serious cell and DNA damage (reviewed by Ott et al. 2007). Therefore, a major challenge for animals in an ever-changing environment is to keep tissue oxygen levels high enough to sustain adequate energy production, but not so high as to result in oxidative stress. In order to achieve this balance, animals have evolved general cellular mechanisms to "sense and respond" to local oxygen levels. Specifically, modulation of certain transcription factors and protein kinases by molecular oxygen regulates processes from cell death to local cardiovascular control to metabolism (Nakayama 2009; Gore et al. 2010). These responses are generally aimed at restoring oxygen homeostasis but may not be sufficient on their own and sometimes lead to pathological conditions. Therefore, the task of oxygen homeostasis at the organismal level falls on the nervous system, which regulates all essential homeostatic functions. Through the use of specialized sensory neurons, termed chemoreceptors, environmental and tissue oxygen levels are monitored and fluctuations encoded as specific activity patterns. These chemoreceptor signals adjust activity in neural regions involved in respiratory rhythm generation and modulation, thus producing respiratory behaviour that satisfies an organism's oxygen demands. A great deal is known about oxygen sensing in vertebrates and the reader is directed towards a few excellent reviews (Milsom and Burleson 2007; López-Barneo et al. 2008; Gonzalez et al. 2010; Nurse 2010). However, much less is known regarding mechanisms of oxygen sensing in invertebrates. Recently, we have made tremendous strides towards the identification and elucidation of mechanisms underlying peripheral oxygen chemoreceptors (POC's) and have begun characterizing their connectivity patterns with rCPG neurons. These novel findings are the focus of this review.

2.2 Oxygen Sensing in Invertebrates

The first multicellular metazoa to evolve out of the Precambrian oceans were invertebrates. This group has been so successful that it remains unmatched in its sheer diversity of form and function and currently comprises 97% of all animal species on earth (Zhang and Shear 2007). However invertebrates, like all aerobes, are constrained by their requirement for molecular oxygen and must maintain sufficient tissue oxygen tensions. This task can be problematic for invertebrates, as many among them experience environmental fluctuations between hypoxia and hyperoxia on a seasonal or even daily basis, especially in aquatic habitats. Although many have mechanisms to tolerate these extreme conditions, most continue to change their behaviour and physiology in an effort to maintain oxygen homeostasis. We hope that by the end of this review, it will become clear to the reader that these simple organisms have not only evolved sophisticated oxygen sensing mechanisms, but that the fundamental mechanisms by which POC's drive central respiratory network activity appears to be conserved in both vertebrate and invertebrate species.

For comparison, we will first provide a brief review of the mammalian respiratory network, which is composed of central neural elements, lungs and sensory input. Rhythm generation originates from the respiratory central pattern generator (rCPG) in the ventrolateral medulla, which drives coordinated activity of muscles surrounding the lungs (Smith et al. 1991, 2009). The rCPG activity is primarily modulated by neural populations in the pons and medulla, and by chemoreceptors. Although oxygen chemoreceptors were first described for the mammalian carotid body by Heymans et al. (1930) subsequent research has identified them in birds, lower vertebrates (reptiles and fish), and many invertebrate groups (Ito et al. 1999; Ishii et al. 1985; Jonz et al. 2004; Milsom and Burleson 2007; and see below). In the carotid body, chemoreceptors depolarize during hypoxia leading to excitation of rCPG neurons and rapid increases in ventilation (Heymans et al. 1930; Blain et al. 2009). In addition, central oxygen chemoreceptors have been identified in mammals, however the mechanisms of oxygen sensing in these neurons and their role in rhythm modulation remains uncertain (reviewed by Neubauer and Sunderram 2004). The invertebrate studies detailed below and summarized in Table 2.1, show that fundamental features of the vertebrate respiratory network are conserved in these simple organisms, especially their ability to sense and respond to environmental oxygen.

Perhaps the most effective way of dealing with extreme environmental conditions is to simply avoid them. It is not surprising then that many invertebrates show active avoidance of hypoxic conditions. Specifically, sea anemones (Cnidaria), *C. elegans* (Nematoda), *Drosophila* larvae (Arthropoda) and Penaeid shrimp (Arthropoda) have all been found to increase locomotor behaviour during hypoxia and seek out normoxic conditions (Fredericks 1976; Dusenbery 1983; Wingrove and O'Farrell 1999; Wu 2002). Furthermore, hypoxic crustaceans are known to elevate their bodies above the water and ventilate their gills using well oxygenated surface water (Taylor et al. 1973). Avoidance of hypoxia however, is not always possible for invertebrates, especially those that are small bodied or sessile. Under conditions of mild to moderate hypoxia, increasing ventilation does however help these animals increase oxygen acquisition. For example, marine polychaete worms respond to hypoxia with increased peristaltic movements, which draws water into their sediment burrows and over the gills (Wohlgemuth et al. 2000). Crustaceans, marine gastropods and many bivalves increase respiratory pumping frequency during hypoxia to draw more water over their gills and increase oxygen extraction (Taylor and Brand 1975; Massabuau and Burtin 1984; Kanz and Quast 1992). Terrestrial invertebrates also show respiratory changes in response to hypoxia. For example, flying insects respond to hypoxia by prolonging spiracle opening and increasing body contractions to enhance convective transport (Case 1956; Harrison et al. 2006). Finally, pulmonate gastropods, which have a simple lung cavity, increase aerial breathing in response to hypoxia (Jones 1961, and described in the next section). The behaviours described here strongly suggest that chemoreceptors facilitate oxygen sensing in most invertebrates. Certainly, detection of oxygen gradients provides support for oxygen sensing, but it could be argued that respiratory responses are initiated by other mechanisms, such as carbon dioxide or pH sensing. While it is probable that these play a role in invertebrate respiration, this does not rule out the involvement of oxygen chemoreceptors *per se*. Firstly, because carbon dioxide is highly soluble in water, it is unlikely that small aquatic invertebrates would accumulate high levels in their tissues. Secondly, since oxygen chemoreceptors have been identified in aquatic and terrestrial species of Nematoda, Mollusca and Arthropoda, this most likely represents an ancestral characteristic that plays a significant role in their respiration.

Among the first invertebrate oxygen chemoreceptors to be tentatively identified were sensory neurons located on the gill spines of Crustaceans. The structure and position of these neurons suggests a role in oxygen sensing (Laverack and Saier 1993). In addition, the hypoxia-sensitive branchiocardiac veins, carrying oxygenated haemolymph to the heart, express dense granule containing cells, support cells and nerve terminals that form a complex structure resembling the carotid body (Ishii et al. 1989;

Table 2.1 Key studies providing evidence of oxygen sensing in various invertebrate phyla

Group	Species	Observations	Reference
Cnidaria	Sea anemone *A. elegantissima*	Actively avoid hypoxia and move to normoxic water	Fredericks (1976)
Nematoda	*C. elegans*	Increase reversal behaviour during hypoxia	Dunsenbery (1983)
		POC's underlying oxygen sensing have been identified	Zimmer et al. (2009)
Annelida – Polychaeta	Lugworm *A. marina*	Increase peristaltic ventilation of burrows during hypoxia	Wohlgemuth et al. (2000)
Arthropoda – Crustacea	Penaeid shrimp *M. ensis*	Actively avoid hypoxia and move to normoxic water	Wu (2002)
	Lobster *N. norvegicus*	Sensory-like neurons enclosed in Type II gill spines are proposed to act as POC's	Laverack and Saier (1993)
	Crayfish *A. leptodactylus*	Increase frequency of respiratory pumping when gills are perfused with hypoxia	Massabuau and Burtin (1984)
		Perfusion of branchiocardiac veins with hypoxia increases branchial nerve activity	Ishii et al. (1989)
		Proposed POC's are located in oxygen sensitive areas of the branchiocardiac veins	Kusakabe et al. (1991)
Arthropoda – Insecta	Flies *Diptera spp.*	Spiracles open more frequently and for longer duration in hypoxia	Case (1956)
	Fruit Fly *D. melanogaster*	Larvae exposed to hypoxia stop feeding and start exploratory behaviour	Wingrove and O'Farrell (1999)
		Terminal sensory cones may contain POC's	Vermehren-Schmaedick et al. (2010)
Mollusca – Bivalvia	Clam *A. islandica*	Increase ventilation of gills during hypoxia	Taylor and Brand (1975)
Mollusca – Gastropoda	Pond snail *L. stagnalis*	Increase duration of aerial respiration in response to hypoxia	Jones (1961)
		POC's underlying hypoxia sensing have been identified	Bell et al. (2007)
	Sea Hare *A. californica*	Increase rate of respiratory pumping during hypoxia	Kanz and Quast (1992)

Kusakabe et al. 1991). In pulmonate gastropods POC's drive aerial respiration in response to hypoxia, but the oxygen sensing mechanisms are poorly understood (Bell et al. 2007). POC's in *C. elegans* have been definitively identified as BAG and URX neurons, which sense decreases and increases in oxygen respectively (Zimmer et al. 2009). In these neurons, oxygen sensing appears to be controlled by atypical soluble guanylyl cyclases (sGC's), which mediate cellular processes by catalyzing cGMP synthesis, have many isoforms and are activated by oxygen. These enzymes are also essential for the hypoxia response of *Drosophila* larvae (Vermehren-Schmaedick et al. 2010). Although POC's have not been conclusively identified in *Drosophila*, some studies elude towards their existence in the terminal sensory cones, which contain cells expressing sGC's (Morton et al. 2008). To date, the existence of central oxygen chemoreceptors in invertebrates remains largely unknown.

An exhaustive review of the cellular and molecular mechanisms underlying invertebrate oxygen chemoreception is beyond the scope of this review, therefore the remainder of this article will focus on the well-studied model organism: *Lymnaea stagnalis*. In this model, peripheral and central elements of hypoxia-driven respiratory behaviour have been identified and well characterized *in vivo* and *in vitro*.

2.3 *Lymnaea* as a Model System

2.3.1 *Aerial Respiratory Behaviour*

The pond snail, *Lymnaea stagnalis,* is a member of the Pulmonata, a unique group characterized by the presence of a simple, internal lung cavity used for aerial respiration. While oxygen is acquired cutaneously (30–40%) under normoxic conditions, aerial respiration helps increase oxygen uptake despite the nature of hypoxic or anoxic water that *Lymnaea* inhabits. *Lymnaea* spend most of their time submerged and only visit the water surface periodically to breathe air, thereby giving rise to an episodic form of respiration analogous to that of aquatic mammals (whales and seals), turtles and crocodilians. Jones (1961) provided a preliminary description of this stereotypical respiratory behaviour, which was later examined in detail by Syed et al. (1991a). Briefly, the animal migrates to the water surface and exposes a small opening, the pneumostome, which connects to the internal lung cavity (Fig. 2.1a). As the pneumostome opens, 'stale' air is expired from the lung cavity via contraction of surrounding muscles. Passive re-inflation draws fresh air into the lung where it is subsequently trapped by penumostome closure. Oxygen extraction then occurs through the highly vascularized lung surface into the haemolymph. The frequency with which aerial respiration is performed in *Lymnaea* increases as environmental oxygen declines, indicating that this is a hypoxia-driven behaviour (Syed et al. 1991a).

2.3.2 *The Respiratory Central Pattern Generator*

Aerial respiration in *Lymnaea* is readily observable thereby offering an excellent opportunity for experimentation in both freely behaving animals and semi-intact preparations. Furthermore, the rhythmical motor control that occurs during aerial respiration makes this behaviour an attractive model for the study of neural mechanisms underlying respiratory pattern generation.

The *Lymnaea* central nervous system is relatively simple, containing approximately 20,000 neurons organized into a ring of nine ganglia that are connected by short commissures. These neurons have relatively large cell bodies (40–90 μm), can be individually identified based on size, color and location and their projections easily traced through the nervous system to peripheral tissues. For these reasons, early research helped identify three neurons of interest in *Lymnaea* aerial respiratory behaviour: the giant dopamine cell "Right Pedal Dorsal 1" (RPeD1) and the two interneurons: "Visceral Dorsal 4" (VD4) and "Input 3 Interneuron" (IP3I; Fig. 2.1b). To determine if these neurons comprise the rCPG underlying respiratory pattern generation, simultaneous intracellular recordings were performed *in vivo*. Together, these three neurons produced patterned, rhythmical activity indicative of the respiratory rhythm. By exploiting the ease of invertebrate cell culture, and the ability of these neurons to reform their synapses in culture, Syed et al. (1990) reconstructed this neural circuit *in vitro*. Electrophysiological recordings from this preparation confirmed that indeed RPeD1, VD4 and IP3I are necessary and sufficient to produce the aerial respiratory rhythm. Specifically, the three neurons reformed their appropriate synapses to produce patterned activity that was not quantitatively different than that recorded *in vivo* (Fig. 2.1c). Interestingly, positive current injection into RPeD1 (the "command neuron") is necessary to initiate patterned activity in both the *in vivo* and *in vitro* preparations demonstrating that this circuit does not contain pacemaker neurons, but rather relies on synaptic connections. These findings confirm that RPeD1, VD4 and IP3I comprise an rCPG, which produces aerial respiration by driving coordinated activity in the motorneurons innervating the pneumostome musculature (Syed and Winlow 1991; Fig. 2.1d). Because of the ease with which *Lymnaea* neurons can be studied, the cellular and synaptic properties of this rCPG have been characterized at all levels from isolated neurons to semi-intact preparations.

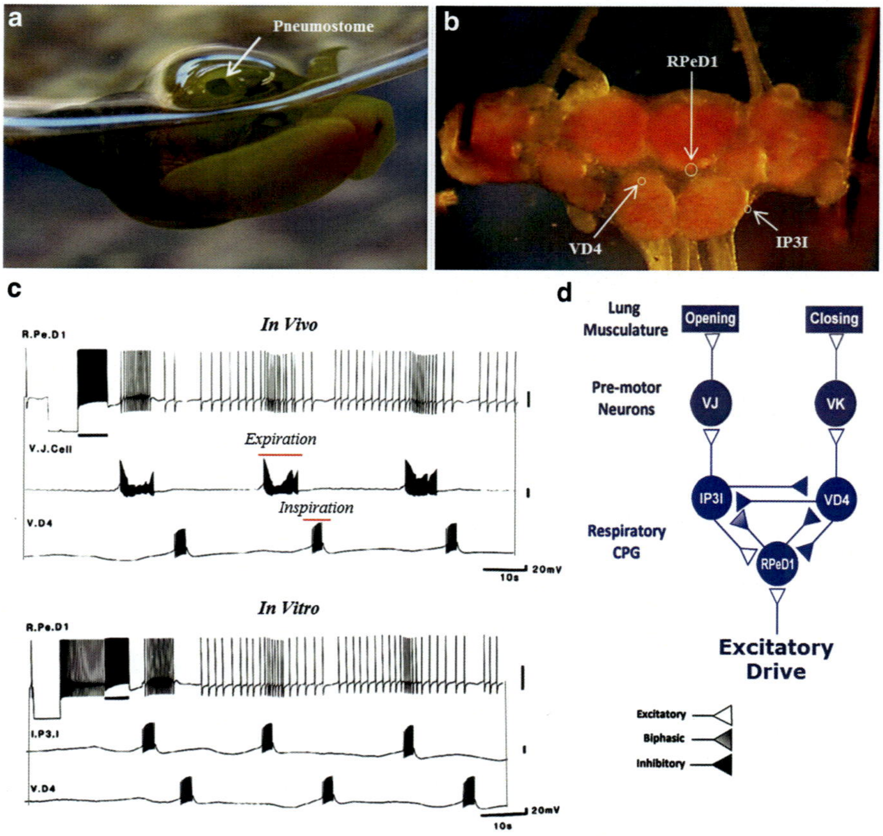

Fig. 2.1 The *Lymnaea* model system. (**a**) During aerial respiration, *Lymnaea* migrates to the water surface and exposes the pneumostome (*arrow*) to atmospheric air. Rhythmic motor activity of the pneumostome and lung musculature is controlled by a respiratory central pattern generator (rCPG). (**b**) Photograph showing the dorsal surface of the *Lymnaea* central ring ganglia. The three neurons: RPeD1, IP3I and VD4 comprising the rCPG are identified based on size, color and location. (**c**) Electrophysiological recordings performed *in vivo* (*top traces*) demonstrate that positive current injection into RPeD1 initiates patterned, rhythmic activity in the rCPG (note that the V.J. pre-motor neuron is used as an indicator of IP3I activity *in vivo*). This rhythmical activity is preserved when RPeD1, IP3I and VD4 are co-cultured *in vitro* and allowed to reform their synaptic connections (*bottom traces*). (**d**) Diagrammatic representation of the *Lymnaea* respiratory network. Excitatory drive causes bursting in RPeD1 leading to rhythmical rCPG activity. IP3I and VD4 alternately excite pre-motor neurons leading to cycles of expiration and inspiration, respectively

2.3.3 Peripheral Oxygen Chemoreceptors Drive Aerial Respiration in Lymnaea

Under normoxic conditions, positive current injection into RPeD1 is necessary to initiate patterned activity in the rCPG suggesting that RPeD1 must receive excitatory input from oxygen chemoreceptors during hypoxia. Although the existence of invertebrate oxygen chemoreceptors is established, none had yet been described in *Lymnaea*. To locate these putative oxygen chemoreceptors, it was first necessary to determine where hypoxic excitation originates both *in vivo* and *in vitro*. In a semi-intact preparation (central nervous system connected to all peripheral organs) where the central ring ganglia and peripheral regions (pneumostome, lung, etc.) were compartmentalized, it was revealed that perfusion of the peripheral compartment with hypoxia (67 Torr O_2) led to patterned, rhythmical activity in a previously silent rCPG. Conversely, no rCPG activity was observed when the compartment containing the central ring ganglia was made hypoxic (Inoue et al. 2001; Fig. 2.2a). These results suggested that the hypoxic drive to the rCPG originates at the periphery, however the precise source

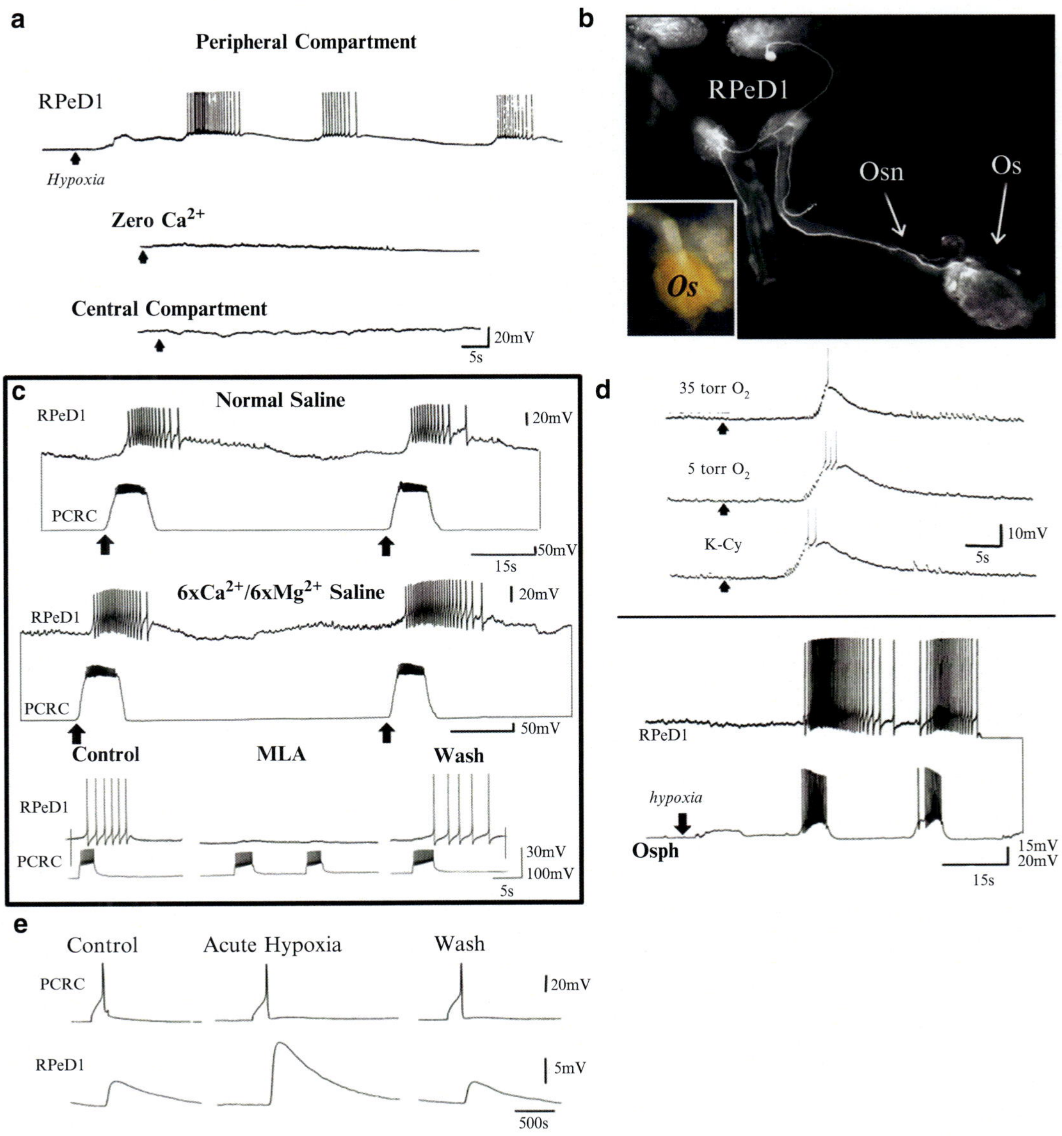

Fig. 2.2 Identification of peripheral oxygen chemoreceptors in *Lymnaea*. (**a**) Demonstration that the hypoxia-induced excitatory drive in RPeD1 originates at the periphery. Perfusion of the peripheral compartment with hypoxic saline (67 Torr O_2) results in rhythmic bursting activity recorded from the rCPG neuron, RPeD1. This activity is abolished by perfusion with zero Ca^{2+} saline and absent when the central compartment is made hypoxic. (**b**) *Lymnaea* preparation showing the central ring ganglia and the osphradial ganglion (Os) connected by a peripheral nerve. Injection of fluorescent Lucifer Yellow dye into RPeD1 labels a peripheral axonal projection that terminates at the osphradial ganglion. Insert: the osphradial ganglion shown at higher magnification. (**c**) Stimulation of an osphradial neuron (PCRC) in a semi-intact preparation produces bursting activity in a previously quiescent RPeD1. This excitation persists in the presence of high Ca^{2+}/high Mg^{2+} saline indicating a monosynaptic connection. Furthermore, perfusion of the synapse reconstructed *in vitro* with the cholinergic antagonist mecamylamine (MLA) reversibly blocks synaptic transmission. (**d**) Osphradial neurons are genuine peripheral oxygen chemoreceptors. Isolated osphradial neurons respond to acute hypoxia with dose-dependent increases in bursting activity. This hypoxia response is mimicked by the addition of potassium cyanide (K-Cy), which inhibits mitochondrial oxidative phosphorylation. Further, excitation of osphradial neurons with hypoxia *in vitro* leads to simultaneous excitation of RPeD1 (*bottom traces*). (**e**) Stimulation of osphradial ganglionic neurons during normoxia results in 1:1 excitatory post-synaptic potentials in RPeD1. However, perfusion of this synapse with hypoxia (65 Torr O_2) transiently increases the EPSP amplitude, indicating that the synapse exhibits short-term synaptic plasticity

of this drive remained unknown. Previously, a peripherally located structure, termed the osphradial ganglion, had been described in *Lymnaea* and proposed to contain oxygen sensitive neurons (Kamardin 1976; Sokolov and Kamardin 1977). Because RPeD1 is the only rCPG neuron with peripheral projections, it was hypothesized that this cell may either directly respond to hypoxia via peripheral branches or receive excitatory input from the putative oxygen sensitive neurons located in the osphradial ganglion (Fig. 2.2b). To test this possibility, synaptic transmission was blocked by co-perfusion of the peripheral compartment with hypoxia and a zero Ca^{2+} saline solution. This treatment effectively blocked hypoxia-induced rhythmical activity in RPeD1 (and hence the rCPG neurons) demonstrating that RPeD1 does not directly sense hypoxia but rather has its activity modulated through synaptic connections with osphradial neurons (Inoue et al. 2001; Fig. 2.2a). To support this postulate further, direct intracellular recordings were performed on the putative POC's located in the osphradial ganglion and RPeD1 in a semi-intact preparation. Indeed, direct stimulation of the osphradial POC's triggered spiking activity in RPeD1 indicating an excitatory synapse between the two neurons (Fig. 2.2c, top traces). Moreover, perfusion of the periphery with high divalent cation saline ($6xCa^{2+}/6xMg^{2+}$) and zero Ca^{2+}/high Mg^{2+} saline demonstrated that this synapse was both monosynaptic and chemical (Fig. 2.2c, middle traces). The identity of this synapse was further investigated by culturing the osphradial POC's with RPeD1 and testing the resulting synapse during application of various neurotransmitter antagonists. The cholinergic antagonist mecamylamine was found to reversibly block synaptic transmission between these two neurons (Bell et al. 2007; Fig. 2.2c, bottom traces). Together, these studies provided the first direct evidence of synaptic connections between the *Lymnaea* rCPG and its peripheral counterparts.

To provide unequivocal evidence that the identified osphradial neurons are indeed genuine oxygen chemoreceptors, these cells were isolated in cell culture and exposed to various hypoxic stimuli. The osphradial POC's responded to hypoxia (ranging from 35 to 5 Torr O_2) and potassium cyanide with bursting activity in a manner analogous to that of vertebrate carotid body oxygen chemoreceptors (Fig. 2.2d, top traces). To further demonstrate that the identified osphradial POC's would cause excitation of RPeD1 during hypoxia, the two neurons were cultured together and subjected to a hypoxic challenge (35 Torr O_2). The perfusion of osphradial POC's with hypoxic saline led to concurrent bursting in RPeD1, while isolated RPeD1 showed no response to hypoxia (Bell et al. 2007; Fig. 2.2d, bottom traces). Together, these studies demonstrate that excitation of *Lymnaea* POC's by hypoxia provides significant excitatory drive to the rCPG via RPeD1. However, the precise cellular and molecular mechanisms underlying this oxygen sensing remain largely unknown.

Interestingly, the perfusion of the osphradial POC-RPeD1 synapse with acute hypoxia (10 min, 67 Torr O_2) resulted in short-term synaptic plasticity. Specifically, synaptic strength was measured by stimulating the POC and recording the resulting excitatory post-synaptic potential (EPSP) amplitude from RPeD1. Compared with normoxia, acute hypoxia produced a significant and transient increase in EPSP amplitude indicative of short-term potentiation (Bell et al. 2007; Fig. 2.2e). Therefore, in addition to driving *Lymnaea* aerial respiration, hypoxia also appears to modulate synaptic strength between POC's and the rCPG. Similar observations have been made in the mammalian chemoreflex pathway suggesting that the fundamental effects of hypoxia on respiratory synapses may follow similar pathways, though the underlying cellular mechanisms remain largely unknown in vertebrates (Dwinell and Powell 1999).

2.4 Future Perspectives

In *Lymnaea*, not only the rCPG but also its peripheral counterparts have been identified and characterized both *in vivo* and *in vitro*. In this preparation, it is now feasible to reconstruct the respiratory network in order to determine the ion channel activity and synaptic mechanisms underlying respiratory rhythm

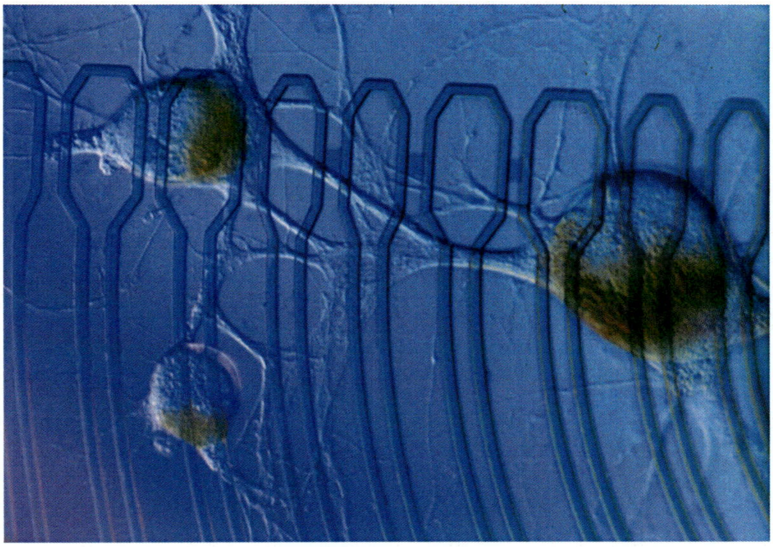

Fig. 2.3 The *Lymnaea* rCPG reconstructed on a silicon chip. The chip surface is embedded with electrodes onto which neurons are cultured. Non-invasive stimulation and simultaneous recording permits long-term monitoring of neural activity with excellent spatial and temporal resolution

generation at a resolution that is unapproachable elsewhere. Specifically, we could now use our novel Planar Patch Chip based on patch clamp recording techniques to decipher not only the ion channels that are sensitive to hypoxia but also investigate their long-term modulation (Py et al. 2010). Similarly, using our novel silicon chip technology, we now have an excellent opportunity to reconstitute the entire *Lymnaea* POC, rCPG and the motor neurons in cell culture to ask fundamental questions vis-à-vis oxygen sensing and central respiratory rhythm generation at an unprecedented level of resolution (Fig. 2.3). These studies will shed significant light on the fundamental mechanisms of central respiratory rhythm generation from the involvement of POC's to rCPG and the motor output. This information could then be used to elucidate fundamental principles of respiratory rhythm generation in vertebrates.

Acknowledgements This work was supported by a Canadian Institutes of Health Research (CIHR) grant awarded to N.I.S. The authors would like to acknowledge the excellent technical support provided by Mr. Wali Zaidi.

References

Bell HJ, Inoue T, Shum K, Luk C, Syed NI (2007) Peripheral oxygen sensing cells directly modulate the output of an identified respiratory central pattern generating neuron. Eur J Neurosci 25:3537–3550

Blain GM, Smith CA, Henderson KS, Dempsey JA (2009) Contribution of the carotid body chemoreceptors to eupneic ventilation in the intact, unanesthetized dog. J Appl Physiol 106:1564–1573

Case JF (1956) Carbon dioxide and oxygen effects on the spiracles of flies. Physiol Zool 29:163–171

Dusenbery DB (1983) Chemotactic behavior of nematodes. J Nematol 15:168–173

Dwinell MR, Powell FL (1999) Chronic hypoxia enhances the phrenic nerve response to arterial chemoreceptor stimulation in anesthetized rats. J Appl Physiol 87:817–823

Fredericks CA (1976) Oxygen as a limiting factor in phototaxis and in intraclonal spacing of the sea anemone *Anthopleura elegantissima*. Mar Biol 38:25–28

Gonzalez C, Agapito MT, Rocher A, Gomez-Niño A, Rigual R, Castañeda J, Conde SV, Obeso A (2010) A revisit to O_2 sensing and transduction in the carotid body chemoreceptors in the context of reactive oxygen species biology. Respir Physiol Neurobiol 174:317–330

Gore A, Muralidhar M, Espey MG, Degenhardt K, Mantell LL (2010) Hyperoxia sensing: from molecular mechanisms to significance in disease. J Immunotoxicol 7:239–254

Harrison J, Frazier MR, Henry JR, Kaiser A, Klok CJ, Rascón B (2006) Responses of terrestrial insects to hypoxia or hyperoxia. Respir Physiol Neurobiol 154:4–17

Hedges S, Blair J, Venturi M, Shoe J (2004) A molecular timescale of eukaryote evolution and the rise of complex multicellular life. BMC Evol Biol 4:2

Heymans C, Bouckaert JJ, Dautrebande L (1930) Sinus carotidien et réflexes respiratoires. II. Influences respiratoires réflexes de l'acidôse de l'alcalose, de l'anhydride carbonique, de l'ion hydrogéne et de l'anoxémie: Sinus carotidiens et échanges respiratoires dans le poumons et au delá des poumons. Arch Int Pharmacodyn 39:400–448

Inoue T, Haque Z, Lukowiak K, Syed NI (2001) Hypoxia-induced respiratory patterned activity in *Lymnaea* originates at the periphery. J Neurophysiol 86:156–163

Ishii K, Ishii K, Kusakabe T (1985) Electrophysiological aspects of reflexogenic area in the chelonian, *Geoclemmys reevesii*. Respir Physiol 59:45–54

Ishii K, Ishii K, Massabuau JC, Dejours P (1989) Oxygen-sensitive chemoreceptors in the branchio-cardiac veins of the crayfish, *Astacus leptodactylus*. Respir Physiol 78:73–81

Ito S, Ohta T, Nakazato Y (1999) Characteristics of 5-HT-containing chemoreceptor cells of the chicken aortic body. J Physiol 515:49–59

Jones JD (1961) Aspects of respiration in *Planorbis corneus*(L.) and *Lymnaea stagnalis*(L.) (gastropoda: pulmonata). Comp Biochem Physiol 4:1–29

Jonz MG, Fearon IM, Nurse CA (2004) Neuroepithelial oxygen chemoreceptors of the zebrafish gill. J Physiol 560:737–752

Kamardin NN (1976) Structure and cellular organization of the osphradium of *Limnea stagnalis* (L.). Arkhiv Anatomii Gistologii I Embriologii 71:87–90

Kanz JE, Quast WD (1992) Respiratory pumping behavior in the marine snail *Aplysia californica* as a function of ambient hypoxia. Physiol Zool 65:35–54

Kusakabe T, Ishii K, Ishii K (1991) Dense granule-containing cells in the wall of the branchio-cardiac veins of a fresh water crayfish (*Astacus leptodactylus*). Anat Embryol 183:553–557

Laverack MS, Saier B (1993) Morphology and ultrastructure of a presumptive oxygen receptor in the gills of the Norway lobster *Nephrops norvegicus* (Decapoda). J Crustac Biol 13:504–510

López-Barneo J, Ortega-Sáenz P, Pardal R, Pascual A, Piruat JI (2008) Carotid body oxygen sensing. Eur Respir J 32:1386–1398

Massabuau JC, Burtin B (1984) Regulation of oxygen consumption in the crayfish *Astacus leptodactylus* at different levels of oxygenation: role of peripheral O_2 chemoreception. J Comp Physiol B 155:43–49

Milsom WK, Burleson ML (2007) Peripheral arterial chemoreceptors and the evolution of the carotid body. Respir Physiol Neurobiol 157:4–11

Morton DB, Stewart JA, Langlais KK, Clemens-Grisham RA, Vermehren A (2008) Synaptic transmission in neurons that express the *Drosophila* atypical soluble guanylyl cyclases, Gyc-89Da and Gyc-89Db, is necessary for the successful completion of larval and adult ecdysis. J Exp Biol 211:1645–1656

Nakayama K (2009) Cellular signal transduction of the hypoxia response. J Biochem 146:757–765

Neubauer JA, Sunderram J (2004) Oxygen-sensing neurons in the central nervous system. J Appl Physiol 96:367–374

Nurse CA (2010) Neurotransmitter and neuromodulatory mechanisms at peripheral arterial chemoreceptors. Exp Physiol 95:657–667

Ott M, Gogvadze V, Orrenius S, Zhivotovsky B (2007) Mitochondria, oxidative stress and cell death. Apoptosis 12:913–922

Py C, Denhoff MW, Martina M, Monette R, Comas T, Ahuja T, Martinez D, Wingar S, Caballero J, Laframboise S, Mielke J, Bogdanov A, Luk C, Syed N, Mealing G (2010) A novel silicon patch-clamp chip permits high-fidelity recording of ion channel activity from functionally defined neurons. Biotechnol Bioeng 107:593–600

Smith JC, Ellenberger HH, Ballanyi K, Richter DW, Feldman JL (1991) Pre-Bötzinger complex: a brainstem region that may generate respiratory rhythm in mammals. Science 254:726–729

Smith JC, Abdala APL, Rybak IA, Paton JFR (2009) Structural and functional architecture of respiratory networks in the mammalian brainstem. Proc Trans R Soc B 364:2577–2587

Sokolov VA, Kamardin NN (1977) The relation of impulse frequency in the osphradial nerve to the concentration of oxygen and inulin in liquid passing over the osphradium of the pond snail. Vestnik Leningr Univ Biol 1:87–90

Syed NI, Winlow W (1991) Respiratory behavior in the pond snail *Lymnaea stagnalis* II. Neural elements of the central pattern generator (CPG). J Comp Physiol 169:557–568

Syed N, Bulloch A, Lukowiak K (1990) *In vitro* reconstruction of the respiratory central pattern generator of the mollusk *Lymnaea*. Science 250:282–285

Syed NI, Harrison D, Winlow W (1991) Respiratory behavior in the pond snail *Lymnaea stagnalis*I. Behavioral analysis and the identification of motor neurons. J Comp Physiol A 169:541–555

Taylor AC, Brand AR (1975) A comparative study of the respiratory responses of the bivalves *Arctica islandica* (L.) and *Mytilus edulis*(L.) to declining oxygen tension. Proc R Soc Lond B Biol 190:443–456

Taylor EW, Butler PJ, Sherlock PJ (1973) The respiratory and cardiovascular changes associated with the emersion response of *Carcinus maenas* (L.) during environmental hypoxia, at three different temperatures. J Comp Physiol 86:95–115

Thannickal VJ (2009) Oxygen in the evolution of complex life and the price we pay. Cell Mol Biol 40:507–510
Vermehren-Schmaedick A, Ainsley JA, Johnson WA, Davies SA, Morton DB (2010) Behavioral responses to hypoxia in *Drosophila* larvae are mediated by atypical soluble guanylyl cyclases. Genetics 186:183–196
Wingrove JA, O'Farrell PH (1999) Nitric oxide contributes to behavioral, cellular, and developmental responses to low oxygen in *Drosophila*. Cell 98:105–114
Wohlgemuth SE, Taylor AC, Grieshaber MK (2000) Ventilatory and metabolic responses to hypoxia and sulphide in the lugworm *Arenicola marina* (L.). J Exp Biol 203:3177–3188
Wu RSS (2002) Hypoxia: from molecular responses to ecosystem responses. Mar Pollut Bull 45:35–45
Zhang Z-Q, Shear WA (2007) Linnaeus tercentenary and invertebrate taxonomy: an introduction. Zootaxa 1668:7–10
Zimmer M, Gray JM, Pokala N, Chang AJ, Karow DS, Marletta MA, Hudson ML, Morton DB, Chronis N, Bargmann CI (2009) Neurons detect increases and decreases in oxygen levels using distinct guanylate cyclases. Neuron 61:865–879

Chapter 3
Peripheral Chemoreceptors in Air- Versus Water- Breathers

Michael G. Jonz and Colin A. Nurse

Abstract Among the vertebrates, peripheral chemoreceptors have evolved to play a key role in matching oxygen delivery to the metabolic needs of the body cells and tissues. Specialized neuroepithelial cells (NECs) distributed within the gill filaments and/or lamellae of water-breathers appear to subserve this function by initiating an increase in ventilation in response to lowering of blood or water PO_2 (hypoxia). It is only recently, however, that these cells have become amenable for detailed investigations using electrophysiological tools. By contrast, the well-studied specialized neuroendocrine cells (i.e. glomus or type I cells) located principally in the carotid body of air-breathers initiate a similar reflex ventilatory response to hypoxia so as to maintain blood PO_2 homeostasis. In some species, however, the carotid body is immature and relatively insensitive to hypoxia at birth; it is during this period that their sympathoadrenal counterparts in the adrenal medulla act as key PO_2 receptors, critical for the proper transition to air-breathing life. It is becoming increasingly clear that in general these chemoreceptors act as polymodal receptors, i.e. capable of detecting several sensory modalities including high CO_2/H^+ or acid hypercapnia. Given the phylogenetic and ontogenetic evidence pointing to homology between the mammalian carotid artery and the first gill arch of teleosts, the question arises whether the mechanisms of chemosensing are conserved among these cell types. This review examines some of the anatomical and functional similarities among these peripheral chemoreceptors, while raising the possibility that the fundamental mechanisms of O_2 and CO_2/H^+ sensing arose first in water-breathers and are conserved among the vertebrates.

Keywords Fish gill • O_2 and CO_2/H^+ receptors • Carotid body • Adrenal medulla • K^+ current

3.1 Introduction

Vertebrates of both aquatic and terrestrial environments respond to hypoxic and hypercapnic challenges. In mammals, compensatory cardioventilatory responses include hyperventilation and tachycardia, while in fish and amphibians hyperventilation and bradycardia will occur. Comparative studies have

M.G. Jonz (✉)
Department of Biology, University of Ottawa, 30 Marie Curie Pvt, Ottawa, ON K1N 6N5, Canada
e-mail: mjonz@uottawa.ca

C.A. Nurse
Department of Biology, McMaster University, 1280 Main St. W, Hamilton, ON L8S 4K1, Canada
e-mail: nursec@mcmaster.ca

C.A. Nurse et al. (eds.), *Arterial Chemoreception: From Molecules to Systems*, Advances in Experimental Medicine and Biology 758, DOI 10.1007/978-94-007-4584-1_3,

demonstrated that these reflexes are mediated, in part, by peripheral chemoreceptors responsive to changes in O_2 and CO_2/pH. As will be described in this review, the location and distribution of peripheral chemoreceptors in air-breathers (e.g. mammals) and water-breathers (fish and larval amphibians) bear significant homology and suggest that the diffuse distribution of respiratory chemoreceptors in fish was antecedent to the evolution of the mammalian carotid body (Milsom and Burleson 2007). In addition, these differences across vertebrate groups may also have arisen as adaptations due to the specific solubility, or availability, of O_2 and CO_2 in aqueous *vs.* aerial environments (Gilmour 2001). The present paper discusses environmental and arterial chemoreceptors of air and water-breathers that detect changes in O_2 and CO_2/pH and initiate respiratory and cardiovascular reflexes. We have placed a particular focus on the carotid (and aortic) bodies of mammals, and the chemosensitive gills of fish and developing amphibians. In addition, we address similarities with adrenomedullary chromaffin cells, which have a similar embryonic origin as carotid body chemoreceptors. These chromaffin cells act as critical O_2 and CO_2/pH sensors in the neonates of some species including rodents and man, aiding in the transition to the air-breathing, extra-uterine environment. Though pulmonary neuroepithelial bodies (NEBs) have been well studied as sensors of airway PO_2 in the perinatal period, they have been the subject of recent reviews (Cutz et al. 2009), and will not be considered here.

3.2 Phylogeny of the Aortic Arches

In all embryonic vertebrates, the anterior arterial system is formed by a series of bilateral aortic arches that branch from the ventral aorta to the paired dorsal aortae. In fish and developing amphibians, these aortic arches course through the pharyngeal arches and provide blood to the gills once they develop (Weichert 1967). Although most vertebrates have six pairs of aortic arches during embryogenesis, this number is reduced in adults. In teleost fish (e.g. goldfish, trout, zebrafish), the first two arches (mandibular and the hyoid) degenerate and leave the remaining aortic arches to develop with the gills. This arrangement is also found in aquatic or larval stages of amphibians. By contrast, in mammals and adult amphibians the fifth aortic arch is also lost. This leaves the third and fourth aortic arches that go on to form the internal carotid artery and the aortic arch, respectively. Furthermore, these structures will eventually form the carotid and aortic bodies in mature adults. An important phylogenetic comparison can then be made. The first gill arch in fish and the mammalian carotid body are thus homologous structures and derivatives of the third embryonic aortic arch, while the second gill arch and the aortic bodies are homologues and derivatives of the fourth embryonic aortic arch. As will be discussed in the following sections, these regions are important because they all represent sites of O_2 and/or CO_2/pH sensing.

3.3 O_2 and CO_2 Receptors in the Gills of Water Breathers

3.3.1 Teleost Fish

The gills of teleost fish are organized into four bilateral arches that give rise to numerous primary gill filaments and secondary lamellae. Both the filaments and lamellae are covered by a thin epithelial layer composed of several cell types that provides a diffusible boundary between the external environment and the extracellular fluid (Evans et al. 2005). The respiratory chemoreceptors of the gill, called neuroepithelial cells (NECs), are found within the filament epithelium (Fig. 3.1a) and are thus poised to detect changes in O_2 and CO_2 from the external environment, as water passes over the gills during

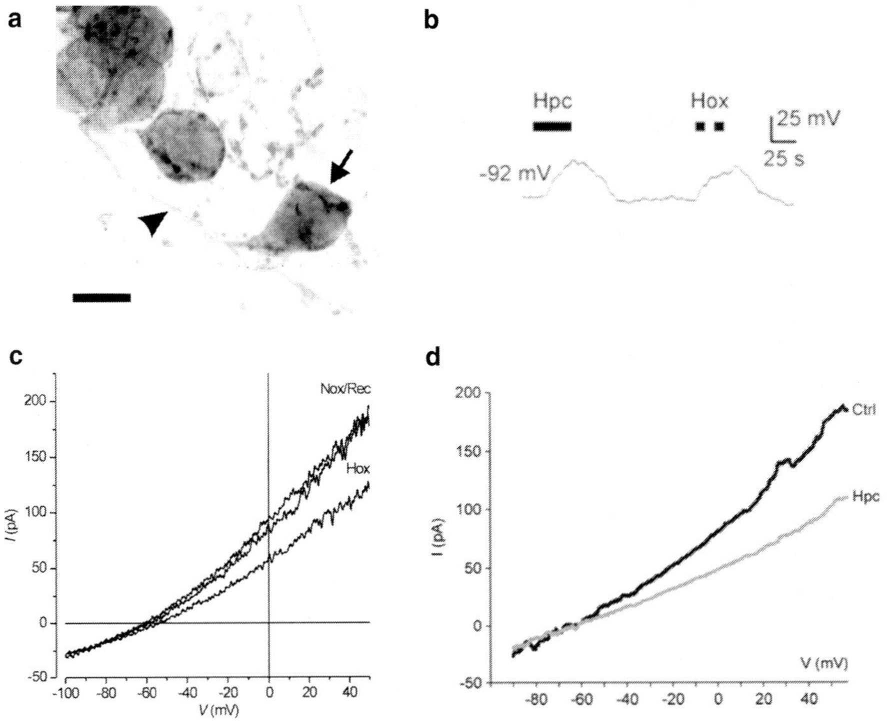

Fig. 3.1 Gill neuroepithelial cells (NECs) from zebrafish respond to hypoxia and hypercapnia. (**a**): A confocal image from a single gill filament showing NECs (*arrow*) and innervating nerve fibres (*arrowhead*). NECs and nerve fibres were immunolabelled with anti-serotonin and anti-zn-12, respectively. Scale bar 5 μm. (**b**): Current-clamp recording from a NEC *in vitro* demonstrating that both hypercapnia (Hpc) and hypoxia (Hox) induce membrane depolarization in the same cell. (**c**), (**d**): Voltage-clamp recordings indicate that hypoxia and hypercapnia reduce outward K^+ current (Panels b and d were reproduced from Qin et al. (2010), and panel c from Jonz et al. (2004). With permission, John Wiley and Sons)

respiration, or from the nearby arterial gill circulation. Although changes in water PO_2 are expected to primarily drive cardioventilatory reflexes, fish do respond to relatively small changes in PCO_2 (Gilmour 2001; Milsom and Burleson 2007). Early comparative studies on a variety of species (e.g. trout, catfish, dogfish) demonstrated that gill NECs resembled carotid body glomus or type I cells at the ultrastructural level (reviewed by Evans et al. 2005; Milsom and Burleson 2007). NECs possessed characteristics of neurosecretion, such as cytoplasmic dense-cored vesicles and storage of serotonin, and received innervation. Unlike type I cells, however, gill NECs are not found in clusters but are usually organized as solitary cells dispersed throughout the filaments on all gill arches. Furthermore, while some species (e.g. trout) have NECs confined to the gill filaments, others (e.g. zebrafish, goldfish) have NECs in the filaments and respiratory lamellae (Saltys et al. 2006). This example of species specificity suggests that the distribution of NECs in the gills may serve an adaptive role. In addition, NECs exhibit plasticity that may underlie acclimatization to aquatic environments with variable oxygen levels. There is evidence in zebrafish that morphological changes in NECs occur, such as proliferation and hypertrophy, following long-term exposure to hypoxia (Jonz et al. 2004), while hyperoxia induces a decrease in NEC density with an associated reduction in the ventilatory response to acute hypoxia (Vulesevic et al. 2006).

Innervation of the gills in fish arises from the glossopharyngeal nerve for the first pair of gill arches, and from the glossopharyngeal and vagus nerves for the remaining three pairs. NECs form synapses with catecholaminergic, serotonergic and nitrergic fibres (Evans et al. 2005). The specific characteristics of sensory nerve terminals that contact NECs are not yet known, but NEC innervation by nerve fibres of the gill arch was correlated with the development of hypoxic hyperventilation in zebrafish, suggesting a sensory component (Jonz and Nurse 2006). While NECs described in most studies are innervated, non-innervated NECs have been reported and these may exert paracrine effects on the surrounding vascular tissues (Coolidge et al. 2008).

In addition to serotonin, in some species NECs may contain acetylcholine, neuropeptides, and the enzymes tyrosine hydroxylase and nitric oxide synthase, which synthesize catecholamines and nitric oxide, respectively (Milsom and Burleson 2007). However, despite the wealth of biochemical evidence from NECs, release of a neurotransmitter from these cells following hypoxic or hypercapnic stimulation has not been demonstrated directly. Therefore, the primary neurotransmitter involved in O_2 or CO_2 sensing in fish remains elusive. An important series of studies did show, however, that in trout a variety of neurochemicals applied by perfusion or exogenous application had stimulatory effects on gill nerve activity (reviewed by Milsom and Burleson 2007). These studies demonstrated that acetylcholine and serotonin increased spike frequency in single afferent fibre recordings of the glossopharyngeal nerve, as did hypoxia. These studies also confirmed the notion that there are O_2-sensitive receptors in the gills oriented towards the external environment and those that are oriented towards the arterial blood supply. The role of each of these receptors is highly species specific but, generally, changes in heart rate are mediated by externally-oriented O_2 receptors, while both cardiac and ventilatory reflexes are controlled by internally-oriented O_2 receptors (Milsom and Burleson 2007). CO_2 receptors of the gill, on the other hand, appear to be predominantly externally-oriented and mediate changes in ventilation (Gilmour 2001). Thus, chemosensing in the gill appears to utilize both aquatic and arterial environments, and may involve a suite of neurotransmitters.

To date, there are four studies that have examined the physiological response to chemical stimuli of NECs isolated from the gills of three fish species: zebrafish (Jonz et al. 2004; Qin et al. 2010), catfish (Burleson et al. 2006), and trout (Zhang et al. 2011). In patch-clamp studies of NECs acutely isolated from the gills of zebrafish, a decrease in extracellular PO_2 below 80 mmHg reduced activity of plasma membrane K^+ channels, and this trend continued in a PO_2-dependent manner (Jonz et al. 2004). An increase in PCO_2 to 7.5 mmHg also reduced K^+ channel activity in NECs (Qin et al. 2010). This response occurred under conditions of constant extracellular pH, suggesting that NECs are responsive directly to CO_2. However, the magnitude of the CO_2 response was affected by carbonic anhydrase activity, suggesting a role for intracellular acidification. In zebrafish, the O_2/CO_2-sensitive conductance was pharmacologically characterized as that of voltage-independent or background K^+ channels (Fig. 3.1c, d). In addition, K^+ channel inhibition by hypoxia or hypercapnia led to membrane depolarization. Moreover, a subset of NECs was found to be responsive to both O_2 and CO_2 (Fig. 3.1b; Qin et al. 2010). Thus, a working model of chemotransduction in gill NECs is similar to that proposed for transduction in carotid body type I cells (López-Barneo 2003), in which membrane depolarization via K^+ channel inhibition leads to activation of voltage-dependent Ca^{2+} channels and neurosecretion. The identity of the O_2 and/or CO_2 sensor in NECs, however, has not been resolved. Ca^{2+} channels have not yet been studied directly in NECs, but evidence from studies in trout indicates that membrane depolarization in NECs induced by high external [K^+] leads to an increase in cytosolic Ca^{2+}, presumably via voltage-gated Ca^{2+} entry (Zhang et al. 2011). This study also provided the first evidence in NECs of the sensing of ammonia (NH_4^+), a strong stimulant of ventilation in fish. Thus, as with type I cells, gill NECs in fish have emerged as polymodal chemosensors, capable of mediating cardiorespiratory reflexes in response to a variety of chemicals.

3.3.2 *Larval Amphibians*

Amphibians undergo complex morphological changes associated with metamorphosis. This includes the transition from gas exchange by internal gills in larvae to ventilation of fully developed lungs in adults. A carotid body-like structure, called the carotid labyrinth, is the primary O_2-sensing organ in adult amphibians. In larvae, however, the carotid labyrinth has not yet developed and peripheral O_2 chemoreception appears to take place in the gills. In *Xenopus laevis*, the gills of pre-metamorphic larvae are organized in a similar manner to those of fish and contain serotonergic NECs that receive cranial nerve innervation (Saltys et al. 2006). As in fish, the distribution of NECs in amphibian gills suggests that they may sense PO_2 changes in water or arterial blood. In nerve recordings from the first gill arch of the bullfrog, an increase in discharge frequency was associated with the change in superfusate from a hyperoxic to hypoxic solution (Straus et al. 2001), indicating the presence of functional gill chemoreceptors. Because studies on the physiological effects of hypoxia on isolated NECs of the amphibian gill are lacking, the mechanism of chemosensory transduction remains to be determined in this model.

3.4 Peripheral O_2 and CO_2/H^+ Chemoreceptors in Air Breathers

3.4.1 Mammalian Carotid and Aortic Bodies

In response to a decrease in arterial PO_2 (hypoxia), air-breathing mammals respond with an increase in ventilation mediated via a reflex pathway that ultimately leads to a restoration of blood PO_2 (Gonzalez et al. 1994). This reflex depends critically on O_2-sensitive peripheral chemoreceptors located principally in the carotid bodies (CB). These bilaterally-paired organs are located near the bifurcation of the common carotid artery and receive a sensory innervation via the carotid sinus nerve whose parent cell bodies reside in the petrosal ganglion. The central projections of petrosal chemoafferent neurons provide input to the central pattern generator in the brainstem, which ultimately regulates breathing. In the CB, innervated chemoreceptor glomus or type I cells are organized in clusters and are intimately associated with glial-like type II cells. Evidence for extra-carotid chemoreceptors emerged from studies of Comroe in the 1930s (Comroe 1939) on CB-denervated animals, in which anoxemia elicited residual respiratory responses that were abolished by sectioning the vagus and aortic nerves. This led to the subsequent discovery of CB-like glomus cells around the aortic arch, located in small "paraganglia-like" structures known as aortic bodies (ABs). Several studies have shown that AB paraganglia are irregularly distributed throughout the cardiac vasculature and surrounding peripheral nerves, with a particularly high density at nerve branch points (McDonald and Blewett 1981). In contrast to the CB which contains several thousand glomus cells in the rat, a typical AB consists of cluster of 5 to >50 glomus cells and associated sustentacular (type II) cells in a ratio of approximately 4:1 (Dahlqvist et al. 1994; McDonald and Blewett 1981). Whereas a few autonomic neuronal cell bodies occur among CB glomus cell clusters, a greater proportion of local neurons appear in association with AB glomus clusters at the vagus-recurrent laryngeal nerve bifurcation (Piskuric et al. 2011). At least some of these local neurons appear to innervate AB glomus cells though it remains to be determined whether or not they subserve a sensory relay function in mediating local cardiovascular reflexes during hypoxemia. Regardless of their location glomus cells express tyrosine hydroxylase (TH), the rate limiting enzyme in catecholamine (CA) biosynthesis, and are thought to be predominantly dopaminergic (Gonzalez et al. 1994; Piskuric et al. 2011; Nurse 2010). However, the presence of other neurotransmitters and neuromodulators, including ACh and serotonin, has been noted in immunocytochemical studies in both the CB and AB of several species (Nurse, 2010; Piskuric et al. 2011).

There is strong evidence that ATP acts as the principle excitatory neurotransmitter during hypoxia chemotransduction in the CB, though other neurochemicals (e.g. ACh, 5-HT, GABA, dopamine, adenosine, histamine) may help shape the final sensory output as reflected by the carotid sinus nerve discharge (Nurse 2010). Though controversies still persist, cellular studies in several laboratories over

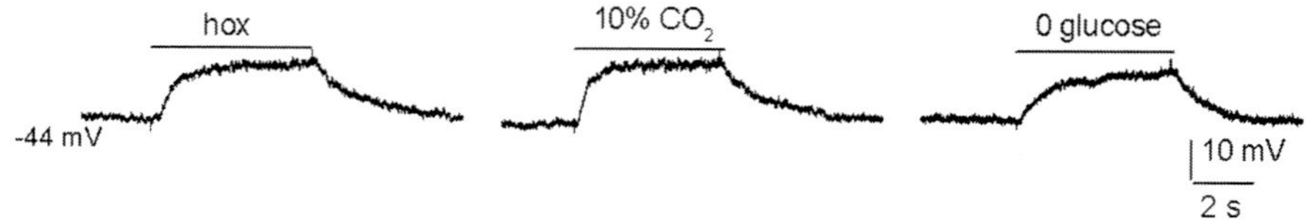

Fig. 3.2 Polymodal chemosensing in a ~10 day-old rat carotid body glomus cell that was a member of a cell cluster. Figure illustrates a glomus cell responding to each of three stimuli, i.e. hypoxia (PO_2 ~15 mmHg), isohydric hypercapnia (10% CO_2; pH = 7.4) and aglycemia (0 mM glucose), with membrane depolarization (Courtesy of Min Zhang)

the last ~25 years have resulted in significant progress towards our understanding of the transduction mechanisms by which CB glomus cells sense hypoxia (López-Barneo 2003; Buckler 2007). However, the molecular identity of the PO_2 sensor still remains arguable, though it is possible that CB glomus cells may use more than one sensor for hypoxia signaling (Buckler 2007; see paper by Evans et al. in this volume). By contrast, little is known about the mechanisms by which AB glomus cells sense hypoxia, and it was only recently that a dissociated cell culture model of isolated AB glomus cells and local neurons was developed (Piskuric et al. 2011).

3.4.2 *Mammalian Glomus Cells as Polymodal Chemoreceptors*

It is clear that hypoxia is only one of several sensory modalities that can excite CB glomus cells, and hence they might best be considered as polymodal chemoreceptors. For example, blood CO_2 and H^+ ions are also under homeostatic control, and though central chemoreceptors in the brainstem are the main CO_2/pH sensors, CB glomus cells can also sense acid hypercapnia and contribute to acid–base balance via reflex control of breathing (Gonzalez et al. 1994; Kumar and Bin-Jaliah 2007; Nurse 2010). Acid hypercapnia depolarizes CB glomus cells, leading to voltage-gated Ca^{2+} entry and neurosecretion. The mechanisms involve intracellular or extracellular pH regulation of acid sensitive TASK and/or ASIC channels, and the effects of elevated CO_2 are mediated principally by cytoplasmic acidification catalysed by intracellular carbonic anhydrase (Gonzalez et al. 1994; Nurse 2010). A more recent stimulus that has been shown to excite the CB is low blood glucose or hypoglycemia, and thus the CB may also be considered a peripheral glucosensor similar to the ones in the liver and portal vein (López-Barneo 2003; Kumar and Bin-Jaliah 2007; Nurse 2010). Given the marked sensitivity of brain neurons to oxygen and glucose deprivation, the CB appears strategically positioned to integrate signals resulting from small reductions in these variables and initiate counter-regulatory autonomic responses (López-Barneo 2003). While there is some debate as to whether glucosensing by the CB *in situ* is direct or indirect (López-Barneo 2003; Kumar and Bin-Jaliah 2007; Nurse 2010), there is evidence that at least some isolated glomus cells depolarize and release neurotransmitters following exposure to acute hypoglycemia (López-Barneo 2003; Nurse 2010). These data suggest that glomus cells may act as general metabolic sensors capable of detecting blood-borne chemicals, e.g. hypoxia, acid hypercapnia, and low glucose, that are the products of metabolically active tissues. An example of a rat CB glomus cell, within an isolated cell cluster, responding to hypoxia, isohydric hypercapnia (10% CO_2; pH = 7.4), and aglycemia with membrane depolarization is shown in Fig. 3.2. The polymodal nature of CB chemoreceptors is further emphasized by the fact that other physico-chemical stimuli in blood including osmolarity, K^+ concentration, and temperature also appear to be sensed by this organ (Kumar and Bin-Jaliah 2007).

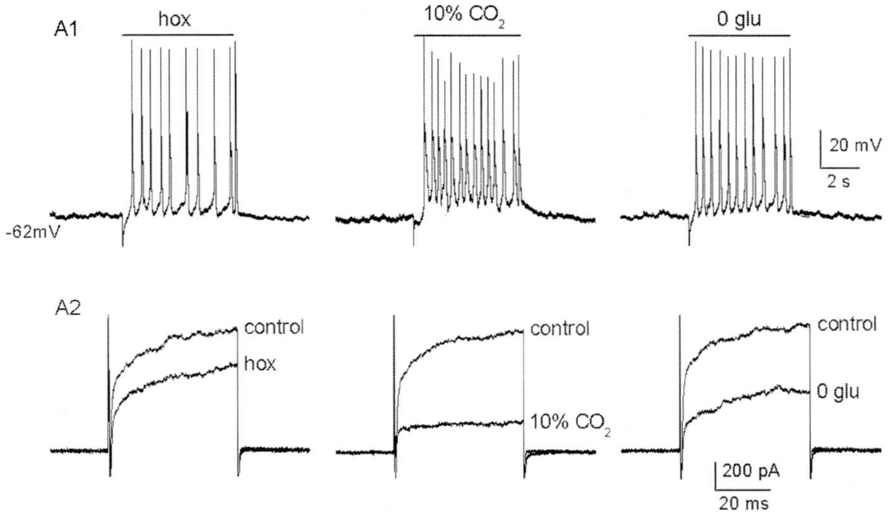

Fig. 3.3 Polymodal chemosensing in a neonatal rat adrenal chromaffin cell. Current clamp recordings showing membrane depolarization and action potential firing during hypoxia, aglycemia, and isohydric hypercapnia (A1). Under voltage clamp, each stimulus caused inhibition of outward K^+ current during a voltage step to +30 mV (A2); holding potential was −60 mV (Courtesy of Min Zhang)

3.4.3 Perinatal Adrenal Chromaffin Cells as Polymodal Chemoreceptors

Adrenomedullary chromaffin cells (AMC) have a similar developmental origin from the neural crest as CB glomus cells and though they both function as peripheral O_2 and CO_2/H^+ chemoreceptors, they do so with different developmental profiles. At birth, the CB of humans and rodents is immature and relatively insensitive to hypoxia and acid hypercapnia (Kim et al. 2011). Interestingly, it is around this period that AMC act as important peripheral O_2 and CO_2/H^+ sensors. In particular, it has been known for some time that catecholamine (CAT) release from neonatal AMC in response to natural asphyxial stressors associated with uterine contractions and postnatal apneas, i.e. hypoxia and acid hypercapnia, plays a critical role in the adaptation of the newborn to extrauterine life (Slotkin and Seidler 1988; Nurse et al. 2009). This CAT secretion is required for initiation of lung respiration via stimulation of surfactant secretion and fluid reabsorption, and its impairment results in a loss of hypoxia tolerance and increased neonatal mortality (Slotkin and Seidler 1988). Because splanchnic sympathetic innervation of the adrenal medulla is immature or absent in the neonates of some species including rat and man, the vital CAT surge in response to asphyxial stimuli occurs via a *direct* 'non-neurogenic' mechanism (Slotkin and Seidler 1988). Similar to CB glomus cells, neonatal AMC sense acute hypoxia via K^+ channel inhibition and membrane depolarization, though the particular subtypes of K^+ channels involved may differ (Fig. 3.3 A1, A2; Nurse et al. 2009). CO_2 sensing by neonatal AMC is mediated by activation of a resting cation conductance and inhibition of K^+ channels coupled to carbonic anhydrase activity (see Nurse et al. 2009); these combine to trigger membrane depolarization and catecholamine secretion during hypercapnia (Fig. 3.3 A1, A2). Interestingly, both the direct hypoxia and CO_2/H^+ sensing mechanisms in AMC are suppressed postnatally along a time course that parallels the development of functional innervation of the adrenal medulla (Slotkin and Seidler 1988; Thompson et al. 1997; Nurse et al. 2009), and conversely, this timing coincides with the functional maturation of the CB as the principal peripheral O_2 and CO_2/H^+ sensor.

The suppression or loss of direct O_2 and CO_2/H^+ sensing by chromaffin cells with postnatal development provides a unique opportunity to address factors that regulate chemosensitivity. These findings, together with the observation that chemosensitivity returns after denervation of the adult adrenal gland (Levinsky and Lopez-Barneo 2009), suggest that chemical signals released from the splanchnic nerve may be directly involved. Given that the splanchnic innervation is cholinergic, it is possible that activation of postsynaptic nicotinic ACh receptors (AChR) on AMC is a contributing factor. Indeed, activation of these receptors by chronic exposure to nicotine *in utero* and *in vitro* resulted in the selective suppression of hypoxic, but not CO_2, chemosensitivity in these cells (Buttigieg et al. 2008). Moreover, the signaling pathway leading to the suppression of hypoxic sensitivity involved a role for α7 nicotinic AChR, and a hypoxia inducible factor 2α (HIF-2α)-dependent upregulation of K_{ATP} channels (Buttigieg et al. 2009). These K_{ATP} channels open during acute hypoxia in neonatal AMC and appear to serve a protective role by limiting the membrane depolarization caused by inhibition of other K^+ channel subtypes. The importance of this pathway, and clinical relevance to mothers who smoke during pregnancy, was demonstrated in experiments where pre-treatment of nicotine-exposed pups *in vivo* with the K_{ATP} channel blocker glibenclamide led to increased survival of the offspring following exposure to a hypoxic stress (Buttigieg et al. 2009). The preservation of CO_2 chemosensitivity in these nicotine-exposed neonatal AMC implied that other mediators are likely to participate in the developmental loss of CO_2 sensitivity. Thus, similar to CB glomus cells, neonatal AMC also appear to act as polymodal chemosensors. This idea is further supported by recent evidence indicating that neonatal chromaffin cells can sense low glucose and this property also appears to be developmentally regulated (Fig. 3.3A1,A2; see paper by Livermore et al. this volume).

Acknowledgements MGJ is supported by the Natural Sciences and Engineering Research Council of Canada (NSERC). CAN received support from the Canadian Institutes of Health Research, NSERC, and Heart and Stroke Foundation of Ontario.

References

Buckler KJ (2007) TASK-like potassium channels and oxygen sensing in the carotid body. Respir Physiol Neurobiol 157:55–64

Burleson ML, Mercer SE, Wilk-Blaszczak MA (2006) Isolation and characterization of putative O_2 chemoreceptor cells from the gills of channel catfish (*Ictalurus punctatus*). Brain Res 1092:100–107

Buttigieg J, Brown ST, Zhang M, Lowe M, Holloway AC, Nurse CA (2008) Chronic nicotine *in utero* selectively suppresses hypoxic sensitivity in neonatal rat adrenal chromaffin cells. FASEB J 22:1317–1326

Buttigieg J, Brown S, Holloway AC, Nurse CA (2009) Chronic nicotine blunts hypoxic sensitivity in perinatal rat adrenal chromaffin cells via upregulation of K_{ATP} channels: Role of α7 nicotinic acetylcholine receptor and hypoxia-inducible factor 2α. J Neurosci 29:7137–7147

Comroe JJ (1939) The location and function of the chemoreceptors of the aorta. Am J Physiol 127:176–191

Coolidge EH, Ciuhandu CS, Milsom WK (2008) A comparative analysis of putative oxygen-sensing cells in the fish gill. J Exp Biol 211:1231–1242

Cutz E, Pan J, Yeger H (2009) The role of NOX2 and "novel oxidases" in airway chemoreceptor O_2 sensing. Adv Exp Med Biol 648:427–438

Dahlqvist A, Neuhuber WL, Forsgren S (1994) Innervation of laryngeal nerve paraganglia: an anterograde tracing and immunohistochemical study in the rat. J Comp Neurol 345:440–446

Evans DH, Piermarini PM, Choe KP (2005) The multifunctional fish gill: dominant site of gas exchange, osmoregulation, acid–base regulation, and excretion of nitrogenous waste. Physiol Rev 85:97–177

Gilmour KM (2001) The CO_2/pH ventilatory drive in fish. Comp Biochem Physiol A Mol Integr Physiol 130:219–240

Gonzalez C, Almaraz L, Obeso A, Rigual R (1994) Carotid body chemoreceptors: from natural stimuli to sensory discharge. Physiol Rev 74:829–898

Jonz MG, Nurse CA (2006) Ontogenesis of oxygen chemoreception in aquatic vertebrates. Respir Physiol Neurobiol 154:139–152

Jonz MG, Fearon IM, Nurse CA (2004) Neuroepithelial oxygen chemoreceptors of the zebrafish gill. J Physiol 560:737–752

Kim D, Papreck JR, Kim I, Donnelly DF, Carroll JL (2011) Changes in oxygen sensitivity of TASK in carotid body glomus cells during early development. Respir Physiol Neurobiol 177:228–235

Kumar P, Bin-Jaliah I (2007) Adequate stimuli of the carotid body: more than an oxygen sensor? Respir Physiol Neurobiol 157:12–21

Levinsky KL, López-Barneo J (2009) Developmental change in T-type Ca^{2+} channel expression and its role in rat chromaffin cell responsiveness to acute hypoxia. J Physiol 58:1917–1929

López-Barneo J (2003) Oxygen and glucose sensing by carotid body glomus cells. Curr Opin Neurobiol 13:493–499

McDonald DM, Blewett RW (1981) Location and size of carotid body-like organs (paraganglia) revealed in rats by the permeability of blood vessels to Evans blue dye. J Neurocytol 10:607–643

Milsom WK, Burleson ML (2007) Peripheral arterial chemoreceptors and the evolution of the carotid body. Respir Physiol Neurobiol 157:4–11

Nurse CA (2010) Neurotransmitter and neuromodulatory mechanisms at peripheral arterial chemoreceptors. Exp Physiol 95:657–667

Nurse CA, Buttigieg J, Brown S, Holloway AC (2009) Regulation of oxygen sensitivity in adrenal chromaffin cells. Ann N Y Acad Sci 1177:132–139

Piskuric NA, Vollmer C, Nurse CA (2011) Confocal immunofluorescence study of rat aortic body chemoreceptors and associated neurons in situ and in vitro. J Comp Neurol 519:856–873

Qin Z, Lewis JE, Perry SF (2010) Zebrafish (*Danio rerio*) gill neuroepithelial cells are sensitive chemoreceptors for environmental CO_2. J Physiol 588:861–872

Saltys HA, Jonz MG, Nurse CA (2006) Comparative study of gill neuroepithelial cells and their innervation in teleosts and *Xenopus* tadpoles. Cell Tissue Res 323:1–10

Slotkin TA, Seidler FJ (1988) Adrenomedullary catecholamine release in the fetus and newborn: secretory mechanisms and their role in stress and survival. J Devel Physiol 10:1–16

Straus C, Wilson RJ, Remmers JE (2001) Oxygen sensitive chemoreceptors in the first gill arch of the tadpole, *Rana catesbeiana*. Can J Physiol Pharmacol 79:959–962

Thompson RJ, Jackson A, Nurse CA (1997) Developmental loss of hypoxic chemosensitivity in rat adrenomedullary chromaffin cells. J Physiol 498:503–510

Vulesevic B, McNeill B, Perry SF (2006) Chemoreceptor plasticity and respiratory acclimation in the zebrafish *Danio rerio*. J Exp Biol 209:1261–1273

Weichert CK (1967) Elements of chordate anatomy. McGraw-Hill, New York

Zhang L, Nurse CA, Jonz MG, Wood CM (2011) Ammonia sensing by neuroepithelial cells and ventilatory responses to ammonia in rainbow trout. J Exp Biol 214:2678–2689

Chapter 4
Sex-Specific Effects of Daily Gavage with a Mixed Progesterone and Glucocorticoid Receptor Antagonist on Hypoxic Ventilatory Response in Newborn Rats

Stéphanie Fournier, Van Diep Doan, and Vincent Joseph

Abstract We tested the hypothesis that daily gavage with mifepristone, a mixed progesterone/glucocorticoid receptor antagonist would alter hypoxic ventilatory response (HVR) in newborn male and female rats. Rats were treated with mifepristone (40μg/g/day), or vehicle between postnatal days 3–12, and used at 10–12 days of age to record baseline ventilatory and metabolic values using whole body plethysmography. HVR was tested by exposing the animals to 14% and 12% O_2 for 20 minutes each. HVR was enhanced by mifepristone treatment, mainly due to an effect on tidal volume that remained higher in mifepristone treated rats during both levels of hypoxic exposure. This effect was sex-specific being apparent only in male rats. In Vehicle treated rats, HVR was higher in females than in males, which was also due to a higher tidal volume in hypoxia (at 14 and 12% O_2). We conclude that the activity of the progesterone and/or glucocorticoid receptors modulates respiratory control in rat pups, and that these effects are different in males and females.

Keywords Newborn rats • Gavage • Mifepristone • Steroid receptor antagonist • Progesterone • Glucocorticoid • Hypoxic ventilatory response • Whole body plethysmography

4.1 Introduction

Progesterone is an effective respiratory stimulant, acting both on the central nervous system and on peripheral chemoreceptors to enhance resting minute ventilation, ventilatory and carotid sinus nerve responses to hypoxia (Soliz and Joseph 2005). This has potential far-reaching implications since progesterone-based hormone replacement therapy reduces the occurrence of sleep apneas in postmenopausal women (Shahar et al. 2003). In newborn rats, chronic low-dose administration of progesterone through the lactating mother enhances the hypoxic ventilatory response (HVR) and reduces apnea frequency under normoxia or hypoxia (Lefter et al. 2008), and it has been suggested that progesterone might potentially be used as a therapy against apneas in preterm neonates (Finer et al. 2006). The glucocorticoid corticosterone is a candidate as a potent modulator of respiratory function. In adult

S. Fournier • V.D. Doan • V. Joseph (✉)
Department of Pediatrics, Laval University, Centre de Recherche CHUQ, Hôpital St-François d'Assise, Local D0-711, 10 rue de l'Espinay, Quebec, QC, G1L 3L5, Canada
e-mail: joseph.vincent@crsfa.ulaval.ca

C.A. Nurse et al. (eds.), *Arterial Chemoreception: From Molecules to Systems*, Advances in Experimental Medicine and Biology 758, DOI 10.1007/978-94-007-4584-1_4,

male rats, chronic implantation of corticosterone pellets (300 mg/14 days) almost doubles HVR (Fournier et al. 2007). Contrastingly, deprivation of maternal stimuli in rat pups (neonatal maternal separation) activates stress response and enhances plasma corticosterone level. While it does not affect HVR in pups (Gulemetova and Kinkead 2011), neonatal maternal separation has long term effects on respiratory control resulting in enhanced HVR and higher apnea frequency in adult rats (Kinkead et al. 2009), suggesting a long-term effect of glucocorticoid receptor activation during a specific period of postnatal development.

In newborn mammals, the gonads and adrenals are not fully developed and circulating levels of both progesterone and corticosterone are low (Galeeva et al. 2010), but local steroid synthesis might occur within the central and peripheral nervous system (Zwain and Yen 1999). We have reported dense staining for progesterone receptor, and cytochrome P450 side-chain-cleavage, an enzyme necessary for steroid synthesis, in the carotid bodies of foetal, newborn, and adult male rats (Joseph et al. 2006). Furthermore, both progesterone (Haywood et al. 1999; Behan and Thomas 2005) and glucocorticoid receptors (Galeeva et al. 2010) (both members of the nuclear hormone receptor superfamily of ligand-activated transcription factors) are found in brain areas involved in respiratory control (hypothalamic and brainstem nuclei, phrenic motoneurons). These observations raise the intriguing possibility that endogenous steroids such as progesterone or corticosterone might modulate respiratory control during postnatal development. Mifepristone, a synthetic steroid compound, binds with very high affinity to the rat progesterone and glucocorticoid receptors (Schreiber et al. 1983), forming an inactive complex, and is one of the few available antagonists for these receptors. Mifepristone also binds weakly to the androgen receptor (9% of testosterone binding activity), and might also have weak anti-androgenic activity (Schreiber et al. 1983). In the present study, we tested the effects of daily treatment with mifepristone (40 µg/g/day, gavage) between postnatal days 3–12 in newborn rats. We then evaluated their normoxic ventilation and hypoxic ventilatory response using whole body plethysmography at 12 days of age.

4.2 Methods

Adult male and female rats were ordered from Charles-River (St. Constant, Québec) and used for mating. At birth, all litters were reduced to 12 pups to avoid specific effects on growth, and pups were randomly assigned to a vehicle (5% ethanol in propylene glycol) or a mifepristone (mifepristone – 40 µg/g/day (Lonstein et al. 2001)) group (daily oral gavage (0.05 ml/10 g) from postnatal days 3 to12). We used a total of five litters for this study. All protocols have been reviewed and approved by our local committee on animal use and care.

Respiratory recordings were performed in 10–12-day-old rat pups using whole-body flow-through plethysmography (Emka technologies, Paris, France) as previously described (Lefter et al. 2008, 2007). Airflow through the chamber was set at ~100 ml/min, and the temperature inside the chamber was fixed at 30 C using a temperature control loop. Oxygen and CO_2 levels were analysed for the calculation of O_2 consumption ($\dot{V}_{O_2}$) and CO_2 production ($\dot{V}_{CO_2}$). All signals were stored on a computer and used to calculate respiratory parameters ($\dot{V}E$, f_R and V_T), O_2 consumption, and CO_2 production. After a period of habituation in the plethysmograph chamber (10–15 min), the chamber was opened, and rectal temperature was measured. The animal was then returned to the chamber and allowed an additional 10 min before the start of the experiments. After baseline recordings for 10–15 min under normoxia, the inlet air was mixed with a flow of nitrogen to reach 14% O_2, maintained for 20 min. Then the O_2 level was decreased from 14% to 12%, and maintained at this level for another 20 min. At the end of the exposure, the chamber was opened, and the rectal temperature was immediately measured.

Table 4.1 Respiratory and metabolic data recorded in normoxia in vehicle (Veh) and mifepristone-treated (Mif) male rat pups

	Males		Females	
	Veh (n = 13)	Mif (n = 14)	Veh (n = 11)	Mif (n = 15)
Body weight (g)	26.8 ± 0.5	23.8 ± 0.6*	25.7 ± 0.5	23.8 ± 0.3*
Rectal temp. (°C)	34.2 ± 0.1	33.9 ± 0.1	33.9 ± 0.1	34.2 ± 0.1
fR (breaths/min)	172 ± 5	178 ± 3	170 ± 4	174 ± 4
V_T (ml/100 g)	0.84 ± 0.03	0.96 ± 0.04*	0.87 ± 0.06	0.95 ± 0.04*
$\dot{V}E$ (ml/min/100 g)	144 ± 7	173 ± 10*	149 ± 13	165 ± 8*
$\dot{V}E$ (ml/min/$g^{0.47}$)	8.2 ± 0.5	8.4 ± 0.7	8.8 ± 0.4	9.2 ± 0.6
$\dot{V}_{O_2}$ (ml/min/$g^{0.52}$)	0.18 ± 0.01	0.19 ± 0.01	0.17 ± 0.01	0.17 ± 0.01
$\dot{V}_{CO_2}$ (ml/min/$g^{0.52}$)	0.11 ± 0.01	0.13 ± 0.01	0.11 ± 0.01	0.12 ± 0.01

Values for minute ventilation, oxygen consumption and CO_2 production have been corrected using allometric scaling to take into consideration the lower body weight of Mif pups. All values are mean ± sem
*p < 0.05 for group effect

All values are reported as means±sem. Baseline values were analyzed with a two-way ANOVA using treatment and sex as grouping variables, followed by one-way ANOVA for treatment effect within each sex (no sex-effect appeared in the analysis). Responses to hypoxia were analyzed with a two-way ANOVA for repeated measures at each level of hypoxia. If a significant effect ($p<0.05$) for sex or treatment, or interaction with hypoxia appeared, we used an ANOVA for repeated measures to perform relevant comparisons by splitting the data table (treatment effect for each sex, sex effect for each treatment). All analysis were performed using statview 5.0.1

4.3 Results

Mifepristone-treated pups had reduced body weight compared to vehicle for both males and females, and slightly higher tidal volume and minute ventilation (Table 4.1). To take into account the difference in body weight we applied allometric scaling (Lefter et al. 2008). When minute ventilation, $\dot{V}_{O_2}$ and $\dot{V}_{CO_2}$ are adequately corrected for body weight they are similar in mifepristone compared to vehicle pups.

Respiratory frequency increased rapidly at 14% O_2 followed by a much more modest increase at 12% O_2 (Fig. 4.1). This effect was not affected by mifepristone treatment or sex.

Tidal volume declined after 5 min of exposure to 14% O_2 in Veh males, but remained close to baseline in mifepristone treated males under 14% (treatment effect $p=0.0005$) and 12% O_2 ($p=0.0005$). This effect of mifepristone was not observed in females ($p=0.2$ and 0.4 for treatment at 14% and 12% O_2).

Minute ventilation increased in response to 14% O_2 in Veh males, then returned to baseline level after 5–6 min hypoxia. Male pups treated with mifepristone maintained higher ventilation at 14% O_2 (treatment effect $p=0.038$), and a similar pattern occured at 12% O_2 (treatment effect $p=0.027$). In females there was an interaction between hypoxia at 14% O_2 and mifepristone treatment ($p=0.017$), essentially due to an increased response during the early phase (peak between the 4th and 6th minute of exposure).

The general ANOVA analysis indicated a sex effect for tidal volume and minute ventilation responses to hypoxia. To further analyse this effect, we focused on steady state responses to hypoxia by averaging values between 10 and 15 min for each level, when differences are more obvious. The results in Fig. 4.2 indicate that, compared to vehicle males, females had a higher response of minute ventilation (at 12% O_2) and maintained tidal volume close to baseline values both under 14% and 12% O_2. These effects were not apparent in mifepristone treated pups.

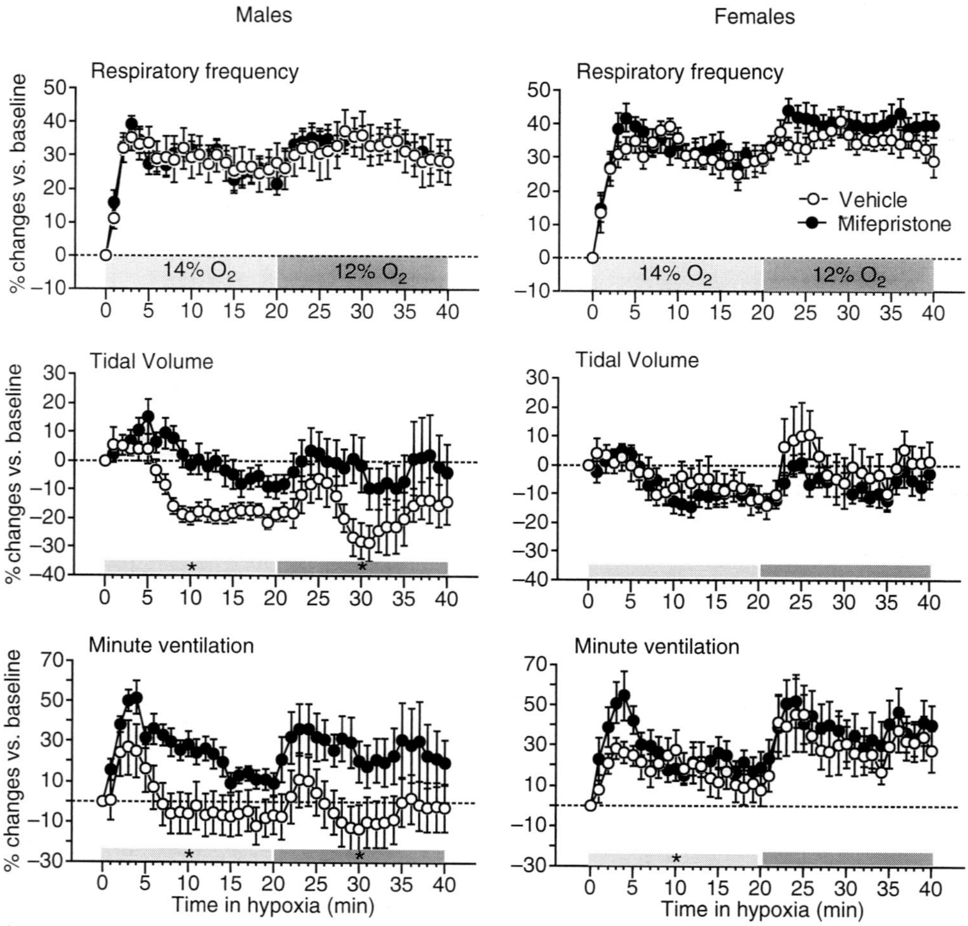

Fig. 4.1 Normalized, minute-by-minute, hypoxic responses for respiratory frequency, tidal volume, and minute ventilation, in 10–12 day old male (*left*) and female (*right*) rats treated by daily mifepristone or vehicle gavage from postnatal day 3. All values are presented as % changes from baseline. O_2 levels (14% and 12%) are indicated in *grey bars* at the *bottom* of each graph. All values are mean ± sem. *: $p<0.05$ for group effect

Oxygen consumption and CO_2 production rate measured at the end of each hypoxic level dropped to similar levels in mifepristone pups compared to vehicle (Fig. 4.3). At the end of the hypoxic exposure, rectal temperature was similar in all four groups (Male vehicle: 33.0±0.3°C, male mifepristone: 33.4±0.3°C, female vehicle: 33.3±0.3°C, female mifepristone: 33.3±0.3°C).

4.4 Discussion

This study indicates that in 12 day-old rats, 10 days of gavage with mifepristone, an antagonist of the progesterone and glucocorticoid nuclear receptors does not alter baseline respiratory parameters, but enhances the ventilatory response to hypoxia. This effect was sex-specific, being observed in males only. Furthermore, vehicle male rats displayed lower hypoxic ventilatory response than females, owing to a reduction of tidal volume below baseline values during hypoxia exposure. Male and female rats treated with mifepristone also displayed lower body weight, presumably indicating a growth promoting effect of steroid receptors activity. At the dose that we used (40 μg/g/day) mifepristone has

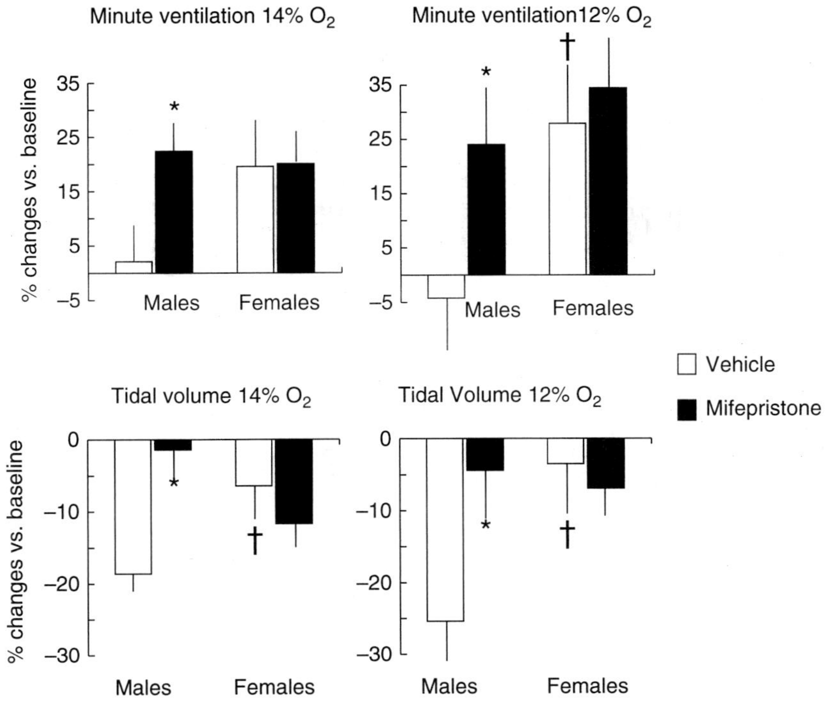

Fig. 4.2 Steady state responses to hypoxia (averaged between 10 and 15 min of exposure at 14% and 12% O_2) for tidal volume and minute ventilation, in 10–12 day old male and female rats treated by daily mifepristone or vehicle gavage from postnatal day 3. All values are presented as % changes from baseline. All values are mean ± sem. *: $p<0.05$ for group effect. †: $p<0.05$ for sex effect

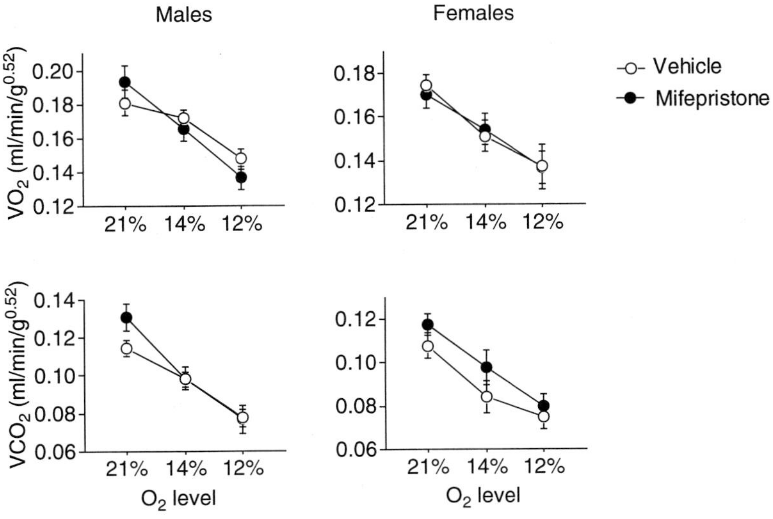

Fig. 4.3 Oxygen consumption (VO_2, ml/min/$g^{-0.52}$) and CO_2 production rate (VCO_2, ml/min/$g^{-0.52}$) recorded in normoxia and in response to 14% and 12% O_2 in 10–12 day old male (*left*) and female (*right*) rats treated by daily mifepristone or vehicle gavage from postnatal day 3. All values are presented as % changes from baseline. All values are mean ± sem

been reported to impair sexual behaviour in adult male rats treated between postnatal days 1 and 10 (Lonstein et al. 2001), and prevent the facilitation of sexual behaviour induced by relevant stimuli in adult female rats (Auger et al. 1997). Because mifepristone is a steroid, it readily crosses the blood–brain barrier, affecting both peripheral and central components of the respiratory control system.

While some studies have reported that pre-pubertal female rats have a higher hypoxic ventilatory response than males (Mortola and Saiki 1996), in most of our recent studies in rat pups this difference was either not found (Niane et al. 2009)_or was only a trend that lacked statistical significance (Julien et al. 2008). Under the experimental and recording conditions used here, we have found a clear sex-specific effect with females showing higher HVR; this effect was mostly due a less important roll-off phase of the HVR in females compared to males. Accordingly, future studies should aim at determining whether this effect is linked to a sex-specific distribution and/or effects of progesterone or glucocorticoid receptor, and their potential interactions with inhibitory mechanisms involved in the roll-off phase of HVR. Interestingly, circulating levels of corticosterone are more elevated in males compared to females between postnatal days 3 and 15 (Galeeva et al. 2010).

Unfortunately, commonly available progesterone receptors (mifepristone and onapristone) are also potent glucocorticoid antagonist. However, KO mice for the progesterone receptors are available and we are currently using them to specifically question the role of this receptor on respiratory control during postnatal development and in adult mice. KO mice for the glucocorticoid receptor are also available but 90% die at birth due to a severe retardation of lung development, and those surviving have altered brain functions (reflected as spatial learning deficits) as adults (see for references (Cole et al. 2001)). Characterization of respiratory control has not been performed in these mice.

While previous studies have reported that progesterone exposure in newborn and adult rats increases minute ventilation and hypoxic ventilatory response (Soliz and Joseph 2005; Lefter et al. 2008) (see also the contribution by Hishri et al. in this volume), the effects reported in this study likely point towards an important contribution of the glucocorticoid receptor on HVR and respiratory control that warrants further investigation.

Study founded by CIHR grant to VJ (MOP 102715).

References

Auger AP, Moffatt CA, Blaustein JD (1997) Progesterone-independent activation of rat brain progestin receptors by reproductive stimuli. Endocrinology 138(1):511–514

Behan M, Thomas CF (2005) Sex hormone receptors are expressed in identified respiratory motoneurons in male and female rats. Neuroscience 130(3):725–734

Cole TJ et al (2001) GRKO mice express an aberrant dexamethasone-binding glucocorticoid receptor, but are profoundly glucocorticoid resistant. Mol Cell Endocrinol 173(1–2):193–202

Finer NN et al (2006) Summary proceedings from the apnea-of-prematurity group. Pediatrics 117(3 Pt 2):S47–S51

Fournier S et al (2007) Chronic corticosterone elevation and sex-specific augmentation of the hypoxic ventilatory response in awake rats. J Physiol 584(Pt 3):951–962

Galeeva A et al (2010) Postnatal ontogeny of the glucocorticoid receptor in the hippocampus. Vitam Horm 82:367–389

Gulemetova R, Kinkead R (2011) Neonatal stress increases respiratory instability in rat pups. Respir Physiol Neurobiol 176(3):103–109

Haywood SA et al (1999) Fluctuating estrogen and progesterone receptor expression in brainstem norepinephrine neurons through the rat estrous cycle. Endocrinology 140(7):3255–3263

Joseph V et al (2006) Expression of sex-steroid receptors and steroidogenic enzymes in the carotid body of adult and newborn male rats. Brain Res 1073–1074:71–82

Julien C, Bairam A, Joseph V (2008) Chronic intermittent hypoxia reduces ventilatory long-term facilitation and enhances apnea frequency in newborn rats. Am J Physiol Regul Integr Comp Physiol 294(4):R1356–R1366

Kinkead R et al (2009) Neonatal maternal separation disrupts regulation of sleep and breathing in adult male rats. Sleep 32(12):1611–1620

Lefter R, Morency CE, Joseph V (2007) Progesterone increases hypoxic ventilatory response and reduces apneas in newborn rats. Respir Physiol Neurobiol 156:9–16

Lefter R, Doan VD, Joseph V (2008) Contrasting effects of estradiol and progesterone on respiratory pattern and hypoxic ventilatory response in newborn male rats. Respir Physiol Neurobiol 164(3):312–318
Lonstein JS, Quadros PS, Wagner CK (2001) Effects of neonatal RU486 on adult sexual, parental, and fearful behaviors in rats. Behav Neurosci 115(1):58–70
Mortola JP, Saiki C (1996) Ventilatory response to hypoxia in rats: gender differences. Respir Physiol 106:21–34
Niane L, Joseph V, Bairam A (2009) Role of cholinergic-nicotinic receptors on hypoxic chemoreflex during postnatal development in rats. Respir Physiol Neurobiol 169(3):323–332
Schreiber JR, Hsueh AJ, Baulieu EE (1983) Binding of the anti-progestin RU-486 to rat ovary steroid receptors. Contraception 28(1):77–85
Shahar E et al (2003) Hormone replacement therapy and sleep-disordered breathing. Am J Respir Crit Care Med 167(9):1186–1192
Soliz J, Joseph V (2005) Perinatal steroid exposure and respiratory control during early postnatal life. Respir Physiol Neurobiol 149(1–3):111–122
Zwain IH, Yen SS (1999) Neurosteroidogenesis in astrocytes, oligodendrocytes, and neurons of cerebral cortex of rat brain. Endocrinology 140(8):3843–3852

Chapter 5
Age-Dependent Changes in Breathing Stability in Rats

Lalah M. Niane and Aida Bairam

Keywords Respiratory instability • Newborn rat

5.1 Introduction

The respiratory control system rapidly develops during the perinatal period in mammalian species. Therefore, premature birth prevents the completion of important neurological maturation processes, which can cause periodic breathing and apneas. The ventilatory response to hypoxia (HVR) is among the test that have been used to study the immature respiratory control system in human and animal subjects (Carroll 2003; Cohen and Katz-Salamon 2005). This response is biphasic because the early increase in the HVR of newborn mammals is not sustained as in adults but decreases to below baseline values (Bissonnette 2000; Cohen and Katz-Salamon 2005; Niane and Bairam 2011). In a preliminary report, we have suggested that rats at postnatal days 1, 4, 7 and 12 (P1, P4, P7 and P12) may be selected to represent the postnatal changes in the biphasic HVR pattern (peak and steady state) and apnea frequency (Niane and Bairam 2011). To better understand the relationship between the maturation of respiratory control and the decreased apnea frequency with age (Niane and Bairam 2011), we re-analyzed our data in P1-, P4-, P7-, P12-, P21- and P90-day-old rats. The following two parameters were studied as an index of respiratory stability during development: the coefficient of variation of

L.M. Niane
Pediatrics Department, Centre de Recherche CHUQ-HSFA, Laval University, Québec, Québec G1R1N9, Canada

Unité de Recherche en Périnatologie, Département de PédiatrieCentre Hospitalier, Universitaire de Québec, Hôpital Saint-François d'Assise, Université Laval
10, rue de l'Espinay, D0-717, Québec G1L 3L5, Canada
lalah-malika.niane.1@ulaval.ca

A. Bairam (✉)
Unité de Recherche en Périnatologie, Département de PédiatrieCentre Hospitalier, Universitaire de Québec, Hôpital Saint-François d'Assise, Université Laval,
10, rue de l'Espinay, D0-717, Québec G1L 3L5, Canada

Centre de Recherche, D0-717, Hôpital Saint-François d'Assise,
10, rue de l'Espinay, Québec G1L 3L5, Canada
aida.bairam@crsfa.ulaval.ca

C.A. Nurse et al. (eds.), *Arterial Chemoreception: From Molecules to Systems*, Advances in Experimental Medicine and Biology 758, DOI 10.1007/978-94-007-4584-1_5,

minute ventilation and the apnea types (spontaneous vs. post-sigh). The correlation between apnea types and the coefficient of variation of minute ventilation under baseline and steady-state conditions of HVR were assessed.

5.2 Materials and Methods

The studies were performed in Sprague–Dawley male rats born and raised in our animal care facility. The P90-day-old rats were weaned from their mother at P21 and 2 rats were placed in each cage until evaluation. The local Animal Care Committee at Laval University approved the experimental protocol. The respiratory frequency and tidal volume were recorded using whole body flow-through plethysmographs in rats aged from P1 to P21 days (IOX, Emka Technologies, Paris, France). The double-chamber plethysmograph (model PLY 3023, Buxco Electronics, Sharon, CT, USA) was used in adult rats. Animal mating, litter reduction, sources of apparatus, technical settings, such as gas flow through the plethysmograph and humidity, the calibration of gas flowing in and out of the plethysmography room, the ambient temperature and the correction of the tidal volume, which was based on the barometric pressure, room and body temperatures and humidity (BTPS), were similar to those previously described and regularly used in our laboratory (Julien et al. 2010, 2011; Niane et al. 2010; Niane and Bairam 2011). To acclimate the rat to the chamber, it was placed in the plethysmograph 20–30 min before the 10-min recording of baseline ventilation (normoxia, $FiO_2 = 21\%$). The rat was exposed to moderate hypoxia ($FiO_2 = 12\%$, 20 min) by adding nitrogen to the air flowing into the plethysmograph. The body temperature was measured at the end of baseline recording and hypoxic exposure (Julien et al. 2010, 2011; Niane et al. 2010; Niane and Bairam 2011).

While the rat was awake, the respiratory frequency, tidal volume, and oxygen flowing in and out of the plethysmograph were continuously recorded and then collected on a minute-to-minute basis using IOX software (Version 1.8.9 EMKA technology, Paris, France). Minute ventilation (respiratory frequency X tidal volume) was represented as the averaged ventilation of the last 5 min under baseline and steady-state conditions of HVR.

The coefficient of variation of minute ventilation was calculated (CV = standard deviation/mean X 100) (Julien et al. 2011). Baseline spontaneous and post-sigh apnea frequencies during the 10-min baseline recording and the last 10-min recording of hypoxic exposure were analyzed using the standardized criteria of Mendelson (Mendelson et al. 1988), as we have previously described (Julien et al. 2010, 2011; Niane and Bairam 2011). ANOVA one-way test was used to assess the effects of age. When the p-value was less than 0.05, Fisher's test was used to compare the age-related effects. The correlation studies used Spearman's test followed by linear regression analysis. Data were represented as the mean ± SEM.

5.3 Results

Under baseline conditions, minute ventilation, the coefficient of variation of minute ventilation and apnea types significantly decreased (Fig. 5.1a–c) with age. The age-dependent decrease in apnea frequency was positively correlated with the decrease in the coefficient of variation of minute ventilation (Fig. 5.1d). Independently of ages studied, spontaneous apnea (Fig. 5.1c) was the predominant type under baseline and represented approximately 85% of total apnea types (Fig. 5.1c). During hypoxia, the steady-state minute ventilation showed a gradual increase with age (Fig. 5.2a). However, this increase was associated with a decrease in the coefficient of variation of minute ventilation and apnea frequency (Fig. 5.2b, c). In contrast to the baseline results, post-sigh (not spontaneous) apnea was the

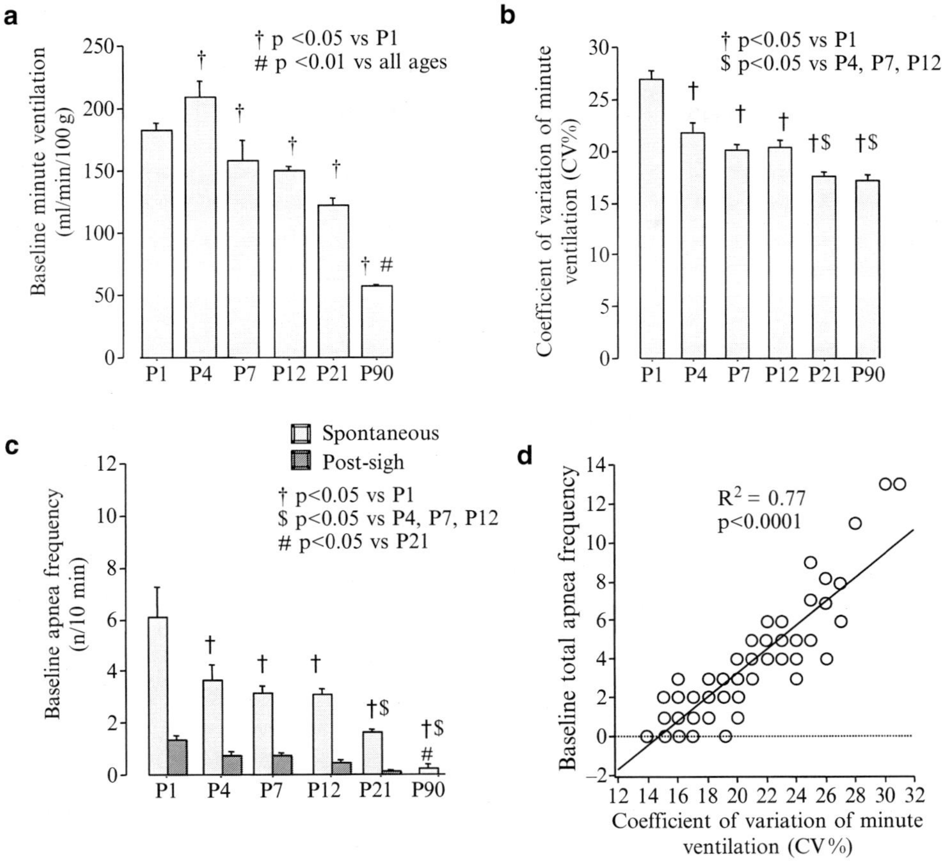

Fig. 5.1 Baseline minute ventilation (**a**), the coefficient of variation of minute ventilation (**b**), spontaneous and post-sigh apnea frequencies (**c**) and the correlation between the total apnea frequency and the coefficient of variation of minute ventilation (**d**) in rats of the studied ages. P: postnatal age. Data are shown as the means ± SEM (*NB*: Fig. 5.1a was reprinted with permission from Niane and Bairam (2011))

predominant type and represented approximately 91% of total apnea types (Fig. 5.2c). The age-dependent decrease in apnea frequency during the steady state of HVR was correlated with the decrease in the coefficient of variation of minute ventilation (Fig. 5.2d).

5.4 Discussion

Our data showed that the age-dependent decrease in apnea frequency was positively correlated with the increase in respiratory stability under normoxic and hypoxic conditions. Spontaneous apnea and post-sigh apnea were the predominant age-independent types of apnea during normoxia and hypoxia, respectively.

Apnea and periodic breathing are the hallmarks of immaturity in the developing respiratory control system, particularly in preterm neonates (Rigatto and Brady 1972a, b; Cohen and Katz-Salamon 2005; Horne et al. 2005). Although the underlying mechanisms are not fully understood, the preterm infants display higher coefficient of variation of minute ventilation than infants who were born at term or adult subjects (Al-Hathlol et al. 2000). This respiratory instability is associated with the larger variability in

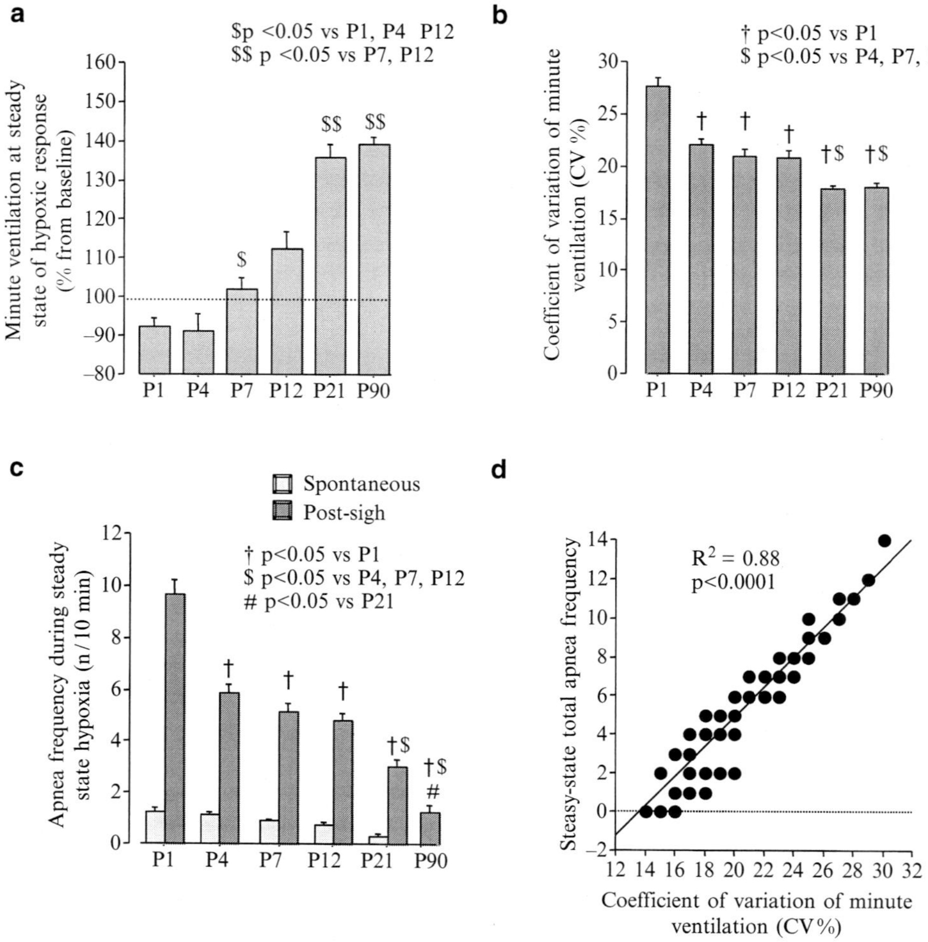

Fig. 5.2 Steady-state minute ventilation in response to mild hypoxia (FiO_2 12%, 20 min) (**a**), the coefficient of variation of minute ventilation during the steady-state hypoxic response (**b**), spontaneous and post-sigh apnea frequencies (**c**) and the correlation between the total apnea frequency and the coefficient of variation of minute ventilation (**d**) in rats of the studied ages. *P* postnatal age (Data are shown as the means ± SEM)

arterial oxygen saturation and alveolar CO_2 (Al-Hathlol et al. 2000). Previous studies have suggested that higher apnea frequencies in preterm infants increase the respiratory instability and alter the developmental course of the respiratory control system (Rigatto and Brady 1972a, 1972b). Paradoxically, an exaggerated hypoxic peripheral chemosensitivity (carotid body) may enhance apnea events that further destabilize breathing (Al-Matary et al. 2004; Nock et al. 2004; Cardot et al. 2007). The present study used rats as a model to confirm the clinical observations. Our results suggest that the analyses of respiratory variability and apnea in developing rats are indices of maturity in the respiratory control system and can be used as variables to describe the developmental changes of this system under normoxic or hypoxic conditions in rats. Furthermore, these variables can be used to study the effects of pharmacological, environmental, genetic or other factors modulating the developmental pattern of respiratory control mechanisms.

Aside from the importance of the HVR in the study of the maturation of breathing control (Bissonnette 2000; Carroll 2003; Niane and Bairam 2011) in animal models, we concluded that the coefficient of variation of minute ventilation and apnea types also represented important variables of the maturational process. These variables can be considered for the conception and refinement of apnea treatments during the neonatal period using rats at 1, 4, 7 and 12 days of age.

Acknowledgements This study was partially supported by the CIHR operating grant MOP-81101 to A. Bairam. We thank Mrs. Melanie Pelletier and Sylvie Viger for animal care.

References

Al-Hathlol K et al (2000) A study of breathing pattern and ventilation in newborn infants and adult subjects. Acta Paediatr 89(12):1420–1425

Al-Matary A et al (2004) Increased peripheral chemoreceptor activity may be critical in destabilizing breathing in neonates. Semin Perinatol 28(4):264–272

Bissonnette JM (2000) Mechanisms regulating hypoxic respiratory depression during fetal and postnatal life. Am J Physiol Regul Integr Comp Physiol 278(6):R1391–R1400

Cardot V et al (2007) Ventilatory response to a hyperoxic test is related to the frequency of short apneic episodes in late preterm neonates. Pediatr Res 62(5):591–596

Carroll JL (2003) Developmental plasticity in respiratory control. J Appl Physiol 94(1):375–389

Cohen G, Katz-Salamon M (2005) Development of chemoreceptor responses in infants. Respir Physiol Neurobiol 149(1–3):233–242

Horne RS et al (2005) Postnatal development of ventilatory and arousal responses to hypoxia in human infants. Respir Physiol Neurobiol 149(1–3):257–271

Julien CA et al (2010) Carotid sinus nerve stimulation, but not intermittent hypoxia, induces respiratory LTF in adult rats exposed to neonatal intermittent hypoxia. Am J Physiol Regul Integr Comp Physiol 299(1):R192–R205

Julien CA et al (2011) Alteration of carotid body chemoreflexes after neonatal intermittent hypoxia and caffeine in rat pups. Respir Physiol Neurobiol 177(3):301–312

Mendelson WB et al (1988) Periodic cessation of respiratory effort during sleep in adult rats. Physiol Behav 43(2):229–234

Niane LM, Bairam A (2011). Selecting representative ages for developmental changes of respiratory irregularities and hypoxic ventilatory response in rats. Open J Mol Integr Physiol 1:1–7 (Website: http://www.scirp.org/journal/OJMIP

Niane LM et al (2010) Ventilatory and carotid body chemoreceptor responses to purinergic P2X receptor antagonists in newborn rats. J Appl Physiol 110:83–94

Nock ML et al (2004) Relationship of the ventilatory response to hypoxia with neonatal apnea in preterm infants. J Pediatr 144(3):291–295

Rigatto H, Brady JP (1972a) Periodic breathing and apnea in preterm infants. I. Evidence for hypoventilation possibly due to central respiratory depression. Pediatrics 50(2):202–218

Rigatto H, Brady JP (1972b) Periodic breathing and apnea in preterm infants. II. Hypoxia as a primary event. Pediatrics 50(2):219–228

Chapter 6
Dose Dependent Effect of Progesterone on Hypoxic Ventilatory Response in Newborn Rats

Oubeidallah Hichri, Jean-C Laurin, Cécile A. Julien, Vincent Joseph, and Aida Bairam

Abstract The effect of progesterone as a respiratory stimulant in newborn subjects is less known than that in adults. This study investigated the dose-response curve (0, 2, 4, and 8 mg/kg, ip) of progesterone on ventilation in non-anesthetized newborn rats at 4- and 12-days old using plethysmography. Progesterone had no effects in the regulation of normoxic ventilation. However, it enhanced the response to moderate hypoxia (FiO_2 12%, 20 min) in 4- but not in 12-days old pups. This response was similar between the dose of 4 and 8 mg/kg. These observations suggested that progesterone enhances in age- and dose-dependent manner the hypoxic ventilatory response in newborn rats.

Keywords Hypoxia • Progesterone • Newborn rat

6.1 Introduction

In adult mammalian species including human, progesterone is a well-known respiratory stimulant, acting peripherally (carotid body) and centrally (brainstem structure) to increase ventilation in response to hypoxia (HVR) (Behan and Wenninger 2008). Recent studies show that progesterone is also a respiratory stimulant in newborn rats. Ten-day-old rats chronically exposed to progesterone through the milk of lactating mother show an increase of HVR with a significant decrease in apnea frequency during

O. Hichri • J.-C. Laurin • C.A. Julien
Unité de Recherche en Périnatologie, Département de Pédiatrie, Centre Hospitalier Universitaire de Québec, Hôpital Saint-François d'Assise, Université Laval, 10, rue de l'Espinay, Québec, QC G1L 3L5, Canada

V. Joseph
Unité de Recherche en Périnatologie, Département de Pédiatrie, Centre Hospitalier Universitaire de Québec, Hôpital Saint-François d'Assise, Université Laval, 10, rue de l'Espinay, Québec, QC G1L 3L5, Canada

CRCHUQ/Hop Laval, 10 de l'espinay Local D0-711, Québec G0A3X0, Canada
e-mail: joseph.vincent@crsfa.ulaval.ca

A. Bairam (✉)
Unité de Recherche en Périnatologie, Département de Pédiatrie, Centre Hospitalier Universitaire de Québec, Hôpital Saint-François d'Assise, Université Laval, 10, rue de l'Espinay, Québec, QC G1L 3L5, Canada

Centre de Recherche, D0-717, Hôpital Saint-François d'Assise, 10, rue de l'Espinay, Québec G1L 3L5, Canada
e-mail: aida.bairam@crsfa.ulaval.ca

C.A. Nurse et al. (eds.), *Arterial Chemoreception: From Molecules to Systems*, Advances in Experimental Medicine and Biology 758, DOI 10.1007/978-94-007-4584-1_6,

hypoxia (Lefter et al. 2007). These studies suggested that progesterone is required for adequate HVR in newborn, but as has been shown in adult subjects (Behan and Wenninger 2008), the chronic use of progesterone could have affected ventilatory control via indirect mechanisms. In light of its stimulatory effects on breathing, a recent report proposed progesterone as an alternative therapy to methylxanthines for apnea treatment in premature infants (Finer et al. 2006). To this day, however, relevant investigations of the effects of progesterone on respiratory activity in newborn are lacking in both animal models and human. To address this issue, we used 4- and 12-days-old rats to evaluate the dose–response curve of progesterone on ventilation and metabolism at baseline and in response to hypoxia (HVR).

6.2 Materials and Methods

Laval University Animal Care Committee approved the experimental protocols. Male Sprague–Dawley rats at P4 and P12 days old were used. All pups (6–8 rats/dose/age; total number 27 at 4-day-old and 30 rats at 12-day-old) were obtained from eight virgin female and male rats (Charles River, St Constant, Québec). Rats were supplied with food and water ad libitum and maintained under standard laboratory conditions (21°C, 12:12 h dark–light cycle: lights on at 07:00 and off at 19:00). Litters were culled to 12 pups on postnatal day 1 with a preference to keep males for experiments whenever possible. Respiratory and metabolic variables were recorded in awake and unrestrained pups using whole body plethysmography as previously described in detail (Lefter et al. 2007; Julien et al. 2008; Niane et al. 2009). Body temperature was measured orally in P4 rats and rectally in P12 rats at the end of baseline recording and hypoxic exposure using a thermocouple for small rodents (Harvard, Holliston, MA, USA). Ambient respiratory gases (oxygen and CO_2), respiratory frequency (Fr) and tidal volume (Vt) were continuously measured during the entire experiment. Then, minute ventilation (Fr X Vt), oxygen consumption [flow x (O_2in-O_2out)-(O_2out x (CO_2out-CO_2in)/(1-O_2out)] and CO_2 production [flow x (CO_2out-CO_2in)- (CO_2out x (O_2in-O_2out)/(1-CO_2out)] were calculated, corrected to STPD and further used for calculating the ratios of ventilation/oxygen consumption or CO_2 production. Each rat was used only for one recording in any of dose or age studied. Each pup received a single intraperitoneal injection of either saline or progesterone at 2, 4, and 8 mg/kg. The calculated dose/body weight (2 μl/g) was administered 30 min before ventilatory assessments. Experiments began with a 10 min recording of baseline ventilation (normoxia: $FiO_2 = 21\%$) followed by exposure to moderate hypoxia ($FiO_2 = 12\%$, 20 min). Progesterone was purchased from Sigma Aldrich Canada (Oakville, ON, Canada), freshly dissolved in saline and stored at 4 C° for use during the 3–4 h following preparation.

Ventilatory variables were measured on a min-by-min basis (IOX software, V1.8.9 EMKA tech., Fr) and then averaged over the last 5 min of baseline recording. To assess the temporal dynamics of the HVR, variables were averaged every 2 min over the first 10 min of hypoxia (early phase). Then, values were obtained every 5 min for the last 10 min (late phase). Each HVR was expressed as percent changes from baseline. Metabolic variables were measured at baseline and at the end of hypoxic exposure. ANOVA for one or multiple factors (dose and age) was used and $p < 0.05$ was considered significant. Data are presented as mean ± SEM.

6.3 Results

6.3.1 *Progesterone Does Not Affect Baseline Ventilation and Metabolism*

In P4 day-old pups, progesterone had no effect on baseline minute ventilation, oxygen consumption or CO_2 production at any of the doses tested compared to saline (0.0 mg/kg) (Fig. 6.1a). Similar results were obtained in P12 days-old rats (Fig. 6.1b).

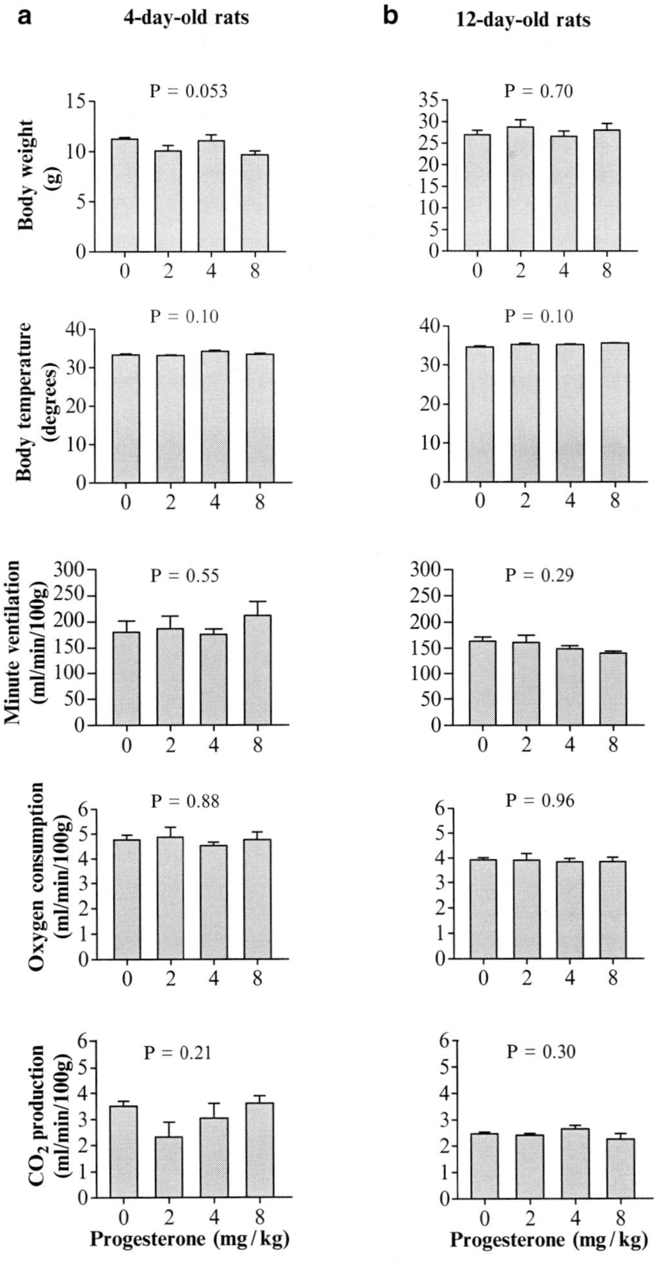

Fig. 6.1 Body weight, body temperature and progesterone dose-response curve on minute ventilation, oxygen consumption and carbon dioxide production in (**a**) 4- and (**b**) 12-days old rats under baseline (normoxic condition). ANOVA results are indicated in the top of each histogram

6.3.2 Dose- and Age-Dependent Effects of Progesterone on HVR and Metabolism

The dynamic response to hypoxia following progesterone administration is shown in Fig. 6.2. In P4 day-old rats, progesterone administration at 4 mg/kg enhanced the early (peak) and the late (steady state) phases of the HVR (% from baseline) above the response measured in saline treated pups; other

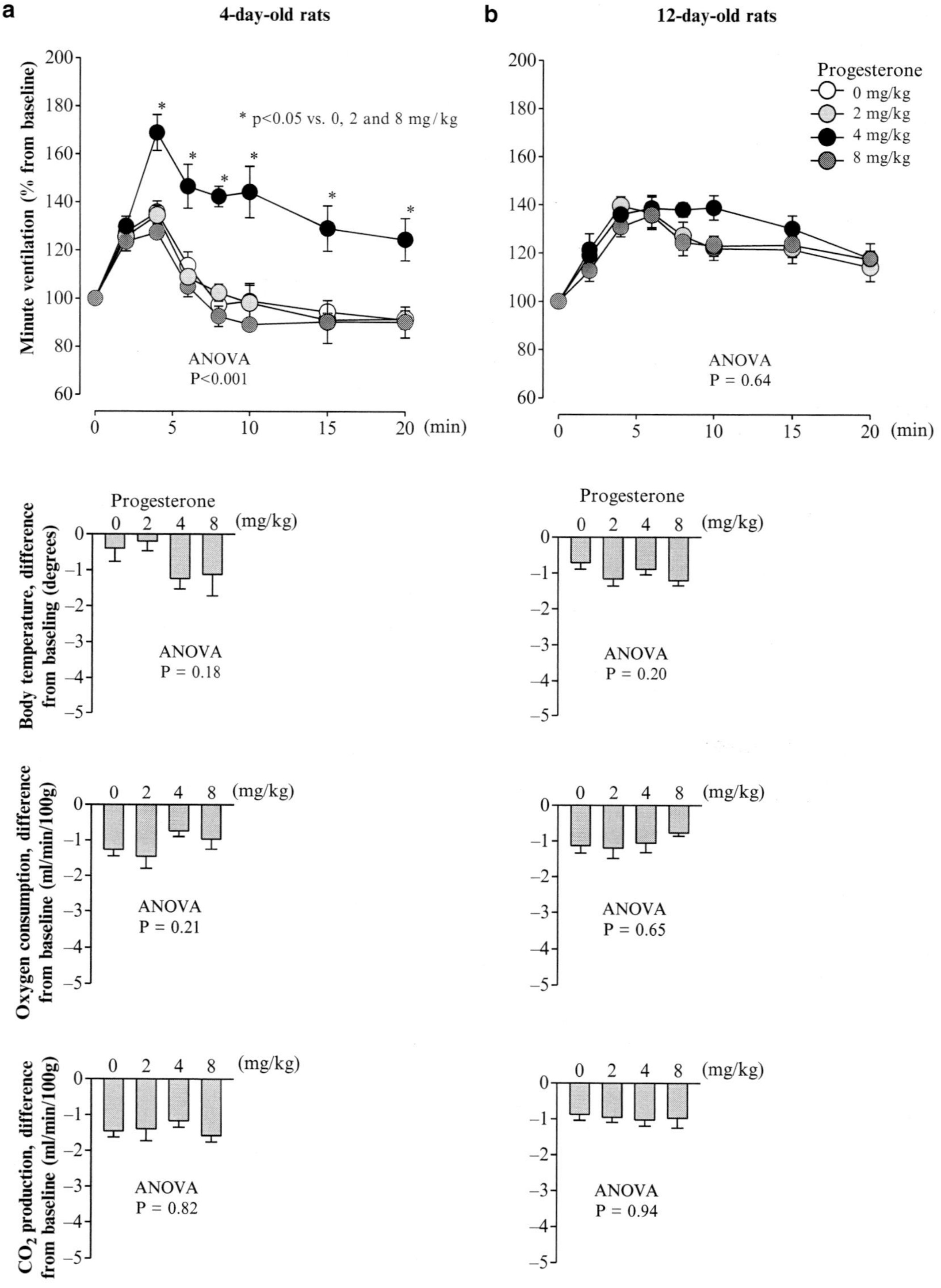

Fig. 6.2 Progesterone dose-response curve on the dynamic response to hypoxia (FiO_2 12%, 20 min) on minute ventilation in 4- and 12 days old rats and on metabolism (body temperature, oxygen consumption and carbon dioxide production) evaluated at the end of hypoxic exposure. ANOVA results analyses are reported in each panel

doses had no effect (Fig. 6.2a). This response contrasted with the effects observed in P12 day-old rats (Fig. 6.2b) in which progesterone had no effect at any of the doses tested. Hypoxia decreased metabolism as indicated by the reduced body temperature, oxygen consumption and carbon dioxide production; however these variables were not affected by any of progesterone doses tested in either P4 or P12 day-old rats (Fig. 6.2a, b).

6.4 Discussion

Our data show that acute progesterone administration in a dose range between 0 and 8 mg/kg did not influence the baseline ventilation or metabolism in both P4- and P12-day-old rats. In contrast, progesterone at 4 mg/kg enhanced HVR in P4 but not in P12-day-old rats.

Interestingly, the acute effects reported here under normoxic and hypoxic conditions corroborate with those obtained in 10-day-old rats in which chronic exposure to progesterone through osmotic mini-pumps to lactating mother from birth failed to increase baseline minute ventilation but enhanced the HVR (Lefter et al. 2007). Progesterone has no effects on baseline ventilation in adult rat either (Yamazaki et al. 2005). Accordingly, we conclude that progesterone has no effect on the tonic regulation of ventilation in either newborn or adult rats.

During hypoxia, however, acute progesterone administration enhanced the HVR in P4 day-old rats only whereas chronic administration enhanced the HVR in 10 days old rats (Lefter et al. 2007). Unfortunately, no data are available on ventilatory effects of chronic progesterone administration in younger animals. The HVR in newborns mammals is bi-phasic (Bissonnette 2000). The first phase corresponds to an initial increase in ventilation that is mainly related to peripheral carotid body chemoreceptor activation. This initial increase is usually followed by a decline to a level lower than baseline values especially during the early days of life (Bissonnette 2000). That decline in ventilation is mainly related to the powerful central inhibitory mechanisms in newborn and to a decreasing in metabolism (Bissonnette 2000). Since progesterone at 4 mg/kg (1) enhanced the first phase, (2) prevented the ventilatory decline during the second phase of the response to bring breathing a level comparable to that observed in P12 rats, and (3) had no effect on metabolism, these data suggest that progesterone stimulates the peripheral and central components of the respiratory control system in very young rats. In rats, progesterone receptors are expressed in the carotid bodies at gestational day 18, postnatal days 0, and 2 and adulthood (Joseph et al. 2006); however, their developmental pattern is not yet determined. To explain the lack of progesterone effect at P12, it is tempting to propose that within the carotid bodies, progesterone receptor expression declines with age. On other hand, the age-dependent increase in inhibitory mechanisms such as dopamine may mask, at least in part, the effects of progesterone (Bairam and Carroll 2005). Indeed, the enhancement of the late phase of HVR in P4 but not in P12 day-old rats with progesterone 4 mg/kg suggested that the role of central progesterone receptors is also more important in early than late days of life. Experiments performed on reduced brainstem preparations from fetuses aged of ages 17–21 days suggest that progesterone act as a regulator of respiratory rhythm generation in early life (Ren and Greer 2006). It is worth mentioning that progesterone receptors are also expressed in the respiratory nuclei of adult rat, see for review (Behan and Wenninger 2008). Taken together, all the above observations suggest that the stimulatory effect of progesterone on the HVR is better expressed during early than late days of age in newborn rats. Also, these effects may be related at least in part to the age-dependent expression of density and/or function of progesterone receptors at peripheral (carotid body) and central (brainstem) levels of breathing control system.

6.5 Conclusions

Progesterone has no effects in the regulation of normoxic ventilation but enhances the HVR in an age- and dose-dependent manner. These observations need to be considered in any further investigations of progesterone on respiratory irregularities and apnea frequency in developing subjects including human infants (Al-Hathlol et al. 2000); See also Niane L. paper in this book.

References

Al-Hathlol K, Idiong N, Hussain A, Kwiatkowski K, Alvaro RE, Weintraub Z, Cates DB, Rigatto H (2000) A study of breathing pattern and ventilation in newborn infants and adult subjects. Acta Paediatr 89:1420–1425

Bairam A, Carroll JL (2005) Neurotransmitters in carotid body development. Respir Physiol Neurobiol 149:217–232

Behan M, Wenninger JM (2008) Sex steroidal hormones and respiratory control. Respir Physiol Neurobiol 164:213–221

Bissonnette JM (2000) Mechanisms regulating hypoxic respiratory depression during fetal and postnatal life. Am J Physiol Regul Integr Comp Physiol 278:R1391–1400

Finer NN, Higgins R, Kattwinkel J, Martin RJ (2006) Summary proceedings from the apnea-of-prematurity group. Pediatrics 117:S47–51

Joseph V, Doan VD, Morency CE, Lajeunesse Y, Bairam A (2006) Expression of sex-steroid receptors and steroidogenic enzymes in the carotid body of adult and newborn male rats. Brain Res 1073–1074:71–82

Julien C, Bairam A, Joseph V (2008) Chronic intermittent hypoxia reduces ventilatory long-term facilitation and enhances apnea frequency in newborn rats. Am J Physiol Regul Integr Comp Physiol 294:R1356–1366

Lefter R, Morency CE, Joseph V (2007) Progesterone increases hypoxic ventilatory response and reduces apneas in newborn rats. Respir Physiol Neurobiol 156:9–16

Niane L, Joseph V, Bairam A (2009) Role of cholinergic-nicotinic receptors on hypoxic chemoreflex during postnatal development in rats. Respir Physiol Neurobiol 169:323–332

Ren J, Greer JJ (2006) Neurosteroid modulation of respiratory rhythm in rats during the perinatal period. J Physiol 574:535–546

Yamazaki H, Haji A, Ohi Y, Takeda R (2005) Effects of progesterone on apneic events during behaviorally defined sleep in male rats. Life Sci 78:383–388

Chapter 7
Postnatal Hyperoxia Impairs Acute Oxygen Sensing of Rat Glomus Cells by Reduced Membrane Depolarization

Insook Kim, David F. Donnelly, and John L. Carroll

Abstract Previous work demonstrated that hyperoxia (30–60% O_2) exposure in the post-natal period reduces the ventilatory response to acute hypoxia and this impairment may continue considerably beyond the period of hyperoxia exposure. Previous work from our laboratory demonstrated that 1–2 weeks of hyperoxia (60% O_2) starting between P1 and P14: reduced the single chemoreceptor unit response to hypoxia, reduced the rise in glomus cell calcium caused by acute hypoxia and reduced hypoxia-induced catecholamine release (Donnelly 05, Donnelly 09). The present study asked whether the impairment extended to hypoxia-induced membrane depolarization, an earlier step in the transduction cascade. Perforated patch, whole-cell recordings were obtained from rat glomus cells exposed to hyperoxia from P0-P8 or P8-P15 and age-matched control groups. In both cases, hypoxia-induced membrane depolarization was significantly less in the hyperoxia treated groups compared to controls, while depolarization to 20 mM K^+ was not significantly affected. Resting membrane potential and input resistance were also not different in the hyperoxia treated groups. Whole carotid body quantitative real time PCR showed that TASK-1, TASK-3 and L-type Ca^{2+} channel expression was significantly down-regulated at Hyper 8–15 compared to controls. We conclude that 1 week of postnatal hyperoxia during the early and late stage of CB maturation impairs organ function by affecting the coupling between hypoxia and glomus cell depolarization. This may be caused by altered expression of TASK1, TASK3 or L-type Ca^{2+} channel gene expression. We speculate that an identification of cellular changes caused by hyperoxia may yield unique insights to the mechanism of oxygen sensing by the carotid bodies.

Keywords Carotid body • Glomus cells • Hyperoxia • Membrane depolarization • Gene expression • qPCR • Postnatal • Impairment

---

I. Kim (✉)
Department of Pediatrics, ACHRI, University of Arkansas for Medical Sciences, 13th Children's Way, Little Rock 72202, AR, USA
e-mail: kiminsook@uams.edu

D.F. Donnelly
Department of Pediatrics, Yale University, School of Medicine, 333 Cedar St., New Haven, CT, USA
e-mail: David.Donnelly@Yale.edu

J.L. Carroll
Department of Pediatrics, ACHRI, University of Arkansas for Medical Sciences, 13th Children's Way, Little Rock 72202, AR, USA

Arkansas Children's Hospital, Slot 512-17, 1 Children's Way, Little Rock, AR 72212, USA
e-mail: carrolljohnl@uams.edu

C.A. Nurse et al. (eds.), *Arterial Chemoreception: From Molecules to Systems*, Advances in Experimental Medicine and Biology 758, DOI 10.1007/978-94-007-4584-1_7,

7.1 Introduction

Chronic hyperoxia (30–60% O_2) exposure for first month of life attenuates the hypoxic ventilatory response (HVR) of adult rats despite a return to normoxia (Ling et al. 1997; Bavis 2005; Fuller et al. 2002) and may occur with shorter durations in the first 2 weeks of life (Bavis et al. 2002; Bisgard et al. 2003). This is reflected by a reduction in the early and later phases of the HVR (Bavis et al. 2010) and a reduction in chemoreceptor afferent nerve activity (Donnelly et al. 2005). These impairments are likely due to cellular alterations within the carotid body – a reduction in the number of O_2-sensitive glomus cells in CB and a loss of chemoreceptor afferent axons in the carotid sinus nerve (CSN) (Erickson et al. 1998; Wang and Bisgard 2005). Furthermore, the remaining glomus cells and chemoafferent neurons in CSN show a significant reduction in their responses to acute hypoxia (Bisgard et al. 2003; Ling et al. 1997).

The alteration in chemoreceptor function is even more pronounced when observed at the end of hyperoxia-treatment period. Large decreases in chemoreceptor function are observed immediately following as little as 4–5day hyperoxia (Donnelly et al. 2009) with more profound decreases following 1–2 weeks of hyperoxia (Bavis et al. 2011; Donnelly et al. 2005). Besides decreases in afferent nerve activity, 1 week of hyperoxia causes a reduction in hypoxia-induced catecholamine release and a reduction of the rise in intracellular calcium produced by acute exposure to hypoxia. The rise in calcium is believed to be due to an inhibition of a 'leak' type K^+ channels (Buckler 1997; Buckler and Vaughan-Jones 1994). Thus, we hypothesized that hyperoxia-treatment alters leak channel expression in glomus cells which underlies the impaired chemoreceptor response to hypoxia caused by hyperoxia exposure. Perforated patch, whole-cell recordings were performed on isolated glomus cells exposed to hyperoxia from P0-P8 or P8-P15 and age-matched control pups. Since previous work identified TASK 1/3 as the 'leak' channel accounting for glomus cell depolarization (Kim et al. 2009, 2011a), we postulated an altered expression of these channels.

7.2 Methods

7.2.1 *Animal Model*

The use of animals in this study was approved by the Animal Care and Use Committee of the University of Arkansas for Medical Sciences. Experiments were conducted on Sprague–Dawley rat pups of both sexes. Hyperoxia exposure was begun at birth (P0) or at postnatal day 8 (P8), which represent the early and late stages of normal post-natal chemoreceptor maturation. For hyperoxia treatment at birth, pregnant rats were placed in environmental chamber (OxyCycler, BioSpherix) 1–2 days prior to expected delivery and were allowed to give birth. Rat pups and dams were maintained in the chamber for 1 week (P0-P8). For hyperoxia treatment starting at P8, the dam and rat pups were placed in the chamber at P8 and maintained for 1 week (P8-P15). The atmosphere within the chamber was maintained at 60% O_2 and chamber CO_2 was maintained under 0.2% by a controlled leak from the chamber. Control animals maintained in normoxia were similarly housed and CB harvested at the same ages.

7.2.2 *Carotid Body Cell Isolation*

Rat pups were anesthetized with isoflurane, decapitated, and the heads placed in ice-cold buffered saline solution (BSS). Whole CBs were dissected out and placed in ice-cold low Ca^{2+}/Mg^{2+} BSS. Isolated CBs were dissociated with an enzyme cocktail containing trypsin and collagenase by incubating at 37°C for 20–25 min. Dissociated CB cells were placed on poly-D-lysine coated glass coverslips and incubated (>3 h) in CB growth medium at 37°C until use.

7.2.3 *Electrophysiological Studies*

Cells were recorded in the current clamp mode (I=0 pA) using perforated patch access. Only cells with access resistance <100MΩ were accepted. The normoxic solution was equilibrated with 21% O_2, 5% CO_2, and balance N_2. The hypoxic responses were elicited by switching the perfusate to saline bubbled with 0% O_2/5% CO_2/bal N_2 and, in some cases, included an oxygen scavenger. Anoxic solutions were achieved by adding an oxygen scavenger, glucose oxidase and catalase (GOC) or 0.5 mM sodium dithionite (Dithio: $Na_2S_2O_4$) into the hypoxic solution. All patch clamp experiments were performed at 35°C.

7.2.4 *Quantitative Real Time PCR (qPCR)*

Isolated whole CBs were stored in RNAlater stabilization reagent (Qiagen) at −80°C. Collected frozen CBs were processed to extract the total RNA using RNeasy Plus Micro Kit (Qiagen). cDNA was synthesized with iScript cDNA synthesis kit (Bio-Rad). Primers for target genes, TASK-1, TASK-3 and L-type Ca^{2+} channels, and reference genes, PPIA and TBP, were designed by Beacon Designer 2.0. qPCR were run on iCycler (Bio-Rad) by using SYBR Green Supermix (Bio-Rad). The relative gene expression ratios and significances between P1 and P15 and between P15 and hyperoxia-treated (P8-15) were calculated and analyzed with REST2009 software (Pfaffl et al. 2004) with two most stable reference genes, PPIA and TBP (Kim et al. 2011b).

7.3 Results

7.3.1 *Effect of Hyperoxia-Exposure on Membrane Depolarization*

The membrane depolarization response to 20 mM KCl, acute hypoxia (0% O_2) and anoxia were studied on dissociated glomus cells of hyperoxia-treated (P0-P8) and hyperoxia-treated (P8-P15) pups and compared to age matched control normoxia pups, P8 and P15 (Fig. 7.1). Anoxia was achieved by adding an oxygen scavenger (GOC or Dithio) to the perfusate reservoir that had been previously bubbled with 0% O_2.

Hyperoxia treatment caused no or slight reduction in the magnitude of depolarization in response to 20 mM K^+ (Fig. 7.1). In contrast, both hypoxia- and anoxia-induced membrane depolarizations were significantly reduced in hyperoxia treated glomus cells, both hyperoxia-treated (P0-P8) (Fig. 7.1b) and hyperoxia-treated (P8-P15) (Fig. 7.1d), compared to age-matched control glomus cells (Fig. 7.1a, c).

7.3.2 *Gene Expression Modulation by Postnatal Hyperoxia Exposure*

The modulation of oxygen-sensitive K^+ channels, TASK-1, TASK-3 and L-type Ca^{2+} channel gene expression were studied by quantitative real time PCR (qPCR). Relative quantification expressed as ratio was analyzed with REST2009 software (Pfaffl et al. 2004) by using two reference genes, PPIA

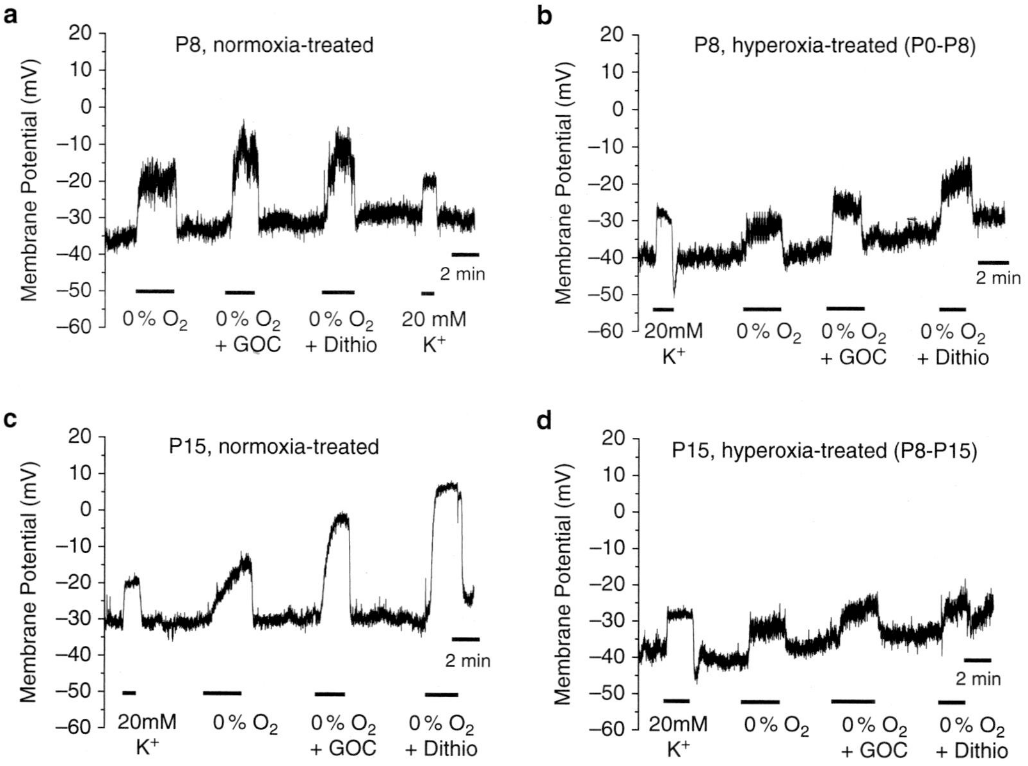

Fig. 7.1 Representative traces of membrane depolarization of normoxia (**a**) and (**c**) and hyperoxia-treated (60% O_2) glomus cells (**b**) and (**d**) in response to 20 mM K^+, 0% O_2 and anoxia (0% O_2+GOC or 0% O_2+dithio)

and TBP, which were selected as the most stable reference genes (Kim et al. 2011b). qPCR with whole carotid body showed that TASK-1, TASK-3 and L-type Ca^{2+} channel expressions were significantly down-regulated at hyperoxia-treated (P8-P15) compared to controls P15 (at $p<0.001$) (Fig. 7.2b), although normal expression of these genes did not change over the P1-P15 normoxia developmental period (Fig. 7.2a).

7.4 Discussion

Our present studies showed that 1 week of hyperoxia exposure starting at P0 or starting at P8 reduced hypoxic depolarization response in CB glomus cells compared to age-matched CB cells. These results are in good agreement with previously reported reductions in HVRs, afferent nerve activities, glomus cell calcium responses and glomus cell secretory responses following exposure to post-natal hyperoxia (Bavis et al. 2011; Donnelly et al. 2009).

Carotid body oxygen sensitivity is low at birth and matures over the first 2 weeks of life (Wasicko et al. 1999). We speculated that the response to hypoxia might be different if started before the maturation process (i.e., P0) compared to a time point at which maturation has largely been accomplished (i.e., P8). Although both of the 1 week hyperoxia exposure periods resulted in reduced glomus cell oxygen responsiveness, the mechanisms could be different. In P8, hyperoxia-treated (P0-P8), the reduced glomus cell hypoxic response might be due to inhibition of CB oxygen sensing development, since its hypoxic depolarization is almost like the same as P0. In P15, hyperoxia-treated (P8-P15), the

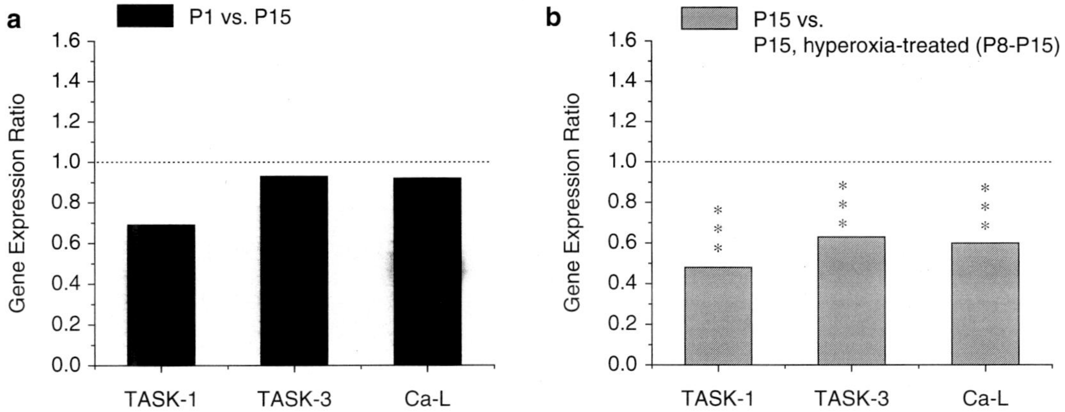

Fig. 7.2 Relative gene expression ratio of TASK-1, TASK-3 and L-type Ca^{2+} (Ca-L) channel were studied by quantitative real time RT-PCR with two reference genes, PPIA & TBP. Significance was marked with *asterisk*, at *p* values <0.001 (***)

reduced glomus cell hypoxic response might be due to free radical species (Gore et al. 2010), since hypoxic sensitivity is largely mature by P8 and hyperoxia treatment caused the response to be below that of P8.

The rise in calcium in response to hypoxia in glomus cells is believed to be due to an inhibition of a 'leak' type K^+ channels (Buckler 1997; Buckler and Vaughan-Jones 1994) and sequentially opening L-type Ca^{2+} channels (Summers et al. 2000). Since previous work identified TASK-1/3 as the 'leak' channel accounting for hypoxia-induced glomus cells depolarization (Kim et al. 2009, 2011a; Buckler 1997), we postulated altered expression of these channels by hyperoxia exposure. qPCR showed that TASK-1 and TASK-3 channel gene expression were reduced after 1 week of hyperoxia exposure at P0 (P0-P7) (Bavis et al. 2011) and P8 (P8-P15), while these TASK channel genes were not modulated during CB maturation in normoxia. The reduced gene expression of oxygen sensitive channels might be one of the important clues which might explain the reduced hypoxic glomus cell membrane depolarization. In addition, the reduced L-type Ca^{2+} channel gene expression might contribute to the reduced calcium response to hypoxia, previously observed.

These data suggest that impairments of CB oxygen sensitivities by postnatal hyperoxia are due to not only post-synaptic, but also pre-synaptic effects. We speculate that elucidation of cellular changes by postnatal hyperoxia may provide unique insights into the mechanism of oxygen sensing by the carotid bodies.

Acknowledgements This study was supported by grants from National Institute of health (HL-073500 (DFD) and HL-05621 (JLC)).

References

Bavis RW (2005) Developmental plasticity of the hypoxic ventilatory response after perinatal hyperoxia and hypoxia. Respir Physiol Neurobiol 149:287–299

Bavis RW, Olson EB Jr, Mitchell GS (2002) Critical developmental period for hyperoxia-induced blunting of hypoxic phrenic responses in rats. J Appl Physiol 92:1013–1018

Bavis RW, Young KM, Barry KJ, Boller MR, Kim E, Klein PM, Ovrutsky AR, Rampersad DA (2010) Chronic hyperoxia alters the early and late phases of the hypoxic ventilatory response in neonatal rats. J Appl Physiol 109:796–803

Bavis RW, Kim I, Pradhan N, Nawreen N, Dmitrieff EF, Carroll JL, Donnelly DF (2011) Recovery of carotid body O_2 sensitivity following chronic postnatal hyperoxia in rats. Respir Physiol Neurobiol 177:47–55

Bisgard GE, Olson EB Jr, Wang ZY, Bavis RW, Fuller DD, Mitchell GS (2003) Adult carotid chemoafferent responses to hypoxia after 1, 2, and 4 wk of postnatal hyperoxia. J Appl Physiol 95:946–952

Buckler KJ (1997) A novel oxygen-sensitive potassium current in rat carotid body type I cells. J Physiol 498(Pt 3): 649–662

Buckler KJ, Vaughan-Jones RD (1994) Effects of hypoxia on membrane potential and intracellular calcium in rat neonatal carotid body type I cells. J Physiol 476:423–428

Donnelly DF, Kim I, Carle C, Carroll JL (2005) Perinatal hyperoxia for 14 days increases nerve conduction time and the acute unitary response to hypoxia of rat carotid body chemoreceptors. J Appl Physiol 99:114–119

Donnelly DF, Bavis RW, Kim I, Dbouk HA, Carroll JL (2009) Time course of alterations in pre- and post-synaptic chemoreceptor function during developmental hyperoxia. Respir Physiol Neurobiol 168:189–197

Erickson JT, Mayer C, Jawa A, Ling L, Olson EB Jr, Vidruk EH, Mitchell GS, Katz DM (1998) Chemoafferent degeneration and carotid body hypoplasia following chronic hyperoxia in newborn rats. J Physiol 509(Pt 2):519–526

Fuller DD, Bavis RW, Vidruk EH, Wang ZY, Olson EB Jr, Bisgard GE, Mitchell GS (2002) Life-long impairment of hypoxic phrenic responses in rats following 1 month of developmental hyperoxia. J Physiol 538:947–955

Gore A, Muralidhar M, Espey MG, Degenhardt K, Mantell LL (2010) Hyperoxia sensing: from molecular mechanisms to significance in disease. J Immunotoxicol 7:239–254

Kim D, Cavanaugh EJ, Kim I, Carroll JL (2009) Heteromeric TASK-1/TASK-3 is the major oxygen-sensitive background K+channel in rat carotid body glomus cells. J Physiol 587:2963–2975

Kim D, Papreck JR, Kim I, Donnelly DF, Carroll JL (2011a) Changes in oxygen sensitivity of TASK in carotid body glomus cells during early postnatal development. Respir Physiol Neurobiol 177:228–235

Kim I, Yang D, Tang X, Carroll JL (2011b) Reference gene validation for qPCR in rat carotid body during postnatal development. BMC Res Notes 4:440

Ling L, Olson EB Jr, Vidruk EH, Mitchell GS (1997) Developmental plasticity of the hypoxic ventilatory response. Respir Physiol 110:261–268

Pfaffl MW, Tichopad A, Prgomet C, Neuvians TP (2004) Determination of stable housekeeping genes, differentially regulated target genes and sample integrity: BestKeeper–Excel-based tool using pair-wise correlations. Biotechnol Lett 26:509–515

Summers BA, Overholt JL, Prabhakar NR (2000) Augmentation of L-type calcium current by hypoxia in rabbit carotid body glomus cells: evidence for a PKC-sensitive pathway. J Neurophysiol 84:1636–1644

Wang ZY, Bisgard GE (2005) Postnatal growth of the carotid body. Respir Physiol Neurobiol 149:181–190

Wasicko MJ, Sterni LM, Bamford OS, Montrose MH, Carroll JL (1999) Resetting and postnatal maturation of oxygen chemosensitivity in rat carotid chemoreceptor cells. J Physiol 514(Pt 2):493–503

Chapter 8
Erythropoietin and the Sex-Dimorphic Chemoreflex Pathway

Jorge Soliz, Hanan Khemiri, Céline Caravagna, and Tommy Seaborn

Abstract During hypoxic or hypoxemic conditions, tissue oxygenation and arterial O_2 carrying capacity are upregulated by two complementary systems, namely the neural respiratory network (central and peripheral) that leads to increased minute ventilation thereby increasing tissue oxygenation, and erythropoietin (Epo) release by the kidney that activates erythropoiesis in bone marrow to augment arterial blood O_2 carrying capacity. Despite the fact that both neural respiratory control and Epo-mediated elevation of red blood cells are responsible for keeping arterial O_2 content optimal, no interaction between these systems has been described so far. Here we review data obtained in our laboratory demonstrating that ventilatory and erythropoietic systems are tightly connected. We found Epo is the key factor mediating this relationship through modulation of the chemoreflex pathway. Moreover, we showed that this interaction occurs in a sex-dependent manner.

Keywords Brainstem • Carotid body • Hypoxia • Hypoxic ventilatory response • Gender • Respiration

8.1 Cerebral Erythropoietin Regulates Hypoxic Ventilation

Erythropoietin (Epo), a 165-amino acid glycoprotein (34 kDa) belonging to type I cytokine superfamily, has low constitutive expression and activity, while being rapidly upregulated under hypoxic stress conditions. Epo expression is regulated by hypoxia-inducible factor-2 (HIF-2), and the strongest hypoxia-related induction occurs in the kidney. Epo has long been thought to be involved exclusively in erythropoiesis, enabling erythroid progenitor cells to survive and mature via its anti-apoptotic action. However, it has been demonstrated during the last 15 years that Epo and its receptor (EpoR) are functionally expressed in a variety of cell types, including cerebral glial cells and neurons, and that Epo modulates neuronal activity (reviewed in (Jelkmann 2007)). Sasaki and co-authors were the first in 1992 to detect presence of Epo and functional EpoR in the neuron-like PC12 and NS6 cell lines

J. Soliz (✉) • H. Khemiri • C. Caravagna • T. Seaborn
Department of Pediatrics, Centre de Recherche de l'Hôpital St-François d'Assise (CR-SFA), Centre Hospitalier Universitaire de Québec (CHUQ), Faculty of Medicine, Laval University,
10 Rue de l'Espinay, Room D0-700, Québec, QC G1L 3L5, Canada
e-mail: jorge.soliz@crchuq.ulaval.ca

C.A. Nurse et al. (eds.), *Arterial Chemoreception: From Molecules to Systems*,
Advances in Experimental Medicine and Biology 758, DOI 10.1007/978-94-007-4584-1_8,

(Sasaki et al. 2001), suggesting that Epo has much broader field of action than previously recognized. In fact, Epo has been shown to play a critical role in neurogenesis and neuroprotection, and to act as a neurotrophic factor (Kumral et al. 2007). Studies in rodents, monkeys and humans showed that Epo exerts a protective function in several ischemic stroke models, traumatic brain injury, spinal cord injury, and perinatal asphyxia (reviewed in (Rabie and Marti 2008; Kumral et al. 2011)). Remarkably, recent studies demonstrated that Epo protects against chronic neurodegenerative diseases and mental disorders (e.g. multiple sclerosis, epilepsy, schizophrenia, Alzheimer's). Mechanisms by which Epo exerts neuroprotective function include several pathways controlling O_2 homeostasis (e.g. by upregulating O_2-free radical scavenger enzymes, by maintaining expression of anti-apoptotic proteins Bcl-2 and Bcl-xl, by activating voltage-gated Ca^{2+} channels). The impact Epo has (directly or indirectly) on the processes involved in O_2 homeostasis, the reported morphological and neurochemical changes associated with hypoxia in the neural respiratory network, and the implication of catecholamines in respiratory response to acute hypoxia at both peripheral (chemoreceptors) and central (respiratory areas) levels (Hilaire et al. 2004), all these evidences lead us to speculate that neuronal Epo modulates hypoxic ventilation by interacting with brainstem respiratory centers.

Our studies performed in wild type (WT) and transgenic mice showed that Epo enhances ventilatory response and acclimatization to acute and chronic hypoxia (Soliz et al. 2005). Furthermore, immunostaining revealed that EpoR in brainstem is expressed in neurons involved in respiratory rhythmogenesis (pre-Bötzinger complex), sensory integration (nucleus tractus solitarius), and catecholaminergic metabolism (A6 and A5 in the pons; A2/C2 and A1/C1 in the medulla oblongata). Since immunoblots provided no evidence for activated Epo-mediated neuroprotective pathways in mice brainstem (Gassmann and Soliz 2009), these results suggested that Epo-related increase in hypoxic ventilation occurs, at least in part, by the modulation of catecholaminergic synthesis in brainstem (Soliz et al. 2005).

Taken together, these observations imply that cerebral Epo is a key factor in the central respiratory network stimulation under conditions of reduced O_2 availability in adult mice. Thus, the next logical question to address was whether Epo impacts also carotid bodies, which are the sensory organs, whose stimulation by hypoxia activates the chemoreflex pathway.

8.2 Carotid Body Mediates the Hypoxic Ventilatory Response

O_2 in the organism is not stored in large amounts because it reacts rapidly with several molecules thereby producing harmful reactive oxygen species. Therefore, O_2 must be continuously supplied to settle the O_2-dependent enzymatic reactions. As a result, mammals are equipped with more than one mechanism to adequately respond to O_2 reduction in arterial blood. The earliest response to occur under the carotid body control, is an increase in ventilation, which enlarges the overall volume of inspired air and thus increases the level of intra-alveolar O_2 (reviewed in (Ortega-Saenz et al. 2007)).

Carotid bodies (Fig. 8.1a) are the main peripheral chemoreceptors identified in mammals and are the most vascularized organs in the body. They sense reduced blood oxygenation and activate chemoreflex pathways by relaying sensory information to brainstem neurons that, in turn, modulate

Fig. 8.1 Erythropoietin (Epo) and its receptor (EpoR) in carotid bodies. (**a**) Carotid body, located at the arterial carotid bifurcation, is the most vascularized organ in the body. Structurally, carotid bodies contain clusters of chemosensitive glomus cells (type I cells) surrounded by sustentacular cells (type II cells). (**b**) Glomus cell is depolarized in response to reduction in arterial O_2 tension. Consequently, voltage-gated Ca^{2+} channels are opened, resulting in an intracellular influx of Ca^{2+}, which in turn promotes the secretion of neurotransmitters held in storage vesicles. Epo and sex hormones also promote neurotransmitter secretion in glomus cells but the signaling pathways underlying this effect are not yet fully deciphered. (**c**) Co-expression of EpoR and tyrosine hydroxylase (TH) in the carotid body (*black* staining) confirms localization of EpoR within chemosensitive glomus cells. (Panel C is adapted from (Soliz et al. 2005), with permission)

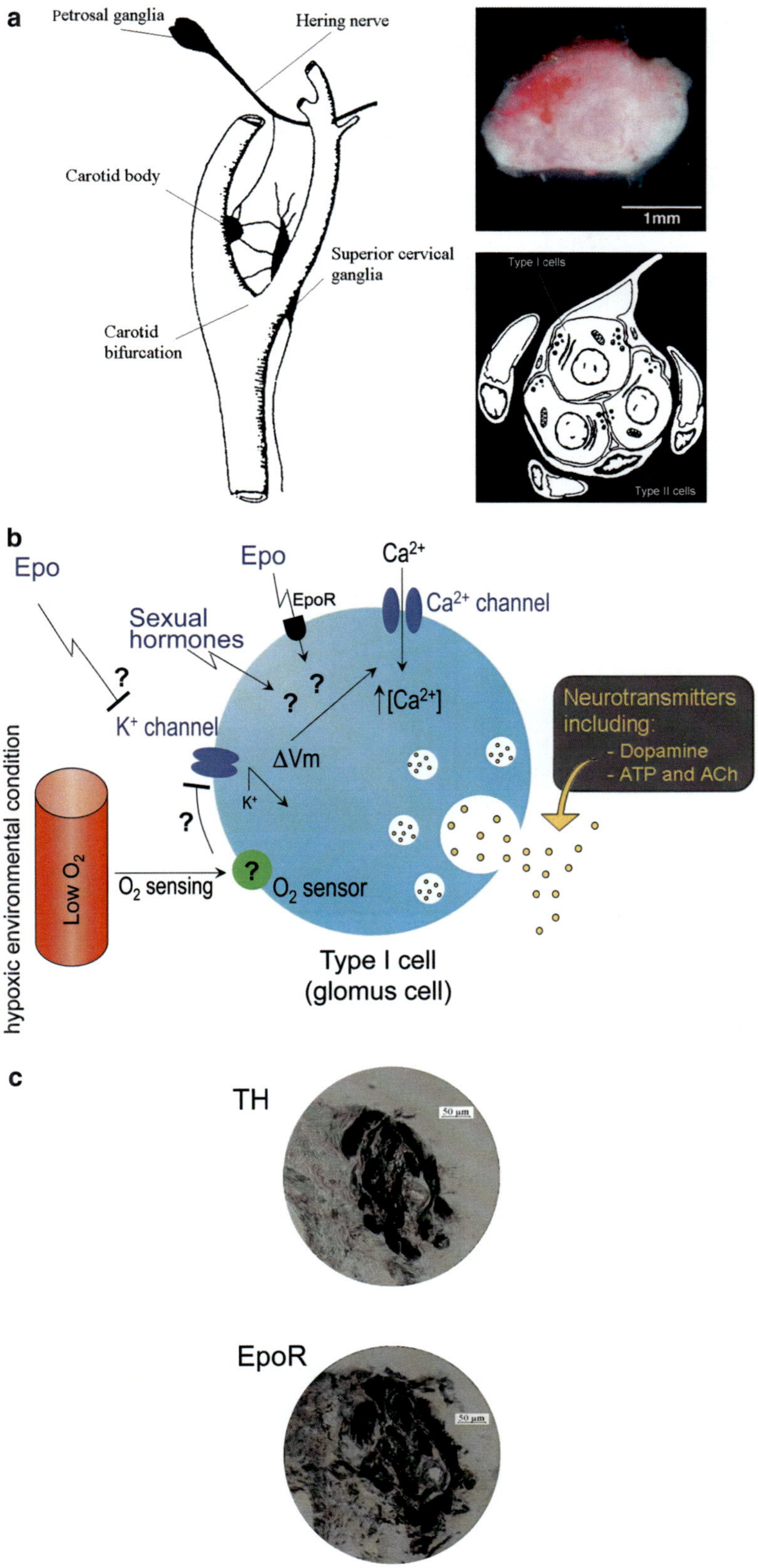
a
Petrosal ganglia
Hering nerve
Carotid body
Superior cervical ganglia
Carotid bifurcation
1mm
Type I cells
Type II cells
b
Epo
Epo
EpoR
Ca2+
Ca2+ channel
Sexual hormones
?
K+ channel
↑[Ca2+]
ΔVm
K+
Neurotransmitters including:
- Dopamine
- ATP and ACh
hypoxic environmental condition
Low O2
O2 sensing
O2 sensor
Type I cell
(glomus cell)
c
TH
50 µm
EpoR
50 µm

corresponding ventilatory adjustments. Increasing ventilation is the most important hypoxic response when mammals are acutely exposed to high altitude, but it also plays a crucial role in acclimatization to chronic hypoxia.

Carotid bodies are composed of innervated clusters of glomus cells which, when activated by reduced O_2 tension, release various neurotransmitters. Dopamine, acetylcholine, adenosine triphosphate (ATP), substance P and nitric oxide (NO) seem to have predominant role in glomus cell signaling. However, glomus cells are essentially catecholaminergic, presenting the highest catecholamine content in the organism.

Several candidates have been proposed as O_2 sensors in glomus cells (reviewed in (Kemp 2006; Lopez-Barneo et al. 2008)), including adenosine monophosphate-activated protein kinase, hemeoxygenase-2, and mitochondria. Whatever the identity of the sensor itself, once the reduction in arterial O_2 is detected, glomus cells rapidly depolarize and generate an influx of Ca^{2+} through voltage-gated Ca^{2+} channels, which leads the release of neurotransmitters (Fig. 8.1b).

8.3 Carotid Body Is a Sex Dimorphic Organ

Only few studies have investigated the impact of sex hormones on carotid bodies (reviewed in (Behan and Wenninger 2008)). Hannhart and co-workers reported that the carotid sinus nerve activity increases following chronic elevation of progesterone in cats, suggesting that carotid body is sensitive to circulating sex hormones. Additional studies suggested this effect might be potentiated by estrogen at the central level. Both estrogen receptor beta and progesterone receptor immunoreactivity have been described in the carotid body of rats (Soliz and Joseph 2005). Moreover, ovarian steroids (i.e. estradiol, estrone, and estriol) are potent respiratory stimulants acting on carotid bodies, thus enhancing basal ventilation and hypoxic responsiveness. Accordingly, several reports have demonstrated sex-related differences in the physiological responses to hypoxia, revealing that women/females show better adaptation to hypoxia than men/males. Indeed, women/females are less susceptible than men/males to a number of hypoxia-associated syndromes, during both infancy and adulthood. In adult mammals, experimental data clearly show that endogenous ovarian steroids act directly on peripheral chemoreceptors, increasing carotid sinus nerve activity, resting minute ventilation ($\dot{V}_E$) and hypoxic chemo-responsiveness. In rats, the stimulatory effects of ovarian steroids on resting $\dot{V}_E$ are mediated by a mechanism involving reduction of dopamine synthesis and signaling on D2 dopamine receptors in the carotid bodies.

8.4 Plasma EPO Interacts with Carotid Body Glomus Cells

Upon sustained hypoxemia, the ventilatory system is essential but insufficient to ensure adequate O_2 supply to cells and tissues. Thus, a second important mechanism increasing O_2 availability is activated when the erythropoietic system augments the blood O_2 carrying capacity.

During erythropoiesis, Epo synthesis in the kidney is accelerated, resulting in increased plasma Epo level. Once in the bone marrow, Epo maintains the availability of erythrocyte progenitor cells, promotes cell division, and increases hemoglobin synthesis, culminating in increased hematocrit level. Despite being complementary, no direct interaction between neural control of ventilation and Epo-mediated elevation in erythrocytes number has been described so far.

Interestingly, Epo exposure in PC12 cells, a cell line considered as a reliable model for the study of carotid body glomus cells, increases their intracellular Ca^{2+} concentration, induces membrane depolarization, and increases cell survival, dopamine release and tyrosine hydroxylase activity (reviewed in (Jelkmann 2007; Rabie and Marti 2008)). Suitably, we suspected that ventilatory and erythropoietic

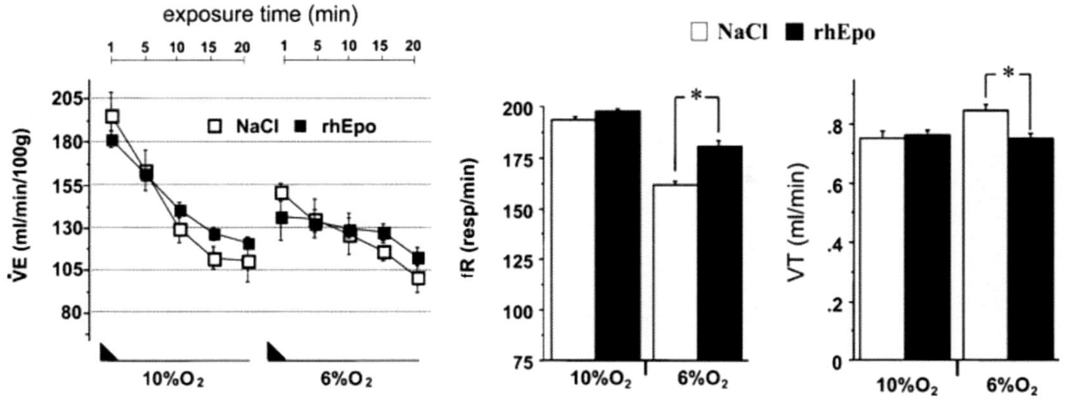

Fig. 8.2 Modulation of hypoxic ventilatory response (HVR) by systemic erythropoietin (Epo). Wild type mice were injected i.v. with recombinant human Epo (rhEpo, 2,000 U/kg) while controls were injected with saline solution. HVR to moderate (10% O_2) and severe (6% O_2) normobaric hypoxia was evaluated by plethysmography. Epo-injected mice showed higher respiratory frequency (fR) but lower tidal volume (V_T) than saline-injected controls when exposed to 6% O_2 hypoxia, suggesting that circulating Epo can modulate peripheral chemoreceptors. (Taken from (Soliz et al. 2005), with permission)

responses were tightly linked. We postulated that higher concentration of Epo in plasma stimulates the hypoxic ventilatory response (HVR) by interacting with carotid body cells, while elevated tissue oxygenation induced by increased ventilation reduces the hypoxic secretion of Epo in renal cells.

Since carotid bodies have embryonic neural origin and glomus cells share some features with PC12 cells, we first addressed the question whether EpoR is present in carotid body glomus cells, as it is the case in brain tissue and neural-like cells. By performing immunostaining on carotid body bifurcations and using tyrosine hydroxylase staining to identify glomus cells, we detected dense staining of EpoR in the carotid body, apparently localized within islets of chemosensitive glomus cells (Soliz et al. 2005) (Fig. 8.1c). This observation implies that circulatory Epo interacts with carotid body cells, most likely by binding the EpoR. We also measured by plethysmography the HVR in mice injected with 2,000 U/kg of recombinant human Epo (rhEpo). We observed that Epo-injected mice showed higher respiratory frequency (fR) but lower tidal volume (V_T) than saline-injected controls when exposed to severe hypoxia (6% O_2 ; Fig. 8.2). Considering that a glycoprotein such as Epo crosses the blood–brain barrier at very high doses only (Banks et al. 2004; Xenocostas et al. 2005), these results suggest that circulating Epo can activate peripheral chemoreceptors.

In addition to our findings, Lam and co-workers demonstrated that protein level of Epo and EpoR increases in carotid body of rats exposed either to intermittent or chronic hypoxia (Lam et al. 2009). These findings suggest that upregulation of Epo and its receptor may be modulated by activation of HIF-1 pathway.

8.5 EPO Modulates the Chemoreflex in a Sex-Dependent Manner

Knowing that both Epo and sex steroids are primary regulators of peripheral chemoreceptors, we tested whether they have functional interactions. We evaluated by plethysmography basal ventilation and HVR to moderate (10% O_2) and severe (6% O_2) normobaric hypoxia in male and female mice following intravenous injection of rhEpo (2,000 U/kg) (Fig. 8.3). Assuming that no or low amounts of Epo cross the blood–brain barrier before 10–12h (Banks et al. 2004; Xenocostas et al. 2005), this experiment allowed us to study the sex-dependent impact of Epo on carotid bodies without the influence of

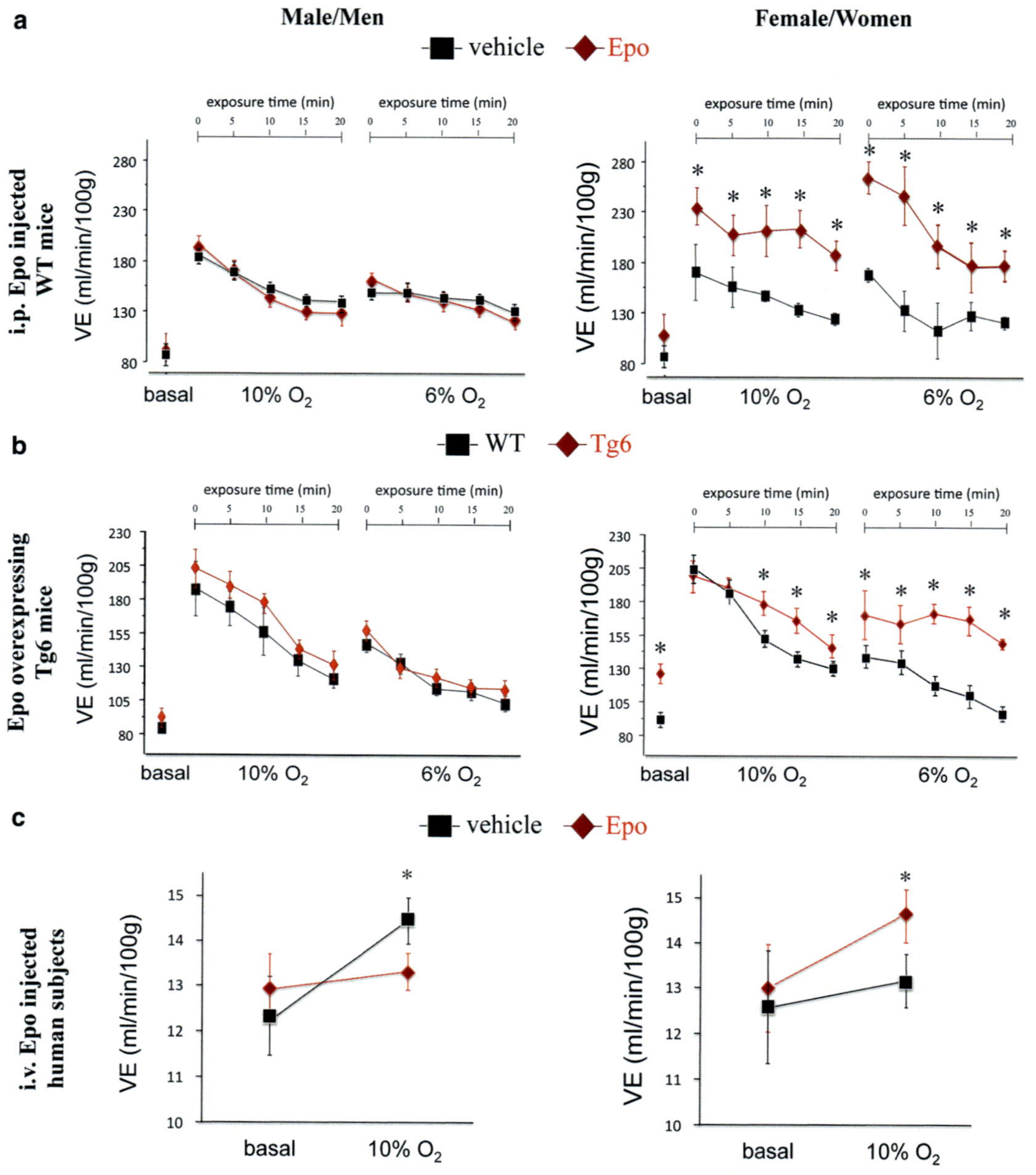

Fig. 8.3 Sex-dependent effect of erythropoietin (Epo) in carotid bodies of mice and humans. Hypoxic ventilatory response (HVR) was determined by plethysmography by measuring minute ventilation ($\dot{V}_E$). (**a**) HVR to moderate (10% O_2) and severe (6% O_2) normobaric hypoxia was evaluated in male and female wild type mice injected with recombinant human Epo (rhEpo) or vehicle. Epo injection increased HVR only in females. (**b**) HVR to moderate (10% O_2) and severe (6% O_2) normobaric hypoxia was evaluated in male and female transgenic mice overexpressing Epo (Tg6). When compared with wild type mice, only Tg6 females showed increased HVR. (**c**) Human subjects were intravenously injected with rhEpo or vehicle and then exposed to 10% O_2 for 15 min. Increased plasma Epo concentration decreased HVR in men while augmenting HVR in women. (Adapted from (Soliz et al. 2009), with permission)

cerebrally-produced Epo. While no difference was observed in male mice during hypoxia, Epo administration to females caused a tremendous increase in HVR (Fig. 8.3a) (Soliz et al. 2009).

By using a transgenic mouse line (Tg6) showing increased Epo levels in brain (26-fold higher than WT) and plasma (12-fold higher than WT), the latter leading to hematocrit values up to 80% (Ruschitzka et al. 2000), we observed that Tg6 males showed similar basal ventilation and HVR to WT control mice (Soliz et al. 2009). Contrastingly, compared with WT, Tg6 females showed increased basal ventilation and augmented HVR under either moderate (10% O_2) or severe (6% O_2) normobaric hypoxia (Fig. 8.3b). In contrast to the data observed in normoxia, this increase was due to significant elevation in fR rather than in V_T (Soliz et al. 2009).

To test whether the situation pictured by our data in mice parallels the situation in humans, we evaluated the HVR in volunteers exposed to 10% O_2 for 15 min upon intravenous injection of rhEpo. We observed that increased plasma Epo concentration altered HVR in all human subjects (Fig. 8.3c). However, while HVR was decreased in men due to a significant decrease in fR, HVR was augmented in women due to a significant increase in V_T. Because we do not expect significant amount of Epo to cross the blood–brain barrier in this short period of time, our results suggest that elevated Epo plasma levels influence peripheral chemosensitivity in both mice and humans (Soliz et al. 2005).

Taken together, these results provide convincing evidence that plasma Epo modulates the carotid body's response to hypoxia in a sex-dependent manner.

8.6 Clinical Implications and Future Research

Subcutaneous treatment with Epo (300 U/kg/dose, 3 times/week) to very premature neonates (gestational age < 30 weeks) has been recently reported to improve both erythropoietic (higher hemoglobin, hematocrit and reticulocytes) and ventilatory functions (reduced need for assisted ventilation and O_2 supplementation) (Tempera et al. 2011). Although a better hematological status clearly helps in stabilizing respiration, the remarkable improvement of ventilatory function caused by Epo cannot be merely explained by a higher number of erythrocytes.

High-dosage of Epo required for neuroprotection (1,000–30,000 U/kg) is well above the range used to treat anemia (500 U/kg) and has prompted concerns about potential adverse effects of Epo. Whereas long-term Epo treatment in adults has been associated with hypertension, seizures, thrombotic events and polycythemia, repeated high-dose Epo treatment in infants and neonates has no discernible adverse effects (Kellert et al. 2007).

Based on the current state of knowledge showing that Epo treatment improves neonatal ventilatory function, we currently investigate the unexplored potential physiological role(s) of Epo in the neuronal control of respiration during postnatal development. Likewise, this work will shed light on the underlying mechanisms of Epo-mediated central regulation of O_2 homeostasis. Furthermore, these data will open new doors for therapeutical use of Epo in neonatal chronic respiratory disorders, which remain significant health problems. Indeed, perinatal asphyxia, neonatal apnea and respiratory distress syndrome in premature infants are major causes of morbidity in neonates. A prophylactic therapy improving the outcomes of respiratory-related diseases in preterm neonates would represent tremendous health benefits. Ultimately, obtained data will help in the future use of Epo in developing novel pharmacological treatments or in the prevention of respiratory instabilities in newborns.

Acknowledgements JS is supported by the Respiratory Health Network of the FRSQ (Fonds de la Recherche en Santé du Québec), the Foundation of Stars for the Children's health research, the CHUQ Research Center direction and axis on Reproduction, perinatal health, and child health, and Faculty of Medicine of Laval University. We apologize to the many researchers who contributed importantly to the field of research and whose works were not cited due to space restriction.

References

Banks WA, Jumbe NL, Farrell CL, Niehoff ML, Heatherington AC (2004) Passage of erythropoietic agents across the blood–brain barrier: a comparison of human and murine erythropoietin and the analog darbepoetin alfa. Eur J Pharmacol 505:93–101

Behan M, Wenninger JM (2008) Sex steroidal hormones and respiratory control. Respir Physiol Neurobiol 164:213–221

Gassmann M, Soliz J (2009) Erythropoietin modulates the neural control of hypoxic ventilation. Cell Mol Life Sci 22:3575–3582

Hilaire G, Viemari JC, Coulon P, Simonneau M, Bévengut M (2004) Modulation of the respiratory rhythm generator by the pontine noradrenergic A5 and A6 groups in rodents. Respir Physiol Neurobiol 143(2–3):187–197

Jelkmann W (2007) Erythropoietin after a century of research: younger than ever. Eur J Haematol 78:183–205

Kellert BA, McPherson RJ, Juul SE (2007) A comparison of high-dose recombinant erythropoietin treatment regimens in brain-injured neonatal rats. Pediatr Res 61:451–455

Kemp PJ (2006) Detecting acute changes in oxygen: will the real sensor please stand up? Exp Physiol 91:829–834

Kumral A, Baskin H, Yesilirmak DC, Ergur BU, Aykan S, Genc S, Genc K, Yilmaz O, Tugyan K, Giray O, Duman N, Ozkan H (2007) Erythropoietin attenuates lipopolysaccharide-induced white matter injury in the neonatal rat brain. Neonatology 92:269–278

Kumral A, Tuzun F, Oner MG, Genc S, Duman N, Ozkan H. (2011). Erythropoietin in neonatal brain protection: The past, the present and the future. Brain Dev. 8:632–43

Lam SY, Tipoe GL, Fung ML (2009) Upregulation of erythropoietin and its receptor expression in the rat carotid body during chronic and intermittent hypoxia. Adv Exp Med Biol 648:207–214

Lopez-Barneo J, Ortega-Saenz P, Pardal R, Pascual A, Piruat JI (2008) Carotid body oxygen sensing. Eur Respir J 32:1386–1398

Ortega-Saenz P, Pascual A, Piruat JI, Lopez-Barneo J (2007) Mechanisms of acute oxygen sensing by the carotid body: lessons from genetically modified animals. Respir Physiol Neurobiol 157:140–147

Rabie T, Marti HH (2008) Brain protection by erythropoietin: a manifold task. Physiology (Bethesda) 23:263–274

Ruschitzka FT, Wenger RH, Stallmach T, Quaschning T, de Wit C, Wagner K, Labugger R, Kelm M, Noll G, Rülicke T, Lindberg RL, Rodenwaldt B, Bauer C, Lüscher TF, Gassmann M (2000) Nitric oxide prevents cardiovascular disease and determines survival in polyglobulic mice overexpressing erythropoietin. Proc Natl Acad Sci U S A 97:11609–11613

Sasaki R, Masuda S, Sasaki R, Masuda S, Nagao M (2001) Pleiotropic functions and tissue-specific expression of erythropoietin. News Physiol Sci 16:110–113

Soliz J, Joseph V (2005) Perinatal steroid exposure and respiratory control during early postnatal life. Respir Physiol Neurobiol 149(1–3):111–122

Soliz J, Joseph V, Soulage C, Becskei C, Vogel J, Pequignot JM, Ogunshola O, Gassmann M (2005) Erythropoietin regulates hypoxic ventilation in mice by interacting with brainstem and carotid bodies. J Physiol 568:559–571

Soliz J, Thomsen JJ, Soulage C, Lundby C, Gassmann M (2009) Sex-dependent regulation of hypoxic ventilation in mice and humans is mediated by erythropoietin. Am J Physiol Regul Integr Comp Physiol 296:R1837–R1846

Tempera A, Stival E, Piastra M, DEL D, Ottaviano C, Tramontozzi P, Marconi M, Cafforio C, Marcozzi P, Rossi N, Buffone E (2011) Early erythropoietin influences both transfusion and ventilation need in very low birth weight infants. J Matern Fetal Neonatal Med 24:1060–1064

Xenocostas A, Cheung WK, Farrell F, Zakszewski C, Kelley M, Lutynski A, Crump M, Lipton JH, Kiss TL, Lau CY, Messner HA (2005) The pharmacokinetics of erythropoietin in the cerebrospinal fluid after intravenous administration of recombinant human erythropoietin. Eur J Clin Pharmacol 61:189–195

Chapter 9
Time-Course of Ventilation, Arterial and Pulmonary CO_2 Tension During CO_2 Increase in Humans

Toru Satoh, Yasumasa Okada, Yasushi Hara, Fumio Sakamaki, Shingo Kyotani, Takeshi Tomita, Noritoshi Nagaya, and Norifumi Nakanishi

Abstract A change of ventilation (VE), $PaCO_2$ (arterial CO_2 tension) and $PvCO_2$ (pulmonary arterial CO_2 tension) with time was not evaluated precisely during exercise or CO_2 rebreathing in humans. In this study, changes of these variables with time were fitted to exponential curves $\{y=Exp(x/T+A)+k\}$ and compared. When exercise pulmonary hemodynamics was examined in 15 cardiac patients to decide therapies, we asked the patients to undergo CO_2 rebreathing using air with supplementation of consumed O_2. Arterial and pulmonary blood was drawn every minute. During exercise, T was 28.2 ± 8.4 and 26.8 ± 12.4, and A was 0.80 ± 0.50 and 0.50 ± 0.90 in VE and $PvCO_2$, respectively, with no statistical differences. During CO_2 rebreathing, T was 18.6 ± 5.8, 41.8 ± 38.0 and 21.6 ± 9.7 and A was 0.39 ± 0.67, 1.64 ± 1.35 and 0.17 ± 0.83 in VE, $PaCO_2$ and $PvCO_2$, respectively, with statistical difference of $PaCO_2$ from other variables, suggesting that VE and $PvCO_2$ showed same mode of change according to time but $PaCO_2$ did not.

Keywords Ventilation • $PaCO_2$ • $PvCO_2$ • CO_2 rebreathing

9.1 Introduction

A change of VE, $PaCO_2$ and $PvCO_2$ with time was not evaluated precisely during CO_2 increasing state like exercise or CO_2 rebreathing in humans. Gelfand and Lambertsen (1973) studied time course of ventilatory change by abruptly adding and removing CO_2 in inhaled air and reported that there were three respiratory components with differing onset lag time and time constant. We did a similar analysis of $PaCO_2$ and $PvCO_2$ as well as VE during CO_2 rebreathing and exercise tests in 15 cardiac patients. We performed exercise test with arterial and pulmonary arterial blood sampling to make therapeutic

T. Satoh (✉)
Cardiology Division, Department of Medicine, Kyorin University School of Medicine, Shinkawa 6-20-2, Mitakashi, Tokyo 181-8611, Japan
tsatoh2008@me.com

Y. Okada
Department of Medicine, Keio University Tsukigase Rehabilitation Center, Amagiyugashima-cho 410-3293, Japan

Y. Hara • F. Sakamaki • S. Kyotani • T. Tomita • N. Nagaya • N. Nakanishi
Division of Cardiology and Pulmonary Circulation, Department of Medicine, National Cardiovascular Center, Suita, Osaka, 565-8565, Japan

C.A. Nurse et al. (eds.), *Arterial Chemoreception: From Molecules to Systems*, Advances in Experimental Medicine and Biology 758, DOI 10.1007/978-94-007-4584-1_9,

decisions for cardiac patients. Then we asked them to undergo CO_2 rebreathing test after explanation of the study purpose to elucidate CO_2 and ventilatory kinetics and know ventilatory control mechanics. We fitted the plotting of $PaCO_2$, $PvCO_2$ and VE with time, to exponential curve. Time constant and fixed constant values of the resultant equations in each variable were calculated and compared each other to see the relation of $PaCO_2$, $PvCO_2$ and ventilation.

The results suggest that the fitted equation of VE with time was statistically different from the fitted equation of $PaCO_2$ with time, but not from the fitted equation of $PvCO_2$ with time. Implication of our results is that VE and $PvCO_2$ are changed identically, but it must await further study that this relation is a cause or a result. We report this result because it may add new insights to ventilation research in terms of CO_2 kinetics.

9.2 Methods

9.2.1 *Study subjects*

The study subjects were 15 patients with cardiac disease, who underwent pulmonary hemodynamic investigations in order to help determine their treatment plans. Eleven had mitral valvular heart disease (4 dominant mitral stenosis, 4 dominant mitral regurgitation and 3 combined mitral stenosis and regurgitation), 2 had dilated cardiomyopathy, and 2 had chronic pulmonary thromboembolism. No patient had ventilatory disorder. Their age was 51 ± 15 years (mean ± S.D.). Eight patients were male and seven were female. The purpose, protocol and risks of the present study were fully explained, and written informed consent was obtained from each patient.

9.2.2 *Protocol*

The patients performed exercise on an upright cycle ergometer 4 h after their usual breakfast and medication, with a Swan-Ganz catheter inserted via the right internal jugular vein into the pulmonary artery and a fine arterial catheter inserted via the left radial artery. They pedaled at a speed of 55 rpm without any added load for 1 min. Then the load was increased by 1 W/ 4 s (15 W/ min) to the symptom-limited maximum. Continuous hemodynamic monitoring, including that of arterial and pulmonary arterial pressures, and expired gas analysis (AE280, Minato Medical Instruments, Osaka) were performed every 6 s throughout the period of exercise. Expired ventilation (VE) was measured by hot-wire flowmeter. Arterial and pulmonary arterial blood samples were collected before exercise and every minute during exercise for blood gas analysis. On the same day, 3 h after lunch, the subjects were tested during CO_2 rebreathing using a bag containing 6 l of air, with the same hemodynamic, expired gas and blood gas analyses as during exercise. Oxygen consumption (VO_2) was determined in advance and an equal amount of O_2 was supplemented into the rebreathing bag to maintain a constant inspired O_2 concentration throughout the rebreathing test.

9.2.3 *Fitting to Exponential Curve*

VE, $PaCO_2$ and $PvCO_2$ were plotted against time. As the relations resembled exponential curve, they were fitted to $y = \mathrm{Exp}\,(x / T + A) + k$. Figure 9.1 demonstrated a representative case during CO_2 rebreathing.

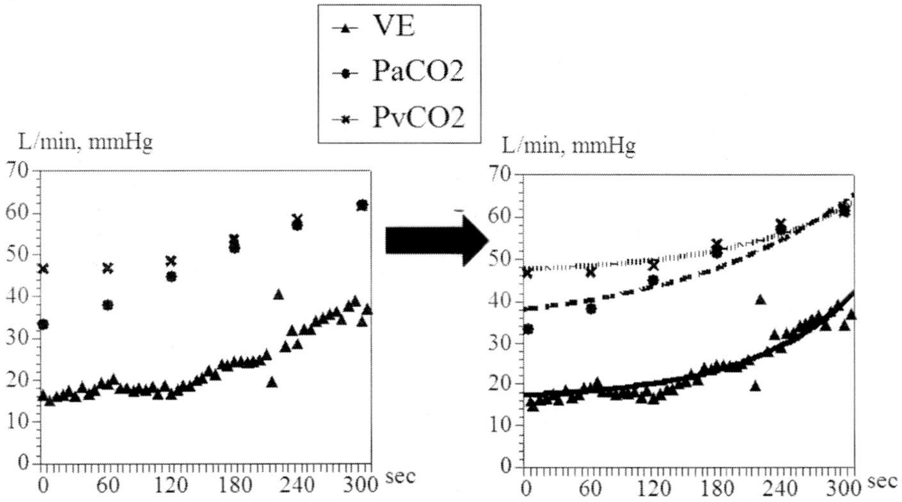

Fig. 9.1 Representative example of fitting to exponential curve. VE, $PaCO_2$ and $PvCO_2$ were determined by averaging forward and backward 5 values. In this example, VE was fitted to y=EXP(x/16.6+0.079)+16.4(R^2=0.87), $PaCO_2$ to y =Exp(x/27.5+1.56)+33.5(R^2=0.92) and $PvCO_2$ to y=Exp (x/19.3+0.114)+46.8(R^2=0.93)

VE was determined by averaging forward and backward 5 values. In this example, VE was fitted to y=Exp (x / 16.6+0.079)+16.4 (R^2=0.87), $PaCO_2$ to y=Exp (x / 27.5+1.56)+33.5 (R^2=0.92) and $PvCO_2$ to y=Exp (x / 19.3+0.114)+46,8 (R^2=0.93).

9.2.4 *Analysis*

T and A were calculated in 15 patients both during exercise and CO_2 rebreathing. T as well as A in VE, $PaCO_2$ and $PvCO_2$ were compared to each other during exercise and CO_2 rebreathing.

9.2.5 *Statistics*

Fitting was carried out by Deltagraph Pro 5.5.1 (SPSS Inc. and Red Rock Software Inc. USA). Comparison of T as well as A in VE, $PaCO_2$ and $PvCO_2$ were done by student-*t* test. Values of $p<0.05$ were considered significant.

9.3 Results

9.3.1 *Changes in Pulmonary Hemodynamics*

VO_2, PaO_2 and arterial pH during exercise are shown in Table 9.1. PaO_2 did not change significantly, as in healthy subjects. Arterial pH was only slightly lowered. Figure 9.2 depicts changes in $PaCO_2$, $PvCO_2$ and VE at rest, and at the RC point in exercise or at the maximal response in CO_2 rebreathing. $PaCO_2$ did not change during exercise (38.3±3.7 to 39.5±5.4 mmHg), but was markedly elevated

Table 9.1 Patient characteristics during exercise

	Rest	RC point	Peak exercise	Statistics
MPA (mmHg)	21 ± 9	50 ± 17		
PCW (mmHg)	12 ± 6	31 ± 12		
VO_2 (ml/min/kg)			18.3 ± 4	
PaO_2 (mmHg)	112 ± 17	107 ± 22		ns
PH	7.37 ± 0.03	7.34 ± 0.05		p=0.02

RC point: respiratory compensation point, when end-tidal CO_2 concentration begins to decrease. Ventilation is linearly related to CO_2 excretion until the RC point. PaO_2: arterial O_2 partial pressure; mPA: mean pulmonary arterial pressure; PCW: pulmonary capillary wedge pressure; VO_2: oxygen consumption; pH: arterial blood pH; ns: not significant; PaO_2 and pH at rest are compared with those at peak exercise by paired *t*-test

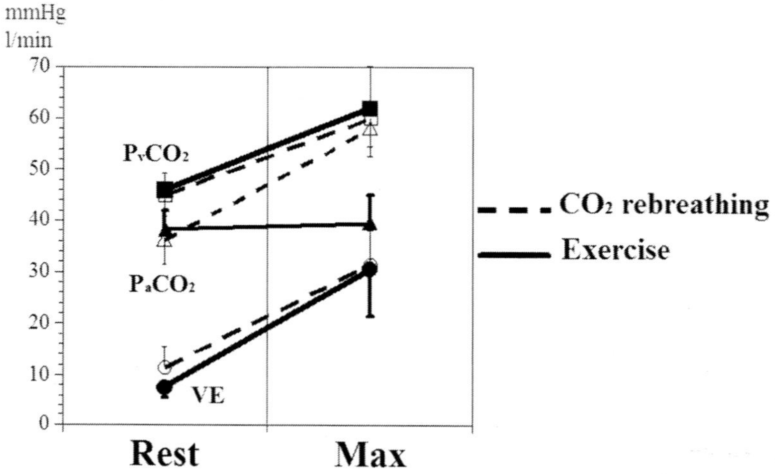

Fig. 9.2 Mean changes in VE, $PaCO_2$ and $PvCO_2$ during exercises and CO_2- breathing for 15 patients. Changes during exercise are shown as *solid lines*. Changes during CO_2 rebreathing are shown as *broken lines*. $PaCO_2$, $PvCO_2$ and VE before both examinations are shown as at rest. Three variables at the RC point in exercise and at the maximal response in CO_2 rebreathing are shown as at Max. See text for details. $PaCO_2$: arterial CO_2 partial pressure; $PvCO_2$: mixed venous CO_2 partial pressure; VE: Ventilation

during CO_2 rebreathing (36.3±4.9 to 58.0±5.5 mmHg). $PvCO_2$ increased greatly during both exercise (45.6±3.3 to 61.9±8.2 mmHg) and CO_2 rebreathing (45.3±4.2 to 60.2±5.5 mmHg). VE increased from 7.5±2.2 to 30.5±9.1 l/min during exercise and from 11.5±4.0 to 31.4±6.7 l/min during CO_2 rebreathing.

9.3.2 *T and A*

T and A in the fitting equation during exercise and CO_2 rebreathing are shown in Tables 9.2 (CO_2 rebreathing) and 9.3 (Exercise). Fitting of 3 variables in all subjects was appropriate because second power of fitness of fitting equations were more than 0.8.

Mean values of T as well as A for VE, $PaCO_2$ and $PvCO_2$ during CO_2 rebreathing are depicted in Fig. 9.3. A and T for $PaCO_2$ were statistically different from A and T for VE and $PvCO_2$.

Mean values of T as well as A for VE and $PvCO_2$ during exercise are depicted in Fig. 9.4. Neither was not statistically different during exercise.

Table 9.2 A and T of fitting equations in each variable during CO_2 rebreathing

	A			T		–: not measured
Patient	VE	$PaCO_2$	$PvCO_2$	VE	$PaCO_2$	$PvCO_2$
1	1.52	1.4	1.56	30.6	34	48.1
2	0.29	2.8	−0.85	16.9	48.5	14.1
3	0.64	2.32	0.2	11.7	36.2	10.3
4	−0.26	2.8	0.86	14.1	52.9	23.8
5	0.96	2.18	−1.35	17.6	25.5	10.2
6	0.99	2.59	−0.54	22.9	52.9	17.5
7	−0.72	1.39	0.15	11.6	27.8	19.9
8	0.22	0.41	0.61	24.1	17	20.5
9	1.5	2.2	1.1	27	44	30
10	−0.06	0.47	−0.03	19.4	17.7	21
11	−0.08	–	1.2	17.1	–	29.4
12	−0.1	−0.3	−0.3	10.7	11.7	14.2
13	0.47	3.8	0.07	20	159	23.5
14	0.02	−0.79	−0.36	16.6	16.5	19.9
15	0.33	1.83	0.02	17.7	40.0	22.5

Table 9.3 A and T of fitting equations in each variable during exercise

	A		T	
Patient	VE	$PvCO_2$	VE	$PvCO_2$
1	0.58	0.39	26.1	23.6
2	0.42	0.76	26	29
3	1.5	−0.03	33.4	24.4
4	1.34	1.3	44.4	56.6
5	1.6	0.94	36.9	32.8
6	1.1	0.25	18.2	16.8
7	0.53	−0.79	22.9	16.6
8	−0.58	−2	15.4	11.9
9	0.49	0.89	38.3	42.4
10	0.8	−0.3	17.6	10.8
11	0.86	0.94	29.7	30.5
12	0.59	0.7	30.2	25.5
13	1.2	0.27	26.1	20
14	0.41	1.37	22.2	35.7
15	1.1	0.7	35.5	34.5

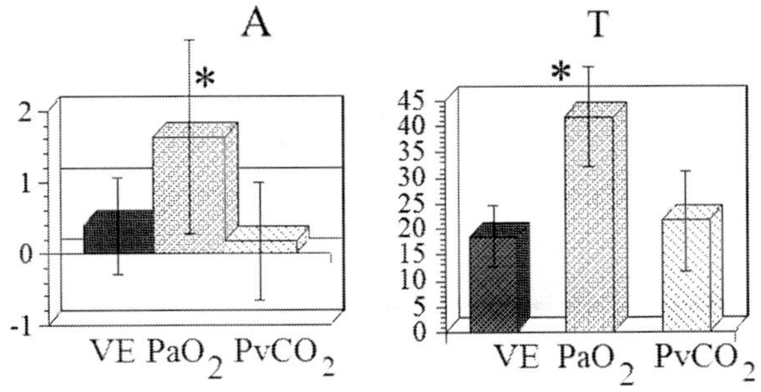

Fig. 9.3 Mean values of A and T in VE, $PaCO_2$ and $PvCO_2$ during CO_2 rebreathing. Mean values of T as well as A in VE, $PaCO_2$ and $PvCO_2$ during CO_2 rebreathing were depicted. A and T in $PaCO_2$ were statistically different from A and T in VE and $PaCO_2$

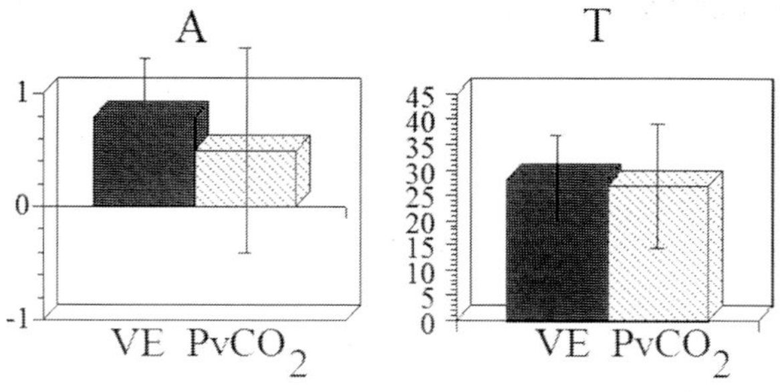

Fig. 9.4 Mean values of A and T in VE and $PvCO_2$ during exercise. Mean values of T as well as A in VE and $PvCO_2$ during exercise were depicted. Both were not statistically different during exercise

9.4 Discussion

This is the first report to analyze changes in VE, $PaCO_2$ and $PvCO_2$ with time and to compare those changes. VE and $PvCO_2$ showed the same mode of change but $PaCO_2$ did not.

9.4.1 Critique of Methods

We studied a diseased population and could be criticized regarding several points, if we extrapolate our results to normal healthy humans. First, patients with heart disease may have additional ventilatory stimuli such as hypoxia or acidosis during exercise. However, the cardiac patients we studied did not show a significant reduction in PaO_2, and demonstrated only a slight decrease in arterial pH. Second, the study patients might have been unique and have had some specific characteristics that led to an unusual conclusion. Among the 15 patients, four had moderate mitral regurgitation, and showed nearly normal hemodynamics during exercise. The results obtained from these four subjects may be representative of normal healthy humans. Other patients had apparently abnormal hemodynamic responses during exercise, but each of them also showed a practically equal change to time between $PvCO_2$ and VE during exercise and CO_2 rebreathing. Therefore, our results may be extended with caution to the normal healthy population, but needs to be confirmed.

Another technical problem may be that the number of sample points for $PvCO_2$ is not sufficient to fit with exponential equation time. However, most of the second power of correlation coefficients for fitting equations were more than 0.9 and at least 0.8, and the present data are considered sufficient for analysis.

9.4.2 Time Course of Respiratory Variants

We fitted the changes in ventilatory parameters (VE, $PaCO_2$ and $PvCO_2$) to exponential equations and compared the time constants and the other constants of the equations between the parameters. In their pioneering work, Gelfand and Lambertsen (1973) confirmed the existence of three different modes of ventilatory components by abruptly adding or stopping CO_2 inhalation. These three components were a peripheral chemoreceptor in the carotid body, a fast central responder to CO_2 increase and a late central responder. We obtained a different time-response equation of the respiratory variables from Gelfand's equation. This is because they inhaled stepwise increase in concentration of CO_2 gas compared to our study in which CO_2 concentration was increased little by little by rebreathing the expired air and the three components of the CO_2 response were not separated. The time constant would be changed according to inhaled CO_2 concentration, individuals, or mode of inspired CO_2 increase.

Our time constant of ventilation was about 20 s during CO_2 rebreathing, and about 30 s during exercise. Gelfand's time constant was 10 s for the fast-responding receptor and 89 s for the slow-responding one. Our time constant was obtained by adding these three components, and is considered to be a reasonable value. Our aim, however, was to compare the differences in time course of three ventilatory variables, VE, $PaCO_2$ and $PvCO_2$.

9.4.3 *Close Coupling of $PvCO_2$ and VE*

The strong coupling of $PvCO_2$ and VE, but not $PaCO_2$ in the pattern of change according to time, suggests two possibilities. One is that VE changes in proportion to $PaCO_2$ but another function intervenes between the two variables, leading to a different pattern of changes in VE and $PaCO_2$ according to time, and $PvCO_2$ is just the result of the VE change. Another possibility is that VE is determined $PvCO_2$, indicating $PvCO_2$ is an important stimulator of ventilation.

If $PvCO_2$ stimulates ventilation, we have to consider the existence of a venous chemoreceptor. This CO_2 chemoreceptor stimulates ventilation both during hypercapnic and eucapnic conditions (CO_2 rebreathing and exercise). Such receptors exist either in the venous system or in the pulmonary artery or pulmonary ventilatory system such as pulmonary stretch receptors (Mitchell et al. 1980; Nilsestuen et al. 1981; Green et al. 1986) or upper airway (Forster et al. 1985). Fedde et al. (1982) reported that pulmonary arterial chemoreceptors for ventilatory control exist in birds. Sheldon and Green (1982) separated the systemic and pulmonary circulation in dogs, controlled CO_2 partial pressure independently in each circuit, and measured VE. They demonstrated that respiratory output was augmented by selectively elevating pulmonary arterial CO_2 partial pressure. However, the existence of venous CO_2 chemoreceptors in mammals has not been proven. Cropp and Comroe (1961) and Sylvester et al. (1973) opposed the theory of the existence of venous CO_2 chemoreceptors in dogs, because infusion of CO_2-equilibrated blood did not initiate ventilatory responses until the infused stimulus reached the systemic arterial circulation. Orr et al. (1988) concluded that venous CO_2 chemoreceptors do not exist in the anesthetized cat, on the basis that venous CO_2 loading did not induce respiratory augmentation in the phrenic neurogram unless $PaCO_2$ was raised.

These reports support the former possibility that VE changes in proportion to $PaCO_2$ but another function intervenes between the two variables, in view of the close coupling of $PvCO_2$ and VE, but not $PaCO_2$. $PaCO_2$ does not directly change ventilation, but some other intervening sensor exists between the variables of $PaCO_2$ and VE. Further study on this issue is needed.

9.5 Summary

VE and $PvCO_2$ showed same mode of change according to time but $PaCO_2$ did not, suggesting that VE and $PvCO_2$ are changed identically with time, but further studies are required to determine whether this relation is a cause or a result. We report this result because it may add new insights to ventilation research in terms of CO_2 kinetics.

References

Cropp GJA, Comroe JHJ (1961) Role of mixed venous blood PCO_2 in respiratory control. J Appl Physiol 16:1029–1033

Fedde MR, Kiley JP, Powell FL, Scheid P (1982) Pulmonary CO_2 receptors and control of breathing in ducks: effects of prolonged circulation time to carotid bodies and brain. Respir Physiol 47:121–140

Forster HV, Pan LG, Flynn C, Bisgard GE, Hoffer RE (1985) Effect of upper airway CO_2 on breathing in awake ponies. J Appl Physiol 59:1222–1227

Gelfand R, Lambertsen CJ (1973) Dynamic respiratory response to abrupt change of inspired CO_2 at normal and high PO_2. J Appl Physiol 35:903–913
Green JF, Schertel ER, Coleridge HM, Coleridge JC (1986) Effect of pulmonary arterial PCO_2 on slowly adapting pulmonary stretch receptors. J Appl Physiol 60:2048–2055
Mitchell GS, Cross BA, Hiramoto T, Scheid P (1980) Effects of intrapulmonary CO_2 and airway pressure on phrenic activity and pulmonary stretch receptor discharge in dogs. Respir Physiol Neurobiol 41:29–48
Nilsestuen JO, Coon RL, Woods M, Kampine JP (1981) Location of lung receptors mediating the breathing frequency response to pulmonary CO_2. Respir Physiol Neurobiol 45:343–355
Orr JA, Fedde MR, Shams H, Roskenbleck H, Scheid P (1988) Absence of CO_2-sensitive venous chemoreceptors in the cat. Respir Physiol 73:211–224
Sheldon MI, Green JF (1982) Evidence for pulmonary CO_2 chemosensitivity: effects on ventilation. J Appl Physiol Respir Environ Exerc Physiol 52:1192–1197
Sylvester JT, Whipp BJ, Wasserman K (1973) Ventilatory control during brief infusions of CO_2-laden blood in the awake dog. J Appl Physiol 35:178–186

Chapter 10
Oxygen Sensitive Synaptic Neurotransmission in Anoxia-Tolerant Turtle Cerebrocortex

Leslie T. Buck, D.W.R. Hogg, C. Rodgers-Garlick, and M.E. Pamenter

Abstract Anoxia rapidly elicits hyper-excitability and cell death in mammal brain but this is not so in anoxia-tolerant turtle brain where spontaneous electrical activity is suppressed by anoxia (i.e. spike arrest; SA). In anoxic turtle brain extracellular GABA concentrations increase dramatically and impact GABAergic synaptic transmission in a way that results in SA. Here we briefly review what is known about the regulation of glutamatergic signalling during anoxia and investigate the possibility that in anoxic turtle cortical neurons $GABA_{A/B}$ receptors play an important role in neuroprotection. Both AMPA and NMDA receptor currents decrease by about 50% in anoxic turtle cerebrocortex and therefore exhibit channel arrest, whereas GABA-A receptor currents increase twofold and increase whole-cell conductance. The increased post synaptic GABA-A receptor current is contrary to the channel arrest hypothesis but it does serve an important function. The reversal potential of the GABA-A receptor (E_{GABA}) is only slightly depolarized relative to the resting membrane potential of the neuron and not sufficient to elicit an action potential. Therefore, when GABA-A receptors are activated, membrane potential moves to E_{GABA} and prevents further depolarization by glutamatergic inputs during anoxia by a process termed shunting inhibition. Furthermore we discuss the presynaptic role of GABA-B receptors and show that increased endogenous GABA release during anoxia mediates SA by activating both GABA-A and B receptors and that this represents a natural oxygen-sensitive adaptive mechanism to protect brain from anoxic injury.

Keywords Glutamate • GABA • Receptors • Reversal potential • Shunting inhibition • Spike arrest • Channel arrest • Metabolic rate

L.T. Buck (✉)
Department of Cell and Systems Biology, University of Toronto,
Toronto, 25 Harbord St, ON M5S 3G5, Canada

Department of Ecology and Evolutionary Biology, University of Toronto, Toronto, ON, Canada
e-mail: les.buck@utoronto.ca

D.W.R. Hogg • C. Rodgers-Garlick
Department of Cell and Systems Biology, University of Toronto,
Toronto, 25 Harbord St, ON M5S 3G5, Canada

M.E. Pamenter
Department of Pediatrics (Division of Respiratory Medicine), University of California San Diego,
La Jolla, CA, USA

C.A. Nurse et al. (eds.), *Arterial Chemoreception: From Molecules to Systems*,
Advances in Experimental Medicine and Biology 758, DOI 10.1007/978-94-007-4584-1_10,

10.1 Introduction

The western painted turtle, red eared slider, common goldfish, and crucian carp are well established vertebrate models of anoxia tolerance that can survive days to months without oxygen (*Chrysemys picta bellii, Trachemys scripta, Carasius auratus,* and *Carasius carasius*, respectively; Bickler and Buck 2007). In many clinical models of anoxia/hypoxia/ischemia tolerance it is often very difficult to discern whether activation of a particular mechanism or pathway is beneficial or detrimental to tissue survival. Indeed, some pathways mediate both protective and deleterious responses to low oxygen stress, depending on the degree to which the mechanism is recruited. As a result, untangling potential cytoprotective mediators in these models can be challenging. Conversely, the strength of anoxia-tolerant models, which have evolved natural adaptations to prolonged low-oxygen stress, is that the mechanisms and cellular pathways activated or inactivated must be protective and can therefore be used to more clearly inform clinical therapeutic targets.

Of these model organisms, the western painted turtle is remarkable in its ability to tolerate up to 4 months of anoxia at 3 C buried in the mud at the bottom of ice covered ponds, a challenge which has been reproduced in the laboratory (Jackson 2002). Temperature is an important factor in lowering metabolic rate in turtles, as increasing temperature to 24°C shortens anoxic survival time to about 24 h. However, after accounting for the effect of temperature on metabolic rate this is the equivalent of increasing anoxic survival of a mammal 100 to 1,000 fold (Bickler and Buck 2007; Jackson 2002). By mammalian standards the degree to which metabolism is depressed in the anoxic turtle is astounding and reflected in various measures of tissue activity. Measurements from laboratory dived (anoxic) animals show that heart rate decreases from 30 beats per minute at 20 C to 1 beat per 10 min at 3°C (Jackson 2002), and electroencephalographic recordings show an 80% decrease in brain electrical activity at 24°C in animals artificially ventilated with nitrogen (Fernandes et al. 1997). Whole animal calorimetry studies confirm a decrease in heat output or metabolic rate of greater than 90%, and experiments with tissue (brain slice) and liver cells show that this response does not require an intact animal, but is intrinsic to isolated tissues and cells (Buck et al. 1993; Jackson 1968; Doll et al. 1994). Remarkably, metabolism is reduced to such a low level that all of the animal's metabolic needs can be met by glycolytically derived ATP. To supply sufficient substrate for prolonged anoxic periods there is an enormous amount of glycogen stored in various tissues, and especially in liver, which contains ten times that of a rat (900 μmoles.g^{-1}.wet^{-1} weight liver; Hochackha and Somero 1984). Correspondingly, there is enhanced blood and tissue pH buffering capacity to accommodate the acid load produced by months of anoxia, which can peak at 200 mmoles lactate•ml^{-1} blood, and a decrease in pH of greater than 1 unit (Jackson 2002). Furthermore, cerebral blood flow (CBF) increases with anoxia to ensure delivery of glycolytic substrate to the brain, removal of anoxic endproducts and to provide adequate pH buffering (Hylland et al. 1994).

In the anoxia-tolerant turtle brain extracellular [glutamate] remains stable or increases slightly (1–4 μM), while [γ-amino butyric acid (GABA)] increases dramatically over 4 h of anoxia (from 0.3 to 27 μM, Nilsson and Lutz 1991; Milton et al. 2002). Despite the lack of a change in [glutamate] the activity (single channel and whole cell currents) of the two major excitatory glutamatergic receptors AMPA and NMDA (α-amino-3-hydroxyl-5-methyl-4-isoxazole-propionate and N-methyl-D-aspartate) decreases by about 50% (Buck and Bickler 1998; Pamenter et al. 2008a, b; Zivkovic and Buck 2010). This also leads to a decrease in spontaneous excitatory post synaptic potential firing, and therefore lowers the level of baseline transmembrane ion leakage. Such decreases in NMDA and AMPA receptor currents provide direct evidence supporting the ion channel arrest hypothesis (CA) which predicts that anoxia tolerant species possess mechanisms to acutely reduce ion movement across the plasma membrane (Hochachka 1986).

More familiar to most is the extreme anoxia sensitivity of the mammalian brain and the cellular events leading to tissue damage (Sattler and Tymianski 2000; Siesjo 1990). In the absence of oxygen cellular [ATP] cannot be maintained, the Na^+/K^+ pump fails and neuronal membrane potential decreases sharply. This process is termed anoxic depolarization (AD) and leads to electrical hyper-excitability, deleterious Ca^{2+} influx through chronically activated NMDA receptors, excitotoxic cell death (ECD), and spreading depression in the penumbra (Anderson et al. 2005). Numerous studies have focused on the role of NMDA receptors in this mechanism and although NMDA receptor blockade prevents ECD (Sattler and Tymianski 2000) it does not prevent AD-mediated injury or post-insult apoptotic cell death (Anderson et al. 2005).

Independent from informative research on anoxia-tolerant turtle brain, research is being conducted on a potential therapeutic alternative to blocking excitatory pathways and that is to up-regulate inhibitory mechanisms mediated by GABA (Blaesse et al. 2009). GABAergic mechanisms are not strongly recruited in ischemic mammalian neurons. In fact, although [GABA] is elevated by ~30% in ischemic murine brain, $GABA_A$ receptor subunit mRNA expression is decreased by ~85%, and GABA-evoked currents and [ATP] run down rapidly, suggesting endogenous GABAergic neuroprotection is transient and ineffective (Erecinska et al. 1984; Allen et al. 2004; Li et al. 1993). Nonetheless, activating GABA receptors pre-insult limits neuronal hyper-excitability in mammalian models of ischemic damage, and AD and cell death are not observed in the afflicted brain region (Galeffi et al. 2000; Costa et al. 2004). Despite these promising results, GABA has received little attention as a clinical stroke intervention and the mode of GABAergic neuroprotection is poorly understood (Ginsberg 2008). In the following section we will give an overview of what is known about oxygen sensitive glutamatergic and GABAergic synaptic neurotransmission in the anoxia-tolerant turtle brain.

10.2 Oxygen Sensitive Glutamatergic Neurotransmission

The ionotropic glutamate receptors (NMDA and AMPA) were the obvious targets of strategies to protect brain from anoxic injury because of their role in permitting intracellular [Ca^{2+}] to rise to excitotoxic levels, which is considered a key point of no return in excitotoxic cell death. Unfortunately, the strategy of inhibiting glutamatergic neurotransmission has thus far been unsuccessful in anoxic/ischemic mammalian brain, largely due to the profound global depression in respiratory and cardiac function. However, in the turtle brain excitotoxicity is not observed during long term anoxia and we have demonstrated that both NMDA and AMPA receptors are regulated in a manner that decreases ionic conductance (Shin and Buck et al. 2005; Pamenter et al. 2008a, b; Zivkovic and Buck 2010).

Whole-cell attached-patch recording of NMDA and AMPA receptor currents from turtle cortical neurons reveal that both receptor currents reversibly decrease by about 50% during a 20 min normoxic to anoxic transition. These experiments are 90 min in length with a 30–40 min anoxic period followed by a normoxic recovery where currents return to pre-anoxic levels. The down-regulation of both types of ionotropic glutamate receptors is linked to the opening of a mitochondrial ATP sensitive K^+ channel (mK_{ATP}; Pamenter et al. 2008a, b; Zivkovic and Buck 2010), which has also been linked to ischemic preconditioning in some animal models, including: cultured rat cortical slices following glutamate toxicity (Kis et al. 2003; 2004), anoxic juvenile mouse brainstem (Muller et al. 2002), and in rat hippocampal and cortical neurons following anoxia/reperfusion injury (Heurteaux et al. 1995; Semenov et al. 2000). The neuroprotective role of the mK_{ATP} channel during anoxia is thought to be due to its ability to regulate Ca^{2+} uptake into the mitochondrial matrix and prevent mitochondrial Ca^{2+} overload. Here, the opening of mK_{ATP} channels increases K^+ conductance and facilitates K^+ influx into the mitochondrion, which activates the K^+/H^+ exchanger and in turn dissipates the mitochondrial proton gradient. This mildly depolarizes the mitochondrial membrane potential and decreases Ca^{2+} uptake into the matrix through the membrane potential dependent Ca^{2+} uniporter (Holmuhamedov et al. 1998). We have demonstrated that a mild increase in cytosolic Ca^{2+} ($[Ca^{2+}]_c$) of about 10–15% occurs during anoxia in the turtle cortex, and that this is linked to mitochondrial Ca^{2+} release via activation of the mK_{ATP} channel (Pamenter et al. 2008b). The anoxia-mediated decrease in both NMDA and AMPA receptor currents is blocked by the inclusion of the mK_{ATP} channel antagonists 5-hydroxydecanoic acid (5HD) and glibenclamide. Alternatively, the mK_{ATP} agonists diazoxide and levcromakalim replicate the decrease in both channel currents during normoxia. The important role of Ca^{2+} in this mechanism is highlighted by the fact that inclusion of a Ca^{2+} chelator (BAPTA – 1,2-bis(*o*-aminophenoxy) ethane-N,N,N′,N′-tetraacetic acid) in the recording electrode abolishes the anoxia mediated decrease in channel activity of both receptors (Shin et al. 2005; Zivkovic and Buck 2010).

Either NMDA or AMPA receptors can be reversibly phosphorylated and in general phosphorylation increases Ca^{2+} and/or Na^{+} currents, whereas dephosphorylation decreases these currents. Both receptors are phosphorylated by a number of protein kinases, these include: protein kinase A, protein kinase C, CaM kinase II, and tyrosine kinase (Chen and Huang 1992; Hajimohammadreza et al. 1995; Wang et al. 2005). The channels are dephosphorylated by the associated protein phosphatases, Ca^{2+}/calmodulin-dependent protein phosphatase 2B, protein-tyrosine phosphatase and protein phosphatase 1&2A (Wang and Salter 1994; Wang et al. 1994; 2005; Tong et al. 1995). Although the transduction mechanism is not yet known, in the case of the NMDAR we previously hypothesized that protein phosphatases 1A (PP1) and 2A dephosphorylate a serine residue on the calmodulin binding site (CBS1) of the NR1 (NMDA receptor subunit 1) C-terminus, permitting Ca^{2+} activated calmodulin to bind to CBS1 (Shin et al. 2005). This facilitates NMDA receptor dissociation from the cytoskeleton and a decrease in NMDA receptor currents. The hypothesis was based on the finding that the calmodulin and phosphatase inhibitors, calmidazolium and cypermethrin, respectively, blocked the anoxia-mediated decrease in NMDA receptor current (Shin et al. 2005).

During prolonged anoxia it is the loss of ion gradients that is ultimately damaging to cells and is characteristically preceded by a large increase in intracellular [Ca^{2+}] and extracellular [K^{+}] (Siesjo 1990; Sattler and Tymianski 2000). Large changes in trans-cellular ion gradients are not observed in anoxia-tolerant species and there is substantial evidence indicating that ion channels undergo channel arrest (CA), including K^{+} channels (Pek and Lutz 1997), Na^{+} channels (Boutilier 2001; Perez-Pinzon et al. 1992), our work on NMDA and AMPA receptors (described above), and our most recent work on Ca^{2+} activated K^{+} channels (unpublished Rodgers-Garlick and Buck). Arresting ion channels will ultimately result in a reduction in the amount of ATP required to maintain transmembrane ion gradients and prolong anoxic survival. It was thought that CA was the underlying cause of the decrease in action potential frequency observed with the onset of anoxia in turtle brain which is termed spike arrest (SA: Sick et al. 1982). However, we will now discuss a recent novel finding by our group demonstrating that SA is mediated by an increase in a specific ion conductance – Cl^{-} conductance through GABA-A receptors.

10.3 Oxygen Sensitive GABAergic Neurotransmission

The link between Ca^{2+} entry via the NMDA receptor and excitotoxicity has overshadowed the possibility that GABA receptors could be a good therapeutic target for anoxic neuroprotection; however, interest in a GABA based strategy is growing. Three different types of GABA receptors have been identified: (a) GABA-A receptor – permeable to Cl^{-} and HCO_3^{-} and found in the postsynaptic density and soma; (b) GABA-B receptor – G-protein coupled receptors that activate an inwardly rectifying K^{+} channel postsynaptically and inhibit a voltage gated Ca^{2+} channel pre-synaptically (Ben-Ari et al. 2007; Kaila et al. 1993);(c) GABA-C or rho receptor – permeable to Cl- and mainly found in the retina but have more recently been characterized in the brain (Rosas-Arellano et al. 2011). In adult mammal brain activation of postsynaptic GABA-A receptors results in hyperpolarization of the membrane potential (V_m) because the reversal potential for Cl- (E_{Cl-}) is more negative than V_m. The hyperpolarized E_{Cl-} results from the relative distribution of Cl^{-} across the cell membrane with the extracellular concentration being much higher than the intracellular. Therefore, when GABA-A receptors are activated the neuron

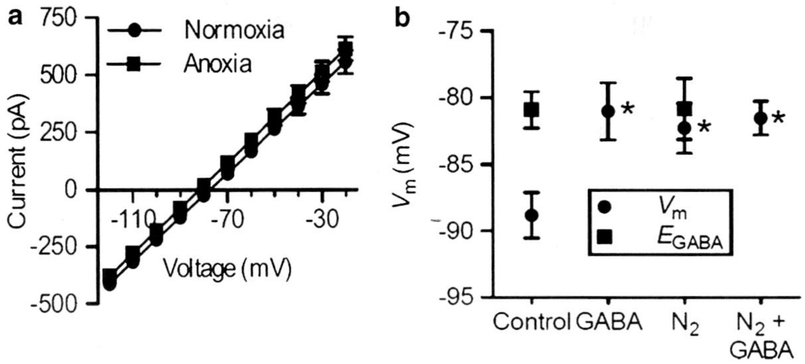

Fig. 10.1 (**a**) Current – Voltage relationship of "pure" GABA currents. Traces are the result of subtracting the background current from the current obtained in the presence of GABA. The reversal or equilibrium potential (E_{GABA}) for the GABA-A receptor is determined by the intersection with the x-axis. There is no change with anoxia but a summary of these data in (**b**) shows that E_{GABA} is slightly depolarized to the resting membrane potential of the cell (Figure modified from Pamenter et al. 2011)

becomes hyperpolarized and more difficult to depolarize by excitatory glutamatergic input. Similarly activation of post-synaptic GABA-B receptors hyperpolarize the cell by permitting an outwardly directed K^+ conductance. Pre-synaptic GABA-B receptor activation results in inhibition of voltage gated Ca^{2+} channels, thereby inhibiting the release of neurotransmitter-containing vesicles. GABA release therefore results in inhibition of action potential generation by three different mechanisms.

Typically intracellular [Cl^-] is much lower than extracellular and the opening of the GABA-A receptor pore results in an inward Cl^- conductance that hyperpolarizes neurons, as described above. In other words the reversal potential for Cl^-(E_{Cl}) is more negative than V_m and opening of the GABA-A receptor drives V_m towards E_{Cl} ($E_{Cl} \approx Cl^-$ reversal potential for GABA-A receptor or E_{GABA}). The [Cl^-] gradient is set by the relative activities of the major Cl^- extrusion transporter KCC2 (K^+/Cl^- cotransporter) and uptake transporter NKCC1 ($N^+/K^+/Cl^-$ cotransporter), which also participate in cell volume regulation (Wang et al. 2003).Our data, in Fig. 10.1, shows that in turtle cortex E_{GABA} is slightly depolarized but not excitatory in relation to V_m; therefore, opening of GABA-A receptor results in an outward Cl^- current (or shunting current, see Fig. 10.4) and V_m depolarizing towards E_{GABA}.

In addition to E_{GABA} being close to resting membrane potential we demonstrate in turtle cortical pyramidal cells a persistent slow GABA-A receptor mediated PSP that occurs at a frequency of about 10 s^{-1} (Fig. 10.2b, normoxic unclamped). A GABA-A receptor mediated current is confirmed by the inhibition of this current by gabazine (specific GABA-A receptor blocker) under both normoxic and anoxic perfusion conditions. The GABA current is also anoxia sensitive as the onset of anoxia causes the current to double in amplitude without a change in frequency (Figs. 10.2 and 10.3). Furthermore, the inclusion of GABA uptake blockers (SKF89976A and (S)-SNAP-5114) causes the current to increase further suggesting that it is an anoxia-mediated increase in presynaptic GABA release that underlies the increase in postsynaptic GABA currents. Preliminary experiments indicate that stellate neurons fire action potentials (APs) in 10 s bursts with 3–4 AP's per burst and that with anoxia, burst activity increases to 6–8 AP's per burst (Hogg, Pamenter, and Buck; unpublished observations). We speculate that the stellate neuron is the oxygen sensor and that the increase in burst activity increases GABA release which then acts postsynaptically to increase GABA-A receptor mediated PSP's in pyramidal cells.

With the onset of anoxia or GABA application action potential frequency (AP_{freq}) decreases from 2 to 0.6 Hz (70%) and whole cell conductance (Gw) increases from about 5–7 nS (40%). Both of these changes could be reversibly blocked by the administration of the GABA-A receptor inhibitor gabazine but only AP_{freq} decreased in the presence of GABA-B receptor specific inhibitor-CGP55845. This result is suggestive of a presynaptic role for GABA-B receptors in anoxia-tolerance through inhibition

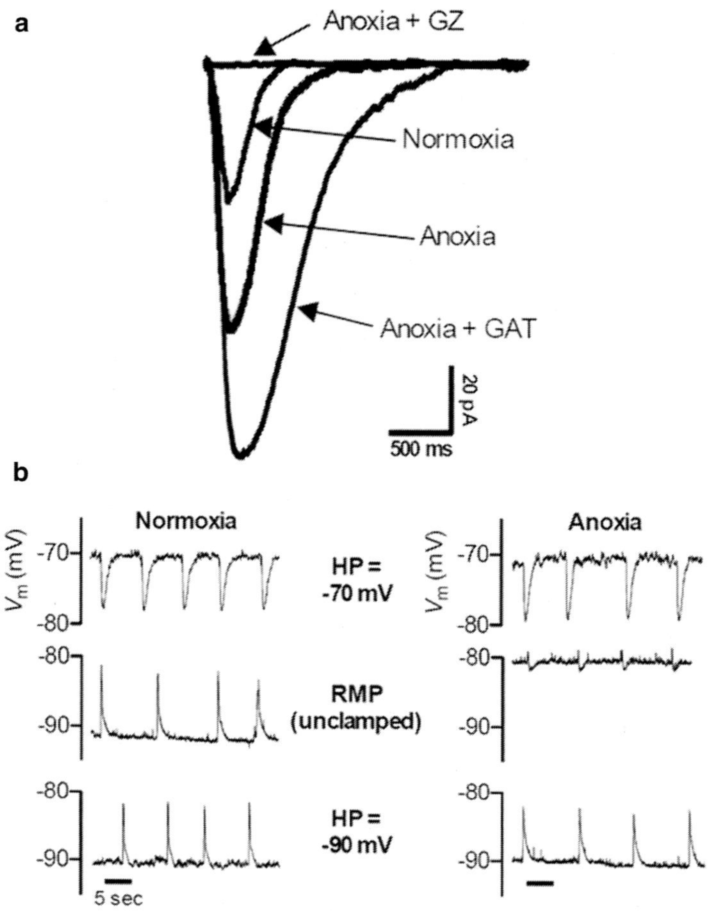

Fig. 10.2 (**a**) Superimposed raw spontaneous GABAergic postsynaptic currents. Traces show the GABA post synaptic currents during: normoxia, an increase with anoxia, inhibition with gabazine and an increase with GABA uptake blockade (SKF 89976A and S-SNAP-5114). (**b**) Sample recordings of post synaptic potentials during normoxia and anoxia while current-clamped at −70 mV, unclamped and −90 mV. GAT refers to GABA uptake blocker (Figure modified from Pamenter et al. (2011) and obtained from a cell-attached gramicidin patch in the presence of NMDA and AMPA receptor blockers)

of neurotransmitter release. Since only inhibition of both receptors causes anoxic depolarization and cell death during anoxia, we conclude that GABA-A and B receptors are necessary for anoxia tolerance.

Our discovery of increased GABA-mediated currents in brain, which is contrary to the CA hypothesis, has raised the possibility that excitable tissue responds differently than non-excitable tissue to anoxia; since we have shown using indirect methods that CA occurs in turtle hepatocytes (Buck and Hochacka 1993).

Summary

Over the last decade we have assessed the oxygen sensitivity of both glutamatergic and GABAergic neurontransmission in the cerebrocortex of the anoxia-tolerant turtle *Chrysemys picta bellii.* Our goal is to understand how excitoxicity is prevented in this species by measuring glutamate receptor

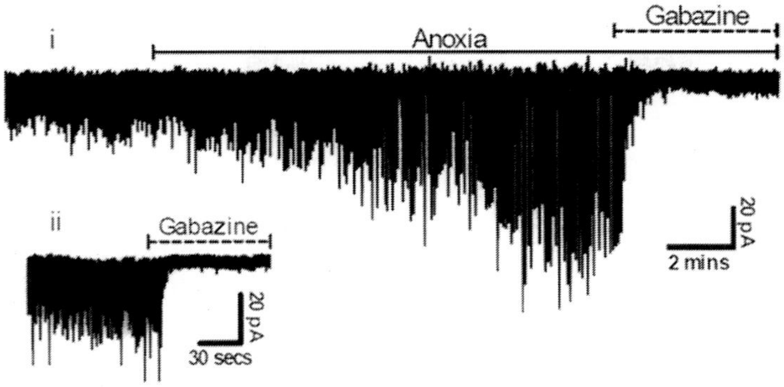

Fig. 10.3 Response of endogenous GABA mediated postsynaptic currents to anoxia and gabazine. The trace is compressed to show the increase in postsynaptic current amplitude with anoxia and inhibition when gabazine is applied under anoxia or normoxia. (Figure modified from Pamenter et al. (2011) and obtained from a cell-attached gramicidin patch in the presence of tetrodotoxin)

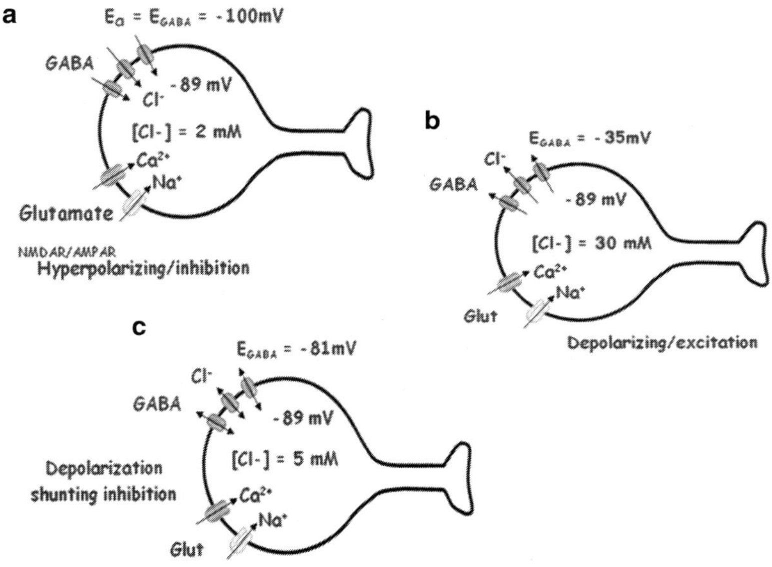

Fig. 10.4 GABAergic neurotransmission and neuronal excitability. Presynaptic GABA release can affect postsynaptic membrane potential (V_m) and excitability in three different ways depending on the relationship between the equilibrium potential (ECl- or EGABA) and V_m. (**a**). Adult brain responds to GABA with membrane hyperpolarization and therefore inhibition of action potentials (AP) resulting from an E_{Cl-} that is hyperpolarized relative to V_m. (**b**). However, in the neonatal brain E_{Cl-} is depolarized relative to V_m; therefore, GABA results in depolarization and neuronal excitation. (**c**). An interesting third possibility is a situation where E_{Cl-} is close to V_m and GABA results in a small shift in V_m that does not significantly hyperpolarize or depolarize the neuron. Thus the GABA-A mediated current (in blue) effectively shunts any positive depolarizing current (NMDA receptor in red; AMPA receptor in yellow) out of the cell and prevents the generation of an action potential. Extracellular [Cl^-] was fixed at 122 mM and V_m was determined using a gramicidin perforated patch. Intracellular [Cl^-] was calculated using the Nernst equation (Model based on results from experiments by Pamenter et al. (2011))

currents (AMPA and NMDA) during a transition from normoxic to anoxic conditions. Glutamate receptor currents in pyramidal neurons are oxygen sensitive since currents decrease by about 50% with the onset of anoxia and return to normal upon re-oxygenation. We also demonstrated that this involves release of Ca^{2+} from the mitochondrion and an increase in cytosolic [Ca^{2+}] of 10–15%. Because the intercerebral concentration of GABA increases dramatically with anoxia we turned our attention to GABAergic neurontransmission and made an interesting discovery. During passive whole-cell recording of electrical activity in pyramidal neurons there is a slow GABAergic PSC that doubles in magnitude with anoxia. Simultaneously the neuron depolarizes slightly so that the membrane potential is at the reversal potential of the GABA-A receptor for Cl^-. The increased Cl^- conductance at its reversal potential effectively becomes a shunting current preventing the neuron from reaching the threshold potential and resulting in spike arrest. This mechanism represents a unique and natural oxygen-sensitive adaptive mechanism to protect brain from anoxic injury.

References

Allen NJ, Rossi DJ, Attwell D (2004) Sequential release of GABA by exocytosis and reversed uptake leads to neuronal swelling in simulated ischemia of hippocampal slices. J Neurosci 24:3837–3849

Anderson TR, Jarvis CR, Biedermann AJ, Molnar C, Andrew RD (2005) Blocking the anoxic depolarization protects without functional compromise following simulated stroke in cortical brain slices. J Neurophysiol 93(2):963–979

Ben-Ari Y, Gaiarsa J-L, Tyzio R, Khazipov R (2007) GABA: a pioneer transmitter that excites immature neurons and generates primitive oscillations. Physiol Rev 87:1215–1284

Bickler PE, Buck LT (2007) Hypoxia tolerance on reptiles, amphibians, and fishes: life with variable oxygen availability. Ann Rev Physiol 69:145–170

Blaesse P, Airaksinen MS, Rivera C, Kaila K (2009) Cation-chloride cotransporters and neuronal function. Neuron 61:820–838

Boutilier RG (2001) Mechanisms of cell survival in hypoxia and hypothermia. J Exp Biol 204:3171–3181

Buck LT, Bickler PE (1998) Adenosine and anoxia reduce N-methyl-D-aspartate receptor open probability in turtle cerebrocortex. J Exp Biol 201:289–297

Buck LT, Hochachka PW (1993) Anoxic suppression of Na^+-K^+-ATPase and constant membrane potential in hepatocytes: support for channel arrest. Am J Physiol 265:R1020–R1025

Buck LT, Land SC, Hochachka PW (1993) Anoxia-tolerant hepatocytes: model system for study of reversible metabolic suppression. Am J Physiol 265:R49–R56

Chen L, Huang L-YM (1992) Protein kinase C reduces Mg^{2+} block of NMDA-receptor channels as a mechanism of modulation. Nature 356:521–523

Costa C, Leone G, Saulle E, Pisani F, Bernardi G, Calabresi P (2004) Coactivation of GABA(A) and GABA(B) receptor results in neuroprotection during in vitro ischemia. Stroke 35:596–600

Doll CJ, Hochachka PW, Hand SC (1994) A microcalorimetric study of turtle cortical slices: insights into brain metabolic depression. J Exp Biol 191:141–153

Erecinska M, Nelson D, Wilson DF, Silver IA (1984) Neurotransmitter amino acids in the CNS. I. Regional changes in amino acid levels in rat brain during ischemia and reperfusion. Brain Res 304:9–22

Fernandes JA, Lutz PL, Tannenbaum A, Todorov AT, Liebovitch L, Vertes R (1997) Electroencephalogram activity in the anoxic turtle brain. Am J Physiol 273:R911–R919

Galeffi F, Sinnar S, Schwartz-Bloom RD (2000) Diazepam promotes ATP recovery and prevents cytochrome c release in hippocampal slices after in vitro ischemia. J Neurochem 75:1242–1249

Ginsberg MD (2008) Neuroprotection for ischemic stroke: past, present and future. Neuropharmacology 55:363–389

Hajimohammadreza I, Probert AW, Coughenour LL, Borosky SA, Marcoux FW, Boxer PA, Wang KKW (1995) Specific inhibitor of calcium calmodulin-dependent protein-kinase-II provides neuroprotection against NMDA-induced and hypoxia hypoglycemia-induced cell-death. J Neurosci 15(5):4093–4101

Heurteaux C, Lauritzen I, Widmann C, Lazdunski M (1995) Essential role of adenosine, adenosine A1 receptors, and ATP-sensitive K+channels in cerebral ischemic preconditioning. Proc Natl Acad Sci U S A 92(10):4666–4670

Hochachka PW (1986) Defense strategies against hypoxia and hypothermia. Science 231:234–241

Hochachka PW, Somero GN (1984) Biochemical adaptation. Princeton University Press, Princeton, p 153

Holmuhamedov EL, Jovanovic S, Dzeja PP, Jovanovic A, Terzic A (1998) Mitochondrial ATP-sensitive K^+ channels modulate cardiac mitochondrial function. Am J Physiol 275:H1567–H1576

Hylland P, Nilsson GE, Lutz PL (1994) Time course of anoxia-induced increased in cerebral blood flow rate in turtles: evidence for a role of adenosine. J Cereb Blood Flow and Metabol 14:877–881

Jackson DC (1968) Metabolic depression and oxygen depletion in the diving turtle. J Appl Physiol 24:503–509

Jackson DC (2002) Hibernating without oxygen: physiological adaptations of the painted turtle. J Physiol 543:731–737

Kaila K, Voipio J, Paalasmaa P, Pasternack M, Deisz RA (1993) The role of bicarbonate in GABA-A receptor-mediated IPSPs of rat neocortical neurones. J Physiol 464:273–289

Kis B, Rajapakse NC, Snipes JA, Nagy K, Horiguchi T, Busija DW (2003) Diazoxide induces delayed pre-conditioning in cultured rat cortical neurons. J Neurochem 87:969–980
Kis B, Nagy K, Snipe JA, Rajapakse NC, Horiguchi T, Grover GJ, Busija DW (2004) The mitochondrial K(ATP) channel opener BMS-191095 induces neuronal preconditioning. Neuroreport 15(2):345–349
Li H, Siegel RE, Schwartz RD (1993) Rapid decline of GABAA receptor subunit mRNA expression in hippocampus following transient cerebral ischemia in the gerbil. Hippocampus 3:527–537
Milton SL, Thompson JW, Lutz PL (2002) Mechanisms for maintaining extracellular glutamate levels in the anoxic turtle striatum. Am J Physiol 282:R1317–R1323
Muller M, Brockhaus J, Ballanyi K (2002) ATP-independent anoxic activation of ATP-sensitive K+channels in dorsal vagal neurons of juvenile mice in situ. Neuroscience 109:313–328
Nilsson GE, Lutz PL (1991) Release of inhibitory neurotransmitters in response to anoxia in turtle brain. Am J Physiol 261:R32–R37
Pamenter ME, Shin DS, Buck LT (2008a) AMPA receptors undergo channel arrest in the anoxic turtle cortex. Am J Physiol 294:R606–R613
Pamenter ME, S-H SD, Cooray M, Buck LT (2008b) Mitochondrial K_{ATP} channels mediate anoxia-induced decreases in NMDAR activity in the turtle cortex. J Physiol 586(4):1043–1058
Pamenter ME, Hogg DW, Ormond J, Shin DS, Woodin M, Buck LT (2011) Endogenous $GABA_A$ and $GABA_B$ receptor-mediated electrical suppression is critical to neuronal anoxia tolerance. Proc Natl Acad Sci USA 108(27):11274–11279
Pek M, Lutz P (1997) Role for adenosine in channel arrest in the anoxic turtle brain. J Exp Biol 200:1913–1917
Perez-Pinzon MA, Rosenthal M, Sick TJ, Lutz PL, Pablo J, Mash D (1992) Downregulation of sodium channels during anoxia: a putative survival strategy of turtle brain. Am J Physiol 262:R712–R715
Rosas-Arellano A, Parodi J, Machuca-Parra AI, Sánchez-Gutiérrez A, Inestrosa NC, Miledi R, Martínez-Torres A (2011) The GABA(A)ρ receptors in hippocampal spontaneous activity and their distribution in hippocampus, amygdala and visual cortex. Neurosci Lett 500(1):20–25
Sattler R, Tymianski M (2000) Molecular mechanisms of calcium-dependent excitotoxicity. J Mol Med 78:3
Semenov DG, Samoilov MO, Zielonka P, Lazarewiscz P (2000) Responses to reversible anoxia of intracellular free and bound Ca2+ in rat cortical slices. Resuscitation 44:207–214
Shin DH, Buck LT (2005) Effect of anoxia and pharmacological anoxia on whole-cell NMDA receptor currents in cortical neurons from the western painted turtle. Physiol Biochem Zool 76(2):532–543
Shin DS, Wilkie MP, Pamenter ME, Buck LT (2005) Calcium and protein phosphatase 1/2A attenuate N-methyl-D-aspartate receptor activity in the anoxic turtle cortex. Comp Biochem Physiol A 142:50–57
Sick TJ, Rosenthal M, La Manna JC, Lutz PL (1982) Brain potassium ion homeostasis, anoxia, and metabolic inhibition in turtles and rats. Am J Physiol 243:R281–R288
Siesjo BK (1990) Calcium, excitotoxins, and brain damage. News Physiol Sci 5:120–125
Tong G, Shepherd DA, Jahr CE (1995) Synaptic desensitization of NMDA receptors by calcineurin. Science 267:1510–1512
Wang YT, Salter MW (1994) Regulation of NMDA receptors by tyrosine kinases and phosphatases. Nature 369:233–235
Wang L-Y, Orser BA, Brautigan DL, MacDonald JF (1994) Regulation of NMDA receptors in cultured hippocampal neurons by protein phosphatases 1 and 2A. Nature 369:230–233
Wang J, Yan Y, Kinter DB, Lytle C, Sun D (2003) GABA-mediated trophic effect on oligodendrocytes requires Na-K-2Cl cotransport activity. J Neurophysiol 90:1257–1265
Wang JQ, Arora A, Yang L, Parelkar NK, Zhang G, Liu X, Choe ES, Limin M (2005) Mechanisms and synaptic plasticity. Mol Neurobiol 32(3):237–249
Zivkovic G, Buck LT (2010) Mitochondrial ATP sensitive K+ channels decrease AMPAR currents in the anoxic turtle cortex. J Neurophysiol 104:1913–1922

Chapter 11
Ion Channel Regulation by the LKB1-AMPK Signalling Pathway: The Key to Carotid Body Activation by Hypoxia and Metabolic Homeostasis at the Whole Body Level

A. Mark Evans, Chris Peers, Christopher N. Wyatt, Prem Kumar, and D. Grahame Hardie

Abstract Our recent investigations provide further support for the proposal that, consequent to inhibition of mitochondrial oxidative phosphorylation, activation of AMP-activated protein kinase (AMPK) mediates carotid body excitation by hypoxia. Consistent with the effects of hypoxia, intracellular dialysis from a patch pipette of an active (thiophosphorylated) recombinant AMPK heterotrimer ($\alpha 2\beta 2\gamma 1$) or application of the AMPK activators AICAR and A769662: (1) Inhibited BK_{Ca} currents and TASK K^+ currents in rat carotid body type I cells; (2) Inhibited whole-cell currents carried by KCa1.1 and TASK3, but not TASK1 channels expressed in HEK293 cells; (3) Triggered carotid body activation. Furthermore, preliminary studies using mice with conditional knockout in type I cells of the primary upstream kinase that activates AMPK in response to metabolic stresses, LKB1, appear to confirm our working hypothesis. Studies on mice with knockout of the catalytic $\alpha 1$ subunit and $\alpha 2$ subunits of AMPK, respectively, have proved equally consistent. Accumulating evidence therefore suggests that the LKB1-AMPK signalling pathway is necessary for hypoxia-response coupling by the carotid body, and serves to regulate oxygen and therefore energy supply at the whole body level.

Keywords Hypoxia • Carotid body • LKB1 • AMPK • BK_{Ca} • TASK • Ventilation • Afferent discharge

A.M. Evans (✉)
Centre for Integrative Physiology, College of Medicine and Veterinary Medicine, University of Edinburgh, Hugh Robson Building, George Square, Edinburgh EH8 9XD, UK
e-mail: mark.evans@ed.ac.uk

C. Peers
Division of Cardiovascular and Neuronal Remodelling, LIGHT, Faculty of Medicine and Health, Garstang Building (level 5), University of Leeds, Clarendon Way, Leeds LS2 9JT, UK
e-mail: c.s.peers@leeds.ac.uk

C.N. Wyatt
Department of Neuroscience, Cell Biology and Physiology, Boonshoft School of Medicine, Wright State University, 3640 Colonel Glenn Hwy, Dayton, OH 45435, USA

P. Kumar
School of Clinical and Experimental Medicine, College of Medical and Dental Sciences, University of Birmingham, Edgbaston, Birmingham B15 2TT, UK

D.G. Hardie
College of Life Sciences, University of Dundee, Dow Street, Dundee DD1 5EH, UK

C.A. Nurse et al. (eds.), *Arterial Chemoreception: From Molecules to Systems*, Advances in Experimental Medicine and Biology 758, DOI 10.1007/978-94-007-4584-1_11,

11.1 Introduction

Initially identified as sensory organs by De Castro (1928), the carotid bodies were first shown to mediate hyperventilation in response to a fall in arterial pO_2 by Heymans in 1930 (Heymans et al. 1930). This pioneering work defined the carotid bodies as the primary peripheral arterial chemoreceptors. The carotid body type I (glomus) cells, which synapse with afferent sensory fibres, were later identified as being responsible for chemosensing within the carotid body (Verna et al. 1975). It is generally accepted that upon exposure to hypoxia a variety of neurotransmitters are released from type I cells by exocytosis, eliciting increased afferent fibre discharge along the carotid sinus nerve (Fig. 11.1) and thereby cardiorespiratory reflexes that compensate for the fall in arterial pO_2 (Gonzalez et al. 1994; Nurse 2010; Iturriaga and Alcayaga 2004; Zhang et al. 2000). However, the question of how type I cells sense and respond to hypoxia remains a contentious issue (Peng et al. 2010; Peers et al. 2010; Evans et al. 2011; Chandel 2010). The mechanism is likely common to all O_2–sensing cells, including carotid body type I cells, pulmonary arterial smooth muscle and endothelial cells, neonatal adrenomedullary chromaffin cells and specialised neurons of the respiratory network within the brainstem; all of which have been defined by their acute sensitivity to "activation" by graded changes in pO_2 lying within the physiological range (Eyzaguirre and Koyano 1965a; Hill et al. 2011; Dipp and Evans 2001; Thompson et al. 1997).

Although different hypotheses have been proposed to account for hypoxia-response coupling in carotid body type I cells (Peers et al. 2010), there is general agreement that hypoxia selectively inhibits specific subtypes of K^+ channels, leading to membrane depolarisation and consequent voltage-gated Ca^{2+} influx that ultimately triggers transmitter release.

11.2 K^+ Channels and the Membrane Hypothesis for O_2 Sensing

Initial insights into the cellular mechanism for hypoxia-response coupling in type I cells was provided in 1988 by the observation that K^+ channels in type I cells are inhibited by acute hypoxia (Lopez-Barneo et al. 1988), a finding subsequently confirmed by others (Delpiano and Hescheler 1989; Stea and Nurse 1991). It is now clear that the response of type I cells from rats is precipitated by inhibition of two different types of K^+ channel. The first to be identified was the large conductance voltage- and Ca^{2+}-activated K^+ current (BK_{Ca}) (Hescheler et al. 1989; Peers 1990). A voltage-independent TASK-like leak K^+ current was subsequently shown to be O_2 sensitive (Buckler 1997), with the TASK3 channel subunit now considered to be critical to the inhibition of this K^+ current by hypoxia (Kim et al. 2009; Ortega-Saenz et al. 2010). However, it is evident that there is a wider variety of O_2-sensitive K^+ channels, with variations in expression of specific channels likely to confer identified species differences (Lopez-Lopez et al. 1993; Perez-Garcia et al. 2004), as well as changes during development that may

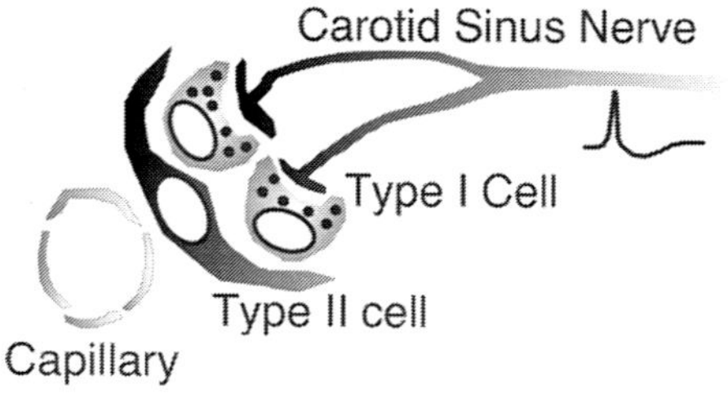

Fig. 11.1 Schematic diagram shows the cellular arrangement within the carotid body. The organ receives a rich arterial blood supply via a dense capillary network. Type II cells encapsulate the transmitter-filled type I (glomus) cells, which are in synaptic contact with afferent fibres of the carotid sinus nerve

contribute to postnatal maturation of O_2 sensitivity (Hatton et al. 1997; Wasicko et al. 2006). Therefore, the ion channels expressed by a given cell type may determine, at least in part, the nature of the response to hypoxia. This led to the proposal that O_2 sensitivity was conferred either by the channels themselves or by accessory subunits (Gonzalez et al. 1995a). However, significant evidence also now points towards the existence of an intracellular signalling cascade and one that is coupled to mitochondrial function (Buckler 2007; Evans et al. 2011; Wyatt and Evans 2007). Thus, the most fundamental questions remain – How does hypoxia mediate K^+ channel inhibition and why does metabolic stress, such as hypoxia, activate carotid body type I cells while suppressing activity in other excitable cell types?

11.3 Mitochondria and O_2 Sensing

The initial clue to a requirement for functional mitochondria in the process of O_2 sensing was provided by the work of Heymans (Heymans et al. 1930), in which it was demonstrated that cyanide mimicked and prevented activation by hypoxia of the carotid body. However, the first direct evidence was provided by spectrophotometric analysis of the respiratory chain redox status, and fluorometric measurement of the NAD(P)H/NAD(P)$^+$ ratio (Mills and Jobsis 1972). By relating outcomes to afferent sinus nerve discharge during hypoxia it was shown that an increase in the NAD(P)H/NAD(P)$^+$ ratio correlated with changes in afferent nerve activity over the physiological range of O_2 levels. At the time it was proposed that mitochondria of most cells may utilise a high affinity (i.e. normal) cytochrome a_3, while the cytochrome a_3 incorporated in mitochondria of O_2 sensing cells may have a low affinity for O_2. However, it is now clear that although mitochondrial activities may vary between cell types, their local environment may be a key regulatory factor. Firstly, intracellular O_2 gradients, and possibly ATP gradients, occur at the cellular level (Jones 1986) and such gradients may vary between and within tissues due to "local" metabolic activity within multicellular preparations (Hill et al. 2011). Moreover, there are tissue-specific differences with respect to mitochondrial function that have been attributed to cell-specific O_2 supply, substrate availability and other intracellular variables (including ADP and ATP demand), which may modulate many aspects of mitochondrial function including the affinity of cytochrome c oxidase for O_2 (Gnaiger et al. 1998; Cooper and Brown 2008; Brown 1992). Thus, as has been suggested previously, the exquisite sensitivity of O_2 sensing cells to a fall in O_2 levels might be attributable to their high rate of O_2 consumption (Gonzalez et al. 1994, 1995b).

Perhaps the strongest evidence in favour of a requirement for functional mitochondria in O_2 sensing comes from recent studies on immortalised neonatal adrenomedullary chromaffin cells that incorporate or lack functional mitochondria (Buttigieg et al. 2008). Cells with functional mitochondria were found to respond to hypoxia and to inhibitors of mitochondrial oxidative phosphorylation. By contrast, those cells lacking functional mitochondria failed to respond to either stimulus. Allied to these findings, many inhibitors of mitochondria (either uncouplers or blockers of specific respiratory chain complexes) mimic hypoxia in their ability to inhibit leak K^+ currents and thereby induce voltage-gated Ca^{2+} entry into carotid body type I cells (Buckler 2007).

11.4 A Set Point for O_2 Sensing Provides the Capacity for Upregulation or Downregulation of Carotid Body Discharge

Discussions on the regulation by O_2 of carotid body function naturally focus on type I cell activation by hypoxia and the consequent increase in afferent fibre discharge (Kumar and Prabhakar 2007). Allied to this, however, it is well documented that hyperoxia attenuates carotid body output

(Biscoe and Pallot 1982). Moreover, early studies on the carotid body have shown that afferent fibre discharge is depressed under anoxic conditions, and so too is the response of the carotid body to mitochondrial inhibition (Eyzaguirre and Koyano 1965b). It is also notable, therefore, that in dorsal root ganglion neurones, which do not serve to monitor O_2 supply, no shift in the NAD(P)H/NAD(P)$^+$ ratio is observed until the pO_2 falls to ~5 mmHg (Duchen and Biscoe 1992a, b), at which point carotid body discharge begins to fail. Thus, it is clear that there is a "set point" around which afferent fibre discharge from the carotid body is modulated, either positively or negatively, by changes in pO_2, and that such modulation occurs across a "pO_2 window" within which the routes of energy supply within carotid body type I cells may support continued exocytosis. This is likely determined by the metabolic status of the type I cells when compared to other cell types, in terms of the balance between energy demand and the capacity for continued energy supply. The AMP-activated protein kinase (AMPK) is central to such metabolic control (Hardie 2007).

11.5 The Emergence of the LKB1-AMPK Signalling Cascade in O_2 Sensing

The ubiquitously expressed AMP-activated protein kinase (AMPK) is a heterotrimer comprising catalytic α and regulatory β and γ subunits, of which there are multiple isoforms (Fig. 11.2a) (Hardie 2007). In response to metabolic stress, AMPK is activated by an increase in the ADP/ATP ratio, which is amplified by adenylate kinase into a greater increase in the AMP/ATP ratio (Hardie et al. 2011). Activation of AMPK (>100-fold) is conferred by phosphorylation at Thr-172 within the α subunit by upstream kinases, which include the tumor suppressor, LKB1, and CaMKK-β. Of these, LKB1 is of primary importance to the response of cells to metabolic stress, while CaMKK-β mediates, when expressed, AMPK activation in response to an increase in cytoplasmic Ca^{2+} concentration ($[Ca^{2+}]_i$). LKB1 appears to phosphorylate Thr-172 constitutively, but displacement of ATP by ADP or AMP at one site on the γ subunit causes a switch to a conformation in which Thr-172 dephosphorylation is prevented (Xiao et al. 2011). Displacement of ATP by AMP, but not ADP, at a second site on the γ subunit causes a further allosteric activation of up to tenfold; the combination of these two effects yielding >1,000-fold activation overall (Fig. 11.2b)(Hardie et al. 2011). In cells subject to a mild energy stress that may still have a high ATP:ADP ratio, the adenylate kinase reaction (ATP + AMP ↔ 2ADP) runs from left to right, keeping AMP concentration much lower than those of ADP and ATP. Under these conditions, AMPK may be regulated primarily by increases in the ADP:ATP ratio, promoting net phosphorylation of Thr-172. However, during a more severe energy stress the fall in ATP:ADP may cause the adenylate kinase to generate AMP, amplifying AMPK activation via the allosteric effect. This dual mechanism ensures that AMPK can respond in a sensitive yet graded manner over a very wide range of different states of energy deprivation. Activation by both mechanisms can be very rapid; when AMPK in T cells is activated by stimulation of the T cell receptor (a process mediated by the Ca^{2+}-CaMKKβ pathway), activation occurs within seconds (Tamas et al. 2006). Once activated, AMPK serves to maintain ATP supply by upregulating catabolic processes (Fig. 11.2c), such as glucose uptake, β-oxidation of fatty acids, inhibition of glycogen synthesis, and by suppressing non-essential ATP-consuming reactions such as protein synthesis (Fig. 11.2d) (Hardie 2007).

As mentioned previously, the balance between these processes of ATP demand and supply could undoubtedly serve to determine the set point for type I cell activation by hypoxia, and maintain the capacity for type I cell exocytosis during hypoxia. However, when coupled to the new concept that AMPK may regulate ion channels and, thereby, cell-specific function (Evans 2006; Evans et al. 2009), it becomes immediately apparent that AMPK has the capability to elicit type I cell activation consequent to the inhibition of mitochondrial oxidative phosphorylation by hypoxia. Thus, the LKB1-AMPK signalling pathway may not only contribute to the regulation of energy (ATP) homeostasis at the cellular level, but may also do so at the whole-body level by regulating O_2 supply.

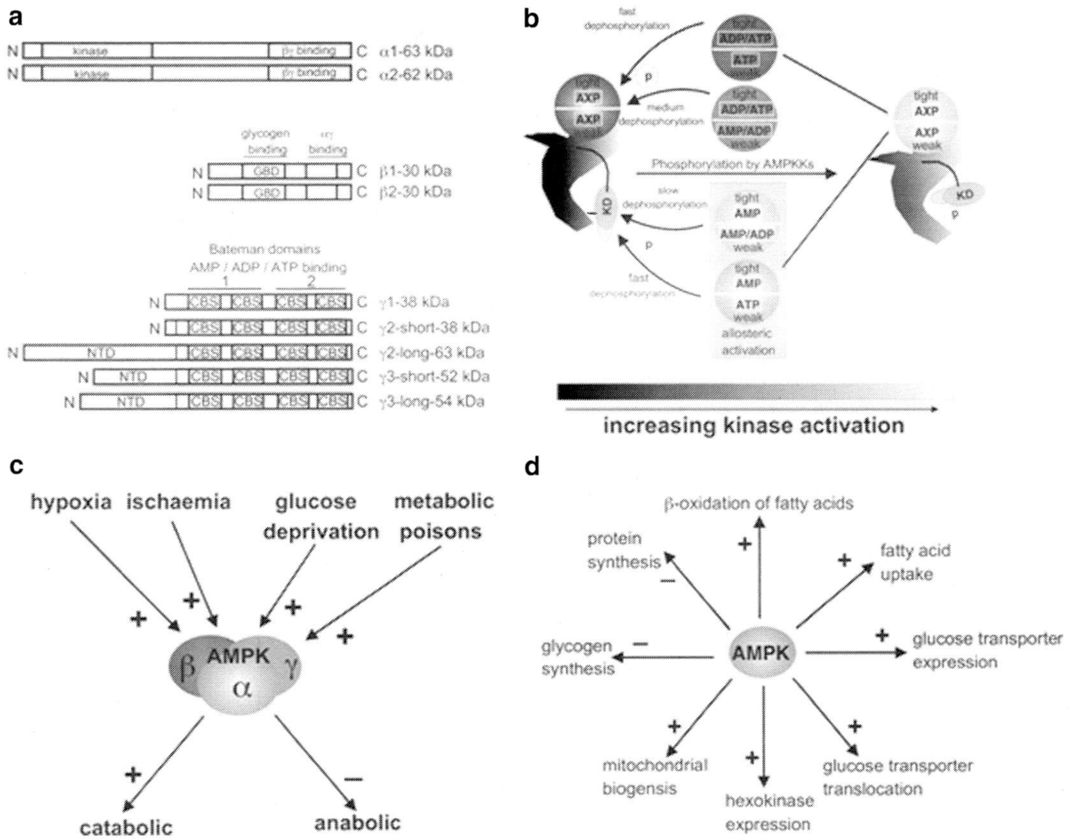

Fig. 11.2 (**a**) The domain structure of AMP-activated protein kinase (AMPK) subunit isoforms, indicating functional regions where known: NTD, N terminal domain; CBS motif x 2 = AMP binding Bateman domain; GBD, glycogen binding domain. (**b**) Schematic describes the regulation by ATP, ADP and AMP via nucleotide association with the two exchangeable sites on the gamma subunit. (**c**) Schematic provides examples of metabolic stresses that may activate AMPK, which then serves to maintain ATP supply by activating catabolic pathways and inhibiting non-essential anabolic pathways. (**d**) Schematic highlights the primary processes regulated by AMPK in order to maintain ATP supply

The first evidence in support of a role for AMPK in O_2 sensing was provided by investigations into its role in hypoxic pulmonary vasoconstriction (Evans et al. 2005). We demonstrated that physiological levels of hypoxia precipitated an increase in the ADP/ATP and AMP/ATP ratio in pulmonary arterial smooth muscle, accompanied by activation of AMPK and phosphorylation of acetyl CoA carboxylase, a well-established marker for AMPK action. We also showed that pharmacological activation of AMPK elicited constriction of pulmonary arteries, and in a manner that mimicked precisely the mechanisms of constriction by hypoxia (for detailed discussion see (Evans et al. 2005, 2011)). The proposal that AMPK may be of general importance to hypoxia-response coupling in all O_2 sensing cells then gained further support from our studies on the carotid body.

11.6 AMPK Mediates Type I Cell Activation by Hypoxia

In our initial investigations into the role of AMPK in carotid body activation by hypoxia, we utilised the pharmacological activator AICAR. This agent is taken up into cells via the adenosine transporters (Gadalla et al. 2004) and converted, by adenosine kinase, to ZMP, which mimics both effects of AMP

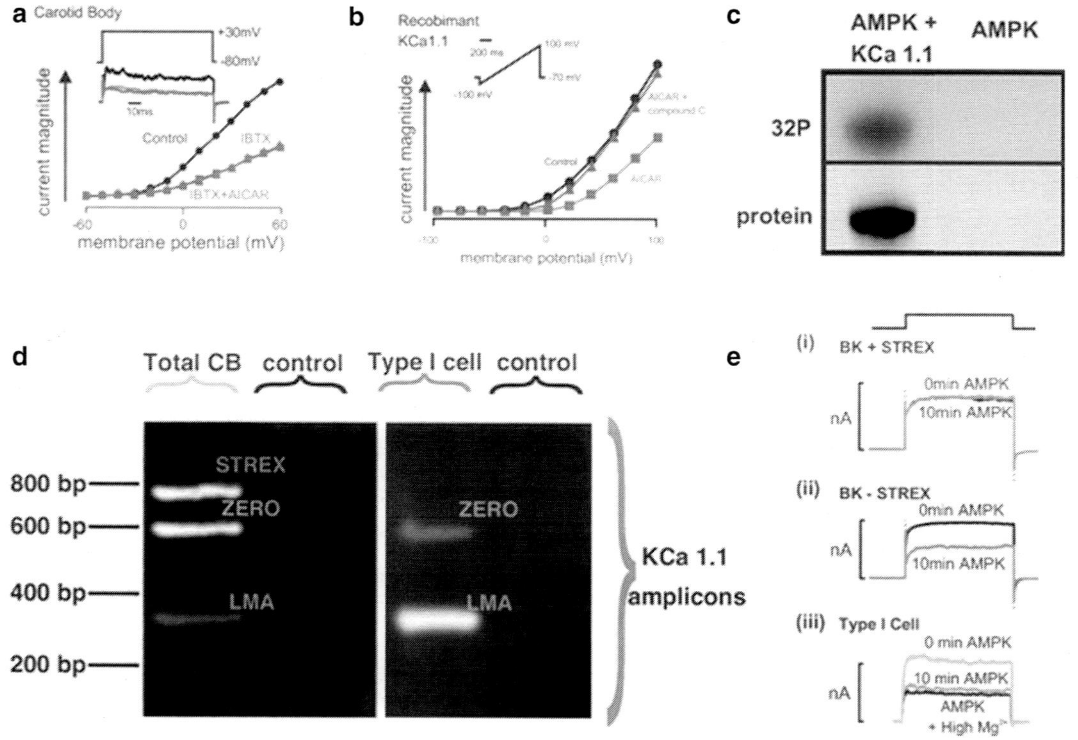

Fig. 11.3 (**a**) Inhibition by AMPK activation of the large conductance Ca^{2+}-activated K^+ current (BK_{Ca}) in acutely isolated type I cells of the carotid body. (**b**) Inhibition by AMPK activation of the large conductance Ca^{2+}-activated K^+ current conferred by recombinant KCa1.1 channels expressed in HEK293 cells. (**c**) AMP-dependent phosphorylation by AMPK of purified KCa1.1. (**d**) Expression of KCa1.1 splice variants in the whole carotid body and in an acutely isolated type I cell. (**e**) Effect of intracellular dialysis of a thiophosphorylated (active), bacterially derived AMPK heterotrimer (α2, β2, γ2) on whole-cell currents carried by (i) KCa1.1 splice variant that incorporates the stress regulated exon when expressed in HEK293 cells, (ii) KCa1.1 splice variant that does not incorporate the stress regulated exon when expressed in HEK293 cells, (iii) BK_{Ca} channels of acutely isolated rat type I cells

on AMPK by binding to the two exchangeable sites on the γ subunits. AICAR elicited a rise in $[Ca^{2+}]_i$ in isolated rat type I cells and increased afferent sensory nerve activity recorded from the carotid sinus nerve (Evans et al. 2005), thus mimicking hypoxia. Furthermore, an AMPK antagonist, compound C, markedly attenuated the hypoxia-induced rise in $[Ca^{2+}]_i$ in type I cells and afferent fibre discharge from the carotid body *in-vitro* (Wyatt et al. 2007).

Importantly, AMPK activation by AICAR, like hypoxia, selectively inhibited both BK_{Ca} (Fig. 11.3a) and leak K^+ currents in rat type I cells (Fig. 11.4), and thereby caused membrane depolarization and voltage-gated Ca^{2+} entry (Wyatt et al. 2007). These observations suggested that hypoxia-response coupling may occur through direct phosphorylation and regulation by AMPK of O_2 sensitive ion channels. This view has gained significant support from our finding that two distinct pharmacological activators of AMPK, AICAR and A-769662, inhibit recombinant BK_{Ca} channels (Fig. 11.3a, b)(Ross et al. 2011; Wyatt et al. 2007) and TASK3, but not TASK1, channels (unpublished observations) stably expressed in HEK293 cells, and in a manner that was blocked by the AMPK antagonist compound C.

These studies have been questioned because of the possible off target effects resulting from pharmacological interventions. However, we have now provided significant support for our original proposals using recombinant AMPK heterotrimers. In these later studies we employed both AMPK purified from rat liver (a mixture of the α1β1γ1 and α2β1γ1 heterotrimers) and a bacterially expressed

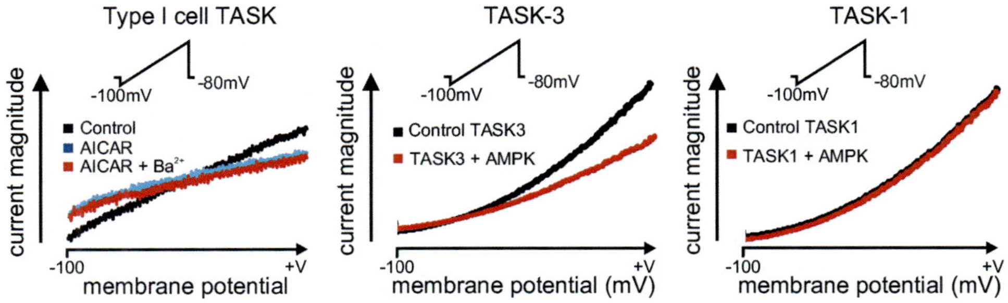

Fig. 11.4 AMPK activation by AICAR inhibits the TASK K^+ current in acutely isolated type I cells of the carotid body (*left panel*). Effect of intracellular dialysis of a thiophosphorylated (active), bacterially derived AMPK heterotrimer (α2, β2, γ2) on whole-cell currents carried by TASK3 (*middle panel*) and TASK1 (*right panel*) stably expressed in HEK293 cells

human α2β2γ1 heterotrimer. Both forms of AMPK phosphorylated recombinant BK_{Ca} channels ($K_{Ca}1.1$) in cell free assays (Fig. 11.3c), and phosphorylation by rat liver AMPK was stimulated by AMP. For experiments with addition of AMPK to intact cells, we used the bacterially expressed α2β2γ1 complex that had been activated by thiophosphorylation of Thr-172 by CaMKKβ using the γ-thiophosphate analogue of ATP. This form is almost as active as the phosphorylated protein, but unlike the latter is not sensitive to cellular phosphatases. Importantly, intracellular dialysis of active, thiophosphorylated AMPK into HEK293 cells inhibited recombinant currents carried by BK_{Ca} channels that lack the stress-regulated exon, but not those carried by BK_{Ca} channels that incorporated this exon (Fig. 11.3e). Crucially, single cell end-point RT-PCR demonstrated that type I cells express only the BK_{Ca} channel variant lacking the stress regulated exon (Fig. 11.3d). Consistent with this, intracellular dialysis of activated, thiophosphorylated human AMPK into rat type I cells inhibited endogenous BK_{Ca} channel currents (Fig. 11.3e) (Ross et al. 2011). Intracellular dialysis of activated AMPK also inhibited whole-cell currents carried by recombinant TASK3 channels, but not TASK1 channels, stably expressed in HEK293 cells. We can conclude, therefore, that AMPK does indeed phosphorylate and inhibit the key O_2-sensitive ion channels that are expressed in rat type I cells.

In marked contrast to its capacity to activate carotid body type I cells, we have recently reported that AMPK may reduce action potential firing frequency in hippocampal neurons by directly phosphorylating and thereby "activating" a different type of O_2 sensitive potassium channel, namely Kv2.1 (Ikematsu et al. 2011). When combined with the above, therefore, our studies suggest that AMPK can increase or decrease cell excitability, in a manner determined by cell-specific expression of members of the K^+ channel superfamily. Thus, this signaling pathway offers great versatility in its capacity to regulate metabolic status at the cellular and whole body levels.

Nevertheless, this left one major question – is the LKB1-AMPK signalling cascade necessary for carotid body activation by, and the ventilatory response to hypoxia?

11.7 The LKB1-AMPK Signalling Cascade and Carotid Body Activation by Hypoxia

To determine whether or not the LKB1-AMPK signalling cascade is required for hypoxia-response coupling by the carotid body we first carried out investigations using pharmacological activators and inhibitors of AMPK. The AMPK antagonist Compound C inhibited basal and hypoxia-evoked afferent fibre discharge from the carotid body *in-vitro* (Fig. 11.5a, b). Moreover, the AMPK activator

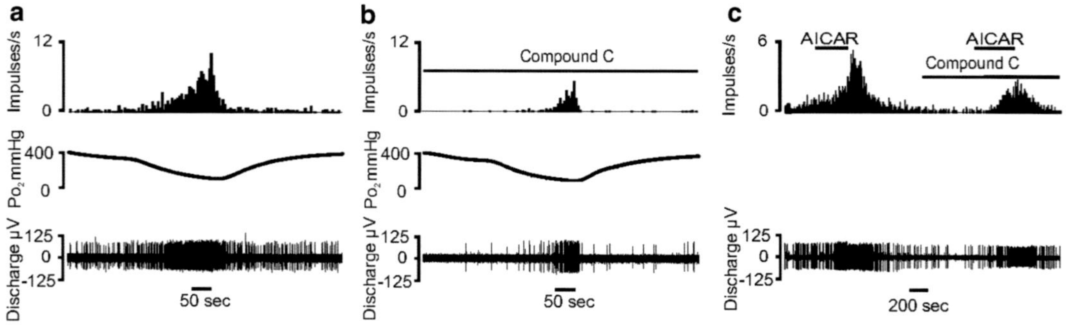

Fig. 11.5 AMPK inhibition attenuates hypoxia and AICAR- induced increases in carotid sinus nerve discharge. (**a**) hypoxia stimulates carotid sinus nerve discharge. (**b**) The AMPK antagonist Compound C reduces basal afferent fibre discharge and inhibits the response to hypoxia. (**c**) AMPK activation by AICAR stimulates carotid sinus nerve discharge in a manner that is inhibited by Compound C

AICAR increased afferent fibre discharge in a manner that was inhibited by Compound C (Evans et al. 2005; Wyatt et al. 2007). This is entirely consistent with the view that the LKB1-AMPK signalling pathway mediates carotid body activation by hypoxia. However, these drugs may have off target effects. Therefore, we have now developed mice in which LKB1 is knocked out in all tyrosine hydroxylase expressing cells, which include carotid body type I cells. Preliminary findings suggest that LKB1 is likely necessary for the initiation of carotid body activation by hypoxia (unpublished observations). This view is supported by preliminary studies on mice lacking either the AMPK α-1 or the α-2 subunit (unpublished observations).

11.8 Summary

Our findings are entirely consistent with the view that a fall in arterial pO_2 leads to inhibition of mitochondrial oxidative phosphorylation in carotid body type I cells, and consequently activates the LKB1-AMPK signalling pathway. Subsequent phosphorylation and inhibition of specific O_2-sensitive K^+ currents by AMPK leads to type I cell depolarisation, voltage-gated Ca^+ entry and exocytosis, ultimately triggering an increase in afferent fibre discharge from the carotid body and cardiorespiratory reflexes that correct for the fall in arterial pO_2. Thus, the LKB1-AMPK signalling cascade regulates energy (ATP) supply at the whole body as well as the cellular level, by regulating respiratory function and thereby O_2 supply as originally proposed (Evans 2006).

Acknowledgements These studies were supported by a Programme Grant from the Wellcome Trust (81195, to AME, CP and DGH).

References

Biscoe TJ, Pallot DJ (1982) The carotid body chemoreceptor: an investigation in the mouse. Q J Exp Physiol 67(4):557–576 (Cambridge, England)

Brown GC (1992) Control of respiration and ATP synthesis in mammalian mitochondria and cells. Biochem J 284(Pt 1):1–13

Buckler KJ (1997) A novel oxygen-sensitive potassium current in rat carotid body type I cells. J Physiol 498(Pt 3):649–662

Buckler KJ (2007) TASK-like potassium channels and oxygen sensing in the carotid body. Respir Physiol Neurobiol 157(1):55–64

Buttigieg J, Brown ST, Lowe M, Zhang M, Nurse CA (2008) Functional mitochondria are required for O2 but not CO2 sensing in immortalized adrenomedullary chromaffin cells. Am J Physiol Cell Physiol 294(4):C945–956

Chandel NS (2010) Mitochondrial complex III: an essential component of universal oxygen sensing machinery? Respir Physiol Neurobiol 174(3):175–181

Cooper CE, Brown GC (2008) The inhibition of mitochondrial cytochrome oxidase by the gases carbon monoxide, nitric oxide, hydrogen cyanide and hydrogen sulfide: chemical mechanism and physiological significance. J Bioenerg Biomembr 40(5):533–539

De Castro F (1928) Sur la structure et l'innervation du sinus carotidien de l'homme et des mammiferes: nouveau faits sur l'innervation et la fonction du glomus caroticum. Trab Lab Invest Biol Univ Madrid 24:330–380

Delpiano MA, Hescheler J (1989) Evidence for a PO2-sensitive K+channel in the type-I cell of the rabbit carotid body. FEBS Lett 249(2):195–198

Dipp M, Evans AM (2001) Cyclic ADP-ribose is the primary trigger for hypoxic pulmonary vasoconstriction in the rat lung in situ. Circ Res 89(1):77–83

Duchen MR, Biscoe TJ (1992a) Relative mitochondrial membrane potential and [Ca2+]i in type I cells isolated from the rabbit carotid body. J Physiol 450:33–61

Duchen MR, Biscoe TJ (1992b) Mitochondrial function in type I cells isolated from rabbit arterial chemoreceptors. J Physiol 450:13–31

Evans AM (2006) AMP-activated protein kinase and the regulation of Ca2+ signalling in O2-sensing cells. J Physiol 574(Pt 1):113–123

Evans AM, Mustard KJ, Wyatt CN, Peers C, Dipp M, Kumar P, Kinnear NP, Hardie DG (2005) Does AMP-activated protein kinase couple inhibition of mitochondrial oxidative phosphorylation by hypoxia to calcium signaling in O2-sensing cells? J Biol Chem 280(50):41504–41511

Evans AM, Hardie DG, Peers C, Wyatt CN, Viollet B, Kumar P, Dallas ML, Ross F, Ikematsu N, Jordan HL, Barr BL, Rafferty JN, Ogunbayo O (2009) Ion channel regulation by AMPK: the route of hypoxia-response coupling in the carotid body and pulmonary artery. Ann N Y Acad Sci 1177:89–100

Evans AM, Hardie DG, Peers C, Mahmoud A (2011) Hypoxic pulmonary vasoconstriction: mechanisms of oxygen-sensing. Curr Opin Anaesthesiol 24(1):13–20

Eyzaguirre C, Koyano H (1965a) Effects of hypoxia, hypercapnia, and pH on the chemoreceptor activity of the carotid body in vitro. J Physiol 178(3):385–409

Eyzaguirre C, Koyano H (1965b) Effects of some pharmacological agents on chemoreceptor discharges. J Physiol 178(3):410–437

Gadalla AE, Pearson T, Currie AJ, Dale N, Hawley SA, Sheehan M, Hirst W, Michel AD, Randall A, Hardie DG, Frenguelli BG (2004) AICA riboside both activates AMP-activated protein kinase and competes with adenosine for the nucleoside transporter in the CA1 region of the rat hippocampus. J Neurochem 88(5):1272–1282

Gnaiger E, Lassnig B, Kuznetsov A, Rieger G, Margreiter R (1998) Mitochondrial oxygen affinity, respiratory flux control and excess capacity of cytochrome c oxidase. J Exp Biol 201(Pt 8):1129–1139

Gonzalez C, Almaraz L, Obeso A, Rigual R (1994) Carotid body chemoreceptors: from natural stimuli to sensory discharges. Physiol Rev 74(4):829–898

Gonzalez C, Vicario I, Almaraz L, Rigual R (1995a) Oxygen sensing in the carotid body. Biol Signal 4(5):245–256

Gonzalez C, Lopez-Lopez JR, Obeso A, Perez-Garcia MT, Rocher A (1995b) Cellular mechanisms of oxygen chemoreception in the carotid body. Respir Physiol 102(2–3):137–147

Hardie DG (2007) AMP-activated/SNF1 protein kinases: conserved guardians of cellular energy. Nat Rev 8(10):774–785

Hardie DG, Carling D, Gamblin SJ (2011) AMP-activated protein kinase: also regulated by ADP? Trends Biochem Sci 29:18–24

Hatton CJ, Carpenter E, Pepper DR, Kumar P, Peers C (1997) Developmental changes in isolated rat type I carotid body cell K+currents and their modulation by hypoxia. J Physiol 501(Pt 1):49–58

Hescheler J, Delpiano MA, Acker H, Pietruschka F (1989) Ionic currents on type-I cells of the rabbit carotid body measured by voltage-clamp experiments and the effect of hypoxia. Brain Res 486(1):79–88

Heymans C, Bouckaert JJ, Dautrebande L (1930) Sinus carotidien et reflexes respiratoires. II. Influences respiratoires reflexes de l'acidose, de l'alcalose, de l'anhydride carbonique, de l'ion hydrogene et de l'anoxemie: sinus carotidiens et echanges respiratoires dans les poumons et au dela poumons. Arch Int Pharmacodyn Ther 39(2)):400–408

Hill AA, Garcia AJ 3rd, Zanella S, Upadhyaya R, Ramirez JM (2011) Graded reductions in oxygenation evoke graded reconfiguration of the isolated respiratory network. J Neurophysiol 105(2):625–639

Ikematsu N, Dallas ML, Ross FA, Lewis RW, Rafferty JN, David JA, Suman R, Peers C, Hardie DG, Evans AM (2011) Phosphorylation of the voltage-gated potassium channel Kv2.1 by AMP-activated protein kinase regulates membrane excitability. Proc Natl Acad Sci U S A 44:18132–18137

Iturriaga R, Alcayaga J (2004) Neurotransmission in the carotid body: transmitters and modulators between glomus cells and petrosal ganglion nerve terminals. Brain Res Rev 47(1–3):46–53

Jones DP (1986) Intracellular diffusion gradients of O2 and ATP. Am J Physiol 250(5 Pt 1):C663–675

Kim D, Cavanaugh EJ, Kim I, Carroll JL (2009) Heteromeric TASK-1/TASK-3 is the major oxygen-sensitive background K+channel in rat carotid body glomus cells. J Physiol 587(Pt 12):2963–2975

Kumar P, Prabhakar N (2007) Sensing hypoxia: carotid body mechanisms and reflexes in health and disease. Respir Physiol Neurobiol 157(1):1–3

Lopez-Barneo J, Lopez-Lopez JR, Urena J, Gonzalez C (1988) Chemotransduction in the carotid body: K+ current modulated by PO2 in type I chemoreceptor cells. Science 241(4865):580–582, New York, NY

Lopez-Lopez JR, De Luis DA, Gonzalez C (1993) Properties of a transient K+ current in chemoreceptor cells of rabbit carotid body. J Physiol 460:15–32

Mills E, Jobsis FF (1972) Mitochondrial respiratory chain of carotid body and chemoreceptor response to changes in oxygen tension. J Neurophysiol 35(4):405–428

Nurse CA (2010) Neurotransmitter and neuromodulatory mechanisms at peripheral arterial chemoreceptors. Exp Physiol 95(6):657–667

Ortega-Saenz P, Levitsky KL, Marcos-Almaraz MT, Bonilla-Henao V, Pascual A, Lopez-Barneo J (2010) Carotid body chemosensory responses in mice deficient of TASK channels. J Gen Physiol 135(4):379–392

Peers C (1990) Hypoxic suppression of K+ currents in type I carotid body cells: selective effect on the Ca2(+)-activated K+ current. Neurosci Lett 119(2):253–256

Peers C, Wyatt CN, Evans AM (2010) Mechanisms for acute oxygen sensing in the carotid body. Respir Physiol Neurobiol 174(3):292–298

Peng YJ, Nanduri J, Raghuraman G, Souvannakitti D, Gadalla MM, Kumar GK, Snyder SH, Prabhakar NR (2010) H2S mediates O2 sensing in the carotid body. Proc Natl Acad Sci U S A 107(23):10719–10724

Perez-Garcia MT, Colinas O, Miguel-Velado E, Moreno-Dominguez A, Lopez-Lopez JR (2004) Characterization of the Kv channels of mouse carotid body chemoreceptor cells and their role in oxygen sensing. J Physiol 557(Pt 2):457–471

Ross FA, Rafferty JN, Dallas ML, Ogunbayo O, Ikematsu N, McClafferty H, Tian L, Widmer H, Rowe IC, Wyatt CN, Shipston MJ, Peers C, Hardie DG, Evans AM (2011) Selective expression in carotid body type I cells of a single splice variant of the large conductance calcium- and voltage-activated potassium channel confers regulation by AMP-activated protein kinase. J Biol Chem 286(14):11929–11936

Stea A, Nurse CA (1991) Whole-cell and perforated-patch recordings from O2-sensitive rat carotid body cells grown in short- and long-term culture. Pflugers Archiv 418(1–2):93–101

Tamas P, Hawley SA, Clarke RG, Mustard KJ, Green K, Hardie DG, Cantrell DA (2006) Regulation of the energy sensor AMP-activated protein kinase by antigen receptor and Ca2+ in T lymphocytes. J Exp Med 203(7):1665–1670

Thompson RJ, Jackson A, Nurse CA (1997) Developmental loss of hypoxic chemosensitivity in rat adrenomedullary chromaffin cells. J Physiol 498(Pt 2):503–510

Verna A, Roumy M, Leitner LM (1975) Loss of chemoreceptive properties of the rabbit carotid body after destruction of the glomus cells. Brain Res 100(1):13–23

Wasicko MJ, Breitwieser GE, Kim I, Carroll JL (2006) Postnatal development of carotid body glomus cell response to hypoxia. Respir Physiol Neurobiol 154(3):356–371

Wyatt CN, Evans AM (2007) AMP-activated protein kinase and chemotransduction in the carotid body. Respir Physiol Neurobiol 157(1):22–29

Wyatt CN, Mustard KJ, Pearson SA, Dallas ML, Atkinson L, Kumar P, Peers C, Hardie DG, Evans AM (2007) AMP-activated protein kinase mediates carotid body excitation by hypoxia. J Biol Chem 282(11):8092–8098

Xiao B, Sanders MJ, Underwood E, Heath R, Mayer FV, Carmena D, Jing C, Walker PA, Eccleston JF, Haire LF, Saiu P, Howell SA, Aasland R, Martin SR, Carling D, Gamblin SJ (2011) Structure of mammalian AMPK and its regulation by ADP. Nature 472(7342):230–233

Zhang M, Zhong H, Vollmer C, Nurse CA (2000) Co-release of ATP and ACh mediates hypoxic signalling at rat carotid body chemoreceptors. J Physiol 525(Pt 1):143–158

Chapter 12
Anoxia Response in Physiological Potassium of the Isolated Inspiratory Center in Calibrated Newborn Rat Brainstem Slices

Araya Ruangkittisakul and Klaus Ballanyi

Abstract Using newborn rat brainstem-spinal cords, we were the first to show that medullary inspiratory networks can generate the neonatal biphasic (initial acceleration-secondary slowing) respiratory response to severe hypoxia causing tissue anoxia. Our findings also indicated that medullary inspiratory interneurons remain functional during sustained anoxia due to effective utilization of anaerobic metabolism. In that previous work by us and related studies by others on respiratory anoxia responses in the above en bloc model or brainstem slices, presumptive recording sites within the pre-Bötzinger complex (preBötC) inspiratory center were not histologically verified. Moreover, preBötC slices were studied in 7–9 mM K^+ to stabilize rhythm which can, however, affect respiratory neuromodulation. Here, we summarize our previous findings on respiratory anoxia responses in the en bloc model in physiological (3 mM) K^+. Using our recently developed 'calibrated' slices, we also exemplify anoxia effects in anatomically identified preBötC cells in physiological K^+ based on recording electrophysiological population activity in conjunction with either membrane potential or cytosolic Ca^{2+}.

Keywords Anoxia • Astrocytes • Breathing • Hypoxia • PBC • pFRG • pre-Botzinger Complex • Respiration • Rhythmogenesis • Two-photon Ca^{2+} Imaging

12.1 Introduction

In newborns, severe arterial hypoxia causing brain anoxia or chemical anoxia, e.g. evoked by cyanide, increases respiratory rate for ~1 min before slowing breathing efforts that persist, at room temperature, for 25–50 min in <1 week-old rats (cf. Ballanyi 2004a, b). It remains unclear to which extent peripheral or central chemoreceptors contribute to this biphasic neonatal hypoxia-anoxia response, and which role ventral respiratory column (VRC) neural networks in the lower brainstem play in this regard (cf. Ballanyi et al. 1999; cf. Ballanyi 2004a, b). Analyzing central respiratory responses to anoxia may also unravel mechanisms of apnea of prematurity, which often requires intensive clinical treatment because apnea events can last >1 min thus evoking arterial hypoxia and, eventually, tissue anoxia (cf. Ballanyi 2004a, b).

A. Ruangkittisakul • K. Ballanyi (✉)
Department of Physiology, Faculty of Medicine and Dentistry, University of Alberta, 750 MSB, Edmonton, AB T6G2H7, Canada
e-mail: klaus.ballanyi@ualberta.ca

C.A. Nurse et al. (eds.), *Arterial Chemoreception: From Molecules to Systems*, Advances in Experimental Medicine and Biology 758, DOI 10.1007/978-94-007-4584-1_12,

Neonatal breathing is initiated and controlled by a dual VRC respiratory center comprising inspiratory pre-Bötzinger complex (preBötC) and expiratory parafacial respiratory group (pFRG) networks (cf. Feldman and Del Negro 2006). The latter review emphasizes that opioids are instrumental for identifying preBötC/pFRG structure-function relationships while our recent work established experimentally-evoked anoxia as another potent tool in this regard (Taccola et al. 2007; Ballanyi et al. 2009). For this research, we use 'en bloc' brainstem-spinal cords with a functional dual respiratory center (cf. Ballanyi et al. 1999) and also transversal preBötC slices that we isolated by defined transection from the en bloc model in our milestone study on the discovery of the inspiratory center (Smith et al. 1991). 'Breathing' preBötC slices are generally studied in solution with 7–9 mM K^+ (which can modulate neural functions) while recording sites in either in vitro model were not histologically identified (cf. Ruangkittisakul et al. 2007; cf. Ballanyi and Ruangkittisakul 2009).

Here, we summarize our published findings on central respiratory anoxia responses, mostly in physiological (3 mM) K^+, in the en bloc model and exemplify present preliminary findings on anoxia effects in physiological K^+ using our anatomically 'calibrated' preBötC slices (Ruangkittisakul et al. 2006, 2007, 2008, 2009, 2012; Ballanyi and Ruangkittisakul 2009).

12.2 Methods

Methods are described in detail in our studies cited in the Results section and our recent review (Ruangkittisakul et al. 2012). In short, brainstem(s) (slices) from 0 to 4 days-old rats were superfused at 5 ml/min and 26–28°C with solution containing (in mM) 120 NaCl; 3–6.2 (en bloc) or 3 (slices) KCl; 1.5–2.4 (en bloc) or 1 (slices) $CaCl_2$; 2 $MgSO_4$; 26 $NaHCO_3$; 1.25 NaH_2PO_4 and 30 (en bloc) or 20 (slices) D-glucose; pH adjusted to 7.4 by gassing with 95% O_2, 5% CO_2.

Suction electrodes for recording extracellular neural population activity (gain 10 k, bandpass-filtering 0.3–3 kHz) were positioned in superficial layers of the ventrolateral slice area containing the VRC with the preBötC. In en bloc models, electrodes were attached to preBötC-driven cervical or hypoglossal (XII) nerve roots or pFRG-driven lumbar or facial (VII) nerve roots. Membrane potential (V_m) was whole-cell recorded (Smith et al. 1991; Schwarzacher et al. 2002; Ballanyi et al. 2009; Panaitescu et al. 2009) using patch pipettes filled with (in mM): 140 K-gluconate, 1 NaCl, 0.5 $CaCl_2$, 2 $MgCl_2$, 1 K_4-BAPTA, 1 Na_2-ATP and 10 Hepes; pH adjusted to 7.4 with KOH; dc resistance 3–8 MΩ. For determining cell location in the rostrocaudal axis, micromanipulator readouts were referred to calibrated slice margins (Panaitescu et al. 2009).

In the en bloc model, the VRC microenvironment was assessed with O_2–, CO_2– and ion-sensitive microelectrodes. In calibrated slices, free cytosolic Ca^{2+} (Ca_i) was monitored with a custom-made multiphoton system comprising an Olympus FV300 confocal microscope attached to a Coherent Ti:Sa laser. Membrane-permeant Ca^{2+} dye was pressure-injected (25–50 mmHg, 10 min) into the preBötC using a broken patch pipette filled with superfusate containing 0.5 mM Fluo-4-AM dissolved in dimethyl sulfoxide containing 20% pluronic acid. Neurons and glia, stained in an area of 150–300 μm diameter, were imaged in 30–75 μm depths for several hours at scanning rates of 1.25–1.43 Hz that are sufficient to resolve peaks of inspiratory-related Ca_i rises (for details, see Ruangkittisakul et al. 2006, 2008, 2009, 2012). Whole-body plethysmography was done in a perfused 100 ml plastic syringe at room temperature using a Validyne DP45-14 differential pressure transducer (Ballanyi 2004a).

12.3 Results and Discussion

In intact perinatal rats, hypoxic anoxia reversibly slows respiration from ~1 breath/s to <1 breath/min with concomitant augmentation of amplitude and duration of single breathing efforts ('anoxic gasps') which persist for >20 min (Fig. 12.1a) (Ballanyi 2004a, b). The typical initial increase in respiratory

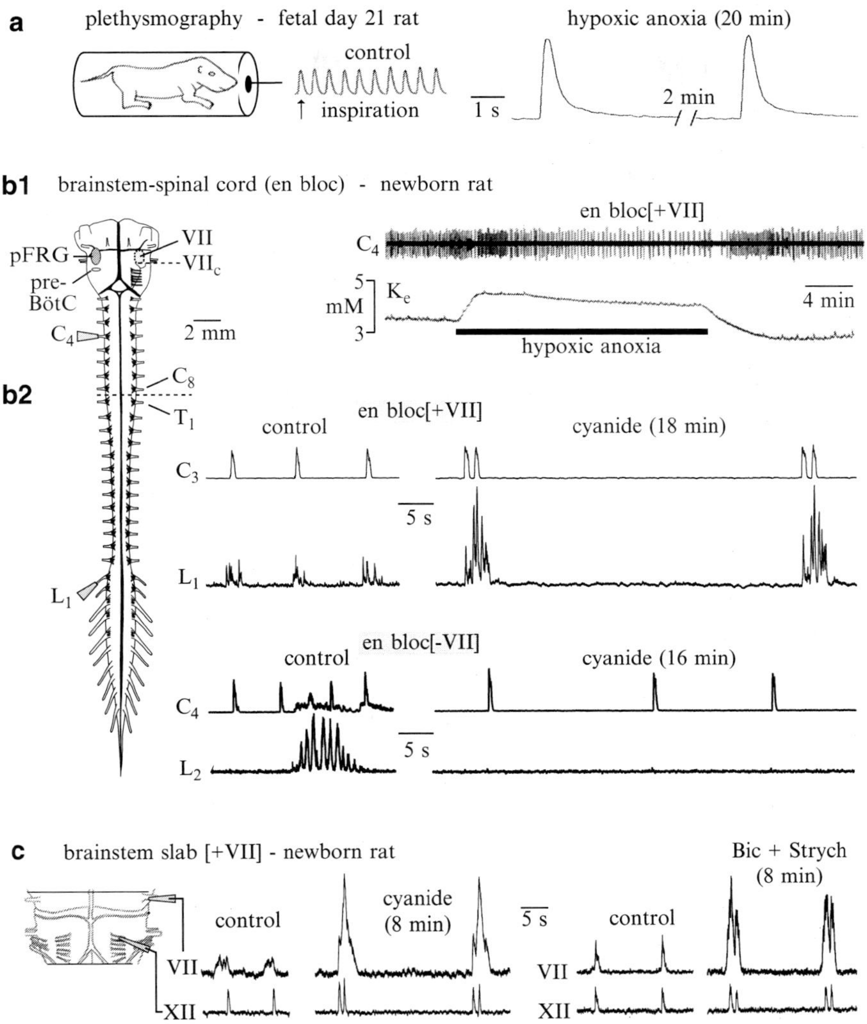

Fig. 12.1 Biphasic respiratory response of neonatal rats to anoxia. (**a**), a rat, isolated on the last embryonic day (E21), responded to hypoxic anoxia with 'anoxic gasps' that were several-fold larger and longer than control breaths while respiratory rate slowed substantially, but breathing persisted during at least 20 min of anoxia. (**b**$_1$) in a newborn rat brainstem-spinal cord preparation transected between cervical (C) and thoracic (T) spinal cord (see dotted line) and containing a complete facial (VII) motor nucleus, hypoxic anoxia in superfusate with 3 mM K^+ and 1.5 mM Ca^{2+} led to initial acceleration and subsequent progressive and persistent slowing of suction electrode-recorded integrated cervical root (C_4) bursting driven by the brainstem pre-Bötzinger complex (preBötC) inspiratory center. This biphasic respiratory anoxia response was accompanied by a modest extracellular K^+ (K_e) rise in the ventral respiratory column (VRC) containing the preBötC. (**b**$_2$) in a different en bloc brainstem model with complete cord and VII nucleus, the upper pair of recordings in 4.5 K/1.5Ca solution shows that chemical anoxia, due to bath-application of 1 mM cyanide, slowed C_3 rhythm with concomitant occurrence of double bursts, whereas pre/post-inspiratory bursting of lumbar root L_1, driven by the expiratory parafacial respiratory group (pFRG) was greatly enhanced in the post-inspiratory phase. In the lower left pair of recordings from a different en bloc preparation with complete cord, but transected at the caudal end of VII nucleus (VII$_c$), L_2 nerve showed only sporadic non-respiratory activity in 4.5 K/1.5Ca which was silenced by chemical anoxia. (**c**) in a newborn rat brainstem 'slab' with complete VII nucleus but no cord, pFRG-driven VII nerve bursting in 3 K/1Ca was enhanced during chemical anoxia contrary to modest amplitude increase of simultaneously recorded inspiratory-related hypoglossal (XII) nerve bursting. Recordings in the right from a different brainstem slab show that bath-application of the $GABA_A$ receptor blocker bicuculline (Bic, 10 μM) plus the glycine receptor blocker strychnine (Strych, 1 μM) mimics anoxic augmentation of VII bursts

rate (cf. Ballanyi 2004a, b) was not resolved with our plethysmographic approach likely due to slow (>1 min) change from room air (plus 5% CO_2) to 100% N_2 (or 95% N_2 plus 5% CO_2) to evoke anoxia. Instead, we found that inspiratory-related spinal cervical nerve bursting in ~40% of en bloc brainstems of 0–4 days-old rats accelerates by up to 100% from ~10 bursts/min in control during the initial 1–3 min of either hypoxic or cyanide-evoked chemical anoxia (Ballanyi et al. 1992, 1996; Ballanyi 2004a, b) (Fig. 12.1b_1). Subsequently, rhythm slows down to <2 bursts/min and persists at this rate for >1 h, typically with occurrence of repetitive spinal and cranial (e.g. XII) nerve bursts (Fig. 12.1b, c). (Ballanyi et al. 1992, 1996, 1999; Ballanyi 2004a, b). If at all, anoxia increases cervical burst amplitude and duration by only ~25%, which is insufficient to account for several-fold augmentation of anoxic gasps in vivo (Fig. 12.1a). During anoxia, functionally expiratory pFRG-driven lumbar or VII nerve bursts show a phase shift from pre/post-inspiratory to inspiratory/post-inspiratory with greatly boosted post-inspiratory activity (Ballanyi 2004a) (Fig. 12.1b_2, c). Blockade of $GABA_A$ (and glycine) receptors may be involved in anoxic amplitude increase of lumbar and VII nerve bursting because this response is mimicked by the selective blockers bicuculline (and strychnine) (Fig. 12.1c). The latter findings in 3 mM K^+ are consistent with our recent data obtained in the en bloc model in 4.5–6.2 mM K^+ (Taccola et al. 2007; Ballanyi et al. 2009). Similarly, we showed previously (Ballanyi 2004a) and more recently (Ruangkittisakul et al. 2007; Taccola et al. 2007) that brainstem transection between the caudal end of VII nucleus (VII_c) and its rostral third abolishes normal and anoxic respiratory lumbar bursting despite different superfusate K^+ levels (Fig. 12.1b_2). In summary, these findings indicate that (*i*) functionally important pFRG neurons are located close to the rostral margin of VII nucleus, (*ii*) anoxia synchronizes the dual respiratory center and (*iii*) modest amplitude increase of repetitive inspiratory nerve bursts, combined with boosted post-inspiratory lumbar activity (resulting in enforced exhalation) contributes to augmented neonatal anoxic gasps in vivo.

In our previous work on en bloc brainstems, anoxia elicited either a stable hyperpolarization, a <10 mV depolarization, or no effect on resting V_m of inspiratory preBötC(−driven) or pre/post-inspiratory pFRG(−driven) VRC neurons (Fig. 12.2) (Ballanyi et al. 1994, 1999; Ballanyi 2004a, b). In ~50% of inspiratory neurons and <40% of pre/post-inspiratory neurons in these studies, bursting persisted during anoxia, independently of effects on resting V_m (Fig. 12.2a, b), whereas the other cells were 'functionally inactivated' (Fig. 12.2c–f). $GABA_A$ and/or glycine receptor-mediated, inspiratory-related hyperpolarization of pre/post-inspiratory neurons could turn into a depolarization during the first ~3 min of anoxia (Fig. 12.2d). Subsequently, such phasic inhibition was suppressed by anoxia, sometimes revealing previously short-circuited excitatory inspiratory input to these cells, similar to effects of bicuculline (Fig. 12.2d–f). Inactivated bursting, but not phasic hyperpolarizing inhibition, could be reactivated by current-evoked countering of anoxic hyperpolarization (Fig. 12.2e). Mechanisms underlying this hyperpolarization are yet unresolved. Adenosine receptors are likely not

Fig. 12.2 (continued) current pulses. (**b**) slowed bursting also continued during anoxic hyperpolarization and concomitant major R_m decrease in inspiratory neuron of a different preparation. (**c**), a pre/post-inspiratory neuron without inspiratory-related inhibition, but showing (non-respiratory) 'ectopic' bursting, responded to anoxia with notable hyperpolarization and R_m decrease and was 'functionally inactivated' due to depression of respiratory-related V_m oscillations and associated action potential discharge. (**d**) in this pre/post-inspiratory neuron anoxia initially reversed hyperpolarizing $GABA_A$ and/or glycine receptor-mediated inspiratory inhibition into excitation and elicited ectopic bursting, but caused functional inactivation 4 min later. (**e**) in this pre/post-inspiratory neuron anoxia greatly depressed pronounced inspiratory hyperpolarizing inhibition and also pre/post-inspiratory synaptic depolarizations and concomitant discharge. In this situation, current-evoked depolarization (+dc) elicited ectopic bursting. After recovery from anoxia, Bic (10 μM) blocked inspiratory inhibition revealing peri-inspiratory and ectopic bursting when the cell was hyperpolarized by current injection (−dc) to the anoxic V_m level. Note that bicuculline evoked tonic C_4 discharge that was not reflected by perturbed cellular bursting. (**f**) a pre/post-inspiratory neuron showed a major anoxic R_m decrease with only modest concomitant hyperpolarization while bursting was abolished. After recovery from anoxia, bath-application of adenosine (ADO, 0.5 mM) did not mimic cellular anoxia effects, but depressed some subthreshold ectopic activity evident as small 'spikes'

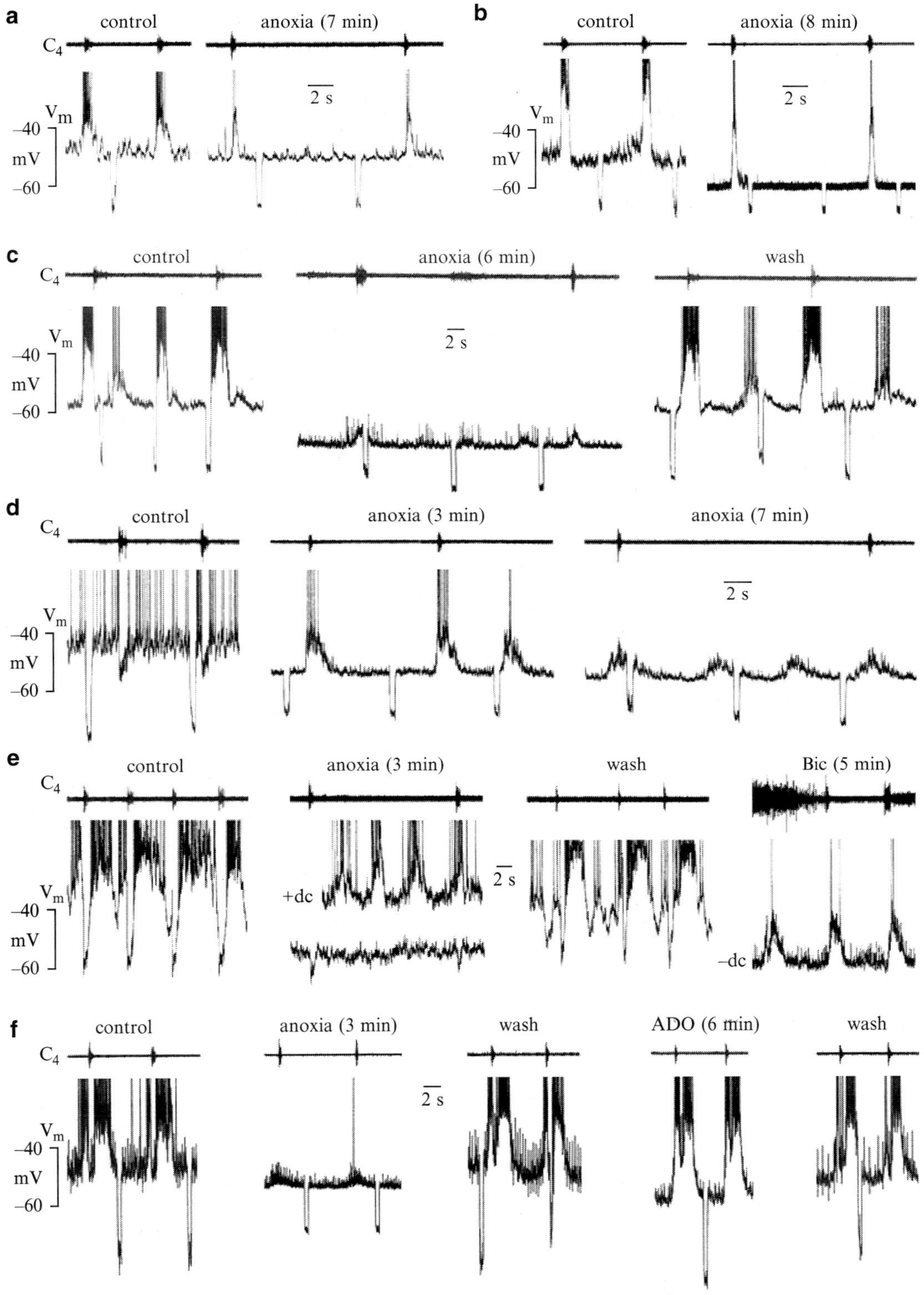

Fig. 12.2 Effect of hypoxic anoxia on preBötC-driven inspiratory and pFRG-driven pre/post-inspiratory VRC neurons of newborn rat en bloc brainstems in 3 K/1.5Ca. (**a**) in an inspiratory neuron slowed cervical bursting persisted during 7 min of anoxia which affected neither resting membrane potential (V_m) nor resistance (R_m) measured by injection of hyperpolarizing

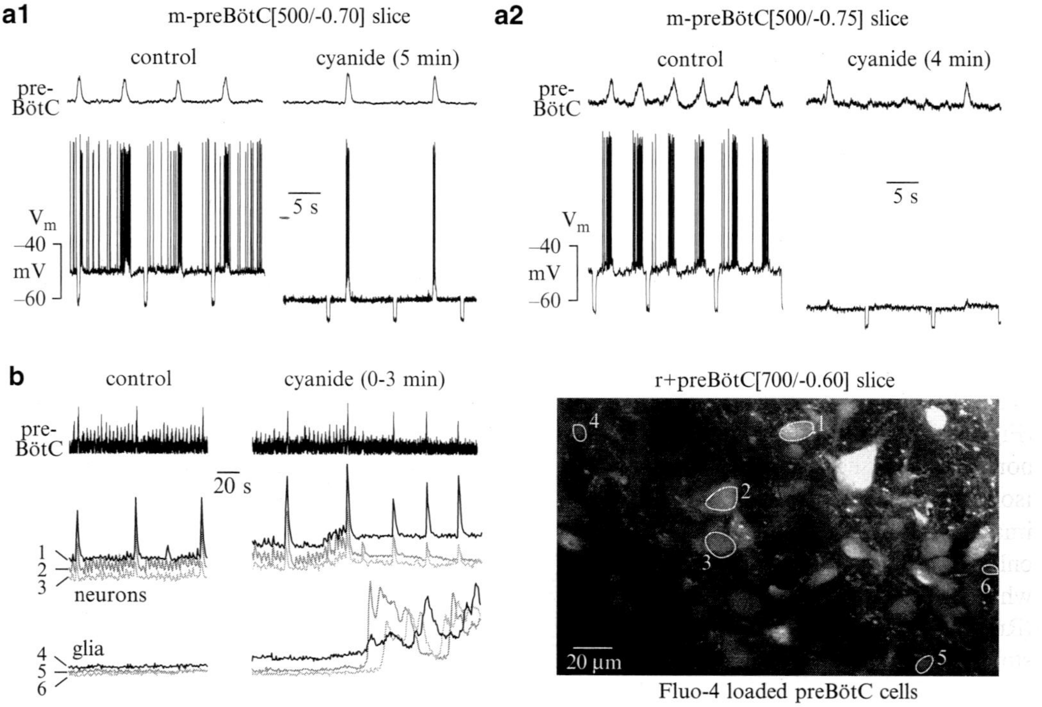

Fig. 12.3 Anoxia effects in 3 mM K^+ on preBötC cells of anatomically calibrated transversal newborn rat brainstem slices. (**a**) in 500 μm thick slices with centered ~200 μm thin preBötC and caudal margin at 0.70 (**a**$_1$) or 0.75 mm (**a**$_2$) caudal to VII_c, cyanide-evoked chemical anoxia hyperpolarized two inspiratory preBötC neurons with a concomitant R_m decrease. While inspiratory-related bursting persisted in the neuron of **a**$_1$, the cell in **a**$_2$ was functionally inactivated. (**b**) in a 700 μm thick slice with caudally exposed preBötC (posterior slice margin 0.60 mm caudal to VII_c) multiphoton imaging of free cytosolic Ca^{2+} (Ca_i) in control solution revealed inspiratory-related Ca_i rises in larger preBötC cells, presumably neurons, whereas Ca_i baseline was unchanged in smaller cells, likely astrocytes. While astrocytes responded to chemical anoxia with a major Ca_i rise, Ca_i baseline only modestly increased in the neurons. Fluorescence intensity traces were measured in regions of interest shown in the image of Fluo-4-AM loaded presumptive preBötC neurons (cells 1–3) and astrocytes (cell 4–6) located at 52 μm depth in that slice

involved because this neuromodulator neither mimics anoxic hyperpolarizations nor slows rhythm (Fig. 12.2f) while (A_1) adenosine receptor blockade does not counter anoxic slowing of rhythm (Ballanyi et al. 1999; Ballanyi 2004a, b). Intrinsically O_2– sensitive K^+ channels do not seem to be involved because hypoxic and chemical anoxia have identical effects on V_m and nerve bursts despite opposite effects on O_2 levels in the VRC (Völker et al. 1995). Our observation that gliclazide partially counters anoxic slowing of inspiratory rhythm in the en bloc model (Ballanyi 2004b) while another sulfonylurea blocker, tolbutamide, depresses anoxic outward currents in neurons of the newborn rat ventrolateral medulla (Ballanyi et al. 1999; Ballanyi 2004b) suggests contribution of ATP-dependent K^+ channels. These latter findings on neurons in unidentified VRC areas were obtained in the en bloc model kept in 3 mM K^+ (and 1.5 mM Ca^{2+}). In contrast, our recent study in 6.2 mM K^+ (and 2.4 mM Ca^{2+}) (Ballanyi et al. 2009) on the same model revealed only modest (~5 mV) anoxic hyperpolarization in pre/post-inspiratory neurons, but no anoxia effect on resting V_m in inspiratory cells within the preBötC, that extends in newborn rats between 0.4 and 0.6 mm caudal to VII_c (Fig. 12.1) as we found using calibrated slices (Ruangkittisakul et al. 2008).

To determine whether the latter lack of occurrence of anoxic hyperpolarization is due to differences between studies in superfusate K^+ (or Ca^{2+}), we analyzed effects of short (~5 min) chemical anoxia on inspiratory preBötC neurons in 500 μm thick calibrated slices kept in physiological K^+ (3 mM) and Ca^{2+} (1 mM) (cf. Ruangkittisakul et al. 2007; cf. Ballanyi and Ruangkittisakul 2009). Four of six neurons responded with a stable hyperpolarization by 4–16 mV to chemical anoxia (Fig. 12.3a), whereas V_m did not change in the other cells. In four of these six neurons, inspiratory drive potentials and concomitant action potential discharge persisted during anoxic slowing of population bursting in the contralateral preBötC aspect (Fig. 12.3a_1), whereas the other two cells were inactivated (Fig. 12.3a_2) similar to ~50% of VRC neurons in our previous reports on the en bloc model (Fig. 12.2) (Ballanyi et al. 1994, 1999; Ballanyi 2004a, b). Our preliminary data on calibrated slices include the observation that, in ~40% of cases, preBötC burst rate accelerates during the first minute of anoxia before rhythm slows down. In summary, our recent findings indicate that the isolated preBötC in newborn rat slices (*i*) is capable of generating a biphasic response to anoxia in physiological K^+ and Ca^{2+} and (*ii*) contains diverse populations of hyperpolarizing *versus* non-hyperpolarizing and inactivating *versus* non-inactivating inspiratory neurons.

A high tolerance to anoxia of the neonatal dual respiratory center due to effective utilization of anaerobic metabolism is also indicated by our previous findings in the VRC of the en bloc model that even >1 h of anoxia does not change extracellular Ca^{2+} and increases extracellular K^+ by only <2 mM (Fig. 12.1b_1) (Ballanyi et al. 1992, 1996; Völker et al. 1995). This modest effect of anoxia on ion homeostasis in the isolated newborn rat VRC, as opposed to massive ion deregulation in the VRC of isolated adult rat brainstems (Ballanyi et al. 1992), is reflected by our preliminary multiphoton imaging data in the preBötC of calibrated slices. Specifically, during 5 min of chemical anoxia Ca_i only rises modestly in inspiratory active cells with 10–25 μm soma diameter likely representing neurons, whereas Ca_i increases notably in small (~10 μm) cells likely representing astrocytes (Fig. 12.3b) (Ruangkittisakul et al. 2009, 2012; Ballanyi et al. 2010). We hypothesized in the latter studies that store-mediated astrocytic Ca_i rises (and eventually resulting gliotransmitter release) contribute to mutual communication between preBötC neurons and neighboring glia. It would be interesting to study if astrocytic activities secondary to their robust Ca_i increase are involved in anoxic slowing of preBötC rhythm.

Acknowledgments This study was supported by Alberta Heritage Foundation for Medical Research (AHFMR), Alberta Innovates Health Solutions (AIHS), Hotchkiss Brain Institute (HBI), Canada Foundation for Innovation (CFI) and the Canadian Institutes of Health Research (CIHR).

References

Ballanyi K (2004a) Neuromodulation of the perinatal respiratory network. Curr Neuropharmacol 2:221–243

Ballanyi K (2004b) Protective role of neuronal K_{ATP} channels in brain hypoxia. J Exp Biol 207:3201–3212

Ballanyi K, Ruangkittisakul A (2009) Structure-function analysis of rhythmogenic inspiratory pre-Bötzinger complex networks in "calibrated" newborn rat brainstem slices. Resp Physiol Neurobiol 168:158–178

Ballanyi K, Kuwana S, Völker A, Morawietz G, Richter DW (1992) Developmental changes in the hypoxia tolerance of the in vitro respiratory network of rats. Neurosci Lett 148:141–144

Ballanyi K, Völker A, Richter DW (1994) Anoxia induced functional inactivation of neonatal respiratory neurons in vitro. Neuroreport 6:165–168

Ballanyi K, Völker A, Richter DW (1996) Functional relevance of anaerobic metabolism in the isolated respiratory network of newborn rats. Eur J Physiol (Pflügers Archiv) 432:741–748

Ballanyi K, Onimaru H, Homma I (1999) Respiratory network function in the isolated brainstem-spinal cord of newborn rats. Prog Neurobiol 59:583–634

Ballanyi K, Ruangkittisakul A, Onimaru H (2009) Opioids prolong and anoxia shortens delay of onset of pre-inspiratory (pFRG) and inspiratory (preBötC) network bursting in newborn rat brainstems. Eur J Physiol (Pflügers Archiv) 458:571–587

Ballanyi K, Panaitescu B, Ruangkittisakul A (2010) Control of breathing by "nerve glue". Sci Signal 3:e41

Feldman JL, Del Negro CA (2006) Looking for inspiration: new perspectives on respiratory rhythm. Nat Rev Neurosci 7:232–242

Panaitescu B, Ruangkittisakul A, Ballanyi K (2009) Silencing by raised extracellular Ca^{2+} of pre-Bötzinger complex neurons in newborn rat brainstem slices without change of membrane potential or input resistance. Neurosci Lett 456:25–29

Ruangkittisakul A, Schwarzacher SW, Secchia L, Poon BY, Ma Y, Funk GD, Ballanyi K (2006) High sensitivity to neuromodulator-activated signaling pathways at physiological [K^+] of confocally-imaged respiratory center neurons in online-calibrated newborn rat brainstem slices. J Neurosci 26:11870–11880

Ruangkittisakul A, Secchia L, Bornes TD, Palathinkal DM, Ballanyi K (2007) Dependence on extracellular Ca^{2+}/K^+ antagonism of inspiratory centre rhythms in slices and en bloc preparations of newborn rat brainstem. J Physiol 584:489–508

Ruangkittisakul A, Schwarzacher SW, Secchia L, Ma Y, Bobocea N, Poon BY, Funk GD, Ballanyi K (2008) Generation of eupnea and sighs by a spatiochemically organized inspiratory network. J Neurosci 28:2447–2458

Ruangkittisakul A, Okada Y, Oku Y, Koshiya N, Ballanyi K (2009) Fluorescence imaging of active respiratory networks. Respir Physiol Neurobiol 168:26–38

Ruangkittisakul A, Secchia-Ballanyi L, Panaitescu B, Bobocea N, Kuribayashi J, Iizuka M, Kantor C, Ballanyi K (2012) Anatomically 'calibrated' isolated respiratory networks from newborn rodents. In: Ballanyi K (ed) Isolated central nervous system circuits. Walz W (ed) Neuromethods series. Springer, New York (in press)

Schwarzacher S, Pestean A, Günther S, Ballanyi K (2002) Serotonergic modulation of respiratory motoneurons and interneurons in brainstem slices of perinatal rats. Neuroscience 115:1247–1259

Smith JC, Ellenberger HH, Ballanyi K, Richter DW, Feldman JL (1991) Pre-Bötzinger complex: a brainstem region that may generate respiratory rhythm in mammals. Science 254:726–729

Taccola G, Secchia L, Ballanyi K (2007) Anoxic persistence of lumbar respiratory bursts and block of lumbar locomotion in newborn rat brainstem spinal cords. J Physiol 585:507–524

Völker A, Ballanyi K, Richter DW (1995) Anoxic disturbance of the isolated respiratory network of neonatal rats. Exp Brain Res 103:9–19

Chapter 13
Hypoxic Redistribution of Iron and Calcium in the Cat Glomus Cells

Mieczyslaw Pokorski, Lidia Faff, and Camillo Di Giulio

Abstract Both iron and calcium are essential for the hypoxia sensing mechanisms in the carotid body. However, trafficking of both ions in chemoreceptor cells in response to hypoxia is unclear. In the present study we seek to determine iron and calcium redistribution patterns in response to hypoxia in the cat chemoreceptor cells. Four cats were used: two each exposed to normoxia ($PaO_2 = 90$ mmHg) and hypoxia ($PaO_2 = 20$ mmHg) for 40 min. Carotid bodies were dissected, 150 nm sections made and processed for the measurements of iron and calcium content in the intracellular organelles of chemoreceptor cells with an energy dispersive X-ray spectroscopy. The results show that iron was distinctly lower in the hypoxic than normoxic chemoreceptor cells' cytoplasm. Conversely, calcium was increased in hypoxia, particularly in the nuclei and the dense-cored vesicles. These results highlight that regional distribution of iron does not coincide with calcium in glomus cells. Redistribution of both ions in response to hypoxia is congruous with their role in hypoxia-sensing. However, the exact determinants of iron/calcium redistribution patterns in glomus cells remain unsettled.

Keywords Calcium • Carotid body • Chemoreceptor cells • Hypoxia • Iron

13.1 Introduction

Both iron and calcium are essential elements for shaping the ventilatory responses to hypoxia in mammals. Ferrous iron and oxygen are co-factors in hypoxia-inducible factor-1 alpha (HIF-1α) degradation in normoxia. In hypoxia, HIF-1α instantly accumulates in cells, enters the nucleus, where it dimerizes with HIF-1β to trigger the transcriptional process of expression of target genes. HIF-1α is now considered

M. Pokorski (✉) • L. Faff
Department of Respiratory Research, Medical Research Center, Polish Academy of Sciences, 5 Pawinskiego St, 02-106 Warsaw, Poland
e-mail: m_pokorski@hotmail.com

C. Di Giulio
Department of Neuroscience and Imaging, University of Chieti-Pescara, 'G. d'Annunzio', Via Dei Vestini, 31 Chieti, Chieti, CH 66100, Italy
e-mail: digiulio@unich.it

C.A. Nurse et al. (eds.), *Arterial Chemoreception: From Molecules to Systems*, Advances in Experimental Medicine and Biology 758, DOI 10.1007/978-94-007-4584-1_13,

a master regulator of an array of genes, expression of which increases when oxygen is insufficient and which function to restore oxygen and energy metabolism (Fandrey and Gassmann 2009). One way to restore oxygen delivery is to increase lung ventilation; for instance, the hypoxic ventilatory response (HVR) being generated by glomus cells of a paired sensory organ of the carotid body (CB). Lack of iron emulates the stabilizing effect on HIF-1α of hypoxia in glomus cells in that the carotid chemosensory discharge, intracellular calcium level, and HIF-1α expression all increase (Roy et al. 2004a). At the functional level, chronic iron chelation appreciably dampens the HVR in conscious rats (Pokorski et al. 2009). The availability of iron is thus liable to influence the carotid body mechanisms. Likewise, calcium has long since been known to be indispensible for these mechanisms. Calcium entry into the glomus cells during hypoxic stimulation is crucial for the subsequent exocytotic catecholamine release and for triggering the subcellular signal transduction cascades (Conde et al. 2006). Yet despite the plausible importance of iron in the carotid chemosensory mechanisms, its intracellular shifts in response to hypoxia are an unknown area. The current study seeks to determine the content of iron and calcium in the cat glomus cells and the redistribution patterns of both elements in response to acute hypoxia.

13.2 Methods

The study was approved by an institutional Ethics Committee and the experiments were performed in compliance with the NIH Guide for Care and Use of Laboratory Animals. Four adult cats (8 carotid bodies) anesthetized with α-chloralose and urethane (35 and 800 mg/kg, i.p., respectively) were used. The cats were tracheostomized, allowed to breathe spontaneously, and the carotid bodies were exposed for later excision. Two cats each were exposed to normoxia ($PaO_2 = 90$ mmHg) or hypoxia ($PaO_2 = 20$ mmHg) for 40 min, after which they were euthanized by perfusion through the left heart with a mixture of 2% paraformaldehyde and 2% glutaraldehyde in 0.1 M cacodylate buffer. Carotid bodies were rapidly dissected, postfixed in the same mixture, and processed routinely for transmission electron microscopy: dehydrated in series of ethanol solutions and embedded in Spurr resin.

The measurements were performed with a Philips 420 scanning transmission electron microscope (TEM) equipped with energy dispersive X-ray spectroscopy (EDS), EDAX DX-4 (Switzerland). The EDS X-ray spectra were used for element identifications and quantifications in an accurately defined organelle viewed in a TEM specimen. For that purpose, the electron beam is focused to spot of interest. The X-rays which are induced due to beam-specimen interaction are recorded and used to form a characteristic spectrum for that specific location. The method detects total calcium and iron, both free and bound, and does not discern between Fe^{2+}/Fe^{3+}.

There were two separate measurement sessions, each consisting of paired 5 normoxic and 5 hypoxic glomus cells. Sections of 150 nm were collected on copper grids, and contrasted with lead hydroxide for 5 min. The grids were placed in a low background holder at an angle of 108° relative to the electron beam. Measurements were done for a 100 s life time and a spot diameter of 50 nm with an accelerating voltage of 80 kV. The only specimens considered suitable for the examination were those which consisted of a clearly identified glomus cell, based on the presence of secretory vesicles and other typical characteristics, covering the visualization area seen under the electron microscope. The primary magnification was 10,500x for all specimens examined. Iron and calcium were assessed in the nucleus, mitochondria, dense-cored vesicles, and the cytoplasm. The mean pooled results for each compartment obtained from the ten cells representing both measurement sessions in either condition were analyzed. Differences in the number of elemental counts were assessed with one-way ANOVA. Results were regarded as statistically significant at $p < 0.05$.

Table 13.1 Subcellular redistribution of iron and calcium elements in cat carotid glomus cells in response to hypoxia

		Normoxia	Hypoxia
Cytoplasm	Fe	401.2±61.3	248.8±13.2*
	Ca	248.5±18.0**	232.5±20.9
Nucleus	Fe	391.5±56.9	226.6±19.7*
	Ca	197.1±23.3**	305.8±81.2
Mitochondria	Fe	459.7±122.5	244.3±20.7
	Ca	233.5±19.5	238.8±29.3
Dense-cored vesicles	Fe	439.7±105.8	255.8±15.3
	Ca	232.1±16.1	309.9±62.2

Summary data for either element in the individual organelles (n=10 for each) in ten glomus cells in either condition. Values are means±SE counts
*$p<0.05$ for the differences between hypoxia and normoxia
**$p<0.05$ for the differences between Ca *vs.* Fe in the normoxic cytoplasm and nuclei

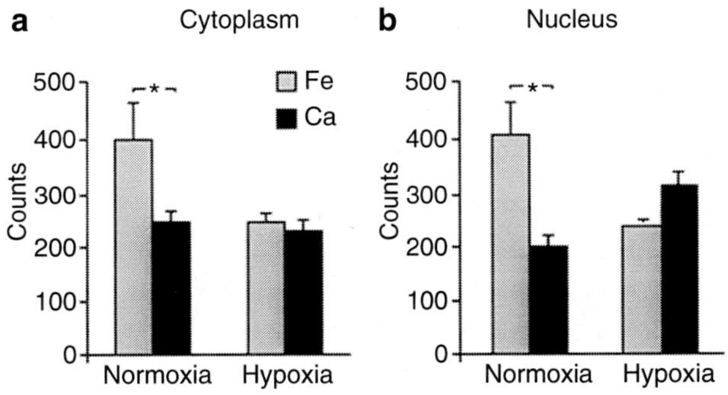

Fig. 13.1 Redistribution of Fe and Ca in cytoplasm and nuclei of cat glomus cells in response to hypoxia; $p<0.05$

13.3 Results

We found that iron content was distinctly lower in the cytoplasm, 248.8±13.2 *vs.* 401.2±61.3 counts, and in the nuclei, 226.6±19.7 *vs.* 391.5±56.9 counts ($p<0.05$), of the hypoxic *vs.* normoxic glomus cells. Iron also was lower in the remaining two organelles examined, albeit the decreases there did not assume statistical significance (Table 13.1). In contrast, calcium content did not change in the cytoplasm and mitochondria, but clearly tended to increase in the nuclei and dense-cored vesicles of the hypoxic glomus cells. However, calcium changes fell short of statistical significance due to data scatter. The opposite alterations in both ions caused that the Fe/Ca ratio, being significantly in favor of the former ion in the normoxic cytoplasm and nuclei, flattened out in hypoxia (Fig. 13.1a, b).

13.4 Discussion

The major objective of the current study was to determine the content of iron in carotid body glomus cells and to trace iron trafficking in response to hypoxia, the natural carotid body stimulus triggering the ventilatory chemoreflex. The presence of iron in glomus cells has never before been directly determined, although iron metabolism is believed to be influential in shaping the ventilatory hypoxic response. The results clearly confirm the presence of iron in glomus cells. Moreover, iron content decreased across the board in response to hypoxia; the most pronounced declines were noted in the cytoplasm and nuclei of glomus cells.

Ferrous iron, along with oxygen, is required for HIF-1α degradation by prolyl hydroxylases in cells. The corollary is that lack of iron would emulate the effects of hypoxia in that HIF-1α stabilizes and, after dimerization with HIF-1β protein, the complex moves to the cell nucleus, whereby the transcription of genes coordinating adaptive responses to hypoxia is upregulated. This sequence of events has been well substantiated in carotid body glomus cells (Baby et al. 2003; Roy et al. 2004b). Chelation of intracellular iron with ciclopirox olamine (CPX) increases the chemosensory discharge and calcium influx in *in vitro* glomus cells, the effects resembling the action of hypoxia (Daudu et al. 2002). Moreover, chronic application of CPX *in vivo* dampens the ventilatory chemoreflex in the awake rat (Pokorski et al. 2009), which may be explicable by the 'psuedo-hypoxic' HIF-1α state created by iron chelation in normoxia, against which background true hypoxia would exert less of an effect. The *in vivo* functional effects of iron chelation are somewhat disputable as other authors have found that desferrioxamine (DFO), an extracellular iron chelator, improves the hypoxic response in the rat, although the improvement evidently tapers off into the second week of DFO administration (Nguyen et al. 2007). The discrepancy in the effects on hypoxic reactivity of chronic iron chelation may be explained by the fact that ferrous iron, interacting with HIF-1α metabolism, belongs mostly to the intracellular free iron pool. DFO, an extracellular chelator, likely acts through different mechanisms. Indeed, DFO decreases chemosensory discharge and calcium influx in *in vitro* carotid body preparation, the effects plainly opposite to those of CPX (Daudu et al. 2002).

From the physiological standpoint, a decrease in iron content in hypoxic glomus cells may be indicative of good reason, as it should facilitate HIF-1α stabilization and thus orchestrate the adaptive ventilatory responses triggered by the carotid body. The cellular determinants of iron decrease by hypoxia are unclear. One plausible mechanism may have to do with hepcidin, a recently described protein which is a powerful negative regulator of iron efflux by causing internalization of ferroportin, the only iron exporter currently known (Kell 2009; Nemeth et al. 2004). Hypoxia suppresses hepcidin (Nicolas et al. 2002), and by doing so it would be liable to foster iron release from a cell. Alternative study designs would be required to explore the exact determinants of cellular iron decrease by hypoxia.

The current results highlight that the regional distribution of iron does not coincide with calcium in glomus cells. The measurement of calcium content in the carotid body was considered referential in the study. Calcium is a key elemental constituent of glomus cells, with an established role in the carotid body mechanisms (Buerk et al. 1997; Faff et al. 2001). In response to the hypoxic stimulus, intracellular calcium increases, participates in transduction cascades (Pokorski et al. 2000; Strosznajder and Pokorski 2000), exocytotic release of neurotransmitters, notably catecholamines (Conde et al. 2006), and is indispensible for glomus cell activation which does not take place in a calcium free environment (Buerk et al. 1997). The present results show rather mild and insignificant alterations in calcium in glomus cells in response to hypoxia. However, the increases in calcium in hypoxic dense-cored vesicles and nuclei are, generally, in concord with the recognized role of calcium signaling in the carotid body.

The study has limitations. The major flaw is that the method employed could not differentiate between bound/unbound ions; nor could it differentiate between ferrous/ferric iron (Fe^{2+}/Fe^{3+}), the latter being insoluble and biologically neutral. These limitations hamper the interpretation of the ionic shifts we noted. However, iron ought to be reduced to be able to enter eukaryotic cells. The intracellular iron pool consists mostly of the labile ferrous iron, shifts of which underlie the response to stimulation. Thus, the patterns of iron shifts we describe should be liable to be operational in generating the glomus cell's response to hypoxia.

Despite the limitations above outlined, we believe we have shown that iron is present in carotid body glomus cells and is elaborated in response to hypoxia, which is congruous with the role of this element, along with calcium, in shaping the carotid body chemoreflex. However, the shifts in both ions noted in hypoxia were rather modest and there was a consistent impression that these shifts took place in but some glomus cells. The highly non-uniform changes in the elemental composition, causing a substantial scatter of data recorded, blur the interpretation of the functional significance of the patterns we describe.

Acknowledgements The X-ray spectroscopy was carried out at the Lab for Electron Microscopy, Leiden University in the Netherlands, where L.F. was a fellow at the time of the study. The authors are thankful to the Dutch collaborators for help in the measurements.

Conflicts of interest The authors declare no conflicts of interest in relation to this article.

References

Baby SM, Roy A, Mokashi AM, Lahiri S (2003) Effects of hypoxia and intracellular iron chelation on hypoxia-inducible factor-1α and -1β in the rat carotid body and glomus cells. Histochem Cell Biol 120:343–352

Buerk DG, Osanai S, Chugh DK, Mokashi A, Lahiri S (1997) Calcium-dependent O2 sensitivity of cat carotid body. Adv Exp Med Biol 411:1–5

Conde SV, Caceres AI, Vicario I, Rocher A, Obeso A, Gonzalez C (2006) An overview on the homeostasis of Ca2+ in glomus cells of the rabbit and rat carotid bodies. Adv Exp Med Biol 580:215–222

Daudu PA, Roy A, Rozanov C, Mokashi A, Lahiri S (2002) Extra- and intracellular free iron and the carotid body responses. Respir Physiol Neurobiol 130:21–31

Faff L, van der Meulen H, Koerten HK, Walski M, Pokorski M (2001) Calcium handling by the cat carotid body – a pyroantimonate study. Acta Hisotochem 103:305–313

Fandrey J, Gassmann M (2009) Oxygen sensing and the activation of the hypoxia inducible factor 1 (HIF-1). Adv Exp Med Biol 648:197–206

Kell DB (2009) Iron behaving badly: inappropriate iron chelation as a major contributor to the aetiology of vascular and other progressive inflammatory and degenerative diseases. BMC Med Genomics 2:2. doi:10.1186/1755-8794-2-2

Nemeth E, Tuttle MS, Powelson J, Vaughn MB, Donovan A, Ward DM, Ganz T, Kaplan J (2004) Hepcidin regulates cellular iron efflux by binding to ferroportin and inducing its internalization. Science 306:2090–2093

Nguyen MVC, Pouvreau S, El Hajjaji FZ, Denavit-Saubie M, Pequignot JM (2007) Desferrioxamine enhances hypoxic ventilatory response and induces tyrosine hydroxylase gene expression in the rat brainstem in vivo. J Neurosci Res 85:1119–1125

Nicolas G, Chauvet C, Viatte L, Danan JL, Bigard X, Devaux I, Beaumont C, Kahn A, Vaulont S (2002) The gene encoding the iron regulatory peptide hepcidin is regulated by anemia, hypoxia, and inflammation. J Clin Invest 110:1037–1044

Pokorski M, Sakagami H, Kondo H (2000) Classical protein kinase C and its hypoxic stimulus-induced translocation in the cat and rat carotid body. Eur Respir J 16:459–463

Pokorski M, Antosiewicz J, Di Giulio C, Lahiri S (2009) Iron chelation and the ventilatory response to hypoxia. Adv Exp Med Biol 648:215–221

Roy A, Li J, Baby SM, Mokashi A, Burek DG, Lahiri S (2004a) Effects of iron-chelators on ion-channels and HIF-1α in the carotid body. Respir Physiol Neurobiol 141:115–123

Roy A, Volgin DV, Baby SM, Mokashi A, Kubin L, Lahiri S (2004b) Activation of HIF-1α mRNA by hypoxia and iron chelator in isolated rat carotid body. Neurosci Lett 363:229–232

Strosznajder RP, Pokorski M (2000) Regulation of phospholipase C activity by calcium ions and guanine nucleotide in the normoxic cat carotid body. Neurochem Res 25:739–743

Chapter 14
Acute Hypoxia Does Not Influence Intracellular pH in Isolated Rat Carotid Body Type I Cells

Ryan L. Shapiro, Barbara L. Barr, Robert W. Putnam, and Christopher N. Wyatt

Keywords Carotid body • Hypoxia • Intracellular pH

14.1 Introduction

In order to interpret data obtained from <u>isolated</u> Type I cells during hypoxia it is critical to know whether parameters such as intracellular pH (pH_i) vary during this challenge. Multiple studies have attempted to address this issue but data are contradictory and methods poorly defined (Wilding et al. 1992; Pang and Eyzaguirre 1993). Wilding et al. used BCECF loaded, isolated Type I cells and demonstrated that 'hypoxia' did not cause a change in pH_i. However, solutions in these experiments were gassed with 2% oxygen and recording chamber PO_2 was not reported. Conversely, Pang and Eyzaguirre used pH-sensitive microelectrodes and demonstrated that single isolated Type I cells did acidify during a hypoxic challenge whereas clusters of Type I cells did not.

These contradictory data were a concern for us as there is agreement that, in intact carotid bodies, pH_i does not change during hypoxia (Garcia-Sancho et al. 1978; Itturiaga et al. 1992). As the majority of work in our laboratory is focused on hypoxia-sensing mechanisms in acutely isolated Type I cells it was essential for us to know if pH_i did indeed vary in our cells during hypoxia.

To address this possibility we exposed acutely isolated rat Type I cells to a level of hypoxia known to initiate calcium signaling and monitored pH_i throughout the exposure. Our data indicate that individual isolated Type I cells maintain their pH_i during an excitatory hypoxic stimulus. Consequently our future experiments addressing the biochemical mechanisms engaged in Type I cells during hypoxia do not have to account for any simultaneous change in pH_i.

R.L. Shapiro • B.L. Barr • R.W. Putnam
Department of Neuroscience, Cell Biology and Physiology, Wright State University, 3640 Colonel Glenn Hwy, Dayton, OH 45435, USA
e-mail: shapirrl@gmail.com

C.N. Wyatt (✉)
Department of Neuroscience, Cell Biology and Physiology, Boonshoft School of Medicine, Wright State University, 3640 Colonel Glenn Hwy, Dayton, OH 45435, USA
e-mail:Christopher.wyatt@wright.edu

C.A. Nurse et al. (eds.), *Arterial Chemoreception: From Molecules to Systems*, Advances in Experimental Medicine and Biology 758, DOI 10.1007/978-94-007-4584-1_14,

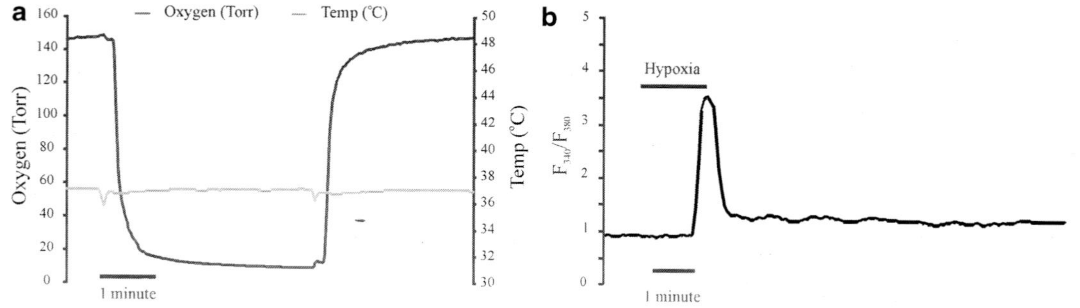

Fig. 14.1 Hypoxic stimuli were sufficient to excite Type I cells. (**a**) Example recording showing the time course of hypoxia and temperature in the perfusion chamber. Solutions were switched at the point where artifacts can be seen in the temperature trace. (**b**) Example recording from a Fura-2 loaded Type I cell indicating that the hypoxic stimulus evoked a rapid and reversible rise in the F_{340}/F_{380} ratio

14.2 Methods

Type 1 cells were isolated from rat (10–21 days old) carotid bodies using standard techniques (Burlon et al. 2009). Cells were then loaded with BCECF or Fura-2 using established protocols (Wilding et al. 1992; Burlon et al. 2009). For Ca^{2+} imaging experiments cells were loaded with 5 μM Fura-2 AM for 30 min at room temperature in the dark. The dye was excited with 340/380 nm light and emitted light was recorded at 510 nm using a Coolsnap HQ CCD camera (Photometrics). Image acquisition was controlled by Metafluor software (Molecular Devices) and cells were visualized using a Nikon TE2000U microscope with a x40 objective. For pH measurements cells were loaded with 5 μM BCECF-AM for 15 min at room temperature in the dark. The dye was excited with 500 and 440 nm light and recorded at 535 nm as previously described. pH calibrations were made using the nigericin technique (Buckler and Vaughan-Jones 1990).

Isolated Type I cells were superfused with bicarbonate buffered Tyrode (95% air, 5% CO_2, 37 C, 8 ml min^{-1}). Solutions were made hypoxic by gassing the Tyrode with 95% N_2, 5% CO_2 and a well around the recording chamber was also filled with hypoxic gas. Perfusion chamber PO_2 was monitored with an Oxymicro (WPI) fiber optic oxygen probe and the excitatory level of hypoxia recorded was <10 Torr (see Fig. 14.1a).

Statistical analysis was performed using two tailed, paired Students t-tests with significance set at 0.05. Data are presented as means ± standard error of the mean.

14.3 Results

10 Torr hypoxia elicited robust, reversible Ca^{2+} signaling in Type I cells (Fig. 14.1b) but had no significant effect on pH_i over a 3 min exposure period (n = 7, Fig. 14.2). These data suggest that acute exposure to excitatory levels of hypoxia does not alter pH_i in isolated Type I cells.

14.4 Discussion

These data indicate that this isolated Type I cell preparation can be used to study hypoxic-signaling mechanisms without having to account for the influence of a simultaneous fall in pH_i. Thus these isolated cells as well as Type I cells in intact carotid bodies (Garcia-Sancho et al. 1978; Itturiaga et al. 1992) do not alter pH_i during acute hypoxia of sufficient severity to evoke a chemosensory response.

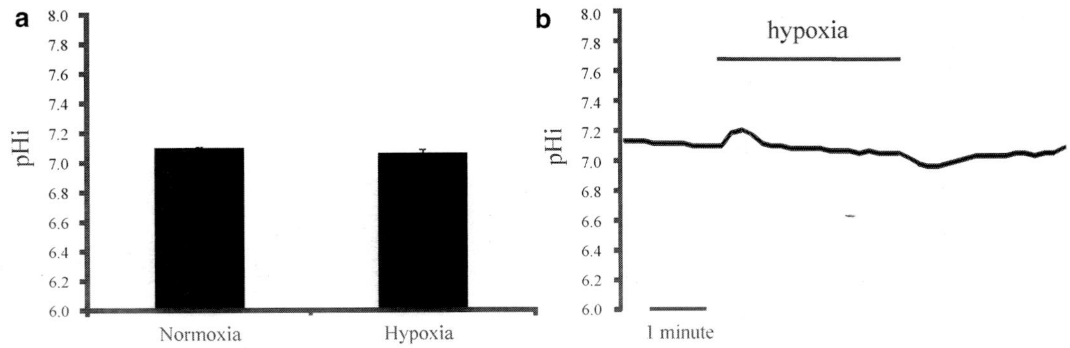

Fig. 14.2 Effect of hypoxia on Type I cell pH_i. (**a**) Bar graph showing average pH_i during normoxia, pH 7.11±0.01 (n=7) and pH_i during the final minute of exposure to hypoxia, pH 7.07±0.03 (n=7, no significant inhibition). (**b**) Example pH imaging trace showing the response to hypoxia (3 min). Observed deviations from baseline may be as a result of temperature switching artifacts (see Fig. 14.1a). Due to the lack of pH response to hypoxia the ability of Type I cells to respond to a challenge to pH_i was tested with Tyrode containing 15 mM NH_4Cl, all cells tested responded robustly and reversibly to this challenge (data not shown)

Acknowledgements RLS was privately funded by R. M. Johnson MD, T. Perry MD and R. Laughlin MD. CNW was funded by NIH 1RO1HL091836

References

Buckler KJ, Vaughan-Jones RJ (1990) Application of a new pH-sensitive fluoroprobe (carboxy-SNARF-1) for intracellular pH measurement in small, isolated cells. Pflügers Archiv 417(2):234–239

Burlon DC, Jordan HL, Wyatt CN (2009) Presynaptic regulation of isolated neonatal rat carotid body type I cells by histamine. Respir Physiol Neurobiol 168:218–223

Garcia-Sancho J, Giraldez F, Belmonte C (1978) Absence of apparent cell pH variations during hypoxia in the carotid body chemoreceptors in vitro. Neurosci Lett 10:247–249

Itturiaga R, Rumsey WL, Lahiri S, Spergel D, Wilson D (1992) Intracellular pH and oxygen chemoreception in the cat carotid body in vitro. J Appl Physiol 72(6):2259–66

Pang L, Eyzaguirre C (1993) Hypoxia affects differently the intracellular pH of clustered and isolated glomus cells of the rat carotid body. Brain Res 623(2):349–355

Wilding TJ, Cheng B, Roos A (1992) pH regulation in adult rat carotid body glomus cells. Importance of extracellular pH, sodium, and potassium. J Gen Physiol 100:593–608

Chapter 15
Hydrogen Sulfide (H_2S): A Physiologic Mediator of Carotid Body Response to Hypoxia

Nanduri R. Prabhakar

Abstract Carotid bodies are sensory organs for monitoring arterial blood O_2 levels. Nitric oxide and carbon monoxide function as inhibitory gasotransmitters in the carotid body. Hydrogen sulfide (H_2S) is another emerging gasotransmitter. The purpose of this article is to review recent studies addressing the role of H_2S in carotid body.Cystathionine γ-lyase (CSE) and cystathionine β synthase (CBS) are the two major enzymes that catalyze the formation of endogenous H_2S. Both CSE and CBS are expressed in glomus cells, the putative site of sensorytransduction in the carotid body. Hypoxia increases H_2S generation in the carotid body. CSE knockout mice displayed absence of hypoxia-evoked H_2S generation and severely impaired sensory excitation by low O_2. Pharmacological inhibitors of CSE as well as CBS showed a similar phenotype in mice and rats. Like hypoxia, H_2S donors stimulated the carotid body sensory activity and this response required Ca^{2+} influx via voltage-gated Ca^{2+} channels. Evidence is emerging implicating Ca^{2+} activated K^+ channels in glomus cells as potential targets of H_2S.

Keywords Hypoxia • Gasotransmitters • K^+ channels • Voltage-activated Ca^{2+} channels • Glomus cells

15.1 Introduction

Mammalian carotid body expresses several classes of neurotransmitters, including biogenic amines, neuropeptides and aminoacids (Prabhakar 2000; Nurse 2010). Nitric oxide (NO) and carbon monoxide (CO) function as gasotransmitters in the nervous system. NO and CO are also synthesized in the carotid body (Prabhakar 1998, 1999; Nurse 2010). For instance, neuronal nitric oxide synthase (nNOS), the enzyme that catalyzes the formation of NO is expressed in the nerve endings that innervate glomus cells and endothelial nitric oxide synthase (eNOS) in blood vessels in the carotid body (Prabhakar 1999; Nurse 2010). NOS inhibitors stimulate and NO donors inhibit the carotid body activity (Prabhakar et al.

N.R. Prabhakar (✉)
Biological Sciences Division, Department of Medicine, Institute for Integrative Physiology and Center for Systems Biology of O_2 Sensing, University of Chicago,
5841 South Maryland Avenue, MC 5068 Room N-711, Chicago, IL 60637, USA
e-mail: nanduri@uchicago.edu

C.A. Nurse et al. (eds.), *Arterial Chemoreception: From Molecules to Systems*,
Advances in Experimental Medicine and Biology 758, DOI 10.1007/978-94-007-4584-1_15,

1993; Chugh et al. 1994; Wang et al. 1995), suggesting that NO functions as an inhibitory gasotransmitter in the carotid body. Recent studies suggest that NO generated by nNOS mediates efferent inhibition of the carotid body sensory activity (Campanucci and Nurse 2007).

Heme oxygenases (HO) catalyze the formation of endogenous CO (Maines 1997). While HO-1 is an inducible enzyme, HO-2 is constitutively expressed in many tissues. HO-2 is localized to glomus cells, and HO inhibitors stimulate the sensory activity of the carotid body (Prabhakar et al. 1995), suggesting that CO is also inhibitory to the sensory activity of the carotid body. Given that molecular O_2 is required for enzymatic activity of nNOS and HO-2, and inhibitors of NOS and HO mimic the effects of hypoxia, it was proposed that stimulation of carotid body activity by hypoxia is in part due to reduced enzymatic generation of CO and NO, thus removing the inhibitory influence on the sensory activity (i.e. disinhibition; Prabhakar 1999).

Recent studies suggest that mammalian cells generate hydrogen sulfide (H_2S) another gas messenger. Like NO and CO, H_2S is a potent vasodilator (Gadalla and Snyder 2010). H_2S has been implicated in mediating hypoxia-evoked relaxation of blood vessels (Olson et al. 2006). The purpose of this article is to review recent studies addressing the role of H_2S in carotid body response to hypoxia.

15.2 H_2S Generating Enzymes in the Carotid Body

Cystathionine γ-lyase (CSE) and cystathionine β-synthase (CBS) are two major enzymes that catalyze the formation of endogenous H_2S. CBS is more abundant in the central nervous system and CSE is predominantly expressed in peripheral tissues (Gadalla and Snyder 2010). CSE-like immunoreactivity was seen in glomus cells of carotid bodies as evidenced by co-localization with tyrosine hydroxylase (TH), an established marker of glomus cells (Peng et al. 2010). CBS expression was also reported in mice (Li et al. 2010) and cat carotid bodies (Fitzgerald et al. 2011). In CSE knockout mice, basal H_2S levels in carotid bodies were half of that seen in wild type mice (Peng et al. 2010). The residual H_2S levels might arise from CBS and/or mitochondrial H_2S generating enzyme(s).

15.3 Effects of Hypoxia on H_2S Levels in the Carotid Body

Hypoxia (PO_2 ~40 mmHg) increases H_2S levels in both mouse and rat carotid bodies (Peng et al. 2010). Remarkably, H_2S generation by hypoxia was completely absent in carotid bodies from $CSE^{-/-}$ mice as well as in rats treated with DL-propargylglycine (PAG), an inhibitor of CSE (Abeles and Walsh 1973; Washtien and Abeles 1977). These observations suggest that CSE is the primary enzyme that contributes to elevated H_2S levels during hypoxia.

15.4 Effects of Inhibition of H_2S Synthesizing Enzymes on Carotid Body Sensory Response to Hypoxia

Li et al. (2010) reported that amino oxyacetic acid (AOAA) and hydroxylamine, inhibitors of CBS reduced mouse carotid body response to hypoxia; whereas PAG and beta-cyano-L-alanine, inhibitors of CSE were without effect. However, Peng et al. (2010) found that carotid bodies from CSE knockout mice exhibit severely impaired sensory response to hypoxia. Similar loss of carotid body hypoxic sensitivity was seen in mice and rats following systemic administration of PAG prior to the experiment (Peng et al. 2010). In striking contrast, acute application of PAG to the *ex vivo* carotid body was ineffective in preventing the hypoxic response, suggesting poor permeability of this compound. The effects of CSE inhibition were selective to hypoxia because carotid body response to hypercapnia was unaffected in CSE knockout mice, and it was augmented in rats treated with PAG, a CSE inhibitor.

15.5 Effects of Inhibition of H_2S Synthesizing Enzymes on Hypoxic Ventilatory Response

Hypoxic ventilatory response (HVR) is a hallmark reflex elicited by the carotid body. $CSE^{-/-}$ mice displayed reduced HVR (Peng et al. 2010). The attenuated HVR was not secondary to changes in metabolism because O_2 consumption (VO_2) and CO_2 production (VCO_2) under hypoxia were comparable between CSE knockout and wild type mice. Unlike hypoxia, ventilatory responses to hypercapnia were unaffected in CSE knockout mice (Peng et al. 2010). Similar reduction in HVR was also seen mice treated with CBS inhibitor (Li et al. 2010).

15.6 Effects of H_2S Donors on Carotid Body Activity

Exogenous application of NaHS, which generates H_2S (H_2S donor), stimulates carotid body sensory activity in mice (Li et al. 2010; Peng et al. 2010) and in rats (Peng et al. 2010). The effects of NaHS were concentration dependent and could be elicited with as little as 30 μM and maximal excitation with 100 μM. NaHS-evoked sensory excitation, like hypoxia, was rapid in onset, occurred within seconds after its application and returned to baseline after terminating the stimulus (Li et al. 2010; Peng et al. 2010). NaHS also stimulated the carotid body sensory activity in CSE knockout mice suggesting that H_2S is a downstream signaling molecule.

Acetylcholine (ACh) and adenosine triphosphate (ATP) are suggested as major excitatory transmitters that mediate sensory excitation by hypoxia (Nurse 2010). Li et al. (2010) reported that NaHS-evoked sensory excitation could be prevented by pyridoxalphosphate-6-azophenyl-2′, 4′-disulfonic acid, an inhibitor of purinergic receptors and hexamethonium, a blocker of nicotinic cholinergic receptors. These observations suggest that H_2S-induced sensory excitation is coupled to release of ATP/ACh from type I glomus cells. However, Fitzgerald et al. (2011) reported that Na_2S, another H_2S donor, inhibits ACh release and had variable effects on ATP release from cat carotid bodies. Pinacidil, an opener of ATP-sensitive K^+ (K_{ATP}) channels mimicked the effects of Na_2S on ACh and ATP release (Fitzgerald et al. 2011). It is intriguing that although H_2S donors stimulate the carotid body sensory activity, they seem to inhibit the release of ACh and ATP, two transmitters implicated in sensory excitation by hypoxia.

15.7 Cellular Targets of H_2S Actions in the Carotid Body

K_{ATP} channels are the major targets of H_2S in blood vessels (Zhao et al. 2001). Glomus cells express K_{ATP} channels (Kim et al. 2011). However, glibenclamide, a potent inhibitor of K_{ATP} channels was ineffective in preventing stimulation of the carotid body by NaHS or hypoxia (Peng et al. 2010). On the other hand, carotid body stimulation by NaHS as well as hypoxia were completely abolished in Ca^{2+}-free medium or in presence of cadmium chloride, a non-selective inhibitor of voltage-activated Ca^{2+} channels (Peng et al. 2010; Li et al. 2010). Like hypoxia, NaHS elevated $[Ca^{2+}]_i$ levels in rat glomus cells and this response was abolished in presence of Ca^{2+}-free medium (V. Makarenko, A.P.

Fox and N.R. Prabhakar, unpublished observations). These findings indicate that H_2S as well as hypoxia facilitate Ca^{2+} influx via voltage-gated Ca^{2+} channels.

Large conductance (maxi K^+) calcium-activated potassium channels (BK) play important role in excitability of many cells and are targets of gasotransmitters (Sitdikova et al. 2010). Li et al. (2010) reported that NaHS, like hypoxia, inhibits BK currents in glomus cells from mouse carotid body, and this effect was prevented by AOAA, an inhibitor of CBS. Although subunits of Kv 3 family of K^+ channels are implicated in glomus cell response to hypoxia in mouse carotid body (Lopez-Lopez and Perez-Garcia 2007), the findings of Li et al. (2010) suggest that H_2S targets BK channels. A study by Telezhkin et al. (2010) reported that H_2S donor inhibits native carotid body and human recombinant BK channels with an IC_{50} of ~275 μM. Inhibition of BK channels by H_2S was rapid and reversible, and does not involve interaction with either the "Ca^{2+} bowl" or residues distal to the Ca^{2+}-sensing domain. These studies suggest that BK channels are potential targets of H_2S and inhibition of these channels might contribute to some if not all the stimulatory effects of H_2S on carotid body sensory activity.

Acknowledgements N.R.P gratefully acknowledges the collaborations of Drs. S.H. Snyder, G. K. Kumar, Y.J. Peng and M.M. Gadalla. Research from author's laboratory is supported by grants HL-76537, HL-90554, and HL-86493 from the National Institutes of Health.

References

Abeles RH, Walsh CT (1973) Acetylenic enzyme inactivators. Inactivation of γ-cystathionase, in vitro and in vivo, by propargylglycine. J Am Chem Soc 95:6124–6125

Campanucci VA, Nurse CA (2007) Autonomic innervation of the carotid body: role in efferent inhibition. Respir Physiol Neurobiol 157:83–92

Chugh DK, Katayama M, Mokashi A, Bebout DE, Ray DK, Lahiri S (1994) Nitric oxide related inhibition of carotid chemosensory nerve activity in the cat. Respir Physiol 97:147–156

Fitzgerald RS, Shirahata M, Chang I, Kostuk E, Kiihl S (2011) The impact of hydrogen sulfide (H2S) on neurotransmitter release from the cat carotid body. Respir Physiol Neurobiol 176:80–89

Gadalla MM, Snyder SH (2010) Hydrogen sulfide as a gasotransmitter. J Neurochem 113:14–26

Kim D, Kim I, Papreck JR, Donnelly DF, Carroll JL (2011) Characterization of an ATP-sensitive K^+ channel in rat carotid body glomus cells. Respir Physiol Neurobiol 177:247–255

Li Q, Sun B, Wang X, Jin Z, Zhou Y, Dong L, Jiang LH, Rong W (2010) A crucial role for hydrogen sulfide in oxygen sensing via modulating large conductance calcium-activated potassium channels. Antioxid Redox Signal 12:1179–1189

López-López JR, Pérez-García MT (2007) Oxygen sensitive Kv channels in the carotid body. Respir Physiol Neurobiol 157:65–74

Maines MD (1997) The heme oxygenase system: a regulator of second messenger gases. Annu Rev Pharmacol Toxicol 37:517–554

Nurse CA (2010) Neurotransmitter and neuromodulatory mechanisms at peripheral arterial chemoreceptors. Exp Physiol 95:657–667

Olson KR, Dombkowski RA, Russell MJ, Doellman MM, Head SK, Whitfield NL, Madden JA (2006) Hydrogen sulfide as an oxygen sensor/transducer in vertebrate hypoxic vasoconstriction and hypoxic vasodilation. J Exp Biol 209:4011–4023

Peng YJ, Nanduri J, Raghuraman G, Souvannakitti D, Gadalla MM, Kumar GK, Snyder SH, Prabhakar NR (2010) H2S mediates 02 sensing in the carotid body. Proc Natl Acad Sci U S A 107:10719–10724

Prabhakar NR (1998) Endogenous carbon monoxide in control of respiration. Respir Physiol 114:57–64

Prabhakar NR (1999) NO and CO as second messengers in oxygen sensing in the carotid body. Respir Physiol 115:161–168

Prabhakar NR (2000) Oxygen sensing by the carotid body chemoreceptors. J Appl Physiol 88:2287–2295

Prabhakar NR, Kumar GK, Chang CH, Agani FA, Haxhiu MA (1993) Nitric oxide in the sensory function of the carotid body. Brain Res 625:16–22

Prabhakar NR, Dinerman JL, Agani FH, Snyder SH (1995) Carbon monoxide: a role in carotid body chemoreception. Proc Natl Acad Sci U S A 92:1994–1997

Sitdikova GF, Weiger TM, Hermann A (2010) Hydrogen sulfide increases calcium-activated potassium (BK) channel activity of rat pituitary tumor cells. Pflugers Archiv 459:389–397

Telezhkin V, Brazier SP, Cayzac SH, Wilkinson WJ, Riccardi D, Kemp PJ (2010) Mechanism of inhibition by hydrogen sulfide of native and recombinant BKCa channels. Respir Physiol Neurobiol 172:169–178

Wang ZZ, Stensaas LJ, Dinger B, Fidone SJ (1995) Nitric oxide mediates chemoreceptor inhibition in the cat carotid body. Neuroscience 65:217–229

Washtien W, Abeles RH (1977) Mechanism of inactivation of γ-cystathionase by the acetylenic substrate analogue propargylglycine. Biochemistry 16:2485–2491

Zhao W, Zhang J, Lu Y, Wang R (2001) The vasorelaxant effect of H2S as a novel endogenous gaseous KATP channel opener. EMBO J 20:6008–6016

Chapter 16
The Retrotrapezoid Nucleus and Breathing

Patrice G. Guyenet, Ruth L. Stornetta, Stephen B.G. Abbott, Seth D. Depuy, and Roy Kanbar

Abstract The retrotrapezoid nucleus (RTN) is located in the rostral medulla oblongata close to the ventral surface and consists of a bilateral cluster of glutamatergic neurons that are non-aminergic and express homeodomain transcription factor Phox2b throughout life. These neurons respond vigorously to increases in local pCO_2 via cell-autonomous and paracrine (glial) mechanisms and receive additional chemosensory information from the carotid bodies. RTN neurons exclusively innervate the regions of the brainstem that contain the respiratory pattern generator (RPG). Lesion or inhibition of RTN neurons largely attenuates the respiratory chemoreflex of adult rats whereas their activation increases respiratory rate, inspiratory amplitude and active expiration. Phox2b mutations that cause congenital central hypoventilation syndrome in humans prevent the development of RTN neurons in mice. Selective deletion of the RTN Phox2b-VGLUT2 neurons by genetic means in mice eliminates the respiratory chemoreflex in neonates.

In short, RTN Phox2b-VGLUT2 neurons are a major nodal point of the CNS network that regulates pCO_2 via breathing and these cells are probable central chemoreceptors.

Keywords Breathing network • Chemoreflexes • Central chemoreceptors • Retrotrapezoid nucleus • Congenital central hypoventilation syndrome • Phox2b • Optogenetics

16.1 Introduction

The central respiratory chemoreflex is the stimulation of breathing caused by an increase in CNS pCO_2. This reflex makes an important contribution to the short-term stability of arterial pCO_2. It provides an especially powerful respiratory stimulus during apnea, airway obstruction and deliberate or accidental inhalation of CO_2. The central respiratory chemoreflex never operates in isolation since CO_2 also controls breathing via the carotid bodies. This essential interaction is discussed elsewhere (Blain et al. 2010).

P.G. Guyenet (✉) • R.L. Stornetta • S.B.G. Abbott • S.D. Depuy • R. Kanbar
Department of Pharmacology, University of Virginia, 1300 Jefferson Park Avenue, 800735, Charlottesville, VA 22908-0735, USA
e-mail: pgg@virginia.edu

C.A. Nurse et al. (eds.), *Arterial Chemoreception: From Molecules to Systems*, Advances in Experimental Medicine and Biology 758, DOI 10.1007/978-94-007-4584-1_16,

The central chemoreflex is presumably triggered by the interaction of protons or, possibly, molecular CO_2 (Huckstepp et al. 2010) with unidentified sensors ((Guyenet et al. 2010) for review). As explained below, the retrotrapezoid nucleus (RTN) is a critical nodal point of the CNS network responsible for the central respiratory chemoreflex and this nucleus seems to derive much of its chemosensitivity from a local effect of CO_2 although the exact location of the proton receptors (RTN neurons, glia or both) is unsettled. The existence of other putative CNS chemoreceptors is discussed elsewhere (Nattie and Li 2009; Corcoran et al. 2009; Guyenet et al. 2010).

16.2 The Retrotrapezoid Nucleus: Definition and Chemosensitivity

The name RTN was coined in 1989 to designate a group of superficial neurons that are located under the facial motor nucleus and innervate the ventral respiratory column (Smith et al. 1989). This thin sheet of cells extends rostrally up to the caudal end of the trapezoid body, hence the name. Smith et al. (1989) speculated that these neurons could have a chemosensory role because their location roughly coincides with a previously identified chemosensitive region of the ventral medullary surface of the cat. Nattie and his colleagues were able to produce reliable albeit modest increases in ventilation in rats (~25% V_E) by artificially acidifying the RTN region (Li and Nattie 2002). These authors produced comparable ventilatory effects by acidifying many other brainstem regions and, on this basis, they concluded that RTN was one among many brain regions that contain central respiratory chemoreceptors (Li and Nattie 2002). This viewpoint is developed elsewhere (see, for example Nattie and Li 2009; Corcoran et al. 2009).

The RTN neurons that regulate breathing remained unidentified for a long time for lack of a histological marker and because this "nucleus" is in reality a sparse collection of cells that are not clearly demarcated from adjoining portions of the reticular formation. Recently, the RTN region was shown to contain a distinctive group of non-aminergic CO_2-sensitive neurons that express VGLUT2 and the transcription factor Phox2b, henceforth called the RTN Phox2b-VGLUT2 neurons (Mulkey et al. 2004; Stornetta et al. 2006). These findings were seminal in two respects. They provided a histological marker for the putative chemoreceptors. The presence of Phox2b enabled investigators to introduce foreign proteins such as channelrhodopsin-2 or the allatostatin receptor relatively specifically into these neurons using lentiviral vectors engineered with PRSx8, a high-efficiency Phox2-activated artificial promoter (Hwang et al. 2001; Abbott et al. 2009). Figure 16.1 depicts the location and projections of the ~2,000 RTN Phox2b-VGLUT2 neurons in the rat.

16.3 How Do RTN Neurons Respond to CO_2?

RTN neurons are very sensitive to hypercapnia (0.5 Hz per 0.01 unit change in pHa), 0.5 Hz representing approximately 5% of their entire dynamic range (Guyenet et al. 2005) (Fig. 16.2a). This high CO_2 sensitivity may result from several mechanisms (Fig. 16.3). The level of evidence supporting each mechanism varies.

Evidence for a cell autonomous pH-sensitivity (Fig. 16.3, mechanism 1) comes from observations that RTN neurons are excited by CO_2 or acidification in slices and that this response persists in the presence of blockers of GABA, glutamate and ATP receptors (Mulkey et al. 2004; Lazarenko et al. 2009) (Fig. 16.2b). This work, performed in neonatal brain slices, has certain limitations. For example, the pH sensitivity of the neurons is only 40% of that observed *in vivo*, even when temperature is factored in (Guyenet et al. 2005) (Fig. 16.2a, b). This may be because this work was performed using

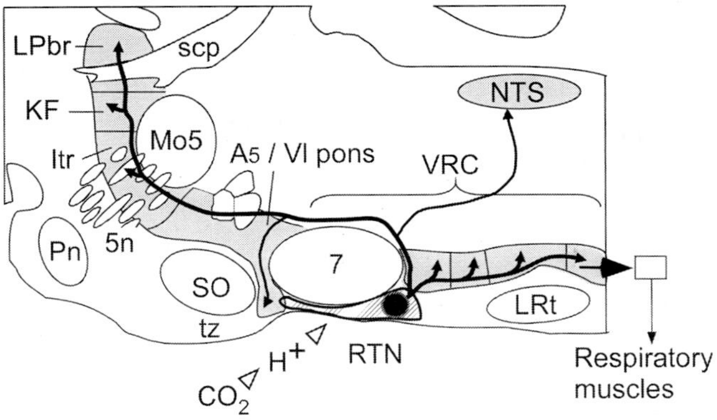

Fig. 16.1 Location and projection of RTN Phox2b-VGLUT2 neurons. Parasagittal schematic section of the lower brainstem of a rodent. The retrotrapezoid nucleus resides at the ventral surface of the medulla oblongata under the facial motor nucleus (7). RTN innervates all four subdivisions of the ventral respiratory column (VRC), the dorsolateral pons (*KF* Kölliker-Fuse nuc., *Itr* intertrigeminal region, *LPbr* lateral parabrachial nuc.), the nucleus of the solitary tract (*NTS*) and the A5 region. Other abbreviations: *LRt* lateral reticular nucleus, *Mo5* trigeminal motor nucleus, *Pn* pontine nuclei, *scp* superior cerebellar peduncle, *SO* superior olive, *tz* trapezoid body, *5n* trigeminal nerve

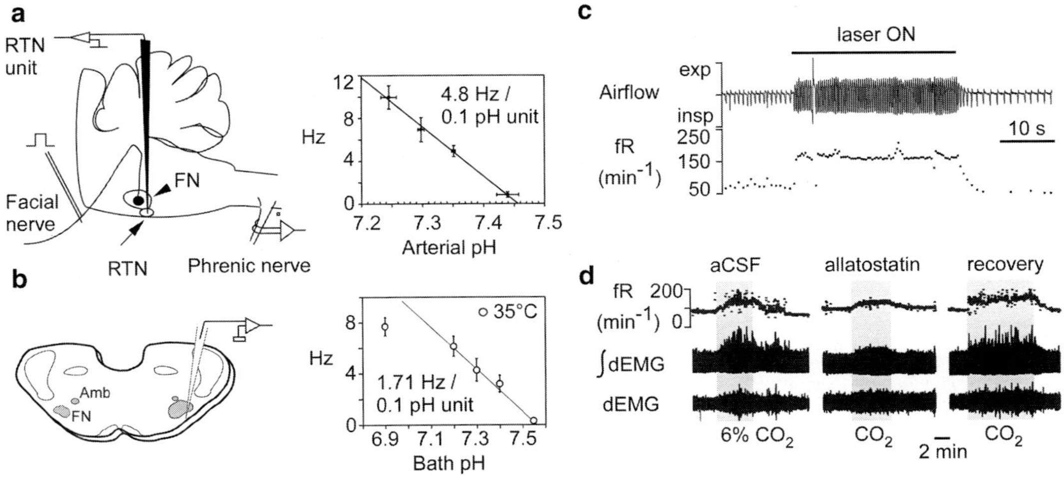

Fig. 16.2 RTN Phox2b-VGLUT2 neurons and the central respiratory chemoreflex. (**a**) *Left*: RTN unit recording *in vivo*. RTN neurons are found below the facial motor nucleus. The lower boundary of this nucleus is identified by recording antidromic field potentials. Right, relationship between discharge rate of RTN units and arterial pH in rats in which kynurenic acid was injected icv to block glutamatergic transmission (Source: Guyenet et al. 2005). (**b**) *Left*: RTN unit recording in slices from neonatal rodents (FN, facial motor nucleus; Amb, nuc. ambiguus). Right, relationship between discharge rate of RTN units and bath pH. Although the recordings were made at 35°C, the pH-sensitivity is only 36% of its value *in vivo* (Source: Guyenet et al. 2005). (**c**) Optogenetic activation of the RTN Phox2b-VGLUT2 neurons elicits massive respiratory stimulation in an awake rat. Channelrhodopsin-2 was selectively transfected into the Phox2b-expressing neurons using a lentiviral vector that expresses its transgene under the control of the artificial promoter PRSx8 (Source: Kanbar et al. 2010). (**d**) Attenuation of the chemoreflex following selective inhibition of the RTN Phox2b-VGLUT2 neurons. The allatostatin receptor was selectively transfected into the Phox2b-expressing neurons using a lentiviral vector that expresses its transgene under the control of the artificial promoter PRSx8 (Source: Marina et al. 2010). In **c** and **d**, the respiratory effects could also be partially due to activation (in **c**) or inhibition (in **d**) of the nearby C1 neurons which also contain Phox2b and therefore also expressed the transgene

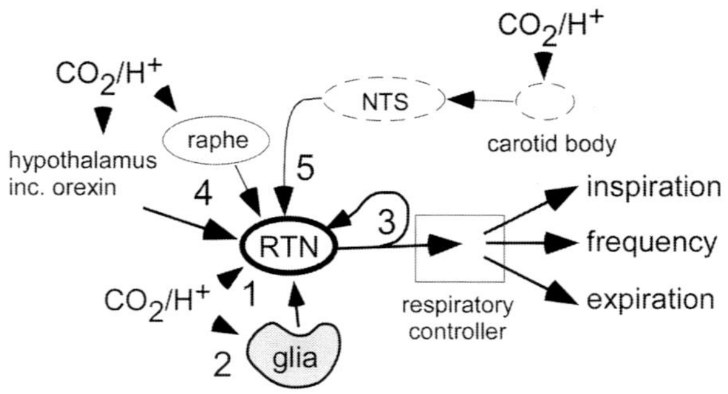

Fig. 16.3 Five mechanisms may cooperate to produce the high sensitivity of RTN Phox2b-VGLUT2 neurons to hypercapnia. Mechanism 1: RTN neurons respond to acidification through an unidentified cell autonomous response that persists in presence of blockers of glutamate and purinergic transmission in slices (Mulkey et al. 2004). Mechanism 2: ventral surface glial cells are depolarized by acidification and probably contribute to the CO_2 sensitivity of the RTN neurons. ATP is the presumed mediator of this paracrine mechanism (Gourine et al. 2010). Mechanism 3: coupling between RTN neurons via gap junctions or recurrent collaterals (depicted) may amplify the cell autonomous effect of pH on RTN neurons (Fortin and Thoby-Brisson 2009). Mechanism 4: RTN neurons receive input from orexinergic or serotonergic neurons. Subsets of these neurons may respond to changes in brain pH/pCO_2 (Guyenet et al. 2010). Mechanism 5: RTN neurons receive a polysynaptic excitatory input from the carotid bodies (Takakura et al. 2006)

tissue from neonate rodents which have weaker chemoreflexes. However, the most glaring hole in the theory that RTN neurons are cell-autonomous CO_2 sensors is that the proton receptor(s) is(are) unidentified (Guyenet et al. 2010).

Evidence for a contribution of surrounding glial cells to the pH response of RTN neurons (Fig. 16.3, mechanism 2) derives from two main observations. The glial cells of the ventral medullary surface are depolarized by acid and, like astrocytes elsewhere, this glia releases ATP (Gourine et al. 2010; Mulkey and Wenker 2011). Secondly, photostimulation of Channelrhodopsin2-transfected RTN glial cells increases breathing, an effect that is attenuated by administration of an ATP receptor antagonist (Gourine et al. 2010). This evidence also has its limitations. A ≤1 kb segment of the GFAP promoter was used to drive expression of ChR2 into the glia using an adenovirus. Short promoters are rarely totally cell-specific. Secondly, the respiratory effects produced by activating glial cells seems too persistent (~30 min) to elicit timely corrective respiratory adjustments to CO_2 changes (Gourine et al. 2010). Finally, blocking ATP receptors has only small effects on the response of RTN neurons to acidification in slices (Mulkey et al. 2004; Wenker et al. 2010).

The exquisite sensitivity of RTN neurons to CO_2 *in vivo* could also be due to the presence of a local circuit (gap junctions, recurrent excitatory collaterals) that amplifies their cell-autonomous response to pH (Fig. 16.3, mechanism 3). During the prenatal period, RTN neurons are coupled by gap junctions that contribute to their group pacemaker activity (Thoby-Brisson et al. 2009).

RTN neurons could also be a convergence point for additional central chemoreceptors, orexin and serotonin being two such candidates (Nattie and Li 2009; Corcoran et al. 2009) (Fig. 16.3, mechanism 4). This hypothesis is still controversial (Depuy et al. 2011) and needs further testing.

The fifth mechanism is a polysynaptic input from the peripheral chemoreceptors (Fig. 16.3, mechanism 5). This input is well-documented and involves a glutamatergic projection from the caudal portion of the nucleus solitary tract to the RTN (Takakura et al. 2006).

16.4 Contribution of the RTN Neurons to the Chemoreflexes

In adult rats, RTN Phox2b-VGLUT2 neurons mediate a large portion of the central respiratory chemoreflex. Selective photoactivation of RTN Phox2b-VGLUT2 neurons transfected with Channelrhodopsin2 produces a massive stimulation of breathing (Kanbar et al. 2010). Relatively selective lesions of the RTN neurons attenuate the central chemoreflex of anesthetized rats according to the degree of cell loss (Takakura et al. 2008). Finally, selective inhibition of the RTN neurons in conscious rats with intact carotid bodies attenuates the response to CO_2 by 60% on average (Marina et al. 2010).

Mice with genetic deletion of RTN Phox2b-VGLUT2 neurons lack central respiratory chemosensitivity at birth (Goridis and Brunet 2010). The first generation of these experiments involved making a heterozygous mouse that uniformly expresses the toxic (27 alanine) mutation of *Phox2b* (Dubreuil et al. 2008). Mice carrying this mutation have no RTN and die within hours of birth from respiratory failure and possibly other undetected deficits. Subsequently, Phox2b was selectively excised from subsets of neurons that depend, like the RTN, on a particular set of transcription factors such as KROX20 (EGR2) or ATOH-1 (Dubreuil et al. 2009) . Each manipulation aborted the development of the RTN Phox2b-VGLUT2 neurons and produced mice that failed to breathe after birth. In all these models, the core of the respiratory pattern generator developed seemingly normally but lacked CO_2 sensitivity *in vitro* revealing that the respiratory network of the neonate is insensitive to moderate acidification (0.2 pH units) in the absence of RTN neurons. This important observation seems to eliminate the possibility that pH-sensitivity is more or less ubiquitous within the respiratory network. The main interpretative limitation of this data is that other neurons besides the RTN Phox2b-VGLUT2 neurons could have been eliminated or impaired by these genetic crosses. In a recent study from the same group, the toxic 27 alanine mutation was selectively introduced in *Egr2*-dependent neurons resulting in an even more specific deletion of the RTN neurons than in prior experiments (Ramanantsoa et al. 2011). These mice have a selective, complete loss of the RTN Phox2b-VGlut2 neurons, but survive into adulthood despite having no central respiratory chemoreflex up to 2 weeks after birth.

In summary, the respiratory chemoreflex is entirely dependent on the integrity of the RTN Phox2b-VGLUT2 neurons at and shortly after birth. Remarkably, selective genetic deletion of the RTN Phox2b-VGLUT2 neurons and loss of breathing stimulation by CO_2 is survivable, apparently because of a compensatory, possibly hypoxic, drive from the carotid bodies (Ramanantsoa et al. 2011). In adulthood at least 60% of the chemoreflex depends on the RTN.

16.5 How Do the RTN Neurons Regulate the Respiratory Network and What Else Do They Control?

RTN Phox2b-VGLUT2 neurons activate the breathing rate, increase the respiratory rhythmic contractions of the diaphragm and activate abdominal expiratory muscles (Janczewski and Feldman 2006; Abdala et al. 2009b; Marina et al. 2010; Kanbar et al. 2010). Further understanding of how these various outflows are controlled will require deciphering the connections of RTN Phox2b-VGLUT2 neurons. This difficult integrative issue has only received partial answers to date. RTN neurons innervate the entire ventral respiratory group, suggesting that they target many types of respiratory neurons (Abbott et al. 2009). The pre-Bötzinger complex is among these targets and RTN neurons likely regulate the respiratory rate at this level (Bochorishvili et al. 2011). RTN also innervates the caudal most part of the respiratory column which contains expiratory premotor neurons (Iscoe 1998; Abbott et al. 2009). This projection likely underpins some of the regulation of active (phase-2) expiration by RTN neurons but more rostral portions of the ventral respiratory column also contribute to this response (Pagliardini et al. 2011). The respiratory network includes many inhibitory neurons that would not be active without an excitatory drive (Smith et al. 2009). These neurons are also plausible targets of the RTN.

There may be several subgroups of RTN Phox2b-VGLUT2 neurons. For example, only two-thirds of the RTN Phox2b-VGLUT2 neurons are *Atoh-1* dependent for their development in mice (Ramanantsoa et al. 2011). Functional heterogeneity between the Phox2b-VGLUT2 neurons is also suggested by the fact that, in neonatal rat brainstem maintained *in vitro*, these neurons are either phasically (preI-post-I neurons) or tonically active (Onimaru et al. 2008). Both types respond to CO_2 however, and their differences in discharge patterns may result from the fact that these neurons are in the process of losing their embryonic group pacemaker properties (Thoby-Brisson et al. 2009). In the adult, RTN Phox2b-VGLUT2 neurons have a variety of stereotyped respiratory modulations (Guyenet et al. 2005) that denote differences in input and, possibly, functional heterogeneity. Finally, the RTN region contains neurons whose activity is restricted to the late expiratory period and become active only at very high levels of CO_2 (Abdala et al. 2009a). These neurons could conceivably be specialized in driving active expiration but evidence that they are Phox2b-VGLUT2 neurons is currently lacking.

Finally, based on their projections, RTN Phox2b-VGLUT2 neurons could regulate other notably CO_2-dependent CNS outflows such as airway tone, CO_2-induced arousal, the stimulatory effect of CO_2 on cardiovascular function, and the urge to breathe.

16.6 Pathophysiology

Non-catecholaminergic Phox2b-positive neurons most likely homologous to the rodent RTN have been identified in man (Rudzinski and Kapur 2010). CCHS (congenital central hypoventilation syndrome) is a developmental disease whose signs and symptoms include attenuated or absent chemoreflexes, sleep apnea and loss of the sensation of dyspnea (Weese-Mayer et al. 2010). Expansion of the 20-alanine tract of Phox2b is the most frequent cause of this syndrome (Weese-Mayer et al. 2010). These mutations produce a dominant negative and toxic version of the transcription factor which, in mice, impairs the development or survival of RTN neurons with considerable specificity as explained above (Dubreuil et al. 2008; Goridis and Brunet 2010). The loss of the RTN has yet to be documented in CCHS. It is a plausible cause of the respiratory deficits experienced by these patients but is unlikely to account for the entire syndrome (Weese-Mayer et al. 2010). The development of the carotid bodies, sympathetic post-ganglionic neurons and the nucleus of the solitary tract are all Phox2b-dependent (Goridis and Brunet 2010). Developmental abnormalities in any of these regions could interfere with autonomic regulations and exacerbate the chemoreflex deficit caused by RTN loss.

16.7 Conclusions

The RTN Phox2b-VGLUT2 neurons are a major nodal point through which lung ventilation is regulated by CO_2. Evidence that these neurons are central respiratory chemoreceptors is considerable but will remain incomplete until the sensor(s) for pH (or pCO_2) are identified and proof is obtained that the RTN Phox2b-VGLUT2 neurons respond to pCO_2 in a cell autonomous manner. At the present time, the CO_2 sensitivity of the RTN neurons is best explained by a mixture of cell-autonomous and paracrine responses to pH/pCO_2.

A superficial analogy between a sensory organ like the carotid body and the RTN is tempting but one should not lose sight of the fact that the RTN Phox2b-VGLUT2 neurons receive synaptic inputs from multiple brain regions (e.g. hypothalamus, the respiratory network and the raphe) and are subject to regulation by polysynaptic inputs from cardiopulmonary receptors.

The RTN Phox2b-VGLUT2 neurons enhance the breathing rate, inspiratory amplitude and active expiration. Subsets of RTN Phox2b-VGLUT2 neurons may mediate the orderly recruitment of various muscle groups when the organism is exposed to incremental levels of CO_2. Finally, the carotid body input to RTN neurons underscores the highly integrated nature of central and peripheral chemoreflexes.

References

Abbott SBG, Stornetta RL, Fortuna MG, Depuy SD, West GH, Harris TE, Guyenet PG (2009) Photostimulation of retrotrapezoid nucleus Phox2b-expressing neurons *in vivo* produces long-lasting activation of breathing in rats. J Neurosci 29:5806–5819

Abdala AP, Rybak IA, Smith JC, Paton JF (2009a) Abdominal expiratory activity in the rat brainstem-spinal cord in situ: patterns, origins, and implications for respiratory rhythm generation. J Physiol 587:3539–3559

Abdala AP, Rybak IA, Smith JC, Zoccal DB, Machado BH, St-John WM, Paton JF (2009b) Multiple pontomedullary mechanisms of respiratory rhythmogenesis. Respir Physiol Neurobiol 168:19–25

Blain GM, Smith CA, Henderson KS, Dempsey JA (2010) Peripheral chemoreceptors determine the respiratory sensitivity of central chemoreceptors to CO(2). J Physiol 588:2455–2471

Bochorishvili G, Stornetta RL, Coates MB, Guyenet PG (2011) Pre-Bötzinger complex receives glutamatergic innervation from galaninergic and other retrotrapezoid nucleus neurons. J Comp Neurol 520(5):1047–1061

Corcoran AE, Hodges MR, Wu Y, Wang W, Wylie CJ, Deneris ES, Richerson GB (2009) Medullary serotonin neurons and central CO2 chemoreception. Respir Physiol Neurobiol 168:49–58

Depuy SD, Kanbar R, Coates MB, Stornetta RL, Guyenet PG (2011) Control of breathing by raphe obscurus serotonergic neurons in mice. J Neurosci 31:1981–1990

Dubreuil V, Ramanantsoa N, Trochet D, Vaubourg V, Amiel J, Gallego J, Brunet JF, Goridis C (2008) A human mutation in Phox2b causes lack of CO2 chemosensitivity, fatal central apnoea and specific loss of parafacial neurons. Proc Natl Acad Sci U S A 105:1067–1072

Dubreuil V, Thoby-Brisson M, Rallu M, Persson K, Pattyn A, Birchmeier C, Brunet JF, Fortin G, Goridis C (2009) Defective respiratory rhythmogenesis and loss of central chemosensitivity in phox2b mutants targeting retrotrapezoid nucleus neurons. J Neurosci 29:14836–14846

Fortin G, Thoby-Brisson M (2009) Embryonic emergence of the respiratory rhythm generator. Respir Physiol Neurobiol 168:86–91

Goridis C, Brunet JF (2010) Central chemoreception: lessons from mouse and human genetics. Respir Physiol Neurobiol 173:312–321

Gourine AV, Kasymov V, Marina N, Tang F, Figueiredo MF, Lane S, Teschemacher AG, Spyer KM, Deisseroth K, Kasparov S (2010) Astrocytes control breathing through pH-dependent release of ATP. Science 329:571–575

Guyenet PG, Mulkey DK, Stornetta RL, Bayliss DA (2005) Regulation of ventral surface chemoreceptors by the central respiratory pattern generator. J Neurosci 25:8938–8947

Guyenet PG, Stornetta RL, Bayliss DA (2010) Central respiratory chemoreception. J Comp Neurol 518:3883–3906

Huckstepp RT, Id BR, Eason R, Spyer KM, Dicke N, Willecke K, Marina N, Gourine AV, Dale N (2010) Connexin hemichannel-mediated CO2-dependent release of ATP in the medulla oblongata contributes to central respiratory chemosensitivity. J Physiol 588:3901–3920

Hwang DY, Carlezon WA Jr, Isacson O, Kim KS (2001) A high-efficiency synthetic promoter that drives transgene expression selectively in noradrenergic neurons. Hum Gene Ther 12:1731–1740

Iscoe S (1998) Control of abdominal muscles. Prog Neurobiol 56:433–506

Janczewski WA, Feldman JL (2006) Distinct rhythm generators for inspiration and expiration in the juvenile rat. J Physiol 570:407–420

Kanbar R, Stornetta RL, Cash DR, Lewis SJ, Guyenet PG (2010) Photostimulation of Phox2b medullary neurons activates cardiorespiratory function in conscious rats. Am J Respir Crit Care Med 182:1184–1194

Lazarenko RM, Milner TA, Depuy SD, Stornetta RL, West GH, Kievits JA, Bayliss DA, Guyenet PG (2009) Acid sensitivity and ultrastructure of the retrotrapezoid nucleus in Phox2b-EGFP transgenic mice. J Comp Neurol 517:69–86

Li A, Nattie E (2002) CO2 dialysis in one chemoreceptor site, the RTN: stimulus intensity and sensitivity in the awake rat. Respir Physiol Neurobiol 133:11–22

Marina N, Abdala AP, Trapp S, Li A, Nattie EE, Hewinson J, Smith JC, Paton JF, Gourine AV (2010) Essential role of Phox2b-expressing ventrolateral brainstem neurons in the chemosensory control of inspiration and expiration. J Neurosci 30:12466–12473

Mulkey DK, Wenker IC (2011) Astrocyte chemoreceptors: mechanisms of H+sensing by astrocytes in the retrotrapezoid nucleus and their possible contribution to respiratory drive. Exp Physiol 96:400–406

Mulkey DK, Stornetta RL, Weston MC, Simmons JR, Parker A, Bayliss DA, Guyenet PG (2004) Respiratory control by ventral surface chemoreceptor neurons in rats. Nat Neurosci 7:1360–1369

Nattie E, Li A (2009) Central chemoreception is a complex system function that involves multiple brain stem sites. J Appl Physiol 106:1464–1466

Onimaru H, Ikeda K, Kawakami K (2008) CO2-sensitive preinspiratory neurons of the parafacial respiratory group express Phox2b in the neonatal rat. J Neurosci 28:12845–12850

Pagliardini S, Janczewski WA, Tan W, Dickson CT, Deisseroth K, Feldman JL (2011) Active expiration induced by excitation of ventral medulla in adult anesthetized rats. J Neurosci 31:2895–2905

Ramanantsoa N, Hirsch MR, Thoby-Brisson M, Dubreuil V, Bouvier J, Ruffault PL, Matrot B, Fortin G, Brunet JF, Gallego J, Goridis C (2011) Breathing without CO_2 chemosensitivity in Conditional Phox2b Mutants. J Neurosci 31:12880–12888
Rudzinski E, Kapur RP (2010) immunolocalization of the candidate human retrotrapezoid nucleus. Pediatr Dev Pathol 13:291–299
Smith JC, Morrison DE, Ellenberger HH, Otto MR, Feldman JL (1989) Brainstem projections to the major respiratory neuron populations in the medulla of the cat. J Comp Neurol 281:69–96
Smith JC, Abdala AP, Rybak IA, Paton JF (2009) Structural and functional architecture of respiratory networks in the mammalian brainstem. Philos Trans R Soc Lond B Biol Sci 364:2577–2587
Stornetta RL, Moreira TS, Takakura AC, Kang BJ, Chang DA, West GH, Brunet JF, Mulkey DK, Bayliss DA, Guyenet PG (2006) Expression of Phox2b by brainstem neurons involved in chemosensory integration in the adult rat. J Neurosci 26:10305–10314
Takakura AC, Moreira TS, Colombari E, West GH, Stornetta RL, Guyenet PG (2006) Peripheral chemoreceptor inputs to retrotrapezoid nucleus (RTN) CO_2-sensitive neurons in rats. J Physiol 572:503–523
Takakura AC, Moreira TS, Stornetta RL, West GH, Gwilt JM, Guyenet PG (2008) Selective lesion of retrotrapezoid Phox2b-expressing neurons raises the apnoeic threshold in rats. J Physiol 586:2975–2991
Thoby-Brisson M, Karlen M, Wu N, Charnay P, Champagnat J, Fortin G (2009) Genetic identification of an embryonic parafacial oscillator coupling to the preBotzinger complex. Nat Neurosci 12:1028–1035
Weese-Mayer DE, Berry-Kravis EM, Ceccherini I, Keens TG, Loghmanee DA, Trang H (2010) An official ATS clinical policy statement: congenital central hypoventilation syndrome: genetic basis, diagnosis, and management. Am J Respir Crit Care Med 181:626–644
Wenker IC, Kreneisz O, Nishiyama A, Mulkey DK (2010) Astrocytes in the retrotrapezoid nucleus sense H + by inhibition of a Kir4.1-Kir5.1-like current and may contribute to chemoreception by a purinergic mechanism. J Neurophysiol 104:3042–3052

Chapter 17
The Interaction Between Low Glucose and Hypoxia in the *in vitro*, Rat Carotid Body

Andrew P.S. Holmes, David Hauton, and Prem Kumar

Abstract A role for the carotid body (CB) in systemic glycaemic control is yet to be fully characterised. Observations made on fasted, anaesthetised cats, rats and dogs in vivo showed that intra-arterial injection of sodium cyanide into the carotid sinus region immediately increased carotid sinus nerve (CSN) discharge frequency and elicited a subsequent significant increase in the systemic arterial glucose concentration, within 2–8 min of drug administration (Alvarez-Buylla and Alvarez-Buylla 1988). These responses were abolished in animals in which both CSNs had been surgically sectioned, demonstrating that the increased arterial glucose concentration detected following CB stimulation was dependent on CSN input into the NTS. Although not directly tested by these authors, it was proposed that low plasma glucose directly stimulated the CB, as the increase in CSN discharge frequency elicited with NaCN was attenuated by direct injection of a hyperglycaemic solution into the common carotid artery (Alvarez-Buylla and Alvarez- Buylla 1988; Alvarez-Buylla et al. 1997). Additionally, in dogs with bilateral CB resection (CBR), the rate of exogenous glucose infusion required to maintain a fixed hypoglycaemic level was significantly higher, whilst the endogenous hepatic glucose production was significantly lower, compared to control (CSN intact) animals (Koyama et al. 2000). These results further suggested a dependence on CB stimulation for the maintenance of a physiologically normal plasma glucose concentration, but again no direct measure of CB response to hypoglycaemia had been made.

Keywords Carotid body • Hypoxia • Glucose • Hypoglycaemia • Stimulus Interaction

17.1 Introduction

Subsequently, electrophysiological recordings of single and few fibre chemoafferents fibres from the CSN, in whole isolated CB *in vitro* preparations, demonstrated that a change in the glucose concentration of the superfusate from 11 mM to either 2 mM or 0 mM failed to elicit any acute increase in neuronal discharge frequency for up to 20 min (Bin-Jaliah et al. 2004, 2005). Further, it was demonstrated that while hypoxia significantly increased the concentration of ATP release from the CB, a 0 mM glucose

A.P. S. Holmes • D. Hauton • P. Kumar (✉)
School of Clinical and Experimental Medicine, The Medical School,
University of Birmingham, Birmingham, West Midlands B15 2TT, UK
e-mail: p.kumar@bham.ac.uk

C.A. Nurse et al. (eds.), *Arterial Chemoreception: From Molecules to Systems*,
Advances in Experimental Medicine and Biology 758, DOI 10.1007/978-94-007-4584-1_17,

solution did not (Conde et al. 2007). Taken together, these studies suggested that low glucose could not be a direct stimulus for the whole *in vitro* CB preparation.

In contrast, in a later study, the absolute amount of ACh released from whole cat CBs was not significantly elevated in either severe hypoxia ($PO_2 < 10$ mmHg) or in a 0 mM glucose solution (Fitzgerald et al. 2009). However, when these results were normalised it was calculated that the percentage of ACh release had increased significantly in both the hypoxic and the 0 mM glucose solution. When the two stimuli were applied simultaneously, the response was greater than that of the addition of the two individual responses, suggesting that 0 mM glucose increased the sensitivity of the CB to PO_2. Interestingly, these authors also showed that 0 mM glucose failed to induce any significant increase in ATP release compared with control (Fitzgerald et al. 2009).

Supporting a role for the CB in direct glucose sensing, data obtained using CB slices had indicated at least a fivefold increase in catecholamine release in severe hypoglycaemic 2 mM and 0 mM glucose solutions (in 150 mmHg PO_2) that was similar to the release observed in hypoxia (Pardal and López-Barneo 2002). A response to a more physiological hypoglycaemia (3.3 mM) was only observed when PO_2 was reduced to 90 mmHg. Single cell patch clamp was used to demonstrate that 0 mM glucose significantly inhibited the outward K^+ current when the cells were held at 0 mV and +20 mV but not at membrane potentials less than −40 mV. No alteration in inward ionic conductance was observed at any holding voltage in 0 mM glucose. It was therefore proposed that stimulation of type I cells in low glucose was a consequence of inhibition of K^+ channel activity and subsequent depolarisation (López-Barneo 2003; Pardal and López-Barneo 2002). Later data from the same group additionally showed, in single type I cells, a significant change in membrane potential stimulated by 0 mM glucose, averaging +9.6 mV, that was not dependent on a reduction in the outward K^+ conductance (García-Fernández et al. 2007). Rather, it was shown that 0 mM glucose significantly increased a background inward Na^+ current at −40 mV and it was therefore hypothesised that 0 mM increased TRP channel activity leading to type I cell depolarisation. Consistent with this was the finding that patch clamp recordings in co-cultures of type I cell clusters with dissociated petrosal neurons showed an increased action potential discharge frequency in a physiological hypoglycaemic (3.3 mM glucose) solution, at a PO_2 of 90 mmHg which was greater than that observed at a higher PO_2 (Zhang et al. 2007). Therefore, the authors advocated a direct sensitivity of type I cells to low physiological glucose in the co-culture preparation and proposed a similar response to low glucose of the whole carotid body in vivo. One confounding factor in attempting to gain consensus between groups appears to lie with the level of concomitant PO_2 during the hypoglycaemic challenge, with a requirement for lower, non-hyperoxic PO_2 seemingly essential for a physiological glucose response. As we had not previously determined glucose sensitivity at varying, acute levels of PO_2, this became the objective of the present study.

17.2 Methods

Tissue was isolated from adult male Wistar rats (100–200 g). Anaesthesia was induced with 4% isoflurane in O_2 and maintained at 1.5–2.0% isoflurane in O_2 via a facemask The carotid bifurcation, the superior cervical ganglion, glossopharyngeal nerve, CSN, and CB were excised together and the tissue immediately placed in a ice-cold bicarbonate buffered extracellular Krebs solution containing, in mM: 125 NaCl, 3 KCl, 1.25 NaH_2PO_4, 5 Na_2SO_4, 1.3 $MgSO_4$, 24 $NaHCO_3$, 2.4 $CaCl_2$, 11 D-glucose, equilibrated with 95% O_2 and 5% CO_2. Animals were immediately killed by exsanguination. The tissue was then continuously superfused in a small volume recording chamber with a bicarbonate buffered extracellular Krebs solution containing, in mM: 125 NaCl, 3 KCl, 1.25 NaH_2PO_4, 5 Na_2SO_4, 1.3 $MgSO_4$, 24 $NaHCO_3$, 2.4 $CaCl_2$, 11 D-glucose, equilibrated with 95% O_2 and 5% CO_2. The temperature was maintained at 37 C. The CSN was identified by its connection into the glossopharyngeal nerve and the whole tissue was partially digested by incubation in a bicarbonate buffered, equilibrated 95% O_2 and 5% CO_2 enzyme solution (0.075 mg/ml collagenase type II, 0.0025 mg/ml dispase type I;

Sigma) at a temperature of 37 C, for 20–30 min. Extracellular recordings of chemoafferent single- or few-fibre activity were made from the cut end of the CSN using extracellular glass suction electrodes. The superfusate PO_2 was measured using an O_2 electrode (OXELP, World Precision Instruments). PO_2 and chemoafferent derived voltage were both recorded using a CED micro1401 (Cambridge Electronic Design) and visualised on a PC with Spike2 software (Cambridge Electronic Design). Offline analysis using Spike2 (Cambridge Electronic Design) allowed for discrimination of electrical activity originating from a single chemoafferent fibre dependent on wavemark voltage frequency, shape and amplitude. Wavemarks from each group were counted and binned into 10 s time intervals. For graded hypoxic responses, the PO_2 in the superfusate was gradually reduced and the PCO_2 was maintained at 40 mmHg. Following hypoxia, superfusate PO_2 was rapidly returned to hyperoxia. Following the initial hypoxic response and a recovery period of >5 min the same CB was superfused with a 0 mM glucose solution. A second hypoxic response was then applied after 15 min of exposure to 0 mM glucose. Following this, the CB was subsequently superfused with an 11 mM glucose solution and then a third hypoxic response was recorded to detect any potential desensitisation of the hypoxic response due to time or repeated exposures to hypoxia. Single chemoafferent fibre discharge frequency (Hz) was plotted against PO_2 and the data was fitted to an exponential decay curve with offset:

$$y = a + be^{-cx}$$

where y is the single-fibre discharge frequency in Hz, x is the PO_2 in mmHg, a is the discharge frequency as the PO_2 tends to infinity (offset), b is the discharge frequency when the PO_2 is 0 mmHg (minus the offset) and c is the rate constant. All curves were fitted using DeltaGraph (version 5), Red Rock Software.

17.3 Results

The whole, freshly isolated *in vitro* CB, in 11 mM glucose, was superfused at a PO_2 of 250 mmHg and hypoxic responses obtained by reducing PO_2 until discharge increased to approximately 10 Hz after which it was rapidly reversed to avoid failure. Three characteristic hypoxic response curves from one animal, in 11 mM control, 0 mM glucose and 11 mM post-control are shown in Fig. 17.1a and it can be observed that the response recorded in 0 mM glucose was significantly and reversibly left shifted. For all animals (n=6), the effect of 0 mM glucose was determined by subtracting the discharge frequency between control and 0 mM glucose over a range of PO_2 (Fig. 17.1b). This showed a significant reduction in discharge in 0 mM glucose that increased with decreasing PO_2. These observations support the proposal that removal of the glucose substrate during hypoxia significantly decreased the ability of the organ to maintain a normal hypoxic response and thus show that lowering PO_2 did not augment glucose sensitivity. Stating the finding in a different way, grouped paired data at a discharge frequency of 5 Hz showed that the superfusate PO_2 in the 0 mM glucose solution was, for each recording, significantly less than that recorded in the control, 11 mM glucose solution; 133.0±7.3 mmHg vs 141.1±8.8 respectively ($P<0.05$; paired *t*-test). Thus, the PO_2 needs to be lower in glucose-free solutions to obtain the same discharge frequency as seen in controls. Taken together these data further suggest that the sensitivity to hypoxia was significantly reduced in the 0 mM glucose solution.

17.4 Discussion

Hypoxic responses in all freshly isolated CBs were observed to be significantly left shifted when superfused with 0 mM glucose. Therefore, in order to stimulate a given discharge frequency during the hypoxic response, the PO_2 was always lower in the 0 mM glucose solution compared with the

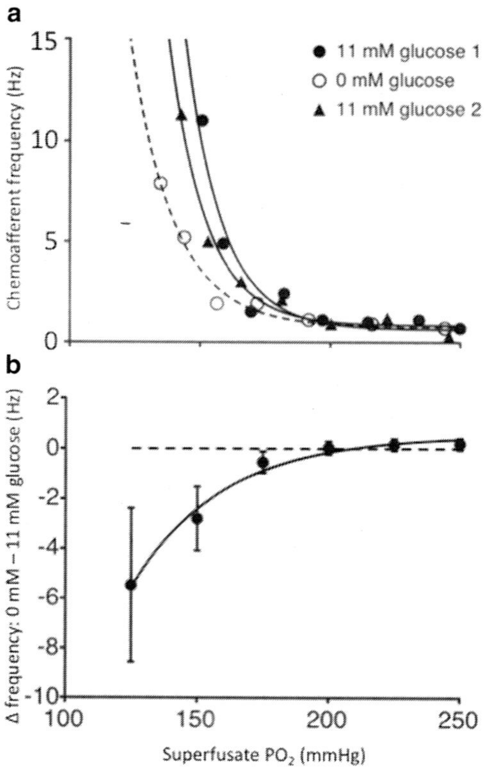

Fig. 17.1 The chemoafferent sensitivity to 0 mM glucose is not augmented with decreasing PO_2. (**a**) Discharge frequency was recorded from a single chemoafferent fibre of a freshly isolated carotid body. Single exponential functions (with offset) were fitted to frequency responses to falls in PO_2 at 11 mM and 0 mM glucose. The hypoxic response curve in 0 mM glucose (*open circles*) was left shifted compared to the pre (*closed circles*) and post-control (*closed triangles*) response curves in 11 mM glucose. (**b**) The carotid body afferent chemodischarge frequency at 6 fixed levels of PO_2 in 11 mM glucose was determined and subtracted from the frequency at these PO_2 levels in 0 mM glucose to show the significant negative impact of glucose removal upon discharge. The *dashed line* indicates no difference from control. Data shown is mean ± SEM (n = 6)

11 mM glucose solution. Due to the diffusion limitations inherent in using a superfusion technique, different individual type I cell clusters in the CB responded over different ranges of the superfusate PO_2. Therefore, it would be unreliable to compare discharge frequency at fixed PO_2 values across different experiments, although it would be acceptable for paired responses from the same preparation. An acute hypoxic stimulus applied to the whole *in vitro* carotid body was shown to significantly increase the rate of glucose uptake and the rate of oxygen consumption (Obeso et al. 1993; Starlinger and Lubbers 1976), suggesting a significant increase in metabolic demand. We suggest therefore that, in a combination of hypoxia and 0 mM glucose, we see a direct consequence of the CB being unable to maintain sufficient energy production during times of excessive metabolic demand. This is also in agreement with earlier observations indicating that the CB CO_2 sensitivity was also reduced by 0 mM glucose (Bin-Jaliah et al. 2005).

However, our data are not compatible with reports from other CB preparations (Pardal and López-Barneo 2002; Zhang et al. 2007) and, at present, there is no clear reason for this discrepancy. However, should the CB be able to detect low glucose, this sensing mechanism appears to involve, like hypoxia, Type I cell depolarisation, with a decrease in K^+ and an increase in Na^+ conductance both being implicated. Thus, the carotid body appears to differ here from orexin-sensitive CNS neurons, where a specific inhibition of background "leak" K^+ channels has been described in response to low glucose (Burdakov et al. 2006). The reason for the various differences between species, between carotid body preparations and between carotid body tissue and other glucose-sensing systems, is an important one to determine. Zhang et al. (2007) suggested that in the experiments of Bin-Jaliah et al. (2004) the relatively high PO_2 deliberately used by these workers to exclude hypoxia-induced excitation, may have prevented the full glucose response from being observed, given the positive interaction previously described between hypoxia and glucopenia (Pardal and López-Barneo 2002; Fitzgerald et al. 2009). Whilst our present data appear to rule out that potential mechanism, an interaction between some element of glucose sensing and oxygen appears to occur in the whole animal and further studies are anticipated to resolve this issue.

Acknowledgements APSH was supported by an A.E. Hills Scholarship and an Alumni Scholarship from the University of Birmingham

References

Alvarez-Buylla R, Alvarez-Buylla E (1988) Carotid sinus receptors participate in glucose homeostasis. Resp Physiol 72:347–360

Alvarez-Buylla R, Alvarez-Buylla E, Mendoza H, Montero S, Alvarez-Buylla A (1997) Pituitary and adrenals are required for hyperglycaemic reflex initiated by stimulation of CBR with cyanide. Am J Physiol 272:R392–R399

Bin-Jaliah I, Maskell P, Kumar P (2004) Indirect sensing of insulin-induced hypoglycaemia by the carotid body in the rat. J Physiol 556:255–266

Bin-Jaliah I, Maskell P, Kumar P (2005) Carbon dioxide sensitivity during hypoglycaemia-induced, elevated metabolism in the anaesthetised rat. J Physiol 563:883–893

Burdakov D, Jensen LT, Alexopoulos H, Williams RH, Fearon IM, O'Kelly I, Gerasimenko O, Fugger L, Verkhratsky A (2006) Tandem-pore K+ channels mediate inhibition of orexin neurons by glucose. Neuron 50:711–722

Conde S, Obeso A, Gonzalez C (2007) Low glucose effects on rat carotid body chemoreceptor cells' secretory responses and action potential frequency in the carotid sinus nerve. J Physiol 585:721–730

Fitzgerald R, Shirahata M, Chang I, Kostuk E (2009) The impact of hypoxia and low glucose on the release of acetylcholine and ATP from the incubated cat carotid body. Brain Res 1270:39–44

García-Fernández M, Ortega-Sáenz P, Castellano A, López-Barneo J (2007) Mechanisms of low-glucose sensitivity in carotid body glomus cells. Diabetes 56:2893–2900

Koyama Y, Coker R, Stone E, Lacy D, Jabbour K, Williams P, Wasserman D (2000) Evidence that carotid bodies play an important role in glucoregulation in vivo. Diabetes 49:1434–1442

López-Barneo J (2003) Oxygen and glucose sensing by carotid body glomus cells. Curr Opin Neurobiol 13:493–499

Obeso A, Gonzalez C, Rigual R, Dinger B, Fidone S (1993) Effect of low O_2 on glucose uptake in rabbit carotid body. J Appl Physiol 74:2387–2393

Pardal R, López-Barneo J (2002) Low glucose-sensing cells in the carotid body. Nat Neurosci 5:197–198

Starlinger H, Lubbers D (1976) Oxygen-consumption of isolated carotid body tissue. Eur J Physiol 366:61–66

Zhang M, Buttigieg J, Nurse C (2007) Neurotransmitter mechanisms mediating low-glucose signalling in co-cultures and fresh tissue slices of rat carotid body. J Physiol 578:735–750

Chapter 18
Do the Carotid Bodies Modulate Hypoglycemic Counterregulation and Baroreflex Control of Blood Pressure In Humans?

Erica A. Wehrwein, Timothy B. Curry, Ananda Basu, Robert A. Rizza, Rita Basu, and Michael J. Joyner

Abstract Peripheral chemoreceptors in the carotid body regulate respiration, sympathetic outflow, and blood pressure in response to hypoxia. The carotid bodies play a role in the counterregulatory response to hypoglycemia in animal models and may interact with the arterial baroreflex. We hypothesized that desensitization of the carotid bodies by hyperoxia in humans would blunt hypoglycemic counterregulation and baroreflex control of blood pressure during hypoglycemia. Seven healthy adults (age 26.7 ± 0.39, BMI 25 ± 0.32, M/4, F/3) each underwent two 180 min hyperinsulinemic (2 mU/kg FFM/min), hypoglycemic (3.33 mmol/L) clamps 1 week apart, randomized to either normoxia (arterial P_{O2} (P_aO_2) 111 ± 6.3 mmHg) or hyperoxia (P_aO_2 345 ± 80.6 mmHg) ($p<0.05$). Plasma glucose concentrations were similar during normoxia and hyperoxia at baseline and during the clamp. The glucose infusion rate was 44.2 ± 3.5% higher ($p<0.01$) during hyperoxia than normoxia during the clamp. Area under the curve values (expressed as % normoxia response) for counterregulatory hormones during hypoglycemia were significantly suppressed by hyperoxia. In addition, mean blood pressure during hypoglycemia was significantly lower with hyperoxia than with normoxia (delta reduction from baseline: −5.4 ± 3.4 mmHg normoxia vs. −13.8 ± 1.9 mmHg hyperoxia, $p<0.05$). The typical baroreflex-mediated rise in heart rate and sympathetic activity with lower blood pressure did not occur when the CB were silenced. These data support the idea that the carotid bodies play a role in the counterregulatory response to hypoglycemia and in baroreflex control of blood pressure in humans.

Keywords Hypoglycemia • Counterregulation • Chemoreceptor • Glucose • Glomus cell

---

E.A. Wehrwein • T.B. Curry • M.J. Joyner (✉)
Department of Anesthesiology, Mayo Clinic, 26 Mayo Park Dr SE, Rochester, MN 55904, USA
e-mail: wehrwein.erica@mayo.edu; Joyner.michael@mayo.edu

A. Basu • R.A. Rizza • R. Basu
Endocrine Research Unit, Division of Endocrinology, Diabetes, Metabolism and Nutrition, Mayo Clinic College of Medicine, Rochester, MN, USA

C.A. Nurse et al. (eds.), *Arterial Chemoreception: From Molecules to Systems*, Advances in Experimental Medicine and Biology 758, DOI 10.1007/978-94-007-4584-1_18,

18.1 Introduction

Peripheral chemoreceptors in the carotid bodies regulate respiration, sympathetic outflow, and blood pressure in response to hypoxia. In addition, cellular and animal studies suggest that the carotid bodies sense and respond to low glucose (Pardal and Lopez-Barneo 2002; Koyama et al. 2000). Given the developing understanding of the carotid bodies as polymodal sensors and that they have a strong influence on sympathetic nervous system, we tested the hypothesis that the carotid bodies play an integrative role in mediating both the counterregulatory hormone response to hypoglycemia as well as the baroreflex control of blood pressure during hypoglycemia. We have recently published a comprehensive report on the potential influence of the carotid chemoreceptors on hypoglycemic counterregulation (Wehrwein et al. 2010). In the current paper we highlight our earlier findings and also report novel data on potential chemoreflex/baroreflex interactions and blood pressure regulation that were generated during our original study (Wehrwein et al. 2010).

We sought to test the role of the carotid bodies in the systemic counterregulatory response to hypoglycemia and blood pressure control in humans. One approach to study this topic in humans is to use hyperoxia to acutely desensitize the carotid bodies. In humans hyperoxia can depress minute ventilation (Downes and Lambertsen 1966), suppresses afferent nerve traffic from the carotid bodies in animals (Fitzgerald and Lahiri 1986), and inhalation of 100% oxygen is thought to substantially reduce peripheral chemoreceptor activity (Lahiri and DeLaney 1975). On separate days we performed paired hyperinsulinemic hypoglycemic clamps in healthy humans exposed to normoxia or hyperoxia. We hypothesized that the glucose infusion rate would be augmented and neuro-hormonal counterregulation blunted during hypoglycemia when the carotid bodies were desensitized by hyperoxia. In addition, we tested the role of the carotid bodies in blood pressure regulation in hypoglycemia under the same experimental conditions and hypothesized that desensitizing the carotid bodies would attenuate baroreflex control of blood pressure.

18.2 Methods

18.2.1 Subjects and Monitoring

Subjects were healthy non-obese (BMI < 28 kg/m^2) persons between 21 and 35 years old.

Prior to starting the glucose clamp, a 20-gauge catheter for blood sampling and blood pressure monitoring was placed in a brachial artery under ultrasound guidance after local anesthesia. Two intravenous catheters were placed in the arm opposite the brachial arterial catheter for infusions. For patient safety, heart rate was monitored with a 5-lead electrocardiogram, respirations via a pneumobelt, and arterial oxygen saturation by a pulse oximeter.

18.2.2 Hypoglycemic Clamps

Intravenous insulin (Novolin®, Novo Nordisk Inc., Princeton NJ) was infused at a constant rate of 2.0 mU/kg FFM/min from protocol time (T) 0 to T180 min, and exogenous glucose (50% Dextrose solution (Hospira, Inc, Lake Forest, IL)) was infused in amounts sufficient to maintain glucose concentrations at hypoglycemic levels (~3.3 mmol/L (60 mg/dL)) (Lecavalier et al. 1989). From T0 until the end of the study, subjects breathed either normoxic or hyperoxic gas via a face mask as described below. Plasma glucose was measured every 5–10 min at the bedside using a glucose oxidase method (Analox Instruments USA Inc., Lunenberg, Massachusetts).

18.2.3 Hyperoxia and Normoxia

During hyperoxia trials subjects breathed 100% oxygen through a face mask connected to a non-rebreathing valve and large meteorological balloon which served as a volume reservoir. On the normoxia day, the identical set up was used except the bag was filled with air (21% oxygen). Normoxia and hyperoxia trials occurred on different days separated by 1 week in random order. Arterial blood gases were measured at baseline and every hour during the clamp.

18.2.4 Analytical Methods

Arterial blood was drawn for glucose and hormone measurements (insulin, c-peptide, glucagon, growth hormone, cortisol, epinephrine, dopamine, and norepinephrine) at T −120, −30, −20, −10, 0 and during the glucose clamp at 60, 120, 150, 160, 170, and 180 min.

18.2.5 Data Analysis and Statistics

Each subject was assessed under both experimental conditions (normoxia and hyperoxia). The time-by-group interaction term was included in the model and supplemental analyses were performed to compare groups at each time period using the paired t-test. For hormones, an area under the curve above baseline was calculated for each hormone in both conditions over time and expressed as percentage of normoxia response. Where relevant, baseline values were calculated as an average of T−30 ,−20,−10 and 0 and clamp values as an average of T150,160,170,180. p-values of <0.05 were considered statistically significant. Sigma Stat 2.03 software was used for analysis.

18.3 Results

18.3.1 Blood Gases

The average P_{aO2} during normoxia was 111 ± 6.3 mmHg and significantly elevated during hyperoxia to 345 ± 80.6 mmHg.

18.3.2 Plasma Glucose, Insulin, and C-Peptide

Plasma glucose concentrations were similar during normoxia and hyperoxia at baseline (5.52 ± 0.15 vs. 5.55 ± 0.13 umol/mL) and during the clamp (3.4 ± 0.05 vs. 3.3 ± 0.05 umol/mL). Baseline and clamp insulin concentrations were no different between normoxia and hyperoxia study days (baseline: 36.4 ± 6.7 vs 32.3 ± 4.9 pmol/L and clamp: 944.2 ± 86.8 vs 916.1 ± 75.9 pmol/L).

18.3.3 Glucose Infusion Rate

Despite similar plasma glucose concentrations during normoxia and hyperoxia, the glucose infusion rate required to maintain hypoglycemia was significantly higher during hyperoxia compared to

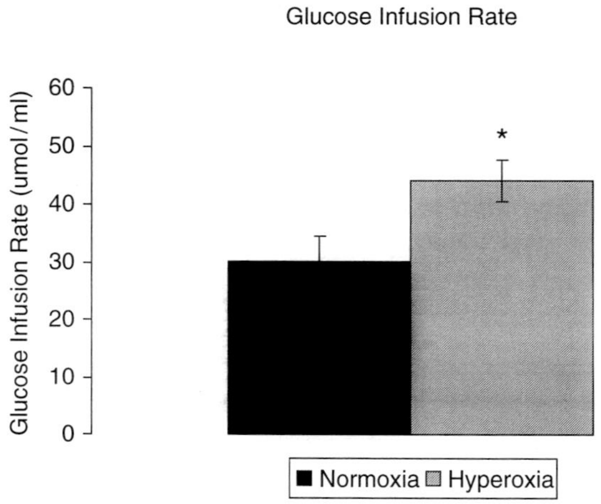

Fig. 18.1 Glucose infusion rate is significantly higher under hyperoxic conditions. Glucose infusion rate (umol/mL) under normoxia and hyperoxia at steady state hypoglycemia. Normoxia: *black bar*, hyperoxia: *gray bar*, Significance denoted as *$p<0.05$. N=7

normoxia (42.7±0.65 vs 28.2±0.15 umol/kg FFM/min; $p<0.01$) by 44.2±3.5% (Fig. 18.1). The higher glucose infusion rate on the hyperoxia study day was observed in all subjects.

18.3.4 Counterregulatory Hormones

Table 18.1 summarizes the data for norepinephrine, epinephrine, cortisol, glucagon, and growth hormone concentrations in response to hypoglycemia. All catecholamines and hormones were significantly lower ($p<0.05$) during hyperoxia vs. normoxia.

18.3.5 Blood Pressure

Mean arterial pressure during steady state hypoglycemia fell from baseline in both normoxia and hyperoxia groups and was significantly lower with hyperoxia than with normoxia (Fig. 18.2; delta from baseline: −5.4±3.4 mmHg normoxia vs. −13.8±1.9 mmHg hyperoxia, $p<0.05$). Systolic and diastolic blood pressures were similarly lower when the carotid bodies were inactivated with hyperoxia compared to normoxia but did not reach statistical significance with these preliminary findings (Data not shown; systolic delta from baseline: 12.0±8.7 mmHg normoxia vs. −3.8±5.8 mmHg, $p=0.1$; diastolic delta from baseline: −8.6±2.0 mmHg vs. −15.3±1.5 mmHg, $p=0.3$). Despite the reduction in all measures of arterial pressure during hyperoxia, the typical baroreflex-mediated rise in heart rate and sympathetic activity that normally occurs with lowered blood pressure did not occur when the CB were silenced. This is evidenced by a similar heart rate under both conditions at steady state clamp conditions (Fig. 18.2b; delta from baseline under normoxic conditions and hyperoxic condition 12.2±3.2 beats per minute vs 7.9±3.3 beats per minute, $p>0.05$) and by a paradoxical reduction in plasma catecholamines under conditions that should induce a rise in sympathetic activity as described above.

Table 18.1 Plasma levels of key counterregulatory hormones in hypoglycemia

Catecholamines and counterregulatory hormones	Percent reduction in hyperoxia from normoxia plasma levels (%)
Norepinephrine	–50.7 ± 5.2%*
Epinephrine	–62.6 ± 3.3%*
Glucagon	–48.6 ± 2.1%*
Growth hormone	–53.1 ± 2.7%*
Cortisol	–63.2 ± 2.1%*

Data is shown as percent reduction in catecholamine and hormone levels in hyperoxic hypoglycemia vs normoxic conditions

*p<0.05. N=7

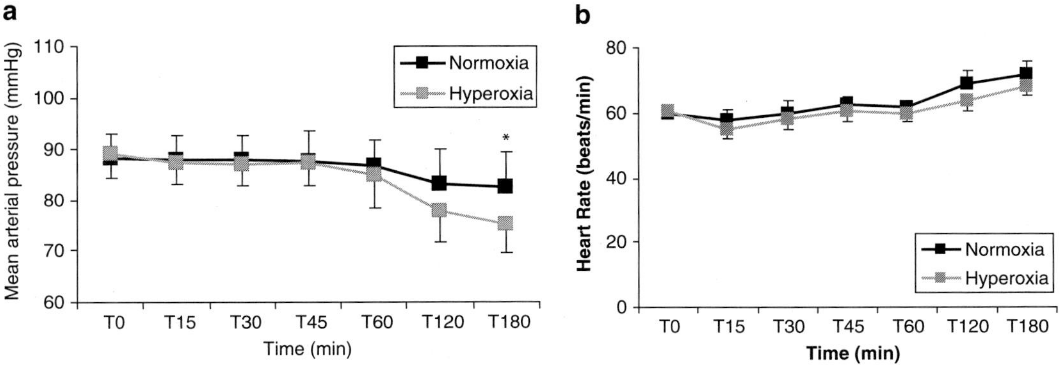

Fig. 18.2 Blood pressure is reduced during hypoglycemia under hyperoxic conditions without a corresponding baroreflex increase in heart rate. (**a**) Mean arterial pressure. Normoxia: *black squares*; hyperoxia: *gray squares*. (**b**) Heart rate is not different between conditions. Significance denoted as *p<0.05 for differences between normoxia and hyperoxia. N=5

18.4 Discussion

The carotid bodies were "inactivated" in humans by exposure to hyperoxia in conjunction with a hypoglycemic clamp. Our study provides evidence that the carotid bodies play an important role in the regulation of integrative physiology including blood glucose homeostasis and baroreflex control of blood pressure in humans: (1) the exogenous glucose infusion rate during the hyperinsulinemic hypoglycemic clamps was ~50% higher during desensitization of the carotid bodies with hyperoxia and hyperoxia during hypoglycemia resulted in a marked blunting of the counterregulatory hormone responses. (2) Sympathetic baroreflex adjustments to reduced arterial pressure during hypoglycemia are blunted during carotid body desensitization. Our data are also consistent with animal data showing the carotid bodies play an integral role in blood glucose homeostasis *in vivo* (Koyama et al. 2000, 2001; Alvarez-Buylla et al. 1997) and reveal for the first time that the carotid bodies are vital in blood pressure maintenance during hypoglycemia in humans.

The present report highlights our previously published findings (Wehrwein et al. 2010) showing that exposure to hyperoxia during hypoglycemia blunted the release of numerous counterregulatory hormones, including the primary counterregulatory hormones to acute hypoglycemia glucagon and epinephrine (Gerich et al. 1979). Consequently, hyperoxia resulted in an increase in glucose infusion rate during the clamp. These findings are consistent with studies in carotid body resected dogs (Koyama et al. 2000, 2001). In addition, we present novel findings in this report demonstrating for the first time that the carotid bodies are integral to the blood pressure responses to hypoglycemia. With the carotid bodies desensitized with hyperoxia, there is a fall in arterial pressure below normoxic conditions and this is not countered by the typical baroreflex response of elevation of heart rate and sympathetic outflow. This supports the notion that the peripheral chemoreceptor afferents integrate in a meaningful way with the arterial baroreflex or that there are barosensitive characteristics in the carotid body organ per se.

18.4.1 Mechanisms

Although we did not directly test the mechanisms by which hypoglycemia interacts with the carotid body glomus cells in humans, a discussion of potential mechanisms is appropriate. Isolated Type I glomus cells from the carotid bodies release neurotransmitter in response to both decreased extracellular glucose and hypoxia (Pardal and Lopez-Barneo 2002). In addition, hypoglycemia has been shown to activate a standing sodium current that depolarizes glomus cells in a dose dependent manner (Garcia-Fernandez et al. 2007). These studies provide a powerful rationale for our overall hypothesis, our specific study design, and the interpretation of our data. In this context, the proposed mechanism for our findings is that hyperoxia desensitized the carotid bodies limiting the ability of the carotid chemoreceptor cells to release neurotransmitter in response to low glucose *in vivo*. Therefore, carotid sinus nerve afferent activation under hyperoxic conditions was reduced with an associated attenuation of activity at the primary central synapse of these sensory fibers in the nucleus tractus solitarii (NTS) (de Campos Cruz et al. 2010). This sequence of events then leads to blunted counterregulatory hormonal responses.

The involvement of the NTS as the first synapse of the carotid body afferent fibers is important in that, in addition to mediating counterregulatory responses, it is a key site for integration of the arterial baroreflex. In addition, there is evidence that the peripheral chemoreflex and arterial baroreflex are highly integrated. Taken together, it is of interest to understand the role that carotid body afferents play in mediating the central integration of the baroreflex. In this study we determined that the baroreflex control of blood pressure during hypoglycemia was attenuated as evidenced by a significant fall in blood pressure under hyperoxia conditions. This drop is arterial pressure would typically be countered by a rise in heart rate and sympathetic activity when the baroreflex pathways are intact; however, in this study we show that there is no increase in heart rate and there is a paradoxical decrease in plasma catecholamines suggesting that the sympathetic outflow is inhibited by the desensitization of the carotid bodies.

18.5 Conclusion

We have provided evidence consistent with the idea that the carotid bodies play a role in the counterregulatory responses to hypoglycemia and in baroreflex control of blood pressure during hypoglycemia in humans. Additionally, it is tempting to speculate that altered carotid body function might have implications for clinical conditions associated with altered regulation of blood glucose and blood pressure including sleep apnea, Type II diabetes and hypertension.

References

Alvarez-Buylla R et al (1997) Pituitary and adrenals are required for hyperglycemic reflex initiated by stimulation of CBR with cyanide. Am J Physiol 272(1 Pt 2):R392–R399

de Campos Cruz J et al (2010) Fos expression in the NTS in response to peripheral chemoreflex activation in awake rats. Auton Neurosci 152(1–2):27–34

Downes JJ, Lambertsen CJ (1966) Dynamic characteristics of ventilatory depression in man on abrupt administration of O. J Appl Physiol 21(2):447–453

Fitzgerald R, Lahiri S (1986) Reflex responses to chemoreceptor stimulation. In: Fishman A (ed) Handbook of physiology-the respiratory system, vol 2. American Physiological Society, Bethesda, pp 313–362, Section 3, Chapter 10

Garcia-Fernandez M et al (2007) Mechanisms of low-glucose sensitivity in carotid body glomus cells. Diabetes 56(12):2893–2900

Gerich J et al (1979) Hormonal mechanisms of recovery from insulin-induced hypoglycemia in man. Am J Physiol 236(4):E380–E385

Koyama Y et al (2000) Evidence that carotid bodies play an important role in glucoregulation in vivo. Diabetes 49(9):1434–1442

Koyama Y et al (2001) Role of carotid bodies in control of the neuroendocrine response to exercise. Am J Physiol Endocrinol Metab 281(4):E742–E748

Lahiri S, DeLaney RG (1975) Relationship between carotid chemoreceptor activity and ventilation in the cat. Respir Physiol 24(3):267–286

Lecavalier L et al (1989) Contributions of gluconeogenesis and glycogenolysis during glucose counterregulation in normal humans. Am J Physiol 256(6 Pt 1):E844–E851

Pardal R, Lopez-Barneo J (2002) Low glucose-sensing cells in the carotid body. Nat Neurosci 5(3):197–198

Wehrwein EA et al (2010) Hyperoxia blunts counterregulation during hypoglycaemia in humans: possible role for the carotid bodies? J Physiol 588(Pt 22):4593–4601

Chapter 19
Shifting from Hypoxia to Hyperoxia to Assess the Peripheral Chemosensory Drive of Ventilation

Patricio Zapata, Carolina Larraín, Edison-Pablo Reyes, and Ricardo Fernández

Abstract The study of the initial effects of a sudden and brief replacement of air by pure oxygen has been proposed as a tool (Dejours' test) to determine the tonic influence that arterial (peripheral) chemoreceptors were exerting upon ventilation under previous normoxic conditions. Therefore, the acute ventilatory response to transient hyperoxia should be used to assess the level of hypoxic chemosensory drive. In spontaneously ventilated pentobarbitone-anesthetized cats, we observed that the degree of ventilatory depression provoked by hyperoxia was correlated to the degree of previous hypoxia. Minimal tidal volumes (V_T) or transient apnea were reached between second to fourth cycles after switching from 5% to 100% O_2 breathing. Continuous recordings of chemosensory discharges from one carotid (sinus) nerve allowed correlation of the falls in frequency of chemosensory discharges to the degree of hyperoxia-induced ventilatory depression and provided an accurate measure of the prevailing chemosensory drive of ventilation exerted during hypoxic steady-state conditions.

Keywords Carotid body • Dejours' test • Hyperoxia • Hypoxia • Ventilatory chemosensory drive

19.1 Introduction

Dejours (1957) argued that mechanisms operating under steady-state conditions can only be demonstrated by the initial (primary) change evoked by a sudden perturbation, since later (secondary) changes compensate for the initial changes, leading to the establishment of a new steady-state. Therefore, he proposed that sudden inhalation of pure oxygen might reveal the existence of O_2 level operating as a stimulus during the previous steady-state condition. This sudden withdrawal procedure was rapidly validated by several experimental studies performed in humans and dogs, showing that an "oxygen (or chemoreflex) drive" for ventilation was operating under eucapnic normoxia (see Dejours 1962; Gautier 2003).

P. Zapata (✉) • C. Larraín • E.-P. Reyes
Facultad de Medicina, Clínica Alemana, Universidad del Desarrollo,
Ave Las Condes 12438, Lo Barnechea, Santiago, 771.0162, Chile
e-mail: pzapata@udd.cl; clarrain@udd.cl; preyes@udd.cl

R. Fernández
Facultad de Ciencias Biologicas y Facultad de Medicina,
Universidad Andres Bello, Santiago, Chile
e-mail: rfernandez@unab.cl

C.A. Nurse et al. (eds.), *Arterial Chemoreception: From Molecules to Systems*,
Advances in Experimental Medicine and Biology 758, DOI 10.1007/978-94-007-4584-1_19,

The transient ventilatory depression evoked by hyperoxia –as well as the hyperventilation elicited by hypoxic challenges– are not longer observed after bilateral section of the buffer nerves: carotid (Hering's sinus) and aortic (Cyon's depressor) nerves (see Dejours 1962). Therefore, O_2 sensing is restricted to peripheral arterial chemoreceptors (carotid and aortic bodies). Leitner et al. (1965) demonstrated that a brief inhalation of pure oxygen induces an almost immediate silencing of carotid nerve chemosensory discharges which –by withdrawing their reflex effects upon ventilation– results in rapid depression of ventilatory volume. Thus, the O_2-related reflex control of ventilation under steady-state conditions is known now as "peripheral chemosensory drive", in opposition to the CO_2-related "central chemosensory drive". Both mechanisms are superimposed upon centrally generated ventilatory output.

The O_2 (hyperoxic, Dejours') test described above has revealed an enhanced peripheral chemosensory drive under hypoxic conditions in humans (Holtby et al. 1988; Sajkov et al. 1997), as well as in cats, guinea-pigs and rats (Fernández et al. 2003). The present work was intended to assess the hyperoxic-induced depression of ventilation, when preceded by different levels of hypoxic stimulation.

19.2 Methods

Data were obtained from 21 experiments performed on male adult cats, anaesthetized with sodium pentobarbitone 40 mg/kg I.P., with supplementary doses given I.V. (12 mg) to maintain a light level of surgical anesthesia (stage III, plane 2). At the end of experiments, animals were euthanized by an overdose of pentobarbitone. Experimental protocols had been approved by the Commission of Bioethics and Biosafety, P. Catholic University of Chile, where the experiments had been performed.

Cats were placed in supine position over a thermoregulated heating pad, maintaining body temperature at 38.0 ± 0.1 C. The right saphenous vein was cannulated for drugs administration. Arterial pressure was recorded from the left femoral artery through a cannula PE-100 filled with heparin 50 UI/mL in saline solution, connected to a Statham pressure transducer.

Cats breathed spontaneously throughout the experiments. The trachea was intubated *per os* with a flexible tube, connected to a heated Fleisch pneumotachograph head (N° 00), and to a Grass volumetric differential pressure transducer for measuring tracheal ventilatory flow ($\delta V/\delta t$), which was converted into tidal volume (V_T) through a Grass 7P10 full-wave integrator. Because of the differences in resistance of the pneumotachograph inner tubings to the flows of O_2 and N_2, the instrument was calibrated separately for flows of different mixtures of O_2 and N_2. Values of V_T reported in this paper are corrected for each different gas mixture. Ventilatory signals were derived to a tachograph (Grass 7P4) to obtain instantaneous respiratory frequency (f_R). The inspired fraction of oxygen (F_IO_2), resulting from mixtures of variable flows of pure O_2 and N_2 delivered from tanks, was determined by a paramagnetic analyzer (Servomex). In some preparations, O_2 saturation of arterial blood (S_aO_2) was measured continuously through a multi-site sensor applied to the cat's tongue and connected to a pulse oximeter II (Criticare Systems, Inc.). A fine tubing (PE-10) was introduced deep into the tracheal cannula for continuous sampling of air and recording of the end tidal CO_2 pressure ($P_{ET}CO_2$) through an infrared gas analyzer (Infrared Industries, Inc). When required, the frequency of chemosensory discharges (f_x) was recorded from one carotid nerve.

Preparations were subjected to 1-min periods of breathing hypoxic mixtures (17.5, 15, 12.5, 10, 7.5 and 5% O_2 in N_2), followed immediately by 1-min periods of breathing 100% O_2. Hyperoxic-induced ventilatory depressions were expressed by their lowest values (nadirs), generally as percentages of respective preceding basal values.

19.3 Results and Discussion

In spontaneously ventilating cats with buffer nerves intact, sudden exposure to 100% O_2 breathing provoked almost immediately transient ventilatory depression, as shown in Fig. 19.1a. Reductions in V_T and f_R were small when cats had been breathing room air, but became more pronounced when they had been breathing hypoxic mixtures. The larger the hypoxia-induced preceding hyperventilation, the larger the ventilatory depression provoked by the sudden shift to 100% O_2 breathing. Furthermore, cats responding more vigorously to hypoxia exhibited larger ventilatory depression in response to the immediately following hyperoxia. These observations demonstrate the enhanced peripheral chemosensory drive achieved during hypoxic breathing.

Figure 19.2 summarizes data collected from experiments in five cats submitted to hypoxic exposures of from 1 min to 10 min, revealing that prolonging hypoxic exposures to 2 min provoked only slighter larger increases in V_T than those provoked by 1 min exposures, and that responses to 2 min, 5 min or 10 min hypoxia did not differ significantly in magnitude of V_T increases. This was valid when exposing cats to either 15%, 10% or 5% O_2 breathing, but increases in V_T were larger for 5% O_2 level exposures than those to 10% O_2 level, and the latter ones larger than those to 15% O_2 level. The right part of this figure shows results of shifting to 100% O_2 breathing after each of these hypoxic levels and at each exposure duration. Here, we observe that the larger the increase in V_T achieved during hypoxia,

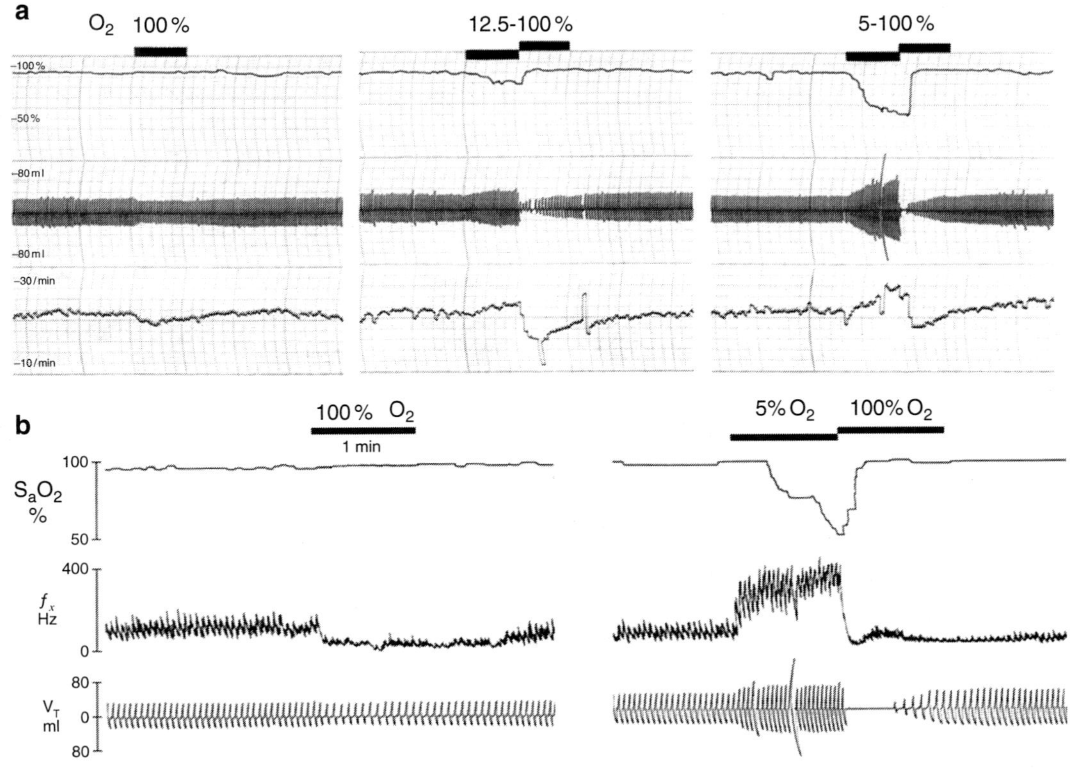

Fig. 19.1 Ventilatory effects of 1-min hyperoxia (*upward bars*) and 1-min hypoxia (*downward bars*) in cats with intact buffer nerves (**a**) and after unilateral carotid neurotomy (**b**). Recordings of arterial O_2 saturation of hemoglobin (S_aO_2), ventilatory tidal volume (V_T), instantaneous respiratory frequency (f_R) and instantaneous frequency of carotid chemosensory discharges (f_χ)

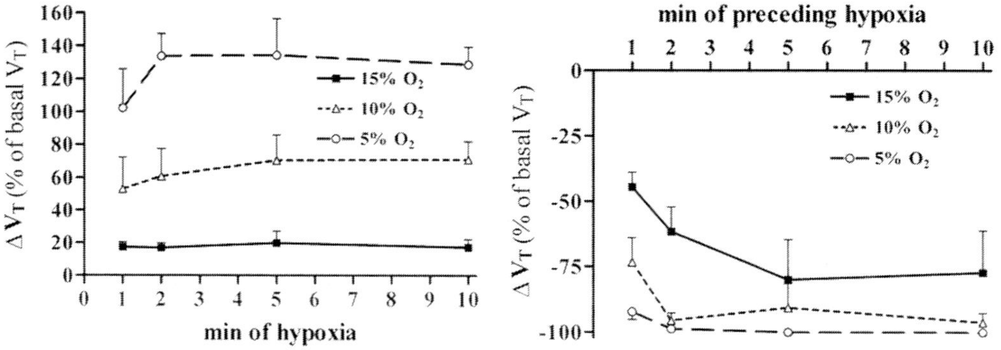

Fig. 19.2 Changes in ventilatory tidal volume (ΔV_T) occurring after 1, 2, 5 and 10 min of breathing either 15, or 10 or 5% O_2 along 10 min (*left panel*) and those observed shortly after subsequent breathing of 100% O_2 for 1 min (*right panel*). Ordinates, changes in V_T expressed as percentages of basal V_T (in normoxia). Abscissae, time of exposure to hypoxic mixtures. Means +/– SD. Data from experiments performed in five cats with intact carotid nerves

the larger the ensuing fall in V_T immediately after hyperoxia. After 2–10 min of 10% or 5% O_2 breathing, the initial response to hyperoxia was commonly a transient respiratory arrest. As comparison, we observed only moderate decreases in V_T and f_R when intense hypoxia (breathing 5% O_2) was followed by sudden readmission of air. These observations indicate that the O_2 test is capable of discriminating between mild, moderate and intense preceding hypoxias.

After unilateral carotid neurotomy, ventilatory responses to hyperoxia following hypoxia were of similar magnitudes, or only slightly less pronounced, than those previously recorded when both carotid nerves were still intact. Thus, the remaining carotid nerve completely or partially replaced the strength of the chemosensory drive formerly provided by the sectioned carotid nerve. This agrees with the enhanced transient ventilatory depression provoked by anesthetic block of a given carotid nerve when the contralateral nerve has been blocked or sectioned (Eugenin et al. 1989).

After sectioning both carotid nerves, hypoxia-induced hyperventilation was only observed upon 5% O_2 breathing, and mild transient ventilatory depression was subsequently evoked by 100% O_2 breathing. No ventilatory responses to either hypoxic stimulation or the following hyperoxic test were observed after sectioning both carotid nerves and both aortic nerves, *i.e.*, total peripheral chemodenervation, confirming their origin from the excitation and inhibition, respectively, of arterial chemoreceptor discharges.

Recordings of the instantaneous frequency of carotid chemosensory discharges (f_x) allowed us to correlate the magnitudes of the changes in chemo-afferent discharges with those in spontaneous ventilation. Figure 19.1b shows that interruption of normoxia by 100% O_2 breathing rapidly decreased f_x to very low levels, but only moderate decreases in V_T were observed. However, the large increases in f_x and V_T while breathing 5% O_2 were followed by intense fall in f_x and respiratory arrest (lasting *ca* 30 s) by sudden shift to breathing pure O_2. Ventilatory fluctuations in f_x were observed under normoxic basal conditions (as described by Black and Torrance 1971), becoming exaggerated during hypoxic-induced hyperventilation.

Simultaneous recordings of ventilation and chemosensory discharges in cats submitted to unilateral carotid neurotomy showed that exposures to 10% O_2 for 1, 2, 5 and 10 min induced higher levels of f_x and V_T, maintained along the maneuver. Shifting to 100% O_2 breathing after these hypoxic levels resulted in immediate nearly silencing of carotid discharges in all cases, and reduced V_T to nearly half control levels when applied after 1 min of hypoxia, but extremely small and slow respiratory cycles (near to respiratory arrest) when applied after 2, 5 and 10 min hypoxic challenges. Thus, in spite of hyperventilation-induced falls in $P_{ET}CO_2$ occurring in our experiments (not illustrated), a powerful chemosensory drive upon

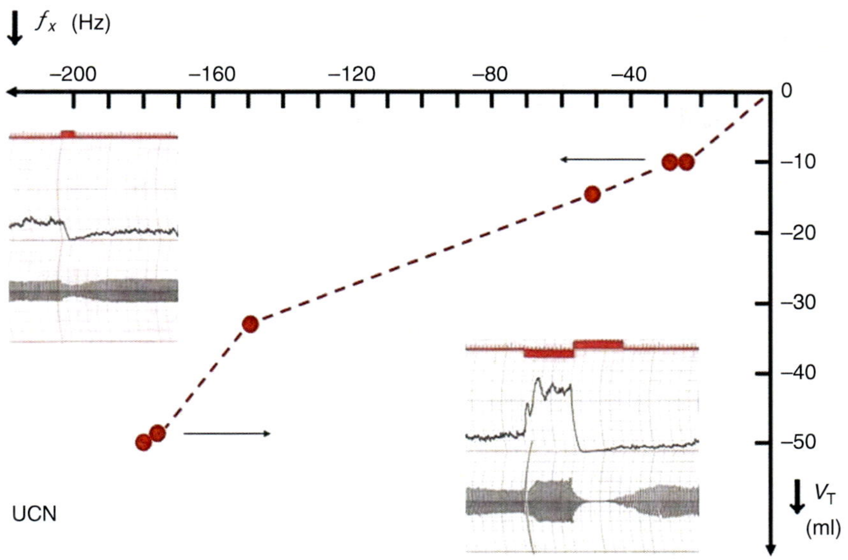

Fig. 19.3 Correlation between maximal reductions in ventilatory tidal volume (↓V_T) and in carotid nerve chemosensory frequency (↓f_χ), upon breathing 100% O_2 immediately after exposure to different levels of hypoxia (from 17.5% to 5% O_2) for 1 min in a cat after unilateral carotid neurotomy (UCN). All hyperoxic tests reduced f_χ down to 0 Hz; thus differences in (↓f_χ) were due to larger levels of f_χ during previous hypoxic exposures. Insets, recordings of f_χ and V_T at conditions indicated by *arrows*

ventilation was maintained all along hypoxic breathing, as revealed by the marked reduction of ventilation elicited by the sudden withdrawal of such tonic reflex influence provoked by O_2 tests.

Breathing 10% CO_2 for 1 min caused increases in V_T and f_R, which dissipated slowly during the following minute upon returning to normal air breathing. A sudden shift from breathing the hypercapnic mixture to 100% O_2 did not induce the immediate ventilatory depression observed when hypoxia was followed by hyperoxia. In fact, no significant differences in ventilatory changes were observed when 10% CO_2 was followed by either 100% O_2 or air breathing. Therefore, the residual ventilatory stimulation occurring immediately after CO_2 exposure should be due to excitation of central (medullary) chemoreceptors, and unaffected by the hyperoxic silencing of arterial chemosensory discharges.

Furthermore, rebreathing from a blind sac for 1 min caused progressive and pronounced increases in f_x, with slightly delayed increases in V_T and f_R. A sudden shift from this asphyxial condition to 100% O_2 breathing reduced immediately f_x to very low levels, but V_T and f_R returned slowly to normal levels, without the initial ventilatory depression response observed by the shift from pure hypoxia to hyperoxia. In fact, no significant differences were observed when rebreathing into the blind sac was followed by either pure O_2 or air inhalation. Therefore, the ventilatory effects of hypoxia plus hypercapnia are not immediately relieved by subsequent hyperoxic breathing, pointing out to the conclusion that central excitation is conspicuous in this condition and responsible for hyperventilation, in spite of the fact that arterial chemoreceptors are also excited.

Taken together, above observations indicate that hypoxia produces an enhanced chemosensory drive upon ventilation only when associated with normocapnia or the resultant hypocapnia provoked by hyperventilation. In other words, a transient pronounced fall in ventilation upon 100% O_2 breathing is a specific response revealing previous hypoxic conditions.

The study in each animal of the concomitant decreases in f_x and V_T provoked by the shift from different levels of hypoxia to the same level of hyperoxia revealed a strong correlation between falls in f_x and reductions in V_T to their respective nadirs (Fig. 19.3). It must be noted that all hyperoxic tests reduced f_χ down to 0 Hz. Thus, the different magnitudes of falls in f_χ were due to the larger levels of f_χ during previous hypoxic exposures.

In the present series of experiments, as well as in other series, each hyperoxic test interrupting basal normoxia usually resulted in complete silencing of chemosensory discharges for a few seconds, but ventilation was transiently depressed by only 20–30% of basal ventilation in normoxia. However, here we report that in cases in which f_χ was not completely silenced by hyperoxia applied upon a hypoxic background, ventilatory arrest could indeed occur. The comparison between these two conditions suggests that the magnitude of ventilatory depression is not commensurate to the level of chemosensory depression provoked by the hyperoxic challenge. Therefore, the previous level of increased peripheral chemosensory drive (replacing "central ventilatory drive") occurring during near steady-state hypoxia is the basic condition for determining the pronounced subsequent fall in ventilation provoked by shifting to hyperoxia.

References

Black AMS, Torrance RW (1971) Respiratory oscillations in chemoreceptor discharge in the control of breathing. Respir Physiol 13:221–237

Dejours P (1957) Intérêt méthodologique de l'étude d'un organisme vivant à la phase initiale de rupture d'un équilibre physiologique. Compt Rend Acad Sci Paris 245:1946–1948

Dejours P (1962) Chemoreflexes in breathing. Physiol Rev 42:335–358

Eugenin J, Larraín C, Zapata P (1989) Correlative contribution of carotid and aortic afferences to the ventilatory chemosensory drive in steady-state normoxia and to the ventilatory chemoreflexes induced by transient hypoxia. Arch Biol Med Exp 22:395–408

Fernández R, Arriagada I, Garrido AM, Larraín C, Zapata P (2003) Ventilatory chemosensory drive in cats, rats and guinea-pigs. Adv Exp Med Biol 536:489–495

Gautier H (2003) Honoring Pierre Dejours: his contribution to the study of the role of the arterial chemoreceptors in the regulation of breathing in humans. Adv Exp Med Biol 536:1–7

Holtby SG, Berezanski DJ, Anthonisen NR (1988) Effect of 100% O_2 on hypoxic eucapnic ventilation. J Appl Physiol 65:1157–1162

Leitner L-M, Pagès B, Puccinelli R, Dejours P (1965) Étude simultanée de la ventilation et des décharges des chémorécepteurs du glomus carotidien chez le chat. I. Au cours d'inhalations brèves d'oxygène pur. Arch Intl Pharmacodyn Thér 154:421–426

Sajkov D, Neill A, Saunders NA, McEvoy RD (1997) Comparison of the effects of sustained isocapnic hypoxia on ventilation in men and women. J Appl Physiol 83:599–607

Chapter 20
CO_2 Signaling in Chemosensory Neuroepithelial Cells of the Zebrafish Gill Filaments: Role of Intracellular Ca^{2+} and pH

Sara J. Abdallah, Steve F. Perry, and Michael G. Jonz

Abstract Adult zebrafish, *Danio rerio*, exhibit hyperventilatory responses to absolute environmental CO_2 levels as low as 1.0 mmHg. The ability of zebrafish to detect and respond to low ambient CO_2 appears to be mediated by chemosensory neuroepithelial cells (NECs) of the gill filaments. Recent electrophysiological characterization of this response revealed that the partial pressure-dependent depolarization of NECs in response to a hypercapnic stimulus is dependent on the rate of acidification associated with the hydration of CO_2 and the inhibition of background K^+ channels. In order to further elucidate the signaling pathway underlying CO_2 chemotransduction in NECs, we used microfluorimetric techniques to study intracellular changes in pH (pHi) and calcium ($[Ca^{2+}]_i$). Using the ratiometric indicators BCECF-AM and fura-2-AM, we found that a hypercapnic stimulus evoked a decrease in pHi and an increase in $[Ca^{2+}]_i$.

Keywords Neuroepithelial cells • Hypercapnia • pH • Calcium • Zebrafish

20.1 Introduction

The ability of an organism to sense O_2 and CO_2 plays an integral role in meeting cellular metabolic demand. It has long been recognized that in response to aquatic hypoxia and hypercapnia fish exhibit pronounced physiological responses including hyperventilation, bradycardia and changes in gill vascular resistance (Burleson et al. 1992). As a consequence of their environment fish achieve arterial (PCO_2) ($PaCO_2$) levels that are near ambient and approximately an order of magnitude lower (~2–3 mmHg) than the $PaCO_2$ achieved in mammals (~40 mmHg). Consequently, unlike mammals, fish have presumably evolved to sense very small deviations in environmental CO_2. In zebrafish, *Danio rerio*, exposure to environmental CO_2 levels as low as 0.13% ($PCO_2 = 1$ mmHg, Vulesevic et al. 2006) elicits an increase in relative breathing amplitude presumably resulting in an elevation of ventilation volume.

In order for ventilation to be increased the hypercapnic signal must be detected by specialized sensory cells (respiratory chemoreceptors) and subsequently transmitted to the effector organs.

S.J. Abdallah • S.F. Perry • M.G. Jonz (✉)
Department of Biology, University of Ottawa, 30 Marie Curie Pvt, Ottawa, ON K1N 6N5, Canada
e-mail: mjonz@uottawa.ca

C.A. Nurse et al. (eds.), *Arterial Chemoreception: From Molecules to Systems*, Advances in Experimental Medicine and Biology 758, DOI 10.1007/978-94-007-4584-1_20,

In mammals, the primary peripheral chemoreceptors that are responsive to hypercapnic stimuli are type I cells of the carotid body. Type I cells depolarize in response to hypercapnic acidosis via inhibition of background K^+ channels leading to voltage-gated Ca^{2+} influx and subsequent Ca^{2+} dependent neurotransmitter release from the cytoplasmic synaptic vesicles (Rigual et al. 1991; Sato 1994; Buckler and Vaughan-Jones 1994; see review by Putnam et al. 2004). In addition to Ca^{2+} influx, hypercapnic stimulation of the carotid body results in intracellular acidification within type I cells (Buckler et al. 1991a; see review by Putnam et al. 2004).

In zebrafish, neuroepithelial cells (NECs) of the gill filament were recently identified as bimodal O_2 and CO_2 sensors (Jonz et al. 2004; Qin et al. 2010). NECs are found on all four gill arches in zebrafish, receive multiple sources of innervation and are enriched with neurotransmitters (Jonz and Nurse 2003; Perry et al. 2009). As in the glomus cells of the carotid body, NECs are responsive to an increase in CO_2 with PCO_2-dependent depolarization via inhibition of background K^+ channels (Qin et al. 2010). The subsequent steps within the chemotransduction cascade of a NEC in response to hypercapnic acidosis are unknown. However, because NECs are biochemically and morphologically similar to the carotid body type I cells, the chemotransduction cascade is predicted to be similar. As such, it was the aim of this study to characterize changes in the concentration of intracellular Ca^{2+} ($[Ca^{2+}]_i$) and pH (pH_i) in response to a hypercapnic stimulus within NECs of the zebrafish gill filament.

20.2 Materials and Methods

20.2.1 Cell Isolation

Isolation procedures of NECs were carried out as previously described by Jonz et al. (2004). Briefly, adult zebrafish were euthanized and gill baskets were removed and rinsed in a phosphate-buffered solution (PBS) containing 2% penicillin-streptomycin (Sigma–Aldrich, Oakville, ON, Canada). Isolated gill filaments were enzymatically digested at 28°C with 0.25% trypsin/EDTA (Sigma-Aldrich) and subjected to mechanical dissociation with forceps. The trypsin reaction was stopped with the addition of 10% fetal calf serum (FCS). The cell suspension was centrifuged and the pellet re-suspended in Leibovitz's culture medium (L-15, Invitrogen, Burlington, ON, Canada) supplemented with 1% penicillin/streptomycin (Sigma-Aldrich) and 3% FCS. Cells were plated on glass bottom culture dishes (MatTek Corp., Ashland, MA, USA) coated with poly-L-lysine (Sigma-Aldrich) and Matri-Gel (BD Biosciences, Mississauga, ON, Canada).

20.2.2 Measurement of $[Ca^{2+}]_i$

Cells that adhered to the culture substrate were incubated in a solution of 5 μM Fura-2-AM (Invitrogen) in normal Ringer's solution (containing in (mM) 135 NaCl, 5 KCl, 2 Mg_2Cl, 2 Ca_2Cl, 10 glucose, 10 Hepes, pH 7.8) at 28°C for 30 min. Subsequently, the Fura-2-AM solution was removed and the cells were incubated in indicator-free Ringer's for an additional 30 min at 28°C. Finally, NECs were identified using 2 mg/ml Neutral Red, a vital marker used to identify 5-HT-containing cells.

Ratiometric Ca^{2+} imaging was performed using a Nikon Eclipse microscope equipped with a Lamda DG-5 high-speed wavelength changer (Sutter Instruments, Novato, CA, USA). Dual images (340 and 380 nm excitation and 510 emission) were collected and the ratiometric data were obtained using Northern Eclipse software (Empix Imaging Inc., Mississauga, ON, Canada).

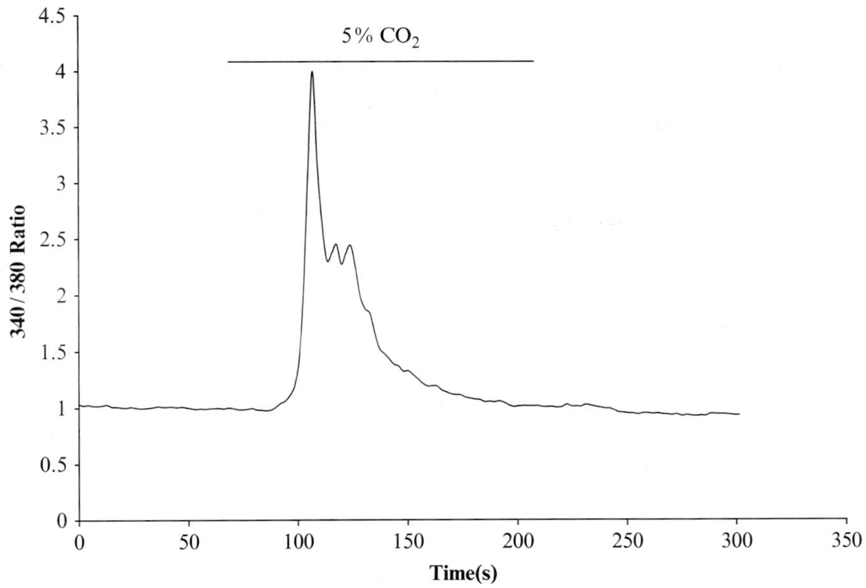

Fig. 20.1 Effect of hypercapnic acidosis (increase in CO_2 to 5%) on $[Ca^{2+}]_i$ of a single isolated neuroepithelial cell (NEC) of the zebrafish, *Danio rerio*, gill filaments. Relative changes in the 340/380 ratio from Fura 2-AM experiments indicate changes in $[Ca^{2+}]_i$. *Bar* represents the duration of the 5% CO_2 application

20.2.3 Measurement of pH_i

The ratiometric indicator 2′,7′-bis-(2-carboxyethyl)-5-(and-6)-carboxyfluorescein (BCECF, Invitrogen) was used to measure changes in pH_i. The same loading procedure was used as described above for Fura-2-AM. Excitation wavelengths of 495 and 440 nm were used, and fluorescence emission was measured at 535 nm.

20.2.4 General Procedures

Dishes containing NECs were fitted with a perfusion chamber and mounted on the microscope stage. Hypercapnia was achieved by bubbling a reservoir of recording solution with a mixture of 5% CO_2 balanced with air until PCO_2 reached equilibrium. The recording chamber was continuously perfused (4 ml min^{-1}) with bath solutions at room temperature (22–24°C). Neutral Red, Fura/BCECF positive cells were selected for imaging, and raw 340/380 or 495/440 ratios were recorded. Data were displayed as ratios to indicate relative changes in $[Ca^{2+}]_i$ or pH_i.

20.3 Results

20.3.1 Effects of Hypercapnic Acidosis on $[Ca^{2+}]_i$

NECs responded to hypercapnia (5% CO_2) with rapid elevation in $[Ca^{2+}]_i$ as represented by the increase in the 340/380 ratio (Fig. 20.1). The $[Ca^{2+}]_i$ response was transient, with recovery beginning before normal Ringer's was returned to the bath.

20.3.2 *Effects of Hypercapnic Acidosis on pH_i*

NECs of zebrafish gill filament acidified in response to the application of 5% CO_2 (Fig. 20.2). NECs exhibited a sharp decrease in pH_i (decease in 495/440 ratio) upon the application of 5% CO_2 that was sustained throughout the exposure. Once the NEC was exposed to normocapnic solution, pH_i returned to baseline levels.

20.4 Discussion

Hypercapnia resulted in intracellular acidification and an increase in $[Ca^{2+}]_i$ within NECs of the zebrafish gill filaments. Although these findings are consistent with the effects of hypercapnia on pH_i and $[Ca^{2+}]_i$ in the rat carotid body type I cells (Buckler et al. 1991a; Buckler and Vaughan-Jones 1993; Sato 1994; see review by Putnam et al. 2004), a few differences are apparent.

Isolated NECs responded to hypercapnic acidosis with a rapid initial rise in $[Ca^{2+}]_i$ followed by a secondary gradual decline back to baseline levels (Fig. 20.1.). Isolated type I cells of the carotid body, on the other hand, exhibit a sustained increase in $[Ca^{2+}]_i$ throughout the acidosis exposure (Buckler and Vaughan-Jones 1993). In fact, averaged responses of type I cells to hypercapnic acidosis showed a rapid initial rise in $[Ca^{2+}]_i$ followed by a secondary adaptation towards a sustained plateau (Buckler and Vaughan-Jones 1993). The distinct calcium profiles exhibited by these two CO_2 sensing structures – transient increase in NECs and a sustained increase in $[Ca^{2+}]_i$ within type 1 cells-may be indicative of a difference in the source of Ca^{2+}. Within type I cells, it has been clearly established that the elevation in $[Ca^{2+}]_i$ in response to hypercapnia is due to an influx of Ca^{2+} via voltage gated L-type Ca^{2+} channels (Buckler and Vaughan-Jones 1994; see review by Putnam et al. 2004). The source of Ca^{2+} contributing to the elevation in $[Ca^{2+}]_i$ in NECs has yet to be elucidated, although preliminary evidence suggests that it may be due, at least in part, to intracellular Ca^{2+} stores (data not shown). Regardless of the source of Ca^{2+}, it is clear that $[Ca^{2+}]_i$ plays a pivotal role in CO_2 sensing within the NECs of the zebrafish gill filament. Ultimately, as in the type I cells of the carotid body, the rise in the $[Ca^{2+}]_i$ within NECs may contribute to neurosecretion.

As CO_2 is sensed, the levels of H^+ and HCO_3 are also affected through the hydration reaction catalyzed by carbonic anhydrase (CA). As such, it was of importance to measure changes in pH_i. As in the type I cells of the carotid body, hypercapnic acidosis within NECs of the gill filament resulted in a sharp decrease in pH_i that was sustained throughout the exposure (Fig. 20.2). Inhibition of CA reduced the response to hypercapnia in type I cells of the rat carotid body (Iturriaga et al. 1991; Buckler et al. 1991b) and NECs of the zebrafish gill filaments (Qin et al. 2010) thereby implicating a role for CA in the decrease in pH_i within the cellular signaling pathway in response to hypercapnic acidosis. Peers and Green (1991) proposed that the decrease in pH_i associated with hypercapnic acidosis inhibits K^+ channels resulting in type I cell depolarization. This depolarization is proposed to lead to Ca^{2+} entry via voltage-gated L-type Ca^{2+} channels and subsequently Ca^{2+} dependent neurosecretion. More recently however, Summers et al. (2002) showed that molecular CO_2 is the primary cause for the augmentation of Ca^{2+} current, that its effects are not secondary to changes in pH and that it is dependent on a protein kinase A mechanism. Ultimately, as suggested by Putnam et al. (2004), the response of type I cells, and likely NECs to hypercapnic acidosis may be dependent on the cumulative effects of pH_i, pH_o and molecular CO_2 on the multiple ion channel targets.

Some pathways within the chemotransduction cascade of CO_2 signaling appear to have been conserved between type I cells of the carotid body and NECs of the zebrafish gill filament. This is evident by the increase in $[Ca^{2+}]_i$ and decrease in pH_i within these two chemoreceptors in response to hypercapnic acidosis. However, major differences between these CO_2 sensing structures do exist.

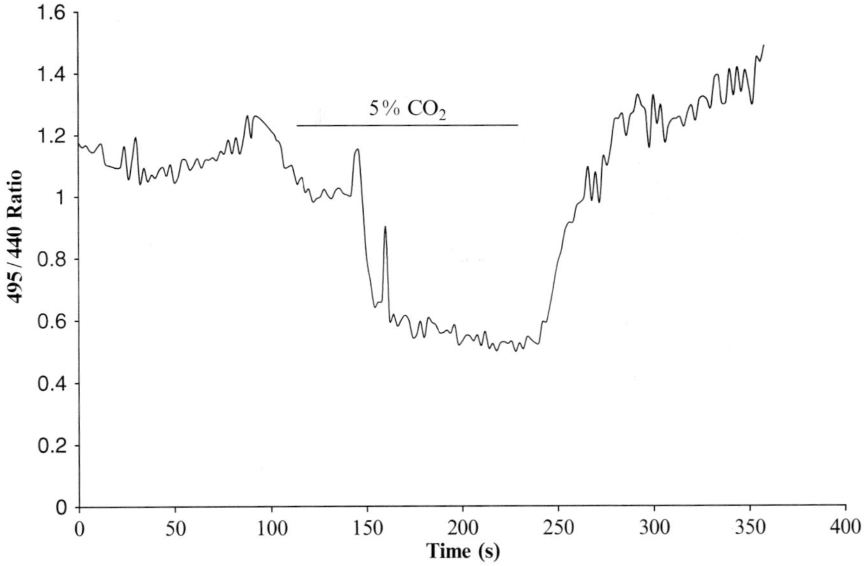

Fig. 20.2 Effect of hypercapnic acidosis (increase in CO_2 to 5%) on pH_i of a single isolated neuroepithelial cell (NEC) of the zebrafish, *Danio rerio*, gill filament. Relative changes in the 495/440 ratio from BCECF experiments indicate changes in pH_i. *Bar* represents the duration of the 5% CO_2 application

Most notable is the ability of NECs to sense relatively small deviations in their environmental CO_2. NECs, unlike type I cells, are able to respond to CO_2 levels as low as 1% with an increase in $[Ca^{2+}]_i$ and decrease in pH_i (Qin et al. 2010 and unpublished data). The variable sensitivity to CO_2 between these two chemoreceptors may reflect an evolutionary adaptation to detect deviations in arterial PCO_2 from normal set points (~40 mmHg in mammals and ~2 to 3 mmHg in fish). Furthermore, the increase in Ca^{2+} in NECs in response to hypercapnic acidosis may be due to the contribution of Ca^{2+} from intracellular stores, whereas in type I cells it is due to an influx of Ca^{2+} from the extracellular environment. Clearly, additional research is required to further elucidate the signal transduction cascade within NECs of the zebrafish gill filaments.

References

Buckler K, Vaughan-Jones R (1993) Effects of acidic stimuli on intracellular calcium in isolated Type-i cells of the neonatal rat carotid-body. Pflugers Arch 425:22–27

Buckler K, Vaughan-Jones R (1994) Effects of hypercapnia on membrane-potential and intracellular calcium in rat carotid-body type-i cells. J Physiol Lond 478:157–171

Buckler K, Vaughan-Jones R, Peers C, Lagadicgossmann D, Nye P (1991a) Effects of extracellular Ph, Pco_2 and Hco3-on intracellular Ph in isolated type-I cells of the neonatal rat carotid-body. J Physiol Lond 444:703–721

Buckler K, Vaughan-Jones R, Peers C, Nye P (1991b) Intracellular pH and its regulation in isolated type-I carotid-body cells of the neonatal rat. J Physiol Lond 436:107–129

Burleson ML, Smatresk NJ, Milsom WK (1992) Afferent inputs associated with cardioventilatory control in fish. In: Hoar WS, Randall DJ, Farrel AP (eds) Fish physiology, vol XIIB. Academic, San Diego, pp 389–426

Iturriaga R, Lahiri S, Mokashi A (1991) Carbonic-anhydrase and chemoreception in the cat carotid-body. Am J Physiol 261:C565–C573

Jonz MG, Nurse CA (2003) Neuroepithelial cells and associated innervation of the zebrafish gill: a confocal immunofluorescence study. J Comp Neurol 461:1–17

Jonz MG, Fearon IM, Nurse CA (2004) Neuroepithelial oxygen chemoreceptors of the zebrafish gill. J Physiol Lond 560:737–752

Peers C, Green FK (1991) Inhibition of Ca2+-activated K+ currents by intracellular acidosis in isolated type-I cells of the neonatal rat carotid-body. J Physiol Lond 437:589–602

Perry SF, Jonz MG, Gilmour KM (2009) Oxygen sensing and the hypoxic ventilatory response. In: Richards JG, Farrell AP, Brauner CJ (eds) Fish physiology, vol 27. Elsevier, New York, pp 193–253

Putnam RW, Filosa JA, Ritucci NA (2004) Cellular mechanisms involved in CO_2 and acid signaling in chemosensitive neurons. Am J Physiol Cell Physiol 287:C1493–C1526

Qin Z, Lewis JE, Perry SF (2010) Zebrafish (Danio rerio) gill neuroepithelial cells are sensitive chemoreceptors for environmental CO_2. J Physiol 588:861–872

Rigual R, Lopezlopez J, Gonzalez C (1991) Release of dopamine and chemoreceptor discharge induced by low pH and high Pco2 stimulation of the cat carotid-body. J Physiol Lond 433:519–531

Sato M (1994) Response of cytosolic Ca2+ to hypercapnic acidosis in cultured glomus cells of the adult rabbit carotid body. Brain Res 662:251–254

Summers BA, Overholt JL, Prabhakar NR (2002) CO_2 and pH independently modulate L-Type Ca^{2+} current in rabbit carotid body glomus cells. J Neurophysiol 88:604–612

Vulesevic B, McNeill B, Perry SF (2006) Chemoreceptor plasticity and respiratory acclimation in the Zebrafish Danio rerio. J Exp Biol 209:1261–1273

Chapter 21
Hyperplasia of Pulmonary Neuroepithelial Bodies (NEB) in Lungs of Prolyl Hydroxylase –1(PHD-1) Deficient Mice

Jie Pan, Herman Yeger, Peter Ratcliffe, Tammie Bishop, and Ernest Cutz

Abstract Pulmonary NEB, widely distributed within the airway mucosa of mammalian lungs, are presumed hypoxia sensitive airway O_2 sensors responding to changes in airway gas concentration. NEB cell hyperplasia has been reported after exposure to chronic hypoxia and in a variety of paediatric and adult lung disorders. Prolyl hydroxylases (PHD 1–3) regulate the stability of hypoxia-inducible factors (HIF's) in an O_2-dependent manner and function as intrinsic oxygen sensors. To determine a possible role of PHD-1in NEB cells we have quantitated NEB's in lungs of neonatal (P2) and adult (2 months) PHD-1-deficient mice and compared them to wild type (WT) control mice. Lung tissues fixed in formalin and embedded in paraffin were processed for immunoperoxidase method and frozen sections for multilabel immunoflourescence using antibodies for NEB markers synaptophysin, synaptic vesicle protein 2 and the peptide CGRP. The frequency and size of NEB in lungs of PHD-1 deficient neonatal mice (P2) and at 2 months was increased significantly compared to WT controls ($p<0.01$). The present data suggests an important role for PHD enzymes in NEB cell biology deserving further studies. Since the PHD-1 deficient mouse appears to be the first animal model showing NEB cell hyperplasia it may be useful for studies of NEB physiology and pathobiology.

Keywords Airway oxygen sensors • Neuroendocrine cells • Oxygen sensing mechanism • Cell proliferation and differentiation • Prolyl hydroxylases • Hypoxia-inducible factor -1

J. Pan • H. Yeger
Division of Pathology, Department of The Paediatric Laboratory Medicine Research Institute, The Hospital for Sick Children, and Department of Laboratory Medicine and Pathobiology, University of Toronto, Toronto, ON, Canada

P. Ratcliffe • T. Bishop
The Henry Wellcome Building for Molecular Physiology, University of Oxford, Oxford, UK

E. Cutz (✉)
Division of Pathology, Department of Pediatric Laboratory Medicine, The Research Institute, The Hospital for Sick Children, 555 University Avenue, Toronto, ON M5G1X8, Canada
e-mail: ernest.cutz@sickkids.ca

C.A. Nurse et al. (eds.), *Arterial Chemoreception: From Molecules to Systems*, Advances in Experimental Medicine and Biology 758, DOI 10.1007/978-94-007-4584-1_21,

21.1 Introduction

Airway mucosa of human and animal lung contains amine (serotonin,5-HT) and peptide producing cells distributed as solitary cells referred to as pulmonary neuroendocrine cells (PNEC) and as innervated clusters called neuroepithelial bodies(NEB). NEB's are widely distributed within the mucosa of intrapulmonary airways and are thought to function as polymodal airway sensors responding to a variety of intraluminal stimuli including gas concentration (pO2, p CO2) and mechanical stretch (Cutz et al. 2007a). PNEC/NEB's appear more numerous in fetal/neonatal lungs and are less conspicuous in adult lungs suggesting developmental regulation and a critical role during the perinatal period. (Cutz et al. 1985, 2007a). NEB cells are excitable and exhibit hypoxia sensing properties owing to the expression of a membrane delimited O2 sensor complex composed of an O2 sensing protein(NAPDH oxidase/NOX2) coupled to an O 2 sensitive K+ channel (Youngson et al. 1993; Wang D et al. 1996; Fu et al. 2000). Although the precise function of NEB is at present unknown, they are considered as a part of a specialized homeostatic oxygen-sensing system of the body (Weir et al. 2005). Hyperplasia of NEB cells has been described in lungs of animals and human exposed to chronic hypoxia and in a variety of pediatric and adult lung disorders (Cutz et al. 2007a; Cutz et al. 2008). The mechanisms and clinical significance of NEB hyperplasia is currently unknown and requires further investigation. It is now well established that hypoxia inducible factor −1 (HIF-1 alpha) is a ubiquitous transcription factor that regulates expression of numerous genes involved in adaptive responses to O2 deprivation (Semenza 2009). Closely linked in mediation of hypoxia responses are prolyl hydroxylase domains enzymes (PHD 1–3)that regulate the stability of HIF's in an O2- dependent manner and hence function as intrinsic O2 sensors (Kaelin and Ratcliffe 2008). Analysis in vivo of PHD expression patterns by immunohistochemistry shows that PHD1 expression is mostly confined to chromogranin A-positive neuroendocrine tissues such as the pancreatic islets, carotid body and adrenal medulla (unpublished observations). Further, PHD-1 was strongly expressed across a range of neuroendocrine tumours, for example, intestinal and lung carcinoids, carotid paragangliomas and phaemochromocytomas. Thus PHD1 is strongly expressed in neuroendocrine tissue, suggesting that it plays an important role in neuroendocrine function. To determine the possible role of PHD- 1 in NEB cells, we examined the frequency, number and the size of NEB's in lungs of PHD-1 deficient mice using NEB cell specific immunomarkers and morphometric methods. Our studies demonstrate a striking hyperplasia of NEBs in lungs of PHD–1 deficient mice during both the perinatal period and in adults suggesting an important role for PHD enzymes in NEB cell biology and pathobiology.

21.2 Methods

Lung tissues were obtained from PHD-1 deficient mice (Egln2 −/−) and wild type (WT) control mice (Aragones et al. 2008). Tissues from neonatal (P1-2, n=12), and adult (2 month, n=7) PHD −1 deficient and age matched WT control mice were fixed in 10% neutral buffered formalin ,embedded in paraffin and processed for immunoperoxidase (IP) labeling method using standard procedures (Cutz et al. 2007b). To identify NEB in formalin fixed paraffin section we used Invitrogen Polymer detection system (*SuperPicTure*™ polymer detection kit, Invitrogen Corp. Camarillo, CA). As primary antibodies we used mouse monoclonal antibody against synaptic vesicle protein 2(SV2 ,Hybridoma Bank, Iowa City, IA ; 1:20 dilution), rabbit monoclonal antibody against synaptophysin (NeoMarkers ,Freemont, CA; 1:100 dilution) and monoclonal antibody against calcitonin gene related peptide (CGRP, Abcam Inc ,Cambridge, MA;1:150 dilution). As a secondary antibody detection procedure we used Invitrogen Polymer detection system with horse radish peroxidase (HRP) polymer-secondary antibody conjugate.

For immunofluorescence (IF)staining and confocal microscopy we used frozen sections of lung tissues(~100um) from PHD-deficient and age matched WT controls (P1 n=4 and P15 n=4 respectively) using tissue preparation and immunolabeling protocols using above primary antibodies as previously reported(Pan et al. 2006). Fluorescent images of PNEC/NEB cells, the airway nerves and smooth muscle actin-FITC (Sigma-Aldrich, St Louis, MI) in the double stained (Texas Red/FITC) whole mounts were obtained using a Leica confocal laser scanning microscope (model TCS-SPE) and LAS-AF software.

Morphometric analysis of NEB frequency and size was performed using methods as previously reported (Cutz et al. 2007b; Pan et al. 2006). The integrated surface area of airways of different sizes, expressed in square millimeters of the section (5 um IP/100 um IF section thickness) was measured by Nikon simple PCI imaging software (IP method) (Cutz et al. 2007b; Pan et al. 2006) and the NIH-Image J program (confocal IF images) standardized by an internal scale bar in each counted image (Pan et al. 2006). For statistical analysis we used Student's *t*-test.

21.3 Results

21.3.1 Immunoperoxidase Method

The overall morphology of lungs from PHD – 1 deficient mice was similar to WT controls with no obvious abnormalities affecting the airways or alveolar compartment. The sections of lungs from neonatal mice (P2) ,both PHD- 1 deficient and age matched WT controls, immunostained for SV 2 showed clusters of positive cells within the airways corresponding to NEB (Fig. 21.1a, b). In addition peribronchial nerve fibers and fine nerve endings were also positive for SV2 (Fig. 21.1d, e). At higher magnification ,NEB in the lungs of WT mice usually consisted of between 5 and 7 cells (Fig. 21.1c, d), while in PHD-1 deficient mice they appeared much larger consisting of 15–30 cells (Fig. 21.1e, f). Some hyperplastic NEB in PHD – 1 deficient mice formed prominent cell aggregates at airway branch points protruding into the airway lumen (Fig. 21.1f). Morphometric analysis confirmed the increased frequency and size of NEB's in lungs of PHD –1 deficient mice compared to their WT counterparts (Fig. 21.2a, b). The mean frequency of NEB (%immunostained area/mm2) in PHD- 1 deficient mice at P2 (2.24 +/–0.4))was increased significantly compared to WT controls(1.62 +/–0.3 ; $p<0.01$) Similarly the size of NEB in PHD –1 deficient l mice (1494.6 +/– 175.3)was close to twice that of WT control mice (865.2 +/–162.6; $p<0.01$). Immunostaining for CGRP, the principal peptide in NEBs of rodent lung, showed a pattern of expression similar to SV2 with almost twofold increase in the number and size of NEB's in lungs of PHD- 1 deficient mice compared to age matched WT controls (Fig. 21.2c, d). In the lungs of 2 month old PHD-1 deficient mice the number and size of NEB assessed by immunostaining for either SV2 or CGRP was increased compared to WT controls, but the difference was at lower level of statistical significance($p<0.05$)(Fig. 21.2a, b, c, d).

21.3.2 Multilabel Immunofluorescence Method

The studies using IF method confirmed the findings of IP study showing that NEB s in the lungs of PHD- 1deficient mice at either P2 or P15 were more numerous and larger compared to WT age matched controls ,and were frequently localized at airway branch points (Fig. 21.3a).At higher magnification, hyperplastic NEBs in lungs of PHD –1 deficient mice sometimes formed an intraepithelial row of up to 20 columnar cells with fine nerve branches entering at the base (Fig. 21.3b). In contrast, NEB in the lungs of WT controls were much smaller (~5–7 cells) but showed a similar pattern of innervation (Fig. 21.3c). Morphometric analysis of sections immunostained for synaptophysin confirmed up to

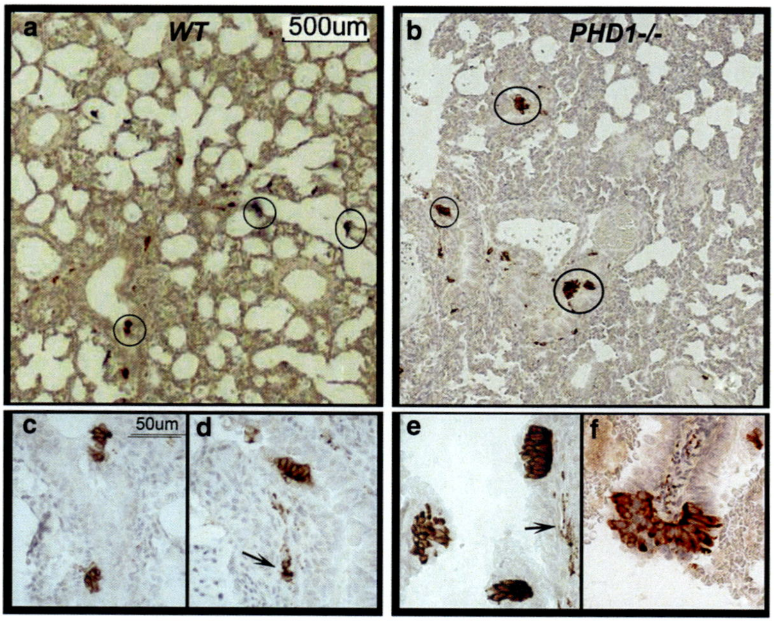

Fig. 21.1 NEB in lungs of WT control and PHD-1 deficient mice visualized using immunoperoxidase method with anti-SV2 antibody. (**a**) Low magnification view of lung section from neonatal (P2) WT mice showing positive NEB in small airways (*circled*). (**b**) Section of lung from neonatal (P2) PHD-1 deficient mouse lung at the same magnification as in (**a**) showing more prominent NEBs (*circled*) in similar distribution. (**c**) Close up of two NEBs consisting of small clusters of immunopositive cells in epithelium of small airway in lung from WT mouse. (**d**) Another small NEB in the same sample as in (**c**) with immunopositive nerve fibers (*arrow*) in submucosa. (**e**) Higher magnification view from sample (**b**) with three prominent NEBs forming compact intraepithelial corpuscles and immunoreactive nerve fibres in submucosa (*arrow*). (**f**) Close up of large, hyperplastic NEB from sample (**b**) situated at airway bifurcation and protruding into airway lumen

twofold increase in the number(58.3 +/−5.2 vs.34.2 +/−4.3) and size(15.3 +/− 3.2 vs 11.2 +/−2.3) of NEB in PHD-1 deficient mice compared to age-matched WT controls (Fig. 21.2e, f, g, h). Although nerve fibers associated with NEB and airway smooth muscle were more clearly visualized, there were no obvious differences in nerve density between PHD-1 deficient and WT mice.

21.4 Discussion

We report striking NEB cell hyperplasia in the lungs of PHD- 1deficient mice. The increased number and size of NEB was observed in lungs of neonates and was maintained into adulthood, suggesting persistence of NEB cell hyperplasia during post natal lung growth and development. Although the precise mechanisms are not known, NEB cell hyperplasia is likely the result of increased recruitment from precursor cells rather than due to increased cell proliferation since in vivo and in vitro labeling studies revealed that NEB cells are a slowly renewing cell population with low cell turnover compared to adjacent airway epithelial cells (Sorokin et al. 1997). In preliminary studies we observed that NEB cells of PHD1 –deficient mice did not show increased labeling with a proliferation marker MIB-1(Ki-67) indicating that the mitotic activity was not affected. During lung development the differentiation of both PNEC and NEB is governed by proneural genes such as a mammalian homolog of the achete-scute complex (Mash-1 in mice; HASH-1 in human). Mice deficient in Mash −1(Mash −1 k/o) lack both PNEC and NEBs (Borges et al. 1997).

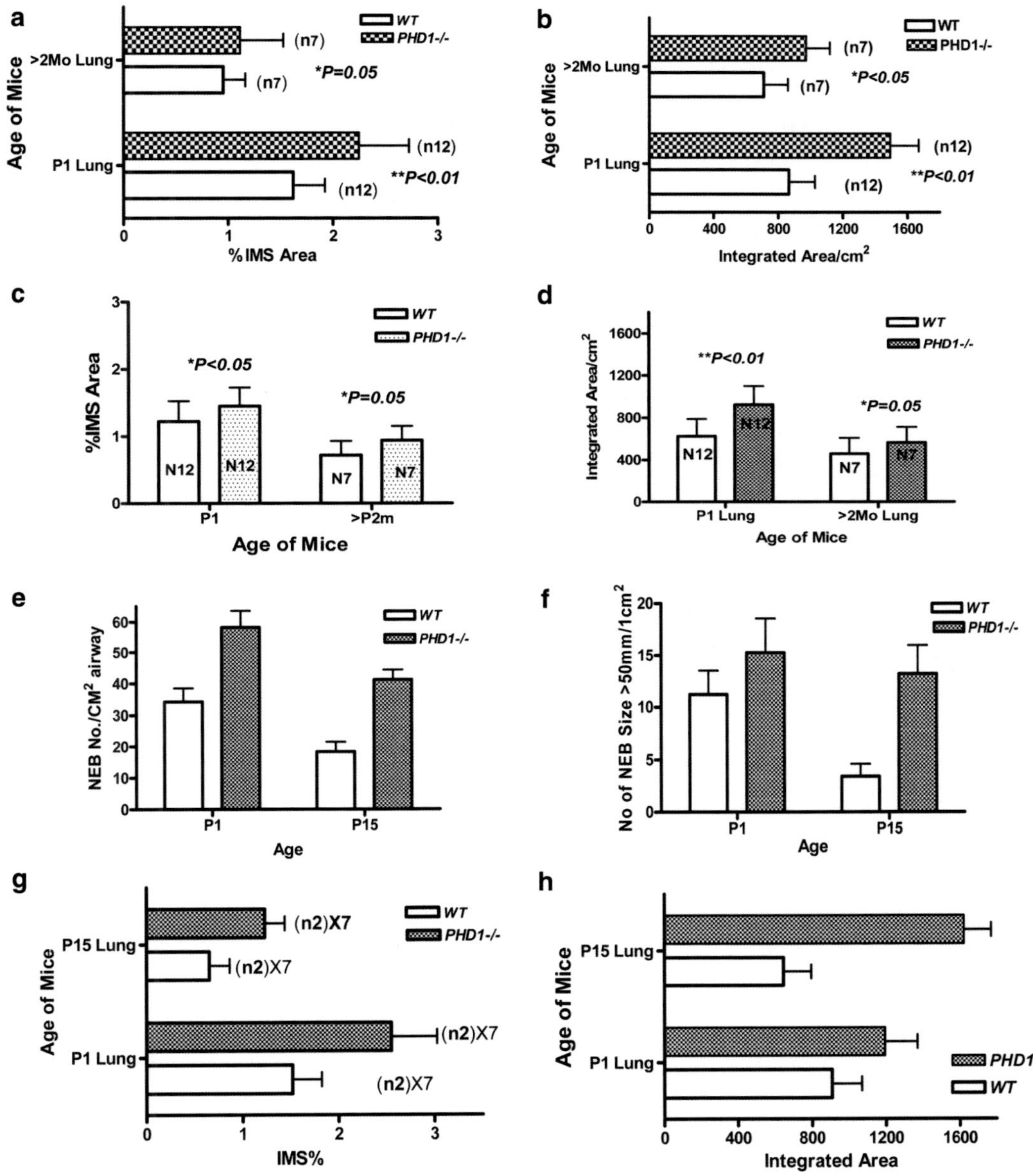

Fig. 21.2 Morphometric assessment of NEB frequency, number and size in lungs of PHD-1deficient and WT control lungs using immunoperoxidase (**a**, **b**, **c**, **d**) and immunofluorescence (**e**, **f**, **g**, **h**) methods. (**a**) Comparison of mean values of %immunostained (SV2 antibody) area (IMS)per mm2 of section from PHD1- deficient and WT controls at different ages (P2 and 2 month). (**b**) Mean values for NEB size from same samples as in a). (**c**) Means of % IMS for NEB immunostained for CGRP as in samples shown in (**a**) & (**b**). (**d**) Means of NEB size in samples immunostained for CGRP as in samples shown in (**a**) & (**b**). (**e**) Comparison of mean number of synaptophysin immunoreactive NEBs/cm2 of airway in lungs of PHD-1 deficient mice and WT controls at P2 and P15. (**f**) Mean values for NEB size from same samples as in (**e**). (**g**) & (**h**) Same samples as in (**e**) & (**f**) with values expressed as IMS %of NEB (**g**) and NEB size /integrated area (**h**)

We have shown recently that during early stages of mouse lung development, pO2 concentration in concert with Mash-1 expression modulates neuroendocrine cell phenotype (McGovern et al. 2010). Organ cultures of fetal mouse lung at E12 maintained in hypoxia (5%O2) showed loss of Mash-1 expression and consequently markedly reduced numbers of PNEC, whereas cultures incubated in

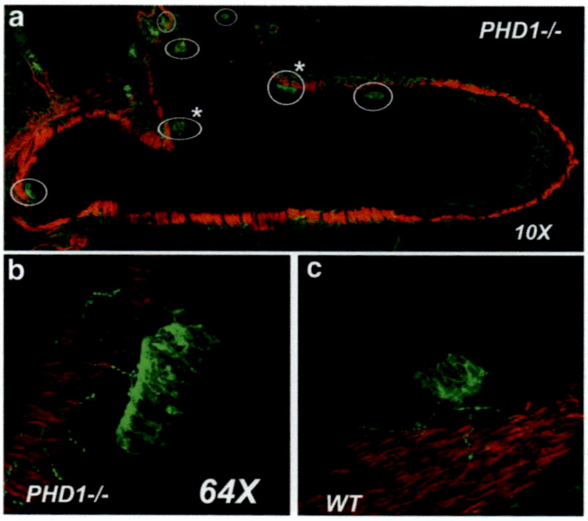

Fig. 21.3 NEBs in lung sections from PHD-1 deficient and WT control mice visualized by multilabel immunoflourescence method and confocal microscopy. (**a**) Low magnification view of cross section of large airway in lung from PHD −1 deficient mouse (P2) immunostained for synaptophysin (*green*) to label NEB and nerve fibers and for smooth muscle actin to outline airway smooth muscle (*red*). Several large NEBs (*circled*) are present, some that are located at airway branch points (*asterisks*). (**b**) Close –up of large NEB from sample a) with closely packed immunoreactive cells and fine, beaded nerve fibres in adjacent smooth muscle and at the base of NEB cell cluster. (**c**) For comparison, NEB in lung of WT mouse is significantly smaller; however submucosal nerve fibres appear similar

"normoxia" (20% O2) showed high levels of Mash-1 expression and numerous PNEC. Switching hypoxia grown cultures to normoxia resulted in a burst of Mash-1 expression followed by PNEC differentiation when cultures were kept under normoxia. In contrast, hypoxia had no effects on Mash-1 expression or PNEC numbers in cultures from later gestation (E16) indicating a locked-in developmental programming. Therefore it seems plausible that hypoxia promotes expansion of precursor cells and that oxygen promotes differentiation of PNEC via Mash-1.

Although the precise mechanisms regulating Mash-1 expression in NEB cells is at present unknown, hypoxia inducible factors (HIFs 1–3) are likely to be involved. Prolyl hydroxylase domain enzymes (PHD1-3) function as cellular O2 sensors via modulation of HIF's expression and thus play a key role in oxygen homeostasis in both health and disease (Kaelin and Ratcliffe 2008; Appelhoff et al. 2004). In terms of hypoxia responses, studies using various tumor cell lines have shown that levels of PHD −1 mRNA were unchanged or decreased under hypoxia,whereas mRNA levels of PHD −2 and PHD −3 were increased, especially PHD- 3 (Appelhoff et al. 2004). The tissue and cell expression of PHD enzymes is variable (Kaelin and Ratcliffe 2008; Appelhoff et al. 2004). For example PHD-3 mRNA is highly expressed in cardiac muscle while PHD −1 mRNA levels are increased in the testis. At the protein level, however, PHD-2 was found to be abundantly expressed in most mouse organs examined (Appelhoff et al. 2004). The expression and distribution of PHD isozymes in the lung (or NEB cells) has not been investigated in detail. In terms of physiological importance, PHD-2 appears to be critical in oxygen sensing under basal conditions, since its inactivation during development is embryo lethal (Kaelin and Ratcliffe 2008). In contrast, PHD- 1 and PHD −3 deficient mice are viable and appear normal at birth (Kaelin and Ratcliffe 2008).

The precise role of PHD −1 in NEB cell hyperplasia and whether it involves the O2 sensing mechanism is at present unknown. In oxidase deficient mice(gp91 phox, NOX2 k/o mice), inactivation of a gene involved in acute O2 sensing, that is HIF-independent, NEB cell responses to acute hypoxia were abrogated without causing NEB cell hyperplasia (Fu et al. 2000; Kazemian et al. 2001).We postulate that in PHD-1 deficient mice, NEB cell hyperplasia could be mediated via a HIF driven up regulation of neurogenic gene Mash-1, which in turn could lead to increased recruitment and differentiation from precursor cells(see above). In this scenario, the lack of PHD −1 may be compensated by over expression of PHD's 2 and 3, especially under conditions of hypoxia since the lung develops in a relatively hypoxic environment (fetal pO2 20–30 mmHg) (Appelhoff et al. 2004). This in turn could lead to increased production of HIF's and downstream activation of the Mash-1 gene. Alternatively, other HIF-independent mechanisms of PHD's function may be involved since in breast cancer cells PHD –1 is induced by estrogen and can stimulate cell proliferation in vitro (Seth et al. 2002).

Hyperplasia of PNEC/NEB cells, suggesting altered function, has been described in a number of pediatric lung disorders including bronchopulmonary dysplasia, cystic fibrosis and asthma (Cutz et al. 2007a).In adults it has been linked to the pathogenesis of tobacco induced lung disease, pulmonary fibrosis and lung carcinogenesis (Cutz et al. 2008). It is of interest to note that the PHD- 1 deficient mouse appears to be a first example of a transgenic animal model with NEB cell hyperplasia. Therefore further studies using the PHD- 1 deficient mouse model may provide mechanistic insights in to the function of NEB cells under normal conditions and in a variety of pulmonary diseases.

Acknowledgements Supported by grants from the Canadian Institutes for Health Research (MOP 15270) and Canadian Cystic Fibrosis to E.C. and H.Y and Wellcome Trust (Programme grant #091857) to P.R. and T.B.

References

Appelhoff RJ, Tian Ya-Min, Raval RR et al (2004) Differential function of the prolyl hydroxylases PHD 1, PHD 2, and PHD 3 in the regulation of hypoxia –inducible factor. J Biol Chem 279:38458–38465

Aragones J, Schneider M, VanGeyte K et al (2008) Deficiency or inhibition of oxygen sensor PHD 1 induces hypoxia tolerance by reprogramming basal metabolism. Nat Genet 40:170–189

Borges MW, Linnoila RI, van de Velde HJ et al (1997) An achaete-scute homologue essential for neuroendocrine differentiation in the lung. Nature 386:852–855

Cutz E, Gillan JE, Bryan AC (1985) Neuroendocrine cells in the developing human lung: morphologic and functional considerations. Pediatr Pulmonol 1:S21–S29

Cutz E, Yeger H, Pan J (2007a) Pulmonary neuroendocrine cell system in pediatric lung disease-recent advances. Pediatr Dev Pathol 10:419–435

Cutz E, Perrin DG, Pan J, Haas EA, Krous FK (2007b) Pulmonary neuroendocrine cells and neuroepithelial bodies in sudden infant death syndrome: potential markers of airway chemoreceptor dysfunction. Ped Dev Pathol 10:106–116

Cutz E, Yeger H, Pan J, Ito T (2008) Pulmonary neuroendocrine cell system in health and disease. Curr Respir Med Rev 4:174–183

Fu XW, Wang D, Nurse CA, Dinauer MC, Cutz E (2000) NADPH oxidase in an O_2 sensor in airway chemoreceptors: evidence from K^+ current modulation in wild type and oxidase deficient mice. Proc Natl Acad Sci U S A 97:4374–4379

Kaelin WG, Ratcliffe PJ (2008) Oxygen sensing by metazoans: the central role of the HIF hydroxylase pathway. Mol Cell 30:393–402

Kazemian P, Stephenson R, Yeger H, Cutz E (2001) Respiratory control in neonatal mice with NADPH oxidase deficiency. Respir Physiol 126:89–100

McGovern S, Pan J, Oliver G, Cutz E, Yeger H (2010) The role of hypoxia and neurogenic genes(Mash-1 and Prox-1) in the developmental programming and maturation of pulmonary neuroendocrine cells in fetal mouse lung. Lab Invest 90:180–195

Pan J, Luk C, Kent G, Cutz E, Yeger H (2006) Pulmonary neuroendocrine cells, airway innervation and smooth muscle are altered in Cftr null mice. Am J Respir Cell Mol Biol 35(3):320–6

Semenza G (2009) Regulation of oxygen homeostasis by hypoxia-inducible factor. Physiology 24:97–106

Seth P, Krop L, Porter D, Polyak K (2002) Novel estrogen and tamoxifen induced genes identified by SAGE (Serial analysis of gene expression). Oncogene 21:836–843

Sorokin SP, Hoyt RF Jr, Shaffer MJ (1997) Ontogeny of neuroepithelial bodies: correlations with mitogenesis and innervation. Microsc Res Tech 37:43–61

Wang D, Youngson C, Wong V et al (1996) NADPH oxidase and hydrogen peroxide-sensitive K+ channel may function as an oxygen sensor complex in airway chemoreceptors and small cell carcinoma cell lines. Proc Natl Acad Sci U S A 93:13182–13187

Weir EK, Lopez-Barneo J, Buckler KJ, Archer SL (2005) Acute oxygen sensing mechanisms. N Engl J Med 353:2042–2055

Youngson C, Nurse C, Yeger H, Cutz E (1993) Oxygen sensing in airway chemoreceptors. Nature 365:153–156

Chapter 22
Precision-Cut Vibratome Slices Allow Functional Live Cell Imaging of the Pulmonary Neuroepithelial Body Microenvironment in Fetal Mice

Kathy Schnorbusch, Robrecht Lembrechts, Inge Brouns, Isabel Pintelon, Jean-Pierre Timmermans, and Dirk Adriaensen

Abstract We recently developed an *ex vivo* lung slice model that allows for confocal live cell imaging (LCI) of neuroepithelial bodies (NEBs) in postnatal mouse lungs (postnatal days 1–21 and adult). NEBs are morphologically well-characterized, extensively innervated groups of neuroendocrine cells in the airway epithelium, which are shielded from the airway lumen by 'Clara-like' cells. The prominent presence of differentiated NEBs from early embryonic development onwards, strongly suggests that NEBs may exert important functions during late fetal and neonatal life. The main goal of the present study was to adapt the current postnatal LCI lung slice model to enable functional studies of fetal mouse lungs (gestational days 17–20).

In vibratome lung slices of prenatal mice, NEBs could be unequivocally identified with the fluorescent stryryl pyridinium dye 4-Di-2-ASP. Changes in the intracellular free calcium concentration and in mitochondrial membrane potential could be monitored using appropriate functional fluorescent indicators (e.g. Fluo-4).

It is clear that the described fetal mouse lung slice model is suited for LCI studies of Clara cells, ciliated cells, and the NEB microenvironment, and offers excellent possibilities to further unravel the significance of NEBs during the prenatal and perinatal period.

Keywords Fetal mouse lung • Vibratome slices • NEBs • Live cell imaging • Calcium imaging • Clara-like cells • 4-Di-2-ASP • Styryl pyridinium dyes

---

This work was supported by the following research grants:
A fellowship from the Agency for Innovation by Science and Technology in Flanders (IWT) to Robrecht Lembrechts (SB 81162), by grants of the Fund for Scientific Research-Flanders (FWO; G.0081.08 to D.A. and I.B., G.0589.11 to D.A. and J-P.T.), and by grants of the University of Antwerp (GOA BOF 2007 to D.A. and KP BOF 2011 to I.B.).

K. Schnorbusch • R. Lembrechts • I. Brouns • I. Pintelon • J.-P. Timmermans • D. Adriaensen (✉)
Department of Veterinary Sciences, Laboratory of Cell Biology and Histology, University of Antwerp, Groenenborgerlaan 171, BE-2020 Antwerp, Belgium
e-mail: dirk.adriaensen@ua.ac.be

C.A. Nurse et al. (eds.), *Arterial Chemoreception: From Molecules to Systems*, Advances in Experimental Medicine and Biology 758, DOI 10.1007/978-94-007-4584-1_22,

22.1 Introduction

Nowadays, *in vitro* fresh lung slices are used for a wide variety of applications, and have become a standard tool for the investigation of lung physiology in health and disease (for recent review see Liberati et al. 2010; Sanderson 2011). Also for physiological studies on neuroepithelial bodies (NEBs), fresh precision-cut lung slices have been reported very useful in several species including mice (Fu et al. 2000). Since NEBs occur as clusters of pulmonary neuroendocrine cells (PNECs) dispersed in the airway epithelium, fresh lung slices are valuable to study NEBs in their 'natural environment', i.e., selectively innervated and surrounded by their neighboring cells and tissues (Adriaensen et al. 2003, 2006; Liberati et al. 2010; De Proost et al. 2008; Pintelon et al. 2005). A few years ago, the combination of fresh lung slices and functional fluorescent indicators in a confocal live cell imaging (LCI) set-up, allowed the design and optimization of a LCI model for NEBs in postnatal mice (Pintelon et al. 2005; De Proost et al. 2008, 2009). In this model, simultaneous changes in physiological parameters of multiple NEB cells, and their interactions with surrounding cells and tissues can be detected in real time, and already allowed to conclude that ATP, released from NEB cells is able to activate the surrounding Clara-like cells (De Proost et al. 2009).

NEBs have been proposed to mediate several functions in the regulation of physiological processes in the lungs during prenatal, perinatal and postnatal life, but the strongest evidence today implicates them in airway oxygen sensing (Cutz and Jackson 1999).

Whereas pulmonary NEBs can nowadays be regarded as prime candidates to serve as sensory end-organs in postnatal airways (Widdicombe 2001; Adriaensen et al. 2006), they are also believed to be involved in fetal lung growth, and epithelial differentiation and repair (Sorokin and Hoyt 1990; Linnoila 2006). Differentiated NEBs are present from early embryonic development onwards. Multiple studies have shown that NEBs emerge as foci of growth in the non-endocrine epithelium in perihilar airways first, and gradually mature in more peripheral branches, assuming roles in airway morphogenesis, cellular growth and maturation (Sorokin et al. 1997). Accordingly, major bioactive substances produced by PNECs, such as calcitonin gene-related peptide (CGRP), have been shown to stimulate growth in fetal lungs (Hoyt et al. 1993). Because of the relatively highest density of NEBs during the perinatal period, the potential importance of NEBs for the transition from fetal to neonatal life has been the subject of many studies and discussions (Pan et al. 2004; Linnoila 2006; Bollé et al. 2000). Although the general impression exists that the number of NEBs declines shortly after birth, this apparent decrease of the NEB population may be mainly caused by a 'dilution effect' because of the major expansion of the lungs after birth, whereby the non-endocrine tissue is highly amplified compared to the endocrine tissue (Gosney 1993). Anyway, many investigators have suggested that NEBs could be most active during the perinatal period, possibly complementing carotid body chemoreceptor function at the time when the latter are not yet fully functional (Van Lommel et al. 1999).

Since conclusive physiological data on the exact function of NEBs in pre- and perinatal life in healthy lungs are still lacking, the main goal of the present study was to further extend the possibilities of the current lung slice model for LCI (De Proost et al. 2008) to the study of fetal mouse lungs. As a proof of principle, vibratome slices of fetal mouse lungs were vitally stained to selectively identify NEBs, and were loaded with a Ca^{2+}-indicator for the real-time visualization of calcium-dependent activation in the NEB micro-environment and the surrounding epithelium. Because morphological data for late fetal lungs were also scarce, the NEB microenvironment was further explored immunocytochemically.

22.2 Methods

22.2.1 Animal Preparation

Lung tissue was obtained from prenatal C57-Bl6 mice (GD17-20; n =11; Janvier, Bio Services, Uden, The Netherlands). National and international principles of laboratory animal care were followed and experiments were approved by the local animal ethics committee of the University of Antwerp.

22.2.2 Immunohistochemical Staining on Lung Cryosections and Fixed Lung Slices

For immunohistochemical staining on lung cryosections, prenatal mouse lungs were fixed in 4% paraformaldehyde (0,1 M phosphate buffer; pH 7.42). Serial 25-μm-thick cryostat sections of the complete lung were processed for immunolabeling. Characteristics and sources of the applied primary antisera are listed in Table 22.1, those of secondary and tertiary antisera in Table 22.2. The combinations of primary and secondary antisera used for multiple immunostaining are listed in Table 22.3. Cryostat sections were routinely incubated overnight with the primary antibodies. To enhance sensitivity, a biotin-conjugated tyramide signal amplification kit (TSA kit NEL700; PerkinElmer LAS, Zaventem, Belgium) was applied. To block endogenous mouse IgG, sections were preincubated with mouse-on-mouse blocker (M.O.M.; Vector Labs, Burlingame, CA). After a final wash, all sections were mounted in Citifluor (Ted Pella 19470, Redding, CA, USA).

After LCI, lung slices (see 2.3) were fixed in 4% paraformaldehyde (0,1 M phosphate buffer; pH 7.42; 2 h) and were incubated overnight with a polyclonal primary antibody raised in rabbit (Rb Pc) against CGRP (1:5000, see Table 22.1). For visualization, the slices were further incubated for 4 h with GAR-Fab-Cy3 (1: 2000, see Table 22.2).

Table 22.1 List of primary antisera used for immunohistochemistry

Primary antisera`			
Antigen	Host	Mc/Pc	Source
Calcitonin gene-related peptide (CGRP)	Mouse	Mc	Sigma C7113, Bornem, Belgium
CGRP	Rabbit	Pc	Sigma C8198
$P2X_3$ receptor ($P2X_3$)	Rabbit	Pc	Chemicon AB5895, Temecula, CA, USA
Clara Cell Secretory Protein (CCSP)	Rabbit	Pc	DakoCytomation A0257, DK-2600, Glostrup, Denmark

Table 22.2 List of secondary antisera and streptavidin complexes used for immunohistochemistry

Secondary antisera, streptavidin complexes	Source	Dilution
Biotinylated Fab fragments of goat anti-rabbit IgG (GAR-Fab-BIOT)	Rockland 811–1602, Gilbertsville, PA, USA	1:500
Cy™3- conjugated Fab Fragments of goat anti-rabbit IgG (GAR-Fab-Cy3)	Jackson ImmunoResearch 111-167-003, West Grove, PA, USA	1:2,000
Cy™3- conjugated Streptavidin (STR-Cy3)	Jackson ImmunoResearch 016-160-084	1:6,000
DyLight™488-conjugated Fab Fragments of goat anti-rabbit IgG (GAR-Fab-DyLight488)	Jackson ImmunoResearch 111-487-003	1:500
DyLight™488-conjugated donkey anti-mouse IgG (DAM-DyLight488)	Jackson ImmunoResearch 715-485-151	1:500
ExtrAvidin™ horseradish peroxidase	Sigma E2886	1:1,000

Table 22.3 List of applied immunocytochemical double and triple staining procedures

Multiple stainings					
Double staining using tyramide signal amplification (TSA)					
Primary antisera antigen 1 (TSA)	*Dilution*	*Visualization*	*Primary antisera antigen 2 (conv.)*	*Dilution*	*Visualization*
$P2X_3$	1:10,000	STR-Cy3	CGRP	1:5,000	GAR-Fab-DyLight488
Conventional double staining					
Primary antisera antigen 1 (conv.)	*Dilution*	*Visualization*	*Primary antisera antigen 2 (conv.)*	*Dilution*	*Visualization*
CGRP	1:2,000	DAM-DyLight488	CCSP	1:200	GAR-Fab-Cy3

22.2.3 *Live Cell Imaging on Fresh Lung Slices*

Preparation of lung slices, live staining of pulmonary NEBs and loading of Ca^{2+} indicators was performed as previously described (Pintelon et al. 2005; De Proost et al. 2008). In short, a standard physiological solution was used throughout, which contained (in mM) NaCl, 130; KCl, 5; $CaCl_2$, 1.2; $MgSO_4$, 1; D-glucose, 11; HEPES, 20; pH 7.42 adjusted with NaOH. Solutions containing a high extracellular potassium concentration ($[K^+]_o$) were prepared by equimolar substitution of KCl for NaCl. All chemicals were purchased from Sigma-Aldrich (Bornem, Belgium), except for the calcium dye Fluo-4 AM (F14201) and the vital styryl pyridinium dye 4-Di-2-ASP (D-289) that were purchased from Molecular Probes (Invitrogen, Merelbeke, Belgium). Lung tissue was stabilized by instillation of a 2% agarose solution (low-melt agarose, A4018, Sigma) via a tracheal cannula. Lung slices (100–150 μm thick) were cut using a vibratome (HM650 V, Microm International, Walldorf, Germany) with cooled tissue bath (4°C).

Living lung slices of prenatal mice were selectively stained for NEBs by incubating for 4 min with 4 μM 4-Di-2-ASP in Dulbecco's modified Eagle's medium/F-12 (DMEM-F-12, Gibco, Invitrogen) and rinsing in DMEM-F-12 in an incubator (37°C; 5% CO_2/95% air). To load living lung slices with a Ca^{2+} indicator, the lung slices were incubated in physiological solution with Fluo-4 AM (10 μM), 100 μM sulfobromophthalein, 0.1% DMSO and 0.02% Pluronic F-127 for 1 h at room temperature. All stimuli were applied to lung slices that were submerged in a tissue bath (2 ml) mounted on the microscope stage, and perfused by a gravity-fed system (flow rate >5 ml/min) with electrically triggered valves that allowed the fast exchange of experimental solutions.

22.2.4 *Microscopic Data Acquisition and Data Analysis*

An inverted microscope (Zeiss Axiovert 200; Carl Zeiss, Jena, Germany), attached to a microlens-enhanced dual spinning disk confocal microscope system (Ultra*VIEW* ERS; PerkinElmer), equipped with a three line (488, 568 and 647 nm) argon-krypton laser, was used for LCI and for all high resolution imaging. Time-lapse experiments were mainly recorded in a single confocal plane with a plan apo 40x/ 1.30 N.A. objective lens (Carl Zeiss) and were analyzed off-line by Volocity software (PerkinElmer).

22.3 Results

The airway epithelium of fresh vibratome slices of fetal mouse lungs was imaged after staining with 4-Di-2-ASP and loading with Fluo-4. Viewed from the luminal side of the sectioned airways, the epithelium displayed a mosaic of polygonal 4-Di-2-ASP/Fluo-4 fluorescent cells, intermingled with almost non-fluorescent rounded cells (Fig. 22.1a, b), the ciliated cells and Clara cells respectively (De Proost et al. 2008). Stimulation of Fluo-4 loaded lung slices with ATP (10 μM; 10s) caused a clearly visible, reversible and reproducible rise in Fluo-4 fluorescence in all Clara cells and to a lesser extent also in ciliated cells (Fig. 22.1b–d).

To explore the NEB microenvironment in fetal mouse airways, multiple immunocytochemical staining was performed. Labeling for calcitonin gene-related peptide (CGRP) in cryostat sections of lungs of prenatal mice, revealed immunoreactivity (IR) in cell groups of the airway epithelium (Fig. 22.2a) that could morphologically be characterized as clustered pulmonary neuroendocrine cells

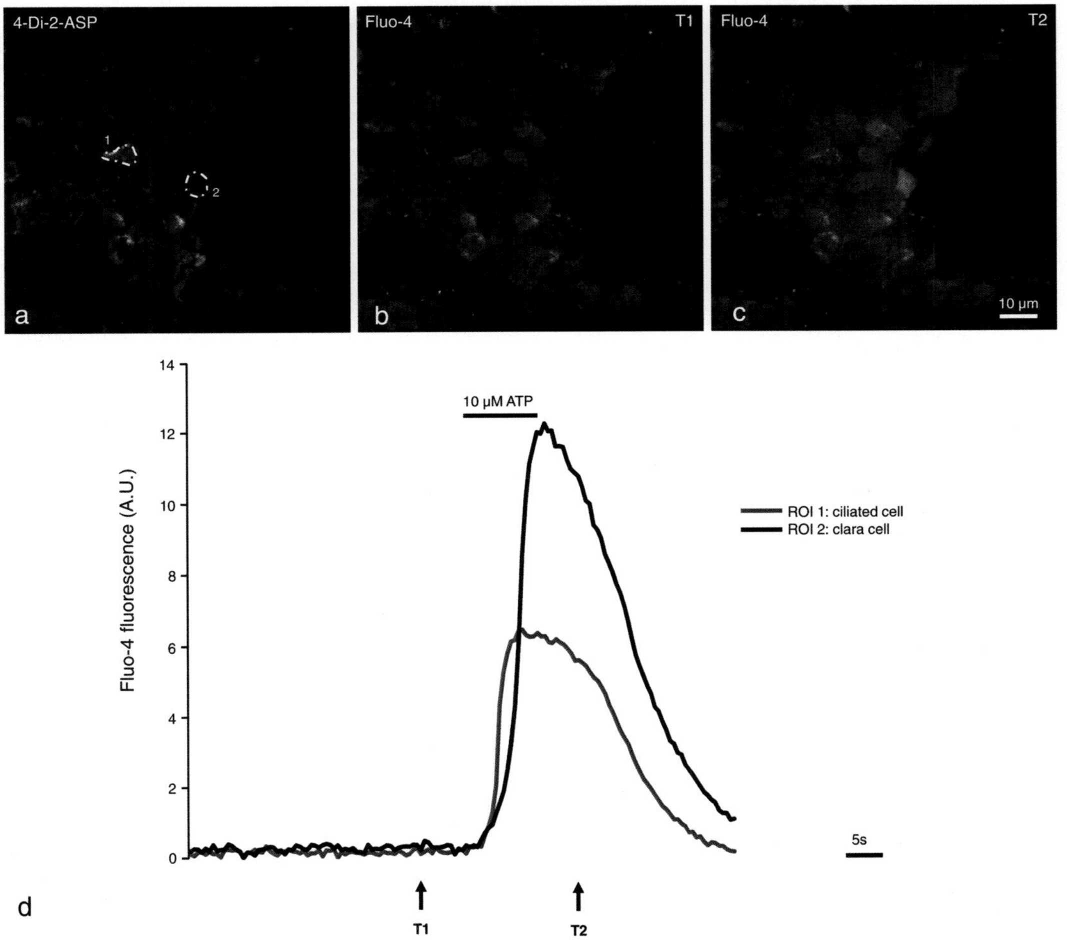

Fig. 22.1 Representative recording of the Fluo-4 fluorescence changes measured in airway epithelium of a 4-Di-2-ASP stained, Fluo-4 loaded fetal murine lung slice, before, during and after stimulation with ATP. (**a**) 4-Di-2-ASP staining allows for discrimination of bright-fluorescent ciliated cells (region of interest (ROI 1)) and non-fluorescent Clara cells (ROI 2). (**b**, **c**) Time-lapse images of Fluo-4 fluorescence in ciliated cells (ROI 1 in a) and Clara cells (ROI 2 in a) at different time points, indicated in the graph as T1 and T2. (**d**) Graph plotting the time course of changes in Fluo-4 fluorescence intensity after a 10s challenge with 10 μM ATP. Stimulation with ATP caused a prominent $[Ca^{2+}]_i$ rise in Clara cells and to a lesser extent in ciliated cells

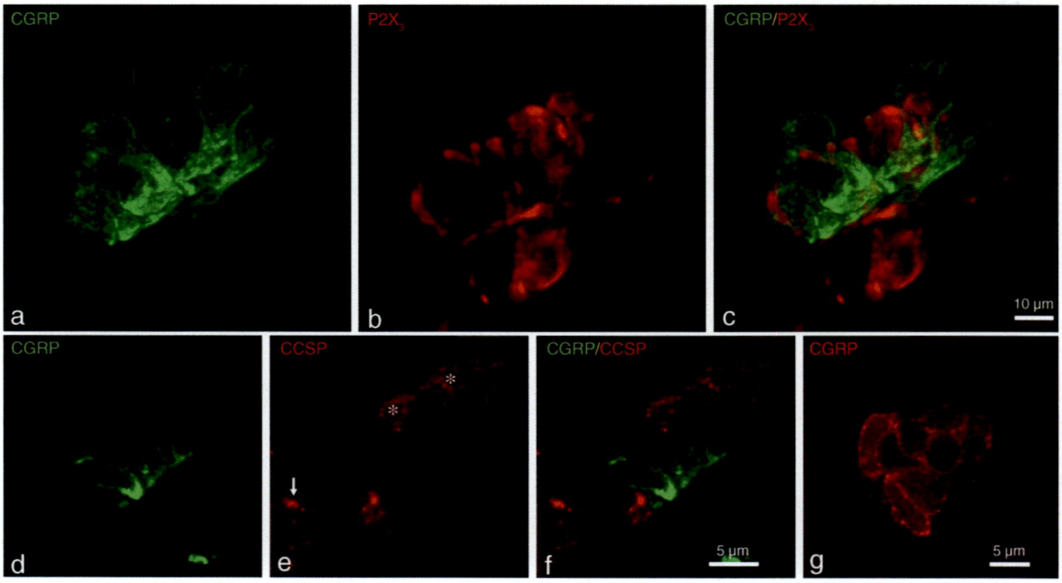

Fig. 22.2 Mouse GD 18. (**a–c**) Double immunocytochemical labeling for CGRP (**a**; *green* Dylight488 fluorescence), a marker for NEB cells in mouse airways, and $P2X_3$ ATP receptors (**b**; *red* Cy3 fluorescence) a marker for one of the subpopulations of the NEB-associated vagal sensory innervation. (**c**). Combination of the *red* and *green* channel shows that $P2X_3$ immunoreactivity (IR) is expressed on laminar intraepithelial nerve terminals that protrude between NEB cells. (**d–f**) Double immunocytochemical labeling for CGRP (**d**; *green* Dylight488 fluorescence) and for CCSP (**e**; *red* Cy3 fluorescence), a marker for Clara (*arrow* in **e**) and Clara-like cells (*asterisks* in **e**) in mouse airways. (**f**) Combination of the *red* and *green* channel shows that CCSP is expressed in Clara-like cells that cover and surround the NEB. (**g**) Immunocytochemical labeling for CGRP (*red* Cy3 fluorescence) revealed IR in prenatal NEBs in fixed murine lung slices

(PNECs). Typically, CGRP IR was more pronounced in the basal part of the PNEC cytoplasm, displaying a granular staining pattern (Fig. 22.2a). Immunostaining for $P2X_3$ ATP receptors, showed that some, but not all of the prenatal NEBs are innervated by a $P2X_3$ receptor-ir subpopulation of vagal sensory nerve terminals that branches between the NEB cells (Fig. 22.2b–c). Double immunostaining for CGRP and Clara cell secretory protein (CCSP) provided a view of the Clara-like cells, that largely

Fig. 22.3 (**a–c**). Representative time-lapse recording of the changes in 4-Di-2-ASP fluorescence measured in NEB cells and surrounding epithelial cells in a fetal murine lung slice, before, during and after stimulation with carbonyl cyanide 4-(trifluoro-methyl) phenylhydrazone (FCCP), a H^+ ionophore that uncouples oxidative phosphorylation and depolarizes mitochondria. (**a**) Time point indicated in the graph (**c**) as T1. Unstimulated fetal mouse lung slice after staining with the fluorescent styryl pyridinium dye 4-Di-2-ASP. A pulmonary NEB (encircled; ROI 1) can be recognized as a compact group of small 4-Di-2-ASP fluorescent cells. The NEB is surrounded by a rim of non-fluorescent Clara-like cells (ROI 3), and are never in contact with 4-Di-2-ASP fluorescent ciliated cells (ROI 2) or non-fluorescent Clara cells (ROI 4). (**b–c**) FCCP (10 μM; 10s) evoked a simultaneous increase in 4-Di-2-ASP fluorescence in all NEB cells and in ciliated epithelial cells. (**c**) Graph plotting the time course of 4-Di-2-ASP fluorescence intensity as observed during administration of FCCP. (**d–h**) Representative recording of the NEB microenvironment after staining with 4-Di-2-ASP and loading with the Ca^{2+}-indicator Fluo-4. (**d**) Image of 4-Di-2-ASP staining recorded prior to the experiment, showing a NEB (encircled ROI 1) which is easily differentiated from the almost non-fluorescent surrounding Clara-like cells (ROI 2), the ciliated cells (ROI 3) and the Clara cells (ROI 4). (**e–g**) Time-lapse images of Fluo-4 fluorescence at different time points, indicated in the graph as T1-T3. (**h**) Graph plotting the time course of changes in Fluo-4 fluorescence intensity. Traces were obtained from the ROI marked on the fluorescence image in (**d**). After a forced depolarization with high $[K^+]_o$ (50 mM; 5 s), NEB cells responded with a clear rise in Fluo-4 fluorescence, a few seconds earlier than Clara-like cells

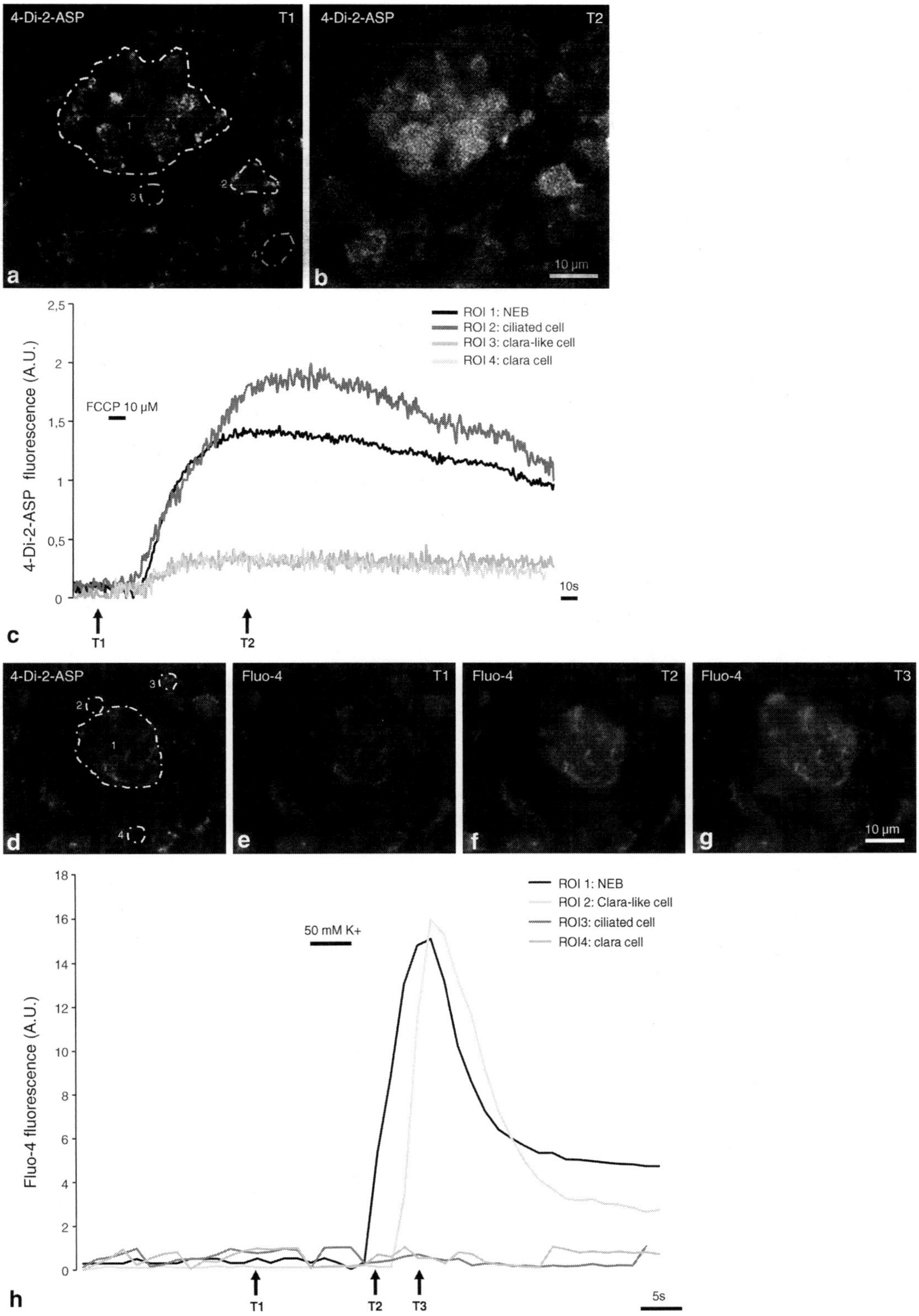
4-Di-2-ASP
T1
4-Di-2-ASP
T2
10 µm
a
b
ROI 1: NEB
ROI 2: ciliated cell
ROI 3: clara-like cell
ROI 4: clara cell
FCCP 10 µM
4-Di-2-ASP fluorescence (A.U.)
2,5
2
1,5
1
0,5
0
10s
T1
T2
c
4-Di-2-ASP
Fluo-4
T1
Fluo-4
T2
Fluo-4
T3
10 µm
d
e
f
g
ROI 1: NEB
ROI 2: Clara-like cell
ROI3: ciliated cell
ROI4: clara cell
50 mM K+
Fluo-4 fluorescence (A.U.)
18
16
14
12
10
8
6
4
2
0
T1
T2
T3
5s
h

shield the PNECs from the airway lumen (Fig. 22.2d–f). Also in lung slices fixed after LCI, immunohistochemical staining revealed CGRP IR in NEB cell groups in the airway epithelium (Fig. 22.2g).

In fresh fetal murine lung slices, pulmonary NEBs could be recognized as clusters of 4-Di-2-ASP-fluorescent small, rounded epithelial cells, surrounded by a continuous layer of non-fluorescent Clara-like cells (Fig. 22.3a). Application to 4-Di-2-ASP-stained lung slices of carbonyl cyanide (trifluoromethoxy)phenylhydrazone (FCCP; 10 μM; 10s), a H^+ ionophore that uncouples oxidative phosphorylation and depolarises mitochondria, could be shown to evoke a simultaneous increase in 4-Di-2-ASP fluorescence in all fetal mouse NEB cells and in ciliated epithelial cells. No alterations in 4-Di-2-ASP fluorescence were observed in the virtually non-stained Clara and Clara-like cells (Fig. 22.3b, c).

Stimulation of a 4-Di-2-ASP stained and Fluo-4 loaded lung slice with the established positive control for postnatal lungs (i.e., 50 mM $[K^+]_o$; 5s), resulted in a robust $[Ca^{2+}]_i$ rise in all prenatal NEB cells, confirming the appropriate Fluo-4 loading and viability of the NEBs. Importantly, the observed $[Ca^{2+}]_i$ changes in prenatal NEB cells were always followed by an increase in Fluo-4 fluorescence in Clara-like cells surrounding the NEBs, but with a delay compared to the rise seen in NEB cells. No other epithelial cells showed a rise in $[Ca^{2+}]_i$ in response to elevation of $[K^+]_o$ (Fig. 22.3d–h).

22.4 Discussion

Although many reports on lung vibratome slices (recent reviews Liberati et al. 2010; Sanderson 2011) have been published, to our knowledge this is the first report of vibratome slices prepared from fetal mouse lungs. Precise agarose filling appears to sufficiently stabilize the tiny lung lobes, enabling the sectioning of 100 μm-thick vibratome slices. Since LCI confirms the functionality of airway epithelial cells, agarose filling does not appear to be harmful. This lung slice technology was developed for fetal mice, because it offers the opportunity to use transgenic models.

A few years ago, a lung slice model was optimized for LCI of neonatal and adult mouse lungs (Pintelon et al. 2005; De Proost et al. 2008). Similar to postnatal lungs, loading of fetal mouse lung slices with the calcium-indicator Fluo-4 or with the styryl pyridinium dye 4-Di-2-ASP resulted in a characteristic mosaic of 4-Di-2-ASP+/Fluo-4+ polygonal ciliated cells and almost non-fluorescent round Clara cells. Also application of 10 μM ATP in Ca^{2+} imaging experiments showed an activation pattern of Clara cells and ciliated cells comparable to our own LCI experiments in postnatal lungs (De Proost et al. 2008), and as described earlier in other airway epithelial models (Knowles et al. 1991).

Immunocytochemical staining of mouse fetal lung cryostat sections, showed that prenatal pulmonary NEB cells are surrounded by a characteristic rim of Clara-like cells. Immunostaining for $P2X_3$ ATP receptors revealed that in mouse fetal lungs some NEBs appear to be already innervated by a vagal sensory component of their innervation. This connection of pulmonary NEBs to the central nervous system long before birth, suggests an important role for NEBs during intrauterine life and/or in neonatal adaptation, as has been proposed for rat fetal NEBs (Brouns et al. 2003).

An important prerequisite to enable the study of pulmonary NEBs in live lung slices, is the possibility to discriminate them in the airway epithelium. For functional LCI in postnatal lung slices, the vital dye 4-Di-2-ASP can be used to identify pulmonary NEBs as 4-Di-2-ASP-fluorescent groups of small epithelial cells (Pintelon et al. 2005; De Proost et al. 2008). 4-Di-2-ASP staining in prenatal lungs showed 4-Di-2-ASP fluorescent NEB cells, surrounded by virtually non-fluorescent Clara-like cells. Since styryl pyridinium dyes are believed to enter selective groups of secretory cells during the recycling of synaptic vesicles (Fukuda et al. 2003), the selective staining of NEB cells in fetal mouse lungs may indicate functional exocytosis of NEB cells in prenatal life. In fetal mice, 4-Di-2-ASP loaded cells, i.e., the NEB cells and ciliated cells, responded to the mitochondrial uncoupler FCCP, pointing out that 4-Di-2-ASP can additionally be used to monitor changes in mitochondrial membrane potential.

Loading of 4-Di-2-ASP stained fetal mouse lung slices with the Ca^{2+} indicator Fluo-4 allowed for detection of the Ca^{2+}-dependent activation of NEB cells. Forced depolarization with high $[K^+]_o$, an established control stimulus for postnatal NEBs, evoked a fast, reversible and reproducible $[Ca^{2+}]_i$ increase in fetal NEB cells. Similar to observations in neonatal lung slices, Clara-like cells displayed the subsequent delayed rise in $[Ca^{2+}]_i$, which is mediated by the release of ATP from activated NEB cells (De Proost et al. 2009). In fetal mouse lung slices, both activation and exocytosis of NEB cells can, therefore, be studied simultaneously.

Following LCI in fetal mouse lung slices, NEB cells could be visualized using antibodies against selective 'marker molecules' such as calcitonin gene-related peptide, allowing the use of this model to link morphology to physiology for fetal mouse NEBs.

The present study has demonstrated the possibility to selectively visualize prenatal NEBs and their responses to applied stimuli in an *ex vivo* mouse lung slice model, using a live cell imaging set-up based on fluorescent indicators of cell function, such as the Ca^{2+} and mitochondrial membrane potential indicators in the present proof-of-concept experiments. The present approach opens interesting new perspectives for unraveling the functional significance of pulmonary NEBs during the prenatal and perinatal period.

References

Adriaensen D, Brouns I, Van Genechten J, Timmermans J-P (2003) Functional morphology of pulmonary neuroepithelial bodies: extremely complex airway receptors. Anat Rec 270A:25–40

Adriaensen D, Brouns I, Pintelon I, De Proost I, Timmermans J-P (2006) Evidence for a role of neuroepithelial bodies as complex airway sensors: comparison with smooth muscle-associated airway receptors. J Appl Physiol 101:960–970

Bollé T, Lauweryns JM, Van Lommel A (2000) Postnatal maturation of neuroepithelial bodies and carotid body innervation: a quantitative investigation in the rabbit. J Neurocytol 29:241–248

Brouns I, Van Genechten J, Burnstock G, Timmermans J-P, Adriaensen D (2003) Ontogenesis of $P2X_3$ receptor-expressing nerve fibres in the rat lung, with special reference to neuroepithelial bodies. Biomed Res 14:80–86

Cutz E, Jackson A (1999) Neuroepithelial bodies as airway oxygen sensors. Respir Physiol 115:201–214

De Proost I, Pintelon I, Brouns I, Kroese ABA, Riccardi D, Kemp PJ, Timmermans J-P, Adriaensen D (2008) Functional live cell imaging of the pulmonary neuroepithelial body microenvironment. Am J Respir Cell Mol Biol 39:180–189

De Proost I, Pintelon I, Wilkinson WJ, Goethals S, Brouns I, Van Nassauw L, Riccardi D, Timmermans J-P, Kemp PJ, Adriaensen D (2009) Purinergic signaling in the pulmonary neuroepithelial body microenvironment unraveled by live cell imaging. FASEB J 23:1153–1160

Fu XW, Wang D, Nurse CA, Dinauer MC, Cutz E (2000) NADPH oxidase is an O_2 sensor in airway chemoreceptors: evidence from K^+ current modulation in wild-type and oxidase-deficient mice. Proc Natl Acad Sci U S A 97:4374–4379

Fukuda J, Ishimine H, Masaki Y (2003) Long-term staining of live Merkel cells with FM dyes. Cell Tissue Res 311:325–332

Gosney JR (1993) Pulmonary neuroendocrine cells in species at high altitude. Anat Rec 236:105–107

Hoyt RF, McNelly NA, Sorokin SP (1993) Calcitonin gene-related peptide (CGRP) as regional mitogen for tracheobronchial epithelium of organ cultured fetal rat lungs. Am Rev Resp Dis 147:A498

Knowles MR, Clarke LL, Boucher RC (1991) Activation by extracellular nucleotides of chloride secretion in the airway epithelia of patients with cystic fibrosis. N Engl J Med 325:533–538

Liberati TA, Randle MR, Toth LA (2010) In vitro lung slices: a powerful approach for assessment of lung pathophysiology. Expert Rev Mol Diagn 10:501–508

Linnoila RI (2006) Functional facets of the pulmonary neuroendocrine system. Lab Invest 86:425–444

Pan J, Yeger H, Cutz E (2004) Innervation of pulmonary neuroendocrine cells and neuroepithelial bodies in developing rabbit lung. J Histochem Cytochem 52:379–389

Pintelon I, De Proost I, Brouns I, Van Herck H, Van Genechten J, Van Meir F, Timmermans J-P, Adriaensen D (2005) Selective visualisation of neuroepithelial bodies in vibratome slices of living lung by 4-Di-2-ASP in various animal species. Cell Tissue Res 321:21–33

Sanderson MJ (2011) Exploring lung physiology in health and disease with lung slices. Pulm Pharmacol Ther 24:452–465

Sorokin SP, Hoyt RF (1990) On the supposed function of neuroepithelial bodies in adult mammalian lungs. News Physiol Sci 5:89–95

Sorokin SP, Hoyt RF, Shaffer MJ (1997) Ontogeny of neuroepithelial bodies: correlations with mitogenesis and innervation. Microsc Res Tech 37:43–61

Van Lommel A, Bolle T, Fannes W, Lauweryns JM (1999) The pulmonary neuroendocrine system: the past decade. Arch Histol Cytol 62:1–16

Widdicombe JG (2001) Airway receptors. Respir Physiol 125:3–15

Chapter 23
Oxygen Sensitivity of Gill Neuroepithelial Cells in the Anoxia-Tolerant Goldfish

Peter C. Zachar and Michael G. Jonz

Abstract In the zebrafish, O_2-chemoreceptive neuroepithelial cells (NECs) of the gill arches detect changes in PO_2 and are believed to initiate cardiorespiratory responses to hypoxia. Goldfish have gill NECs of similar morphology and innervation, yet these animals are naturally tolerant to prolonged periods of anoxia. Whole-cell, voltage-clamp experiments indicated that goldfish NECs express a variety of membrane ion channels, including background and Ca^{2+}-activated K^+ channels. Our initial studies suggest that goldfish NECs do not respond equally to hypoxia (N_2, 25 mmHg) as do those of zebrafish; however, current-clamp recordings indicated that anoxia produced membrane depolarization. In addition, we found that cyanide, which mimics hypoxia, depolarized NECs. Our goal is to further characterize these membrane conductances and determine their potential contribution to O_2 sensing in NECs of the anoxia-tolerant goldfish.

Keywords Goldfish • Neuroepithelial cells • Chemoreceptors • Oxygen sensing • Hypoxia • Anoxia tolerance • Cyanide • Electrophysiology • Patch-clamp • Calcium-activated potassium channels

23.1 Introduction

In fish, the challenges of matching oxygen uptake to metabolic demand are compounded by the limited oxygen availability in their surroundings, making these animals uniquely sensitive to hypoxia and ideal for comparative studies in oxygen sensing. Neuroepithelial cells (NECs) are found in the gills of all fish species investigated and are believed to be the primary sites of peripheral oxygen sensing. NECs may store a variety of neurotransmitters within cytoplasmic synaptic vesicles, although serotonin appears to be most common across species. Gill NECs detect changes in either environmental or arterial oxygen partial pressure (P_{O2}) and initiate reflex cardiorespiratory responses, such as hyperventilation (Perry et al. 2009). These cells are, therefore, regarded as evolutionary precursors of peripheral oxygen chemoreceptors in mammals.

The gills of teleost fish are composed of four pairs of arches that give rise to numerous primary filaments. Each filament in turn produces respiratory lamellae, the sites of gas transfer. The NECs

P.C. Zachar • M.G. Jonz (✉)
Department of Biology, University of Ottawa, 30 Marie Curie, Ottawa, ON K1N 6N5, Canada
e-mail: pzachar@uottawa.ca; mjonz@uottawa.ca

C.A. Nurse et al. (eds.), *Arterial Chemoreception: From Molecules to Systems*,
Advances in Experimental Medicine and Biology 758, DOI 10.1007/978-94-007-4584-1_23,

are located within the primary filament epithelium facing the incident flow of water as it is pumped through the buccal cavity and over the gills during respiration. This places NECs in a prime location to sample the P_{O2} of inspired water or the nearby arterial circulation (Perry et al. 2009). The gills receive a rich supply of sensory innervation from the glossopharyngeal and vagus nerves (Sundin and Nilsson 2002). Furthermore, fibres from multiple neuronal populations contact NECs and may receive pre-synaptic input from these cells during hypoxic stimulation (Jonz and Nurse 2003). The neurochemistry of the NEC-nerve synapse, however, remains largely unresolved (Perry et al. 2009).

The physiological response of NECs to hypoxia at the cellular level has been established in zebrafish (Jonz et al. 2004). A decrease in extracellular P_{O2} inhibits K^+ current through the plasma membrane, representing a mechanism for membrane depolarization and thus transduction of the hypoxic stimulus. This inhibition is insensitive to classical blockers of voltage-activated K^+ channels, such as tetraethylammonium (TEA) and 4-aminopyridine (4-AP), but sensitive to the drug quinidine (Jonz et al. 2004). These observations indicate that the oxygen-sensitive K^+ current is carried by voltage-independent background K^+ (K_B) channels – as is the case in other cellular models of oxygen sensing (Buckler 1997; O'Kelly et al. 1999; Campanucci et al. 2003).

While K_B channels may be responsible for the initial depolarization in response to hypoxia, ion channel content and oxygen sensitivity observed in carotid body type I cells of mammalian models presents a more complicated story. For example, in rat type I cells, K_B channels are oxygen sensitive, yet these cells also express calcium-activated (K_{Ca}) and voltage-dependent (K_V) potassium channels (Buckler 1997; Lopez-Lopez et al. 1997).

In zebrafish, the dominant whole-cell conductance is through K_B channels; however, this may not be the case in other related fish species. Different species of fish exhibit varying tolerances to changes in external P_{O2}, from the relatively hypoxia-intolerant rainbow trout, to the anoxia-tolerant goldfish (Burggren 1982; Bushnell et al. 1984). Differences in tolerance to hypoxia and anoxia may be correlated with different sensitivity or ion channel composition in the NECs of the gill, as NECs of different species retain many morphological similarities regardless of hypoxia tolerance (Saltys et al. 2006). Comparative study across species with varying tolerances is useful in uncovering the relative importance and/or the roles different ion channels might play in modulating the response of NECs to hypoxia. To this end, we are using goldfish (*Carassius auratus*) – an anoxia-tolerant cyprinid fish closely related to zebrafish – to study ion channel expression and oxygen sensitivity in this species.

23.2 Methods

Physiological characterization of the NECs was done using whole-cell patch-clamp electrophysiology on dissociated cells obtained from gill filaments of goldfish following procedures modified from Jonz et al. (2004). Dissociation of goldfish gill filaments was achieved by immersion in 0.25% trypsin (Invitrogen, Burlington, ON, Canada) and mechanical trituration. Once dissociated, cells were plated on glass-bottomed 35 mm culture dishes (MatTek Corporation, Ashland, MA, USA) coated with Poly-L-lysine (Sigma-Aldrich, Oakville, ON, Canada) and matrigel (BD Biosciences, Mississauga, ON, Canada) and cultured in L-15 medium supplemented with 2% penicillin/streptomycin and 2.5% fetal calf serum (Invitrogen, Burlington, ON, Canada). Dishes containing cells in primary culture were used for patch-clamp recordings within 24–36 h. Identification of potential oxygen-sensitive NECs was done by introducing 2 mg/ml Neutral Red, a vital marker taken up by acidic granules in the cytoplasm. This technique is commonly used to identify cells containing amines, such as serotonin (Youngson et al. 1993; Jonz et al. 2004). Electrodes were filled with intracellular recording solution consisting of (in mM): KCl (135), NaCl (5), $CaCl_2$ (0.1), HEPES (10),

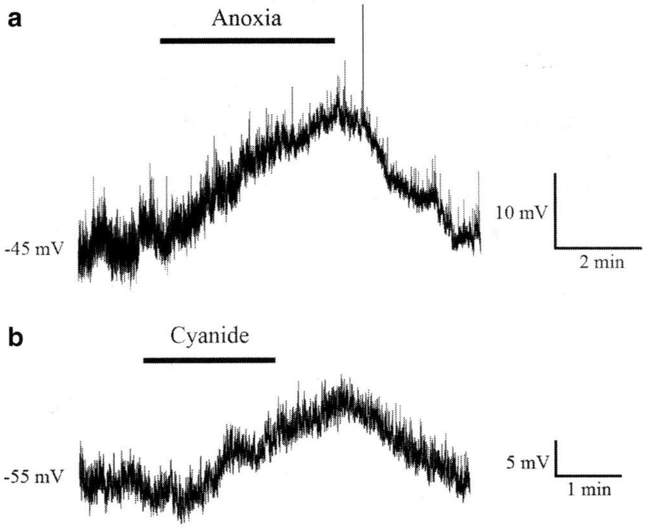

Fig. 23.1 Current-clamp (I = 0) recording of resting membrane potential from goldfish gill neuroepithelial cell (NEC) in primary culture. (a) Membrane depolarization of approximately 15 mV in response to chemical anoxia generated by 2 mM dithionite, (b) depolarization of approximately 10 mV in response to application of 2 mM sodium cyanide. Resting potentials are indicated to the left of each trace. Scale bars indicate the change in potential vs. time

Mg-ATP (2), EGTA (11), and pH adjusted to 7.4 using KOH. Extracellular recording solution consisted of (in mM): KCl (5), NaCl (135), $CaCl_2$ (2), HEPES (10), $MgCl_2$ (2), Glucose (10), and pH was adjusted to 7.8 with NaOH. Recordings were obtained using an Axon Digidata 1440A data acquisition system in conjunction with an Axon Multiclamp 700B microelectrode amplifier (Molecular Devices, Sunnyvale, CA, USA). Signals were sampled at 10 kHz and filtered at 5 kHz. All recordings were corrected appropriately for junction potentials. Hypoxia was generated by bubbling the extracellular solution with 95% N_2. Anoxia was generated by adding 2 mM dithionite (Sigma-Aldrich) to the extracellular solution and maintained by bubbling with N_2. Cyanide (2 mM) was similarly administered. Tetraethylammonium (TEA, 20 mM) and 4-aminopyridine (4-AP, 5 mM) (Sigma-Aldrich) were applied briefly to inhibit voltage-dependent K^+ channels. All solutions were maintained at a constant pH and were applied to the recording bath under constant superfusion at a rate of 2–4 ml/min.

23.3 Results and Discussion

Exposure of Neutral Red-positive NECs from goldfish to levels of hypoxia similar to those used to study zebrafish NECs (25 mmHg; Jonz et al. 2004) yielded no observable changes in membrane currents elicited under voltage-clamp. In preliminary experiments using chemical anoxia generated by the addition of 2 mM dithionite to the extracellular solution, Neutral Red-positive goldfish NECs were observed to depolarize (Fig. 23.1a). Application of 2 mM sodium cyanide elicited a similar response (Fig. 23.1b). In both dithionite and cyanide experiments, NECs repolarized upon wash-out with normal solution.

These observations indicate that, while perhaps not responsive to levels of hypoxia conventionally used to assess oxygen sensitivity in chemoreceptive cells of zebrafish (Jonz et al. 2004), goldfish NECs do respond to more severe drops in P_{O2}. This is not surprising, as the blood oxygen affinity in

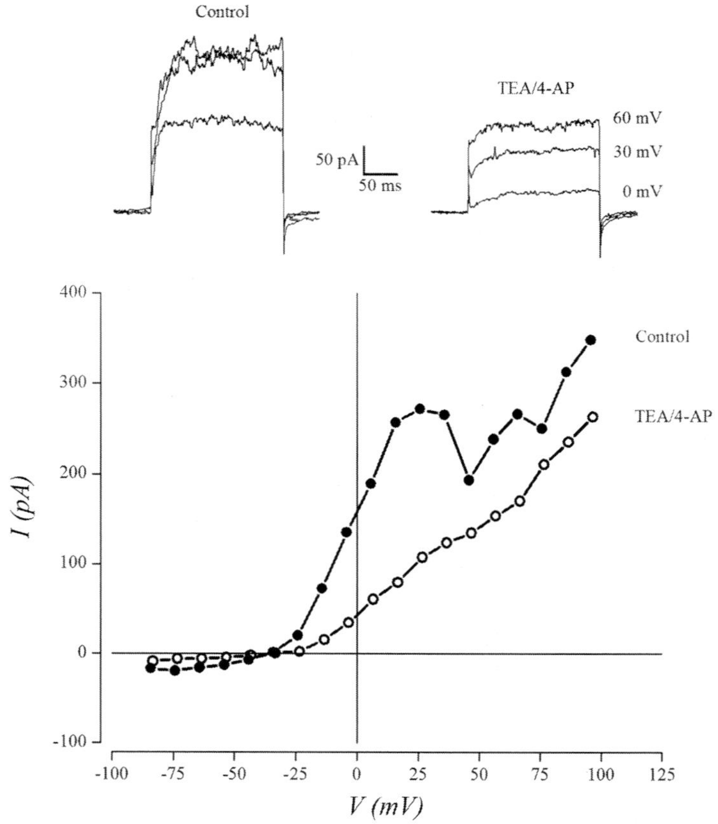

Fig. 23.2 Whole-cell recordings (*inset*) and current–voltage (I-V) relations generated by sequential steps to a range of test potentials from −80 to + 100 mV in 10 mV increments. Cells were held at −60 mV. A characteristic 'shoulder' in the control trace of the I-V indicates the presence of calcium-activated potassium channels (K_{Ca}). Inhibition of voltage-gated potassium channels with TEA and 4-AP reveals the presence of background potassium channels (K_B). *Top panels* show steps to 0, 30, and 60 mV under control conditions and after application of 20 mM TEA and 5 mM 4-AP

the anoxia-tolerant goldfish is relatively high (P_{50} = 2.6 mmHg; Burggren 1982) as compared to other fish, such as sculpins (P_{50} ranging from 22 to 58 mmHg; Mandic et al. 2009) and rainbow trout (P_{50} = 22.9 mmHg; Bushnell et al. 1984), implying that goldfish NECs need not respond until external P_{O2} levels are much lower. In addition, the finding that cyanide, a potent stimulant of the carotid body, depolarizes the NEC membrane suggests that the mitochondria may play a role in the oxygen sensing mechanism (Wyatt and Buckler 2004). This finding is complicated by the nature of cyanide, however, as it is a metabolic inhibitor rather than a chemical known to specifically target oxygen-sensitive cells.

It is evident that a calcium-activated K^+ current (K_{Ca}) is prominent in whole-cell voltage-clamp recordings from goldfish NECs (Fig. 23.2). This type of whole-cell current profile is reminiscent of rat carotid body type I cells (Lopez-Lopez 1997). By contrast, such currents were not reported in zebrafish (Jonz et al. 2004) or channel catfish (Burleson et al. 2006). The present data are also suggestive of a Ca^{2+} conductance across the plasma membrane, as has been reported in gill NECs of trout (Zhang et al. 2011), that precedes activation of K_{Ca} currents. Application of TEA and 4-AP reduced outward current and eliminated the characteristic K_{Ca} 'shoulder' (indicative of K_{Ca} activation) in the whole-cell current–voltage relationship, yielding a profile closely resembling that of an open-rectifier-type background potassium channel (K_B) (Fig. 23.2).

K_{Ca} channels have been observed in rat carotid body type I cells, where they may contribute to the resting membrane potential and provide oxygen sensitivity (Peers and Wyatt 2007). As in type I cells, when the NEC plasma membrane is depolarized by hypoxia (Jonz et al. 2004), voltage-gated calcium channels (Ca_V) would be activated, allowing Ca^{2+} ions to flow down their electrochemical gradient into the cell. Ca^{2+} ions would then activate K_{Ca} channels, increasing K^+ conductance across the plasma membrane. A role for K_{Ca} channels in goldfish, however, is not presently clear. Such a hyperpolarizing current may suggest a role for K_{Ca} channels in blunting the hypoxic response, or promoting repolarization for subsequent hypoxic stimuli. A similar mechanism of negative feedback has been proposed for cardiac and smooth muscle in mammals (Guia et al. 1999; Herrera et al. 2005). Furthermore, future experiments may reveal if K_{Ca} channels in goldfish NECs are oxygen sensitive, as they are in type I cells, and if they may contribute to oxygen chemotransduction.

Experiments to date have demonstrated that goldfish NECs depolarize in response to severe decreases in extracellular O_2, as generated by dithionite. Additionally, TEA and 4-AP resistant (e.g. K_B) channels believed to be primarily responsible for hypoxia-induced depolarization in zebrafish NECs and carotid body type I cells are present in goldfish NECs. The presence of K_{Ca} channels in goldfish NECs, as observed in type I cells, raises the possibility that these channels may be involved in modulation or chemotransduction. These data also indicate that the ensemble of ion channels in peripheral chemoreceptors is species-specific in fish and may correlate with natural history or tolerance to hypoxia or anoxia.

References

Buckler KJ (1997) A novel oxygen-sensitive potassium current in rat carotid body type I cells. J Physiol 498 (Pt 3):649–662

Burggren WW (1982) "Air gulping" improves blood oxygen transport during aquatic hypoxia in the goldfish Carassius Auratus. Physiol Zool 55:327

Burleson ML, Mercer SE, Will-Blaszczak MA (2006) Isolation and characterization of putative O_2 chemoreceptor cells from the gills of channel catfish (Ictalurus punctatus). Brain Res 1092:100–107

Bushnell PG, Steffensen JF, Johansen K (1984) Oxygen consumption and swimming performance in hypoxia-acclimated rainbow trout *Salmo gairdneri*. J Exp Biol 113:225–235

Campanucci VA, Fearon IM, Nurse CA (2003) A novel O2-sensing mechanism in rat glossopharyngeal neurones mediated by a halothane-inhibitable background K+conductance. J Physiol 548:731–743

Guia A, Wan X, Courtemanche M, Leblanc N (1999) Local Ca2+ entry through L-type Ca2+ channels activates Ca2+-dependent K+channels in rabbit myocytes. Circ Res 84:1032–1042

Herrera GM, Etherton B, Nausch B, Nelson MT (2005) Negative feedback regulation of nerve-mediated contractions by KCa channels in mouse urinary bladder smooth muscle. Am J Physiol Regul Integr Comp Physiol 289:R402–R409

Jonz MG, Nurse CA (2003) Neuroepithelial cells and associated innervation of the zebrafish gill: a confocal immunofluorescence study. J Comp Neurol 461:1–17

Jonz MG, Fearon IM, Nurse CA (2004) Neuroepithelial oxygen chemoreceptors of the zebrafish gill. J Physiol 560:737–752

Lopez-Lopez JR, Gonzalez C, Perez-Garcia MT (1997) Properties of ionic currents from isolated adult rat carotid body chemoreceptor cells: effect of hypoxia. J Physiol 499(Pt 2):429–441

Mandic M, Todgham AE, Richards JG (2009) Mechanisms and evolution of hypoxia tolerance in fish. Proc R Soc B 276:735–744

O'Kelly I, Stephens RH, Peers C, Kemp PJ (1999) Potential identification of the O2-sensitive K+current in a human neuroepithelial body-derived cell line. Am J Physiol 276:L96–L104

Peers C, Wyatt CN (2007) The role of maxiK channels in carotid body chemotransduction. Respir Physiol Neurobiol 157:75–82

Perry SF, Jonz MG, Gilmour KM (2009) Oxygen sensing and the hypoxic ventilatory response. In: Richards JG, Farrell AP, Brauner CJ (eds) Hypoxia. Elsevier, New York, p 194

Saltys HA, Jonz MG, Nurse CA (2006) Comparative study of gill neuroepithelial cells and their innervation in teleosts and xenopus tadpoles. Cell Tissue Res 323:1–10

Sundin L, Nilsson S (2002) Branchial Innervation. J Exp Zool 293:232–248
Wyatt CN, Buckler KJ (2004) The effect of mitochondrial inhibitors on membrane currents in isolated neonatal rat carotid body type I cells. J Physiol 556(1):175–191
Youngson C, Nurse C, Yeger H, Cutz E (1993) Oxygen sensing in airway chemoreceptors. Nature 365:153–155
Zhang L, Nurse CA, Jonz MG, Wood CM (2011) Ammonia sensing by neuroepithelial cells and ventilatory responses to ammonia in rainbow trout. J Exp Biol 214:2678–2689

Chapter 24
Interaction of Hypoxia and Core Temperature: Potential Role of TRPV1

Nathaniel Y.W. Yuen, Sandra G. Vincent, Brian Foo, and John T. Fisher

Abstract Hypoxia exposure in small mammals elicits an initial rise in ventilation followed by a reduction to levels that are often less than the normoxic value. The fall in ventilation is matched by a decrease in metabolism rate and a reduction in core body temperature (Tb). The transient receptor potential vanilloid 1 (TRPV1) ion channel has been implicated in thermoregulation (Caterina et al., Science 288:306–313, 2000) and recently shown to exert a tonic effect on Tb in human subjects (Gavva et al., Pain 136:202–210, 2008). We review herein the hypothesis that TRPV1 modulates the Tb response to hypoxia. We provide preliminary evidence that a 24 h hypoxia (FIO_2=0.1) exposure caused an enhanced decrease in Tb in mutant TRPV1$^{-/-}$ mice compared to the TRPV1$^{+/+}$ genotype (Tb was ≈ 1°C lower than TRPV1$^{+/+}$). Further investigation is warranted to determine the extent of TRPV1 ion channel involvement in acute and adaptive responses to hypoxia.

Keywords Hypoxia • TRPV1 • Murine • Thermoregulation •

24.1 Introduction

Thermoregulation is an essential process for homeothermic animals, allowing them to maintain core body temperature (Tb) within a range that aligns with the maintenance of physiologic function and viability (Bicego et al. 2007). However, various environmental factors, such as changes in temperature or oxygen concentration (i.e. hypoxia), may challenge temperature homeostasis (Bicego et al. 2007). Among small mammals, as well as neonates, hypoxia elicits a dual physiologic response; the ventilatory response consists of an acute increase followed by a subsequent reduction to levels that are less than or equal to the normoxic value. The fall in ventilation is matched by a decrease in metabolic rate, which is accompanied by a decrease in body temperature (Teppema and Dahan 2010; Mortola 2004).

N.Y.W. Yuen • S.G. Vincent • B. Foo
Department of Biomedical and Molecular Sciences, Queen's University, 4th Floor Botterell Hall, 18 Stuart Street, Botterell Hall Room 442, Ontario, Kingston K7L 3N6, Canada

J.T. Fisher (✉)
Departments of Biomedical and Molecular Sciences, Paediatrics, Medicine, Queen's University, 4th Floor Botterell Hall, 18 Stuart Street, Botterell Hall Room 442, Ontario, Kingston K7L 3N6, Canada
e-mail: fisherjt@queensu.ca

C.A. Nurse et al. (eds.), *Arterial Chemoreception: From Molecules to Systems*, Advances in Experimental Medicine and Biology 758, DOI 10.1007/978-94-007-4584-1_24,

Compensation for external or internal challenges to the normal homeostatic control of Tb relies on various mechanisms, such as shivering and non-shivering thermogenesis (i.e. metabolism of brown fat) during cold challenge, or activation of mechanisms to enhance heat loss due to elevated temperatures (e.g. sweat and vasodilation) (Bicego et al. 2007). Thermoregulatory pathways embrace a sophisticated control system, which includes sensory receptors, central integration of sensory projections and the eventual recruitment of thermoeffectors to reduce the magnitude or presence of fluctuation of Tb (Bicego et al. 2007; Teppema and Dahan 2010). Gavva and co-workers recently reported an important and unanticipated role for the transient receptor potential vanilloid 1 (TRPV1) ion channel in thermoregulation (Gavva et al. 2007) (for review see (Fisher 2009). This observation coupled with the link between TRPV1 and the regulation of breathing (Lee and Gu 2009; Fisher 2009) raises the question of whether TRPV1 contributes to thermoregulatory control seen in small mammals during hypoxia. We review the rationale for such a link and provide preliminary data to indicate that TRPV1 warrants further attention with respect to the thermoregulatory response to hypoxia.

24.2 The TRPV1 Ion Channel and Thermoregulation

The transient receptor potential vanilloid 1 (TRPV1) ion channel is a non-selective cationic channel belonging to the TRP superfamily of receptors (Clapham 2003; Montell 2005). First described by Caterina and co-workers in the Julius lab (Caterina et al. 1997), the channel is ubiquitously expressed on C-fibre afferents (for review see (Jia and Lee 2007; Fisher 2009). TRPV1 is an important sensory mechanism in the transduction of noxious temperatures (approx. >43°C) and other inflammatory mediators involved in pain (Di Marzo et al. 2002; Van Der Stelt and Di Mazoro (2004); Ross 2003). TRPV1 is well known for its sensitivity to capsaicin, the vanilloid compound that is the pungent component of hot peppers, and activation of the channel by capsaicin elicits a sensation of burning pain (Szallasi and Blumberg 1999). In addition to capsaicin and noxious heat, TRPV1 is also sensitive to the capsaicin-related compounds resiniferatoxin and olvanil, "endovanilloid" compounds such as anandamide, protons, and endogenous inflammatory mediators (Caterina et al. 1997; Caterina and Julius 2001; Pedersen et al. 2005; Ruan et al. 2006). Activation of the ion channel by such diverse stimuli reflects varying sites of action (Julius and Basbaum 2001; Jordt et al. 2000). Activation by a single compound may result in sensitization for other stimuli (Julius and Basbaum 2001; Jordt et al. 2000).

Aside from its role in nociception, recent evidence suggests that TRPV1 is also an important factor in thermoregulation. It is well appreciated that capsaicin administration induces a hypothermic response via TRPV1 to reduce body temperature (Szallasi and Blumberg 1999), which is abolished in mutant TRPV1$^{-/-}$ mice (Caterina et al. 2000). However, a recent study evaluating the analgesic efficacy of a TRPV1 antagonist noted a strong and unanticipated *hyperthermic* effect in human subjects (Gavva et al. 2008). Since the administration of TRPV1 antagonists to TRPV1$^{-/-}$ mutant mice does not elicit a hyperthermia (Caterina et al. 2000), it suggests that TRPV1 is the sole mediator of antagonist-mediated temperature responses. These and other studies led to an appreciation that TRPV1 is tonically active and important in temperature regulation (Caterina 2007; Gavva et al. 2007).

24.3 Effect of Hypoxia On Thermoregulation

In large mammalian species, including humans, hypoxia results in a sustained increase in ventilation, which is the hypoxic ventilatory response (HVR), designed to increase the O_2 flux to the lung (Mortola 2005; Teppema and Dahan 2010). However, smaller mammals (e.g. mice or rats) and neonates (including humans) (Mortola 2005) utilise the control of metabolic rate as an additional mechanism to cope with the reduced O_2 availability associated with hypoxia. Thus, the hypoxic response is typically described as an initial and short term increase in ventilation at the onset of hypoxia, followed by an adaptive or conforming response to the hypoxia through hypometabolism (Mortola 2004). This results in a reduction of metabolic rate as evidenced by decreased VO_2 (Barros et al. 2001; Mortola 2005), inhibition of thermogenesis (Donhoffer and Szelenyi 1967), and increased heat loss (Miller and Miller 1966). The ability to lower metabolism has been suggested to be inversely proportional to the animal size (Frappell et al. 1992).

24.4 Potential Role of TRPV1 During Hypoxia

Ristoiu et al. (2011) identified cellular mechanisms that are activated during hypoxia that alter TRPV1 function. They describe a novel mechanism that sensitizes TRPV1 gating by phosphorylation of serine residues, which in turn relies on the well-described hypoxia-induced activation of hypoxia-inducible factor-1 alpha (HIF-1α) and protein kinase C epsilon (PKCε) (Ristoiu et al. 2011). Thus, both the biophysical and in vivo properties of TRPV1 suggest that it has the potential to not only impact on thermoregulatory control during normoxia, but to influence the thermoregulatory response associated with hypoxia. Preliminary data from our laboratory (see below) support the idea of a role for TRPV1 in the thermoregulatory response to hypoxia.

Despite the finding that TRPV1 antagonism can lead to hyperthermia (Gavva et al. 2008), TRPV1 knockouts do not experience increased body temperature, perhaps reflecting compensatory mechanisms activated during intrauterine and postnatal development (Vincent et al. 2007). Figure 24.1 illustrates the circadian variation of body temperature during a single light–dark cycle of normoxia, followed by 24 h of hypoxia ($FIO_2=0.1$) and 24 h of recovery in normoxia.

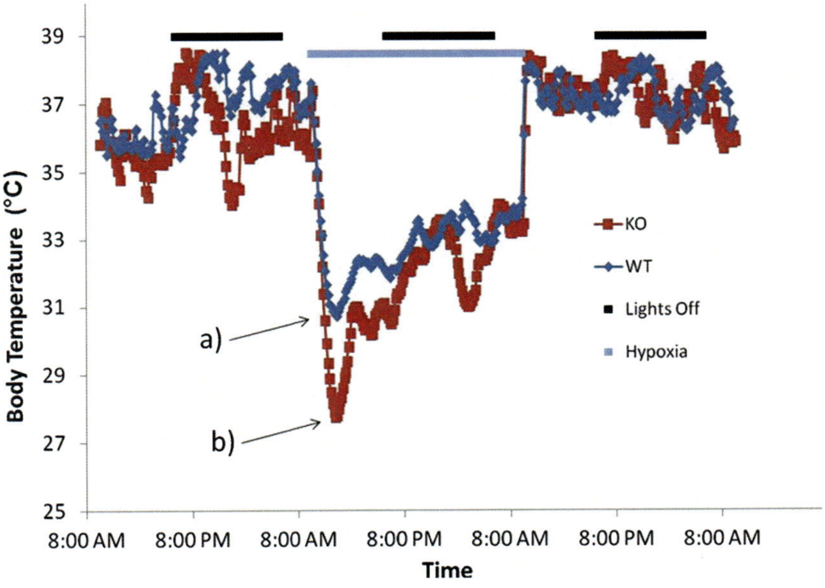

Fig. 24.1 Body temperature response of individual TRPV1$^{-/-}$ (KO) and TRPV1$^{+/+}$ (WT) mice exposed to hypoxia ($FIO_2=0.1$ as indicated). The TRPV1$^{-/-}$ genotype reached a lower Tb compared with the wildtype control. Mice were instrumented with telemetry transmitters (DSI) for the measurement of Tb, HR and activity and allowed to recover for at least 2 weeks. The surgical and exposure protocols were approved by the Institutional University Animal Care Committee and met CCAC guidelines

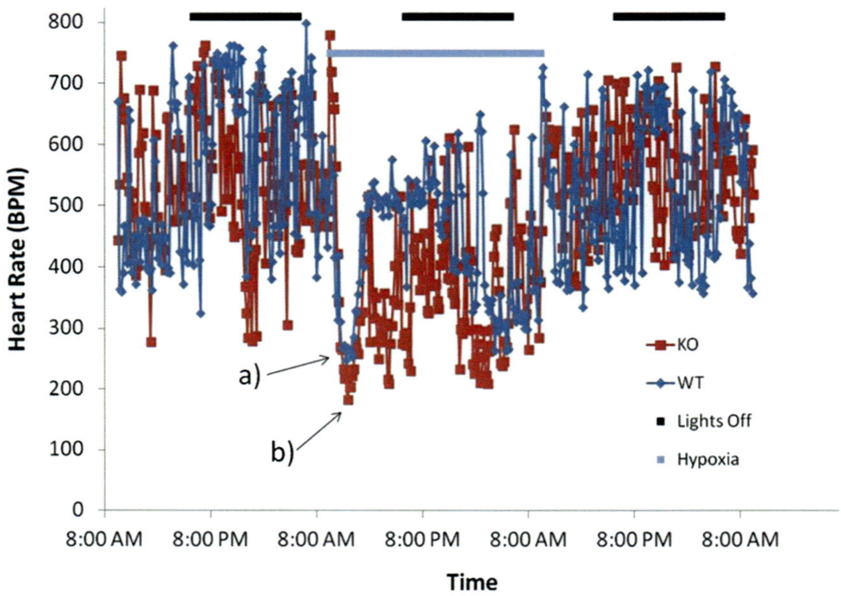

Fig. 24.2 Heart rate response of individual TRPV1$^{-/-}$ (KO) and TRPV1$^{+/+}$ (WT) mice exposed to hypoxia ($FIO_2=0.1$ as indicated). Similar to Tb, the TRPV1$^{-/-}$ genotype reached a lower HR than the wildtype genotype throughout most of the hypoxic exposure

As expected, hypoxia caused a reduction in Tb, which has been shown to be associated with the normal decrease in metabolism (Mortola 2004). However, the TRPV1$^{-/-}$ genotype reached a lower minimum Tb (Fig. 24.1b: 28.8°C; a 20.7% decrease in Tb from the KO normoxic average of 36.38°C) during hypoxia than the wild-type controls (Fig. 24.1a: 29.9°C; a 18.7% decrease in Tb from the WT normoxic average of 36.80°C). This difference in Tb between KO and WT remained throughout the hypoxic exposure. Preliminary statistics did not yield significant differences, however a power analysis suggests that greater numbers would reach statistical significance. All figures depict comparisons between an individual WT and KO mouse but are representative of the preliminary study population. Interestingly, the overall pattern of response to hypoxia is similar for both genotypes, which consists of an initial falling phase, where Tb reaches a nadir followed by a slow recovery phase, as compensatory mechanisms are engaged to maintain homeostasis at a lower set-point (Bicego et al. 2007). The similar trajectory between genotypes indicates that the TRPV1$^{-/-}$ phenotype retains the inherent ability to sense and respond to hypoxia, but the nature of the response is altered.

As with Tb, the decrease in heart rate (Fig. 24.2) for the TRPV1$^{-/-}$ mice reached a lower heart rate (Fig. 24.2b: 201 bpm; a 63.7% decrease in HR from the KO normoxic average of 554 bpm) during hypoxia compared with the wild-type group (Fig. 24.2a: 218 bpm; a 60.7% decrease in HR from the WT normoxic average of 554 bpm). Statistical comparisons reveal no significant differences between the two groups. Telemetry values for activity (Fig. 24.3) were not different between the two genotypes, likely due to the typical response of loss of activity during hypoxia of this magnitude (Brown and Engel 1973).

In summary, our preliminary data suggest that TRPV1 may be implicated in the thermoregulatory responses associated with exposure to hypoxia. Indeed, the *in vivo* responses appear to be consistent with the presence of cellular sensitization of TRPV1 gating, leading to activation of TRPV1 at previously non-functional Tb or endovanilloid concentrations to decrease body temperature. Whether the impact of TRPV1 on Tb relies on a HIF-1-mediated pathway is unclear and further investigation is warranted to determine the full extent of TRPV1 ion channel involvement in acute and adaptive responses to hypoxia.

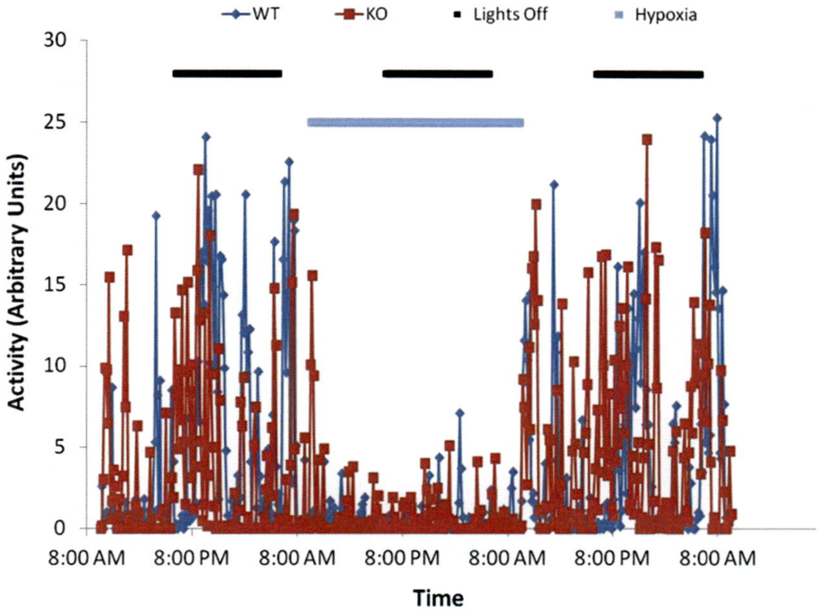

Fig. 24.3 Activity response of individual TRPV1$^{-/-}$ (KO) and TRPV1$^{+/+}$ (WT) mice exposed to hypoxia ($FIO_2 = 0.1$ as indicated). No differences between genotypes were noted

References

Barros RC, Zimmer ME, Branco LG, Milsom WK (2001) Hypoxic metabolic response of the golden-mantled ground squirrel. J Appl Physiol 91:603–612

Bicego KC, Barros RC, Branco LG (2007) Physiology of temperature regulation: comparative aspects. Comp Biochem Physiol A Mol Integr Physiol 147:616–639

Brown R, Engel J (1973) Evidence for catecholamine involvement in the suppression of locomotor activity due to hypoxia. J Pharm Pharmacol 25:815–819

Caterina MJ (2007) Transient receptor potential ion channels as participants in thermosensation and thermoregulation. Am J Physiol Regul Integr Comp Physiol 292:R64–R76

Caterina MJ, Julius D (2001) The vanilloid receptor: a molecular gateway to the pain pathway. Annu Rev Neurosci 24:487–517

Caterina MJ, Schumacher MA, Tominaga M, Rosen TA, Levine JD, Julius D (1997) The capsaicin receptor: a heat-activated ion channel in the pain pathway. Nature 389:816–824

Caterina MJ, Leffler A, Malmberg AB, Martin WJ, Trafton J, Petersen-Zeitz KR, Koltzenburg M, Basbaum AI, Julius D (2000) Impaired nociception and pain sensation in mice lacking the capsaicin receptor. Science 288:306–313

Clapham DE (2003) TRP channels as cellular sensors. Nature 426:517–524

Di Marzo V, Blumberg PM, Szallasi A (2002) Endovanilloid signaling in pain. Curr Opin Neurobiol 12:372–379

Donhoffer S, Szelenyi Z (1967) The role of brown adipose tissue in thermoregulatory heat production in the warm- and cold-adapted adult rat. Acta Physiol Acad Sci Hung 32:53–60

Fisher JT (2009) The TRPV1 ion channel: implications for respiratory sensation and dyspnea. Respir Physiol Neurobiol 167:45–52

Frappell P, Lanthier C, Baudinette RV, Mortola JP (1992) Metabolism and ventilation in acute hypoxia: a comparative analysis in small mammalian species. Am J Physiol 262:R1040–R1046

Gavva NR, Bannon AW, Surapaneni S, Hovland DN Jr, Lehto SG, Gore A, Juan T, Deng H, Han B, Klionsky L, Kuang R, Le A, Tamir R, Wang J, Youngblood B, Zhu D, Norman MH, Magal E, Treanor JJ, Louis JC (2007) The vanilloid receptor TRPV1 is tonically activated in vivo and involved in body temperature regulation. J Neurosci 27:3366–3374

Gavva NR, Treanor JJ, Garami A, Fang L, Surapaneni S, Akrami A, Alvarez F, Bak A, Darling M, Gore A, Jang GR, Kesslak JP, Ni L, Norman MH, Palluconi G, Rose MJ, Salfi M, Tan E, Romanovsky AA, Banfield C, Davar G (2008) Pharmacological blockade of the vanilloid receptor TRPV1 elicits marked hyperthermia in humans. Pain 136:202–210

Jia Y, Lee LY (2007) Role of TRPV receptors in respiratory diseases. Biochim Biophys Acta 1772:915–927

Jordt SE, Tominaga M, Julius D (2000) Acid potentiation of the capsaicin receptor determined by a key extracellular site. Proc Natl Acad Sci U S A 97:8134–8139

Julius D, Basbaum AI (2001) Molecular mechanisms of nociception. Nature 413:203–210

Lee LY, Gu Q (2009) Role of TRPV1 in inflammation-induced airway hypersensitivity. Curr Opin Pharmacol 9:243–249

Miller JA Jr, Miller FS (1966) Interactions between hypothermia and hypoxia-hypercapnia in neonates. Fed Proc 25:1338–1341

Montell C (2005) The TRP superfamily of cation channels. Sci STKE 2005:re3

Mortola JP (2004) Implications of hypoxic hypometabolism during mammalian ontogenesis. Respir Physiol Neurobiol 141:345–356

Mortola JP (2005) Influence of temperature on metabolism and breathing during mammalian ontogenesis. Respir Physiol Neurobiol 149:155–164

Pedersen SF, Owsianik G, Nilius B (2005) TRP channels: an overview. Cell Calcium 38:233–252

Ristoiu V, Shibasaki K, Uchida K, Zhou Y, Ton BH, Flonta ML, Tominaga M (2011) Hypoxia-induced sensitization of transient receptor potential vanilloid 1 involves activation of hypoxia-inducible factor-1 alpha and PKC. Pain 152:936–945

Ross RA (2003) Anandamide and vanilloid TRPV1 receptors. Br J Pharmacol 140:790–801

Ruan T, Lin YS, Lin KS, Kou YR (2006) Mediator mechanisms involved in TRPV1 and P2X receptor-mediated, ROS-evoked bradypneic reflex in anesthetized rats. J Appl Physiol 101:644–654

Szallasi A, Blumberg PM (1999) Vanilloid (Capsaicin) receptors and mechanisms. Pharmacol Rev 51:159–212

Teppema LJ, Dahan A (2010) The ventilatory response to hypoxia in mammals: mechanisms, measurement, and analysis. Physiol Rev 90:675–754

Van Der Stelt M, Di Mazoro V (2004) Endovanilloids. Putative endogenous ligands of transient receptor potential vanilloid 1 channels. Eur J Biochem 271:1827–1834

Vincent SG, Waddell AE, Caron MG, Walker JK, Fisher JT (2007) A murine model of hyperdopaminergic state displays altered respiratory control. FASEB J 21:1463–1471

Chapter 25
Neonatal Intermittent Hypoxia Induces Persistent Alteration of Baroreflex in Adult Male Rats

Cécile A. Julien, Richard Kinkead, Vincent Joseph, and Aida Bairam

Keywords Neonatal apnea • Cardiovascular regulation

25.1 Introduction

Baroreflex is involved in the regulation of arterial blood pressure (BP). An increase in BP activates vagal inhibitory pathways to decrease heart rate; a concomitant decrease in sympathetic discharge reduces vascular resistance. Both responses reduce BP towards normal value. Conversely, a decrease in BP produces opposite effects to increase heart rate and vascular resistance.

We recently showed that adult rats previously exposed to chronic neonatal intermittent hypoxia (n-IH) from postnatal days 3–12 show abnormal cardio-respiratory regulation as indicated by change in ventilation, elevated BP, and attenuated splanchnic nerve activity in response to acute hypoxia (Julien et al. 2010). To explain these results, we proposed that n-IH induces long-term changes of cardiovascular control system that may include a persistent attenuation of the baroreflex. To test this hypothesis, we used the same paradigm of n-IH exposure. At adulthood, we used an anesthetized rat preparation to evaluate the effects of previous exposure to n-IH on (1) BP during changes in peripheral vascular resistances, (2) heart rate (HR) responses to BP changes (cardiac baroreflex) and, (3) splanchnic nerve activity (SNA) in response to BP changes (sympathetic baroreflex).

C.A. Julien (✉) • R. Kinkead • A. Bairam
Department of Pediatrics, Laval University, Centre de Recherche St-François d'Assise, Hôpital Saint-François d'Assise, 10 rue de l'Espinay, Québec, QC G1L3L5, Canada
e-mail: Cecile.Dolbec@crchuq.ulaval.ca

V. Joseph
Department of Pediatrics, Laval University, Centre de Recherche St-François d'Assise, Hôpital Saint-François d'Assise, 10 rue de l'Espinay, Québec, QC G1L3L5, Canada

CRCHUQ/Hop Laval, Québec G0A3X0, Canada

C.A. Nurse et al. (eds.), *Arterial Chemoreception: From Molecules to Systems*, Advances in Experimental Medicine and Biology 758, DOI 10.1007/978-94-007-4584-1_25,

25.2 Materials and Methods

Experimental protocol was approved by the Laval University Animal Care Committee. Ten male Sprague–Dawley rats born in our animal care facility from six virgin females (Charles Rivers St-Constant, QC, CA) were exposed to either neonatal intermittent hypoxia (n-IH, nadir 5% O_2 in 100 s, every 6 min for 1 h followed by 1 h at 21% O_2, repeated 24 h a day) or ambient air (control) for ten consecutive days from postnatal days 3 to 12 as previously described (Julien et al. 2010). At the end of the n-IH exposure, rats were raised under normal conditions until adulthood (3 months old).

For BP, HR, and SNA recordings, adult rats were anesthetized using isoflurane (3.5% in 35% O_2 balanced nitrogen), tracheotomized, and mechanically ventilated (model 683, Harvard Instruments, Holliston, MA, USA). The end-tidal partial pressure of carbon dioxide ($P_{ET}CO_2$) was measured in the expired line of the ventilatory circuit using a capnograph (model 1265, Novametrix, Wallingford, CT, USA) and ventilation was adjusted for $P_{ET}CO_2$ between 35 and 40 mmHg. The left femoral artery was catheterized for BP and HR monitoring and blood sampling for arterial blood gas analysis (PaO_2, $PaCO_2$, pHa, $[HCO_3^-]$ corrected for body temperature (ABL-5, Radiometer, Copenhagen, Denmark). The left femoral vein was catheterized for drug injection. SNA recording was performed as previously described (Julien et al. 2010). The splanchnic nerve signal was amplified (gain = 10,000; model 1,700 AM-System, Everett, WA, USA), band-pass filtered (0.01–10 kHz), and fed to a moving averager (200 ms; model MA-821RSP, CWE incorporated, Ardmore, PA, USA). The signal from the blood pressure transducer was amplified (gain = 100; Transbridge TBM4M-B, World Precision Instrument, Sarasota, FL, USA). Then, all signals were digitized and recorded with a data acquisition system (WinDaq model DI-720, Dataq instruments, Akron, OH, USA). After surgical preparation, the rats were slowly converted from isoflurane to urethane anesthesia (1.6 g/kg iv. in saline, duration 10–20 min) and paralyzed with pancuronium bromide (2.5 mg/kg iv.). The lungs were hyperinflated roughly once per 30 min to prevent alveolar atelectasis.

BP, HR, and SNA amplitude were recorded throughout experimental procedure under 50% O_2. After stabilization, BP, HR, and SNA responses to incrementing doses of vasoactive agents were recorded. Specifically, we used increasing doses of the vasodilator sodium nitroprusside (SNP, 10 mg/ml with infusion from 0.02 to 3.2 ml; duration 1 min each; vasodilation test). This was followed by increasing doses of phenylephrine (PE, 10 mg/ml with infusion from 0.02 to 3.2 ml; duration 1 min each; peripheral vasoconstriction). Between vasodilation and vasoconstriction tests, rats were left undisturbed during at least 20 min. Blood gases were measured before and after each test.

For each dose of SNP or PE, cardiovascular changes were expressed as difference (Δ) with preceding baseline value for BP and HR and as % of preceding baseline value for SNA. Data were expressed as mean ± sem. Due to the small number of rats in each group (n = 5), statistical differences were analyzed using Mann–Whitney non-parametric test. Significance was defined at $p < 0.05$.

25.3 Results

No difference in body weight, blood pressure, heart rate, PaO_2, $PaCO_2$, and pHa was observed in baseline condition between control and n-IH rats (data not shown).

Peripheral vasoreactivity was assessed by measuring the BP response to increasing doses of vasodilator SNP and vasoconstrictor PE in anesthetized control and n-IH adult rats (Fig. 25.1a, b). Vasodilation induced a lower BP decrease in n-IH rats; in this group, the vasoconstriction response was shifted toward higher doses compared to controls. Detailed characteristics of dose–response curve of BP to SNP and PE were determined using sigmoidal regression fit and shown in Table 25.1. Following administration of increasing doses of SNP, the slope of the linear portion of the hypotension response was significantly lower in n-IH than in control rats but the maximal BP response (plateau) was similar.

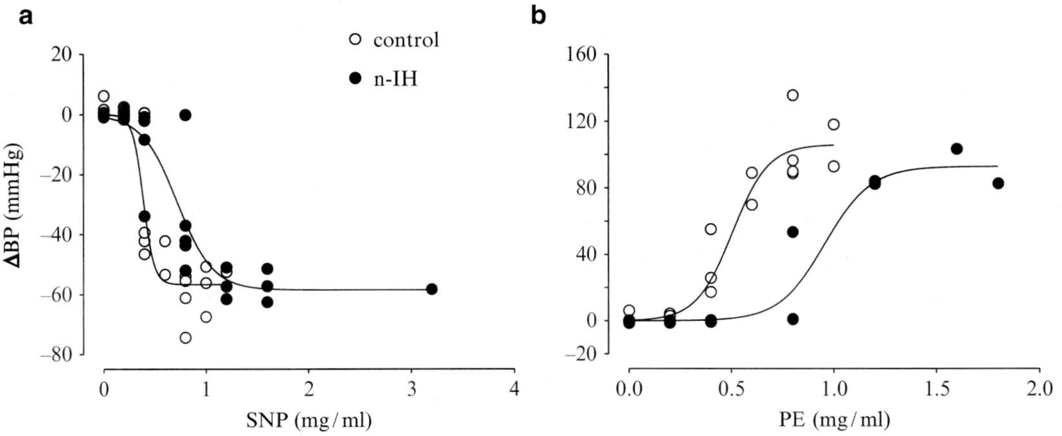

Fig. 25.1 Dose–response curve of blood pressure (BP) to increasing doses of (**a**) sodium nitroprusside (SNP) and (**b**) phenylephrine (PE) in anesthetized control and n-IH adult male rats (n=5 rats per group). BP was expressed as individual difference (Δ) with preceding baseline value for each dose of vasoactive drug

Table 25.1 Descriptive variables for dose-response curves of blood pressure (BP) to sodium nitroprusside and phenylephrine

	Sodium nitroprusside		Phenylephrine	
	Control	n-IH	Control	n-IH
Dose at plateau (mg/ml)	0.84±0.04	1.68±0.39*	0.72±0.10	1.33±0.13*
ΔBP at plateau (mmHg)	−59±3	−57±2	104±8	88±5
Slope (mmHg/mg/ml)	−174±31	−50±12*	150±18	106±12
Fit coefficient, R^2=	0.94	0.92	0.95	0.95

Sigmoidal regression fit was used to determined drug dose at the beginning of the plateau, BP change at plateau and slope of the linear portion in control and n-IH adult male rats. (n=5 per group)
*$p<0.05$ Mann–Whitney *U*-test vs. control rats

During PE injection, the dose at which BP began to increase was greater in n-IH rats than controls; however, the plateau was similar for both groups.

Cardiac baroreflex was evaluated in anesthetized control and n-IH adult rats by representing changes in HR vs. changes in BP (Fig. 25.2a). The index of cardiac baroreflex represented as the mean of individual slopes in n-IH and control groups (right panel), showed a significant increase in cardiac baroreflex in n-IH rats compared to control rats. Thus for a similar BP increase, the bradycardia observed in n-IH rats was greater than controls.

Sympathetic baroreflex was compared between groups by representing the changes in SNA vs. changes in BP (Fig. 25.2b). The index of sympathetic baroreflex represented as the mean of individual slopes (right panel), shows a significant increase in sympathetic baroreflex in n-IH rats compared to control rats, i.e. larger decrease in SNA for similar BP increase in n-IH rats.

25.4 Discussion

This study showed that, in comparison with controls, adult male rats previously exposed to intermittent hypoxia during the neonatal period are characterized by (1) a reduced peripheral vasoreactivity to nitroprusside and phenylephrine, (2) an increased cardiac baroreflex index, and (3) a stronger sympathetic

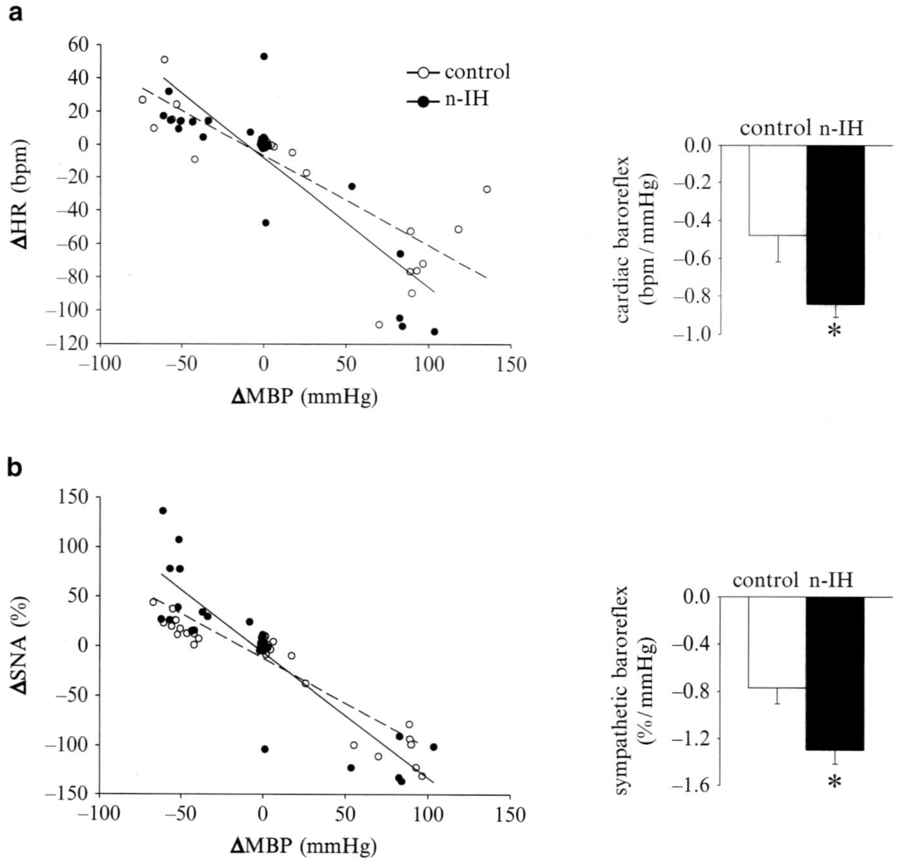

Fig. 25.2 (**a**) Cardiac and (**b**) sympathetic baroreflex in anesthetized control and n-IH adult male rats. Left panel represents individual relationship between changes in heart rate (HR) or splanchnic nerve activity (SNA) and changes in blood pressure (BP). Right panel represents the cardiac or sympathetic baroreflex index calculated in each rat as the slope of the linear relationship between changes in HR or SNA and changes in BP. (n=5 per group).*$p<0.05$ Mann–Whitney U-test vs. control rats

baroreflex index. Taken together, these results indicate that, neonatal intermittent hypoxia induces long term attenuation of peripheral vasoreactivity as well as a persistent increase in both cardiac and sympathetic baroreflex.

Peripheral vasoreactivity attenuation is one of the major factors for cardiovascular deregulation including arterial hypertension. This is observed both in patients with obstructive sleep apnea (OSA) and arterial hypertension (Imadojemu et al. 2002) and in adult animals exposed to chronic intermittent hypoxia (animal model of OSA) (Dematteis et al. 2008; Julien et al. 2003; Tahawi et al. 2001). According to this, the decrease in BP response to both SNP and PE injections may be related to the permanent elevation of BP we previously observed in n-IH adult male rats (Julien et al. 2010). Mechanisms underlying peripheral vasoreactivity attenuation concern regulation of both vasodilation and vasoconstriction. Vasodilation was induced using SNP, a nitric oxide donor that activates the non-endothelium dependent vasodilation pathway via increasing cGMP. Vasoconstriction was induced using PE, an α_1-adrenoceptor agonist that targets smooth muscle cells of the vascular wall and reflects neurogenic activation via the sympathetic system. In n-IH rats, the lower slope of the BP response to SNP as well as the right-shift of the BP response to PE suggest a preferential change in the reactivity rather than in the maximal capacity of the system to react.

In addition to changes in peripheral vasoreactivity in n-IH versus control rats, cardiac and sympathetic baroreflexes were enhanced. The increase in cardiac baroreflex was in agreement with previous observations in awake n-IH rats where the hypoxic-induced BP surge was associated with a larger fluctuation of HR (Julien et al. 2010). Of note, a blunted baroreflex was found in patients with sleep apnea (Carlson et al. 1996; Leung and Bradley 2001) as well as after exposure to chronic intermittent hypoxia either during 30 postnatal days (Soukhova-O'Hare et al. 2006) or in adult age (animal model of OSA) (Dematteis et al. 2008; Fletcher 2001). These opposite results may correspond to differences in experimental setup or period of exposure. Opposite results due to a period of exposure suggests that a small postnatal exposure window such as ten postnatal days, leads to persistent increase in baroreflex. Alteration of baroreflex reflects an impairment of autonomic system either at rest (baroreflex shift) or during BP changes (baroreflex impairment). While the arterial hypertension (Julien et al. 2010) and the vasoreactivity dysfunction (this study) observed in n-IH rats are in accordance with a shift of baroreflex toward higher BP level, the larger fluctuation of HR during hypoxia-induced BP surge (Julien et al. 2010) and the increase in both cardiac and sympathetic baroreflexes are in accordance with baroreflex impairment. In addition, the increased sympathetic baroreflex may reflect a compensatory response to peripheral vasoreactivity alteration. However, we have previously observed that sympathetic nervous response was reduced in n-IH adult rats during chemoreflex challenge using acute hypoxic episodes (Julien et al. 2010). This suggests differential alterations of chemoreflex and baroreflex by n-IH in rats.

In our animal model, exposure to intermittent hypoxia during the 10 consecutive postnatal days produces major persistent cardiovascular alterations that are significant risk factors for cardiovascular disorders such as systemic arterial hypertension, atherosclerosis and/or heart failure at adult age.

Acknowledgments We acknowledge Van Diep Doan for technical assistance, Melanie Pelletier, Sylvie Vigier, and Evelyne Vachon for animal care. Supported by grants from The Hospital for Sick Children's Foundation/Canadian institute on Health Research (Grant #XG07-006), Réseau en Santé Respiratoire du FRSQ.

References

Carlson JT, Hedner JA, Sellgren J, Elam M, Wallin BG (1996) Depressed baroreflex sensitivity in patients with obstructive sleep apnea. Am J Respir Crit Care Med 154:1490–1496

Dematteis M, Julien C, Guillermet C, Sturm N, Lantuejoul S, Mallaret M, Levy P, Gozal E (2008) Intermittent hypoxia induces early functional cardiovascular remodeling in mice. Am J Respir Crit Care Med 177:227–235

Fletcher EC (2001) Invited review: Physiological consequences of intermittent hypoxia: systemic blood pressure. J Appl Physiol 90:1600–1605

Imadojemu VA, Gleeson K, Quraishi SA, Kunselman AR, Sinoway LI, Leuenberger UA (2002) Impaired vasodilator responses in obstructive sleep apnea are improved with continuous positive airway pressure therapy. Am J Respir Crit Care Med 165:950–953

Julien C, Sam B, Patrick L (2003) Vascular reactivity to norepinephrine and acetylcholine after chronic intermittent hypoxia in mice. Respir Physiol Neurobiol 139:21–32

Julien CA, Niane LM, Kinkead R, Bairam A, Joseph V (2010) Carotid sinus nerve stimulation, but not intermittent hypoxia, induces respiratory LTF in adult rats exposed to neonatal intermittent hypoxia. Am J Physiol Regul Integr Comp Physiol 299(1):R192–205

Leung RS, Bradley TD (2001) Sleep apnea and cardiovascular disease. Am J Respir Crit Care Med 164:2147–2165

Soukhova-O'Hare GK, Cheng ZJ, Roberts AM, Gozal D (2006) Postnatal intermittent hypoxia alters baroreflex function in adult rats. Am J Physiol Heart Circ Physiol 290:H1157–1164

Tahawi Z, Orolinova N, Joshua IG, Bader M, Fletcher EC (2001) Altered vascular reactivity in arterioles of chronic intermittent hypoxic rats. J Appl Physiol 90:2007–2013, discussion 2000

Chapter 26
LPS-Induced c-Fos Activation in NTS Neurons and Plasmatic Cortisol Increases in Septic Rats Are Suppressed by Bilateral Carotid Chemodenervation

Edison-Pablo Reyes, Sebastián Abarzúa, Aldo Martin, Jorge Rodríguez, Paula P. Cortés, and Ricardo Fernández

Abstract Lipopolysaccharide (LPS) administered I.P. increases significantly the activation of c-Fos in neurons of the nucleus of the solitary tract (NTS), which in turn activates hypothalamus-pituitary-adrenal axis. The vagus nerve appears to play a role in conveying cytokines signals to the central nervous system (CNS), since -in rodent models of sepsis- bilateral vagotomy abolishes increases in plasmatic glucocorticoid levels, but does not suppress c-Fos NTS activation. Considering that NTS also receives sensory inputs from carotid body chemoreceptors, we evaluated c-Fos activation and plasmatic cortisol levels 90 min after I.P. administration of 15 mg/kg LPS. Experiments were performed in male Sprague–Dawley rats, in control conditions and after bilateral carotid neurotomy (BCN). LPS administration significantly increases the number of c-Fos positive NTS neurons and plasmatic cortisol levels in animals with intact carotid/sinus nerves. When LPS was injected after BCN, the number of c-Fos positive NTS neurons, and plasmatic cortisol levels were not significantly modified. Our data suggest that carotid body chemoreceptors might mediate CNS activation during sepsis.

Keywords LPS • Inflammation • Carotid sinus nerve • c-Fos • NTS

E.-P. Reyes
Facultad de Medicina, Clínica Alemana-Universidad del Desarrollo,
Av. Las Condes 12438, Lo Barrenechea, 7710162 Santiago, Chile

S. Abarzúa • A. Martin • P.P. Cortés
Departamento de Ciencias Biologicas, Facultad de Ciencias Biologicas,
Universidad Andres Bello, Santiago, Chile

J. Rodríguez
Facultad de Medicina, Escuela de Medicina, Universidad Andres Bello, Santiago, Chile

R. Fernández (✉)
Departamento de Ciencias Biologicas, Facultad de Ciencias Biologicas,
Universidad Andres Bello, Av. Republica 252, 8370134 Santiago, Chile
e-mail: rfernandez@unab.cl

C.A. Nurse et al. (eds.), *Arterial Chemoreception: From Molecules to Systems*,
Advances in Experimental Medicine and Biology 758, DOI 10.1007/978-94-007-4584-1_26,

26.1 Introduction

Inflammation induced by lipopolysaccharide (LPS) enhances neural and endocrine response through an increase in release of glucocorticoids and catecholamines, which in turn regulates pro-inflammatory cytokines (Steinman 2004). The effects of LPS may result from either direct actions of pro-inflammatory cytokines on central nervous system (CNS) structures, or mediated by activation of autonomic afferent pathways (Dantzer et al. 2000).

The role of peripheral sensory nerves in immunomodulation is not yet fully understood. Neuroimmune chemosensory transduction would begin at vagal paraganglia glomus cells, innervated by vagal afferent neurons (Berthoud et al. 1995), the somata of which are located in the nodose ganglion, and their central projection ending primarily within the dorsal vagal complex of the *medulla oblongata*. Thus, immunosensory inputs could: (1) initiate local cardiorespiratory reflexes, (2) convey information about the inflammatory process, and (3) activate specific signs of diseases. However, neither tumor necrosis factor (TNF)-α nor interleukin (IL)-1β has significant effects on the frequency of action potentials recorded from vagal nerve paraganglia (Mac Grory et al. 2010). In addition, in rodents exposed to LPS, bilateral subdiaphragmatic vagotomy hinders some sickness symptoms (Bluthe et al. 1994), but failed to prevent the activation –assessed by the induction of c-Fos protein (Hoffman et al. 1993)– of the *nucleus tractus solitarii* (NTS) (Wan et al. 1994; Hermann et al. 2001). Thus, NTS activation and the anti-inflammatory response induced by the administration of LPS could be mediated by the carotid body (CB) chemoreceptors (Fernandez et al. 2008, 2011; Zapata et al. 2011).

The carotid body is the largest paraganglion in the body (Mascorro and Yates 1980), and like other paraganglia, it has specialized glomus cells innervated by the carotid/sinus nerve, which forms its first synapse CNS at the NTS. In consequence, inflammation-derived sensory inputs originated from carotid body chemoreceptors can be differentially processed in the *medulla*. It is noteworthy that central (ascending or descending) projections from the NTS provide a neuronal substrate for interaction between immunosensory signals, hypothalamic-pituitary-adrenal axis, and autonomous nervous system as an immunomodulatory mechanism (Pavlov et al. 2003).

26.2 Materials and Methods

Male Sprague–Dawley rats (100–120 g), anesthetized with Na-pentobarbitone (60 mg/kg intraperitoneally, I.P.) and spontaneously breathing, were placed in supine position on a regulated heating pad. Body temperature was maintained about 37°C using a rectal thermistor probe. Right saphenous vein was cannulated for blood sampling. At the end of experiments, animals were euthanized by an overdose of anesthetic.

A ventral incision in the neck was performed to identify and section both carotid/sinus nerves (BCN, n = 12) close to the carotid body. Thereafter, either saline (Control group) or 15 mg/kg LPS (from *Escherichia coli* Serotype 0127:B8. Sigma-Aldrich Corp, USA) was administered I.P. After 90-min, blood samples were collected in heparin-containing tubes, centrifuged at 1,000 g for 15-min and plasma were stored at −80°C until use. Then, animals were fixed by formaldehyde trans-cardiac perfusion. Same procedure was used for Control animals (Co, n = 14) where both carotid/sinus nerves remained intact.

The brainstems were removed and post-fixed for 24 h before being transferred into sucrose solution. NTS sections were obtained on cryostat and processed for c-Fos protein activation in response to LPS. Briefly, slices were rinsed, blocked, and incubated with primary antibodies against rat c-Fos (Rabbit anti-c-Fos (Ab-5), Calbiochem, Merck, Germany). Then, washed sections were incubated with biotinylated secondary antibodies (Goat ant-Rabbit IgG, Jackson ImmunoResearch Laboratories, Inc., USA) and immunohistochemical detection was revealed with avidin-biotin-peroxidase complex and peroxidase substrate (Vectastain Elite ABC Kit, Vector Laboratories, Inc., USA). Finally, sections were mounted on glass slides and analyzed under light microscopy. c-Fos-labeled nuclei were counted manually (Hermann et al. 2001). Results are presented as number of c-Fos positive cells/mm^2. Dorsal *medulla* areas were identified according to Paxinos and Watson atlas (Paxinos and Watson 2007).

Cortisol plasma levels were measured by ELISA with cortisol EIA Kit, according to manufacturer's instructions (Cayman Chemical Company, USA). Briefly, 50 μL of plasma samples were added to 96-well plate. Non-specific binding (NSB), maximum binding (B_0), and an eight-point standard curve were also included. Then, cortisol tracer and antiserum were added and the plate was incubated overnight at 4°C. Finally, Ellman's reagent was added and the reaction was developed for 90-min. Plate was read at 405 nm. Cortisol concentrations were obtained by subtracting the NSB absorbance from B_0 absorbance (corrected B_0). Then, B/B_0 (sample or standard bound/maximum bound) were calculated subtracting NBS absorbance from the absorbance of the samples or standard curve, and divided by the corrected B_0. Data from standard curve were transformed using Logit, were Logit $(B/B_0) = \ln [(B/B_0) / (1 - B/B_0)]$, and plotted as Logit (B/B_0) *vs.* cortisol concentration.

Data are expressed as Means ± Standard Deviations of the Mean (SD). Statistical analyses were performed using GraphPad Prism© for Windows (GraphPad Software, CA, USA). Statistical differences were considered as significant when $P<0.05$. All protocols were approved by The Commission of Bioethics of the Universidad Andres Bello. Na-pentobarbitone was kindly provided by Dr. Patricio Zapata.

26.3 Results

All medullary brainstem sections containing the NTS and the *area postrema* (AP) were analyzed for the presence of c-Fos stained nuclei. Saline-treated control rats (Co, n=6), with both carotid/sinus nerves intact, showed discrete number of c-Fos labeled neurons in the NTS (Fig. 26.1a). Systemic administration of LPS induced a significant increase in NTS c-Fos labeling (n=6) (Fig. 26.1b). Since the number of c-Fos positive cells after the I.P. administration of both saline and LPS was symmetrical even after bilateral carotid neurotomy (BCN) ($P>0.05$. Paired *t* test), c-Fos count data were pooled (left NTS + right NTS). The graphic representation of the total number of activated neurons (*per* square mm) is displayed in Fig. 26.2, showing fivefold increases in c-Fos expression in response to systemic endotoxin challenge.

BCN decreased the number c-Fos positive cells in saline-treated rats (n=6) (Fig. 26.1c) and suppressed LPS-induced c-Fos increase in the NTS (Fig. 26.1d) (n=6), when BCN was performed prior to endotoxin challenge. Conversely, the number of c-Fos positive nuclei in the *area postrema* was slightly increased in LPS-treated Co and BCN animals, but did not reach statistical significance ($P>0.05$, Kruskal-Wallis test, not shown).

Plasmatic cortisol levels measured 90-min after endotoxin challenge in both saline- (n=7) and LPS-treated (n=7) control rats showed a threefold increase in LPS-treated rats, but no changes were observed in saline-treated rats (Fig. 26.3).

When saline (n=6) was administered after BCN, no significant changes in plasmatic cortisol were observed. Noteworthy BCN (n=6) suppressed the cortisol increases induced by LPS. Moreover, plasmatic levels were even lower than those measured in saline-treated control group (Fig. 26.3).

26.4 Discussion

LPS-induced systemic inflammation is known to evoke NTS activation but bilateral carotid chemodenervation suppresses both NTS activation and plasma cortisol increase in septic rats. Under our experimental conditions, NTS demonstrated a significant increase in c-Fos nuclear protein labeling

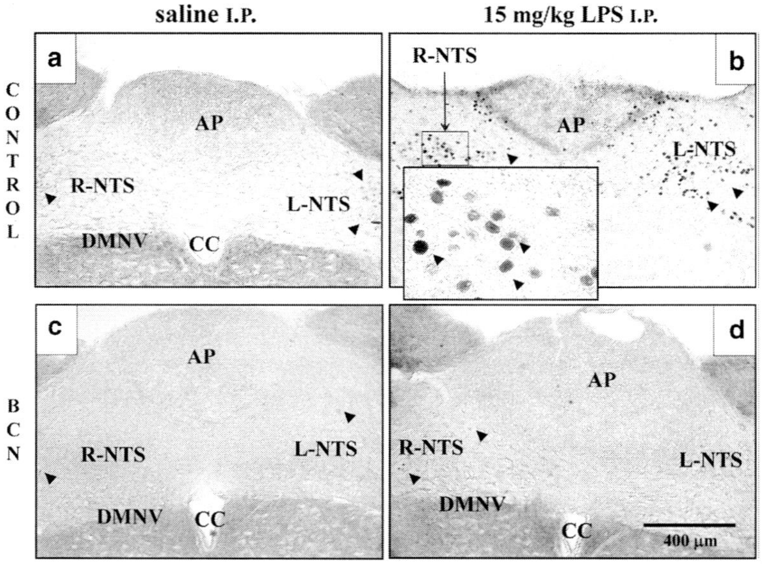

Fig. 26.1 Photomicrographs of coronal histological sections of the NTS at the level of the *area postrema* (AP). c-Fos activation in response to i.p. challenge of either saline (**a, c**) or 15 mg/kg LPS (**b, d**). The latter is characterized by the dark staining cells in **b**. Systemic LPS evoked a substantial, bilateral, and symmetrical increase in c-Fos activation in the NTS excised from control rats, which was suppressed by bilateral carotid neurotomy (BCN). CC, central canal; R, right; L, left; DMNV, dorsal motor nucleus of vagus. Scale bar, 400 μm (10x original magnification). Inset in Co-LPS, 40x original magnification

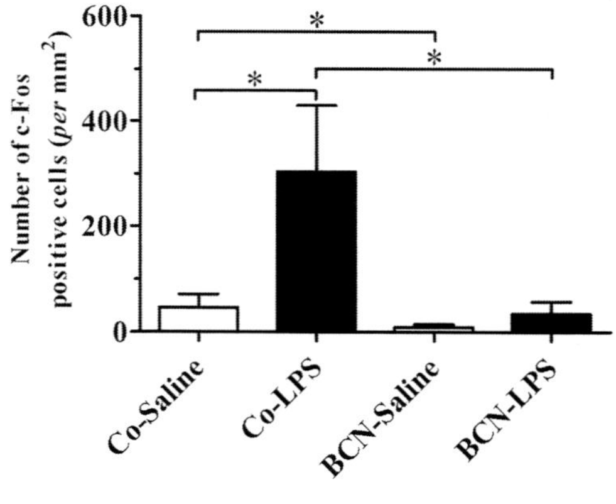

Fig. 26.2 Cell counts of c-Fos-activated neurons in the NTS from control (Co) and bilaterally carotid neurotomized (BCN) rats treated i.p. with either saline (*open bars*) or 15 mg/kg LPS (closed bars). Pooled data showed that LPS increased the number of c-Fos positive neurons in control animals, while in BCN animals, no increase was observed. Values, Means ± SD (n = 6), are expressed *per* mm^2. *$P < 0.05$, Kruskal-Wallis ANOVA followed by Dunn's multiple comparisons test

after systemic challenge of endotoxin in rats with both carotid nerves intact. Vagus nerve appears to be important in conveying inflammation-derived signals to the central nervous system (CNS), but bilateral cervical vagotomy performed in septic rats had a subtle effect on the NTS activation (Hermann et al. 2001). On the contrary, in our experiments we found that bilateral carotid neurotomy (BCN) performed prior to LPS administration suppresses c-Fos activation in the NTS, indicating that

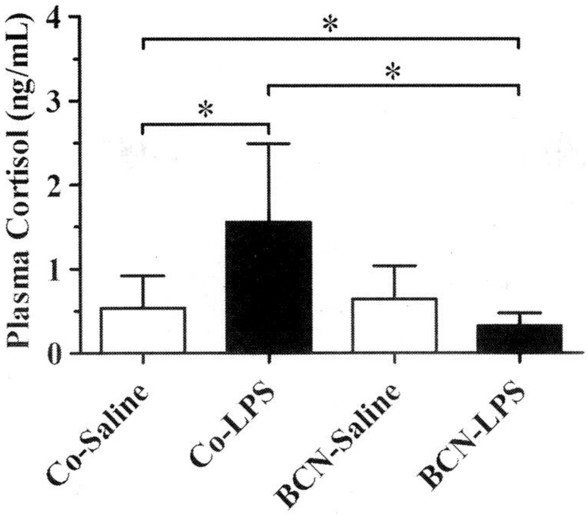

Fig. 26.3 Plasmatic cortisol response to I.P. administration of saline (*open bars*) or 15 mg/kg LPS (*closed bars*) in control (Co) and bilaterally carotid neurotomized (BCN) rats. LPS-induced plasma cortisol increase was suppressed by prior bilateral carotid chemodenervation. Values, Means ± SD (Co, n = 7; BCN, n = 6). *$P < 0.05$, Kruskal-Wallis ANOVA followed by Dunn's multiple comparisons test

carotid body chemoreceptors exert a key effect on the responsiveness of the NTS to LPS and/or immune cells-derived afferent inputs to CNS. It must be mentioned that in this case both vagus nerves remained intact in all animals.

It is also known that chronic hypoxia induces local increase of pro-inflammatory cytokines in the CB, regardless the model studied; intermittent, hypobaric or hypoxic hypoxia (Del Río et al. 2011; Liu et al. 2009; Lam et al. 2008; respectively). We reported that systemic or topical -upon the CB- application of LPS diminished the ventilatory chemoreflex evoked by acute hypoxia, but we did not find any histological evidence of inflammation evoked by that challenge (Fernandez et al. 2008). Considering that apoptosis was not observed, the aforementioned reduction in the chemoreflex may be mediated by release of inhibitory transmitter (De Laurentis et al. 2002). Taken together, above arguments suggest that only a tonical stimulation of the CB induces a local inflammatory response, clearly not enough to produce a systemic inflammatory response syndrome. A possible synergic interaction between a hypoxia stimulated CB and LPS has not been studied yet.

Although nodose neurons may still detect LPS-related signals and convey the information to the CNS through their synaptic inputs to the NTS (Hosoi et al. 2005), we recently described that in septic rats, LPS acts directly upon carotid body chemoreceptors and their sensory ganglion, inducing the local production of TNF-α (Fernandez et al. 2011), which in fact activates NTS neurons (Emch et al. 2000; Hermann and Rogers 2009). Furthermore, in the present work the number of c-Fos positive nuclei in the NTS obtained from BCN rats treated with saline was significantly lower than saline-treated control animals, confirming the tonic discharge from the carotid body chemoreceptors upon NTS.

Carotid body chemoreceptors seem to be an afferent pathway for endocrine response to endotoxemia. In fact, BCN suppresses the cortisol increases induced by LPS. Thus, acting through the NTS, central projections from the carotid body provide a neuronal substrate for interaction between immunosensory signals and hypothalamic-pituitary-adrenal axis, as an immunomodulatory mechanism activated during sepsis.

Acknowledgments Special thanks are due to Mrs. Carolina Larraín for proofreading the manuscript. This work was supported by grant DI-40-11/R (to RF), from the Division for Research of the Universidad Andres Bello (UNAB).

References

Berthoud HR, Kressel M, Neuhuber WL (1995) Vagal afferent innervation of rat abdominal paraganglia as revealed by anterograde DiI-tracing and confocal microscopy. Acta Anat (Basel) 152:127–132

Bluthe RM, Walter V, Parnet P, Laye S, Lestage J, Verrier D, Poole S, Stenning BE, Kelley KW, Dantzer R (1994) Lipopolysaccharide induces sickness behaviour in rats by a vagal mediated mechanism. C R Acad Sci III 317:499–503

Dantzer R, Konsman JP, Bluthe RM, Kelley KW (2000) Neural and humoral pathways of communication from the immune system to the brain: parallel or convergent? Auton Neurosci 85:60–65

De Laurentis A, Pisera D, Caruso C, Candolfi M, Mohn C, Rettori V, Seilicovich A (2002) Lipopolysaccharide- and tumor necrosis factor–induced changes in prolactin secretion and dopaminergic activity in the hypothalamic-pituitary axis. Neuroimmunomodulation 10:30–39

Del Río R, Moya EA, Iturriaga R (2011) Differential expression of pro-inflammatory cytokines, endothelin-1 and nitric oxide synthases in the rat carotid body exposed to intermittent hypoxia. Brain Res 1395:74–85

Emch GS, Hermann GE, Rogers RC (2000) TNF-alpha activates solitary nucleus neurons responsive to gastric distension. Am J Physiol Gastrointest Liver Physiol 279:G582–G586

Fernandez R, Gonzalez S, Rey S, Cortes PP, Maisey KR, Reyes EP, Larrain C, Zapata P (2008) Lipopolysaccharide-induced carotid body inflammation in cats: functional manifestations, histopathology and involvement of tumour necrosis factor-alpha. Exp Physiol 93:892–907

Fernandez R, Nardocci G, Simon F, Martin A, Becerra A, Rodriguez-Tirado C, Maisey KR, Cuna-Castillo C, Cortes PP (2011) Lipopolysaccharide signaling in the carotid chemoreceptor pathway of rats with sepsis syndrome. Respir Physiol Neurobiol 175:336–348

Hermann GE, Rogers RC (2009) TNF activates astrocytes and catecholaminergic neurons in the solitary nucleus: implications for autonomic control. Brain Res 1273:72–82

Hermann GE, Emch GS, Tovar CA, Rogers RC (2001) c-Fos generation in the dorsal vagal complex after systemic endotoxin is not dependent on the vagus nerve. Am J Physiol Regul Integr Comp Physiol 280:R289–R299

Hoffman GE, Smith MS, Verbalis JG (1993) c-Fos and related immediate early gene products as markers of activity in neuroendocrine systems. Front Neuroendocrinol 14:173–213

Hosoi T, Okuma Y, Matsuda T, Nomura Y (2005) Novel pathway for LPS-induced afferent vagus nerve activation: possible role of nodose ganglion. Auton Neurosci 120:104–107

Lam SY, Tipoe GL, Liong EC, Fung ML (2008) Chronic hypoxia upregulates the expression and function of proinflammatory cytokines in the rat carotid body. Histochem Cell Biol 130:549–559

Liu X, He L, Stensaas L, Dinger B, Fidone S (2009) Adaptation to chronic hypoxia involves immune cell invasion and increased expression of inflammatory cytokines in rat carotid body. Am J Physiol Lung Cell Mol Physiol 296:L158–L166

Mac Grory B, O'Connor ET, O'Halloran KD, Jones JF (2010) The effect of pro-inflammatory cytokines on the discharge rate of vagal nerve paraganglia in the rat. Respir Physiol Neurobiol 171:122–127

Mascorro JA, Yates RD (1980) Paraneurons and paraganglia: histological and ultrastructural comparisons between intraganglionic paraneurons and extra-adrenal paraganglion cells. Adv Biochem Psychopharmacol 25:201–213

Pavlov VA, Wang H, Czura CJ, Friedman SG, Tracey KJ (2003) The cholinergic anti-inflammatory pathway: a missing link in neuroimmunomodulation. Mol Med 9:125–134

Paxinos G, Watson CR (2007) The rat brain in stereotaxic coordinates, 6th edn. Academic Press, Burlington, p 456

Steinman L (2004) Elaborate interactions between the immune and nervous systems. Nat Immunol 5:575–581

Wan W, Wetmore L, Sorensen CM, Greenberg AH, Nance DM (1994) Neural and biochemical mediators of endotoxin and stress-induced c-fos expression in the rat brain. Brain Res Bull 34:7–14

Zapata P, Larrain C, Reyes P, Fernandez R (2011) Immunosensory signalling by carotid body chemoreceptors. Respir Physiol Neurobiol 178:370–374

Chapter 27
Developmental Regulation of Glucosensing in Rat Adrenomedullary Chromaffin Cells: Potential Role of the K_{ATP} Channel

Simon Livermore, Nikol A. Piskuric, Shaima Salman, and Colin A. Nurse

Abstract During birth, when the maternal supply of glucose is occluded, there is a drastic fall in blood glucose in the newborn. This stimulus triggers the non-neurogenic release of catecholamines from adrenomedullary chromaffin cells, which restores blood glucose homeostasis. In this report we present preliminary data showing that glucosensing is present in neonatal chromaffin cells from adrenal slices but absent in chromaffin cells from juvenile slices. Moreover, we show that the aglycemia-evoked rise in intracellular Ca^{2+} is robust in neonatal chromaffin cells but blunted in juvenile chromaffin cells. Lastly, we show that the Kir6.2 subunit of the K_{ATP} channel, is upregulated in the adrenal medulla in juvenile animals providing a potential mechanism for the developmental regulation of glucosensing.

Keywords Neonatal • Chromaffin • Amperometry • Development • Fura-2 • Katp • Catecholamines • Western • Aglycemia • Glucose

27.1 Introduction

Occlusion of the umbilical cord at birth prevents the transplacental exchange of O_2, glucose, and CO_2 producing fetal hypoxia, hypoglycemia, and acid hypercapnia (Cuezva et al. 1982; Lagercrantz and Slotkin 1986). These asphyxial stimuli act directly on adrenomedullary chromaffin cells (AMCs) to trigger catecholamine release, which prepares the fetus for extrauterine life by increasing heart contractility, initiating lung fluid absorption, activating lung surfactant secretion, and initiating hepatic glycogen breakdown (Bournaud et al. 2007; Girard et al. 1973; Lagercrantz and Slotkin 1986; Mochizuki-Oda et al. 1997; Munoz-Cabello et al. 2005; Seidler and Slotkin 1986; Thompson et al. 2002, 1997). The sensitivity of AMCs to hypoxia is weak during embryogenesis, peaks in the perinatal period, and is suppressed postnatally coincident with maturation of splanchnic innervation of the adrenal medulla (Bournaud et al. 2007; Lagercrantz and Slotkin 1986; Seidler and Slotkin 1986). The K_{ATP} channel, which opens during hypoxia favouring membrane hyperpolarization, is proposed to underlie this phenomenon because it is highly expressed during embryogenesis and its expression decreases toward birth (Bournaud et al. 2007). Moreover, mimicking splanchnic innervation *in utero* via chronic

S. Livermore (✉) • N.A. Piskuric • S. Salman • C.A. Nurse
Department of Biology, McMaster University, Life Sciences
Building, Rm. 421, 1280 Main St. West, Hamilton, ON L8S 4K1, Canada
e-mail: liverms@mcmaster.ca; piskurn@mcmaster.ca; salmas@mcmaster.ca; nursec@mcmaster.ca

C.A. Nurse et al. (eds.), *Arterial Chemoreception: From Molecules to Systems*,
Advances in Experimental Medicine and Biology 758, DOI 10.1007/978-94-007-4584-1_27,

nicotine exposure abolishes hypoxia-sensing and upregulates the K_{ATP} channel in AMCs (Buttigieg et al. 2009, 2008).

In contrast, there is conflicting evidence regarding the developmental regulation of glucosensing in AMCs. In one study, prolonged hypoglycemia (<4 mM) in adult rats produced a dramatic increase in circulating catecholamines that was abolished by adrenalectomy but not by adrenal denervation, suggesting direct glucosensitivity by adult AMCs (Khalil et al. 1986). In contrast, using an isolated adult rat adrenal gland preparation, hypoglycemia had no effect on catecholamine secretion (de Araujo et al. 1999). In neonatal AMCs, we have recently shown that hypoglycemia/aglycemia activates two opposing pathways, (1) an excitatory pathway favouring membrane depolarization, and (2) an inhibitory pathway via opening of K_{ATP} channels, presumably leading to membrane hyperpolarization (Livermore et al. 2011). Critically, we found that in an immortalized *embryonic* AMC cell line, the primary effect of aglycemia was inhibitory via activation of the K_{ATP} channel (Piskuric et al. 2008). In the present study, we compared the glucosensitivity of neonatal versus juvenile rats, and hypothesized that glucosensitivity was blunted in juvenile AMCs and this developmental difference occurs in parallel with the upregulation of the K_{ATP} channel.

27.2 Materials and Methods

27.2.1 Adrenal Slices

The slicing procedure was done as previously described, with slight modifications (García-Fernández et al. 2007). Adrenals were removed from neonatal (P0-2) or juvenile (P12-14) rats and placed in oxygen-saturated Tyrode's solution (see below) on ice. A single adrenal gland was placed into an agarose solution at 46°C. Once the agarose hardened (15–30 s) extraneous agarose was cut away and the agarose block was glued to the vibratome dish with cyanoacrylate and immersed in bicarbonate-buffered saline (BBS; see below) bubbled with 95% O_2/5% CO_2. Slices were cut 200 μm thick and maintained in BBS bubbled with 95% O_2/5%CO_2 for 15 min to 6 h before experiments were performed. Slices were then placed in a slice chamber (RC-22; Warner Instruments) and pinned to the slide with a slice anchor (SHD-22 L/15; Warner Instruments). The slice platform was placed on the stage of an Axioskop upright microscope (Zeiss) equipped with long distance water immersion objectives. Slices were viewed with differential interference contrast microscopy on a closed-circuit television (CPT-CM 15, Capture, Korea).

27.2.2 Cell Culture

AMCs were cultured as previously described (Thompson et al. 1997). Briefly, adrenal glands were removed from neonatal (P0-2) or juvenile (P12-14) rats and placed in ice-cold L15 medium. The cortex was removed and the remaining medulla was enzymatically dissociated for 1 h then mechanically dissociated with forceps. The cells were then triturated and plated on culture dishes with modified F-12 medium.

27.2.3 Carbon Fiber Amperometry

Catecholamine release was monitored as previously described (Livermore et al. 2011). Briefly, a carbon fiber electrode (Dagan) was placed next to an AMC and polarized to +800 mV. Slices were perfused with BBS (see below) bubbled with 95% air/5% CO_2 at 37°C.

27.2.4 *Fura-2 Ratiometric Ca^{2+} Imaging*

Intracellular Ca^{2+} measurements were performed as previously described (Piskuric et al. 2008). AMCs were loaded with 2.5 μm Fura-2 for 10 min in HEPES-buffered saline (see below) at room temperature and then placed on the stage of an inverted microscope and perfused with BBS (see below) bubbled with 95% air/5% CO_2 at 37°C.

27.2.5 *Western Immunoblotting*

Tissues were homogenized and lysed in Buffer A solution containing: 10 mM HEPES pH 7.6, 10 mM KCl, 0.1 mM EGTA pH 8, 1 mM DTT, protein inhibitors (Complete Mini, Roche, Laval, QC, Canada), 1 mM PMSF, 5 mg/mL Aprotinin, and 5 mg/mL Leupeptin. Protein samples were then placed in boiling water for 5 min. 10 μg of protein samples, measured using the Bradford assay (1:5 dilution reagent and 1 mg/mL BSA), were loaded and resolved on 10% SDS-PAGE gel and transferred onto PVDC membranes. Membranes were then washed and incubated with either primary rabbit polyclonal antibody against K_{ir}6.2 (1:1,000 dilution; Alomone Labs Ltd.) or primary rabbit monoclonal antibody against β-actin (1:10,000 dilution) at 4 °C overnight. Pre-adsorption control was performed in the presence of threefold excess of the respective blocking peptide to determine antibody specificity. Membranes were then washed in PBS, and incubated in a goat anti–rabbit horseradish peroxidase (HRP)–linked secondary antibody (1:10,000 dilution; Jackson Labs, Bar Harbor, ME, USA) for 1 h at room temperature. Immunoreactions were visualized using ECL and exposed to XAR-film.

27.2.6 *Solutions*

The standard BBS contained (in mM): 24 $NaHCO_3$, 115 NaCl, 0–10 glucose, 22–12 sucrose, 5 KCl, 2 $CaCl_2$, and 1 $MgCl_2$, at 37°C, and the pH was kept at 7.4 by bubbling the solution with a 95% air/5% CO_2 gas mixture. HEPES-buffered saline contained (in mM): 135 NaCl, 10 HEPES, 0–10 glucose, 10–0 sucrose, 5 KCl, 2 $CaCl_2$, and 2 $MgCl_2$, at pH = 7.4. The glucose concentration ranged from 10 to 0 mM and the osmolarity was kept constant by the equimolar substitution of sucrose. BBS solutions were made hypoxic (PO_2 ~15 mmHg) by bubbling with 95% N_2/5% CO_2. Tyrode's solution contained (in mM): 148 NaCl, 2 KCl, 3 $MgCl_2$, 10 HEPES, 10 glucose, pH 7.4.

27.3 Results

27.3.1 *Effects of Aglycemia on Catecholamine Secretion from Neonatal and Juvenile AMCs*

We have previously reported that hypoglycemia and/or aglycemia evokes catecholamine secretion from neonatal AMCs *in vitro* (Livermore et al. 2011). In order to study developmental changes in secretory events we made use of an adrenal thin slice preparation that preserves the cellular architecture and functional cholinergic innervation (Barbara et al. 1998). In amperometric recordings a slow upward shift in the baseline during high (30 mM) K^+ exposure corresponded to an increase in bath catecholamines and was used to determine whether or not slices were healthy (Fig. 27.1a–c). When

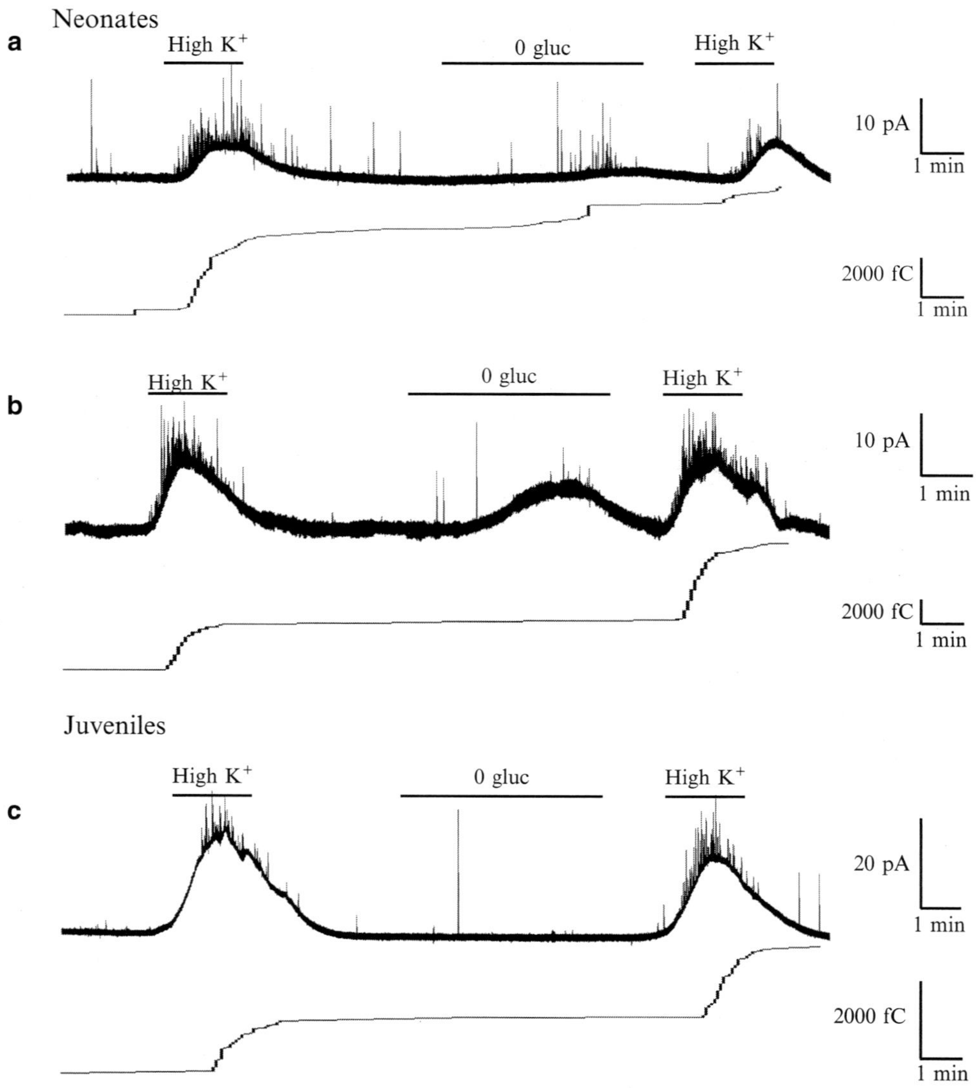

Fig. 27.1 Exemplar traces of aglycemia-evoked responses from juvenile and neonatal adrenomedullary chromaffin cells (AMCs) in thin slices. (**a**) *top*; a recording from a neonatal AMC slice showing quantal release in response to aglycemia (0 Gluc); note the robust high(30 mM)- K^+ evoked secretion, *bottom*; cumulative charge corresponding to the cell above. (**b**) *top*; a trace showing a rise in bath catecholamines during aglycemia, but no quantal release in AMCs from a neonatal slice, *bottom*; cumulative charge corresponding to the cell above. (**c**) *top*; a recording from a juvenile AMC with high K^+-evoked secretion but no aglycemia-evoked secretion, *bottom*; cumulative charge corresponding to the cell above

the carbon fiber microelectrode was close to juvenile or neonatal AMCs quantal catecholamine secretion could be monitored when cells were exposed to high K^+. Likewise, upon switching from control (5 mM glucose) to aglycemia (0 mM glucose), quantal catecholamine secretion could be measured from some neonatal AMCs (Fig. 27.1a). This secretory response was characterized by a ~6 fold increase in frequency of quantal events (5 mM glucose, 3.4 ± 1.9 events/min; 0 mM, 20.0 ± 6.8 events/min; $p < 0.05$, $n = 3$ cells) with no significant change in mean quantal charge. Overall, there was a ~5 fold increase in secretion rate when neonatal AMCs were exposed to aglycemia (5 mM glucose,

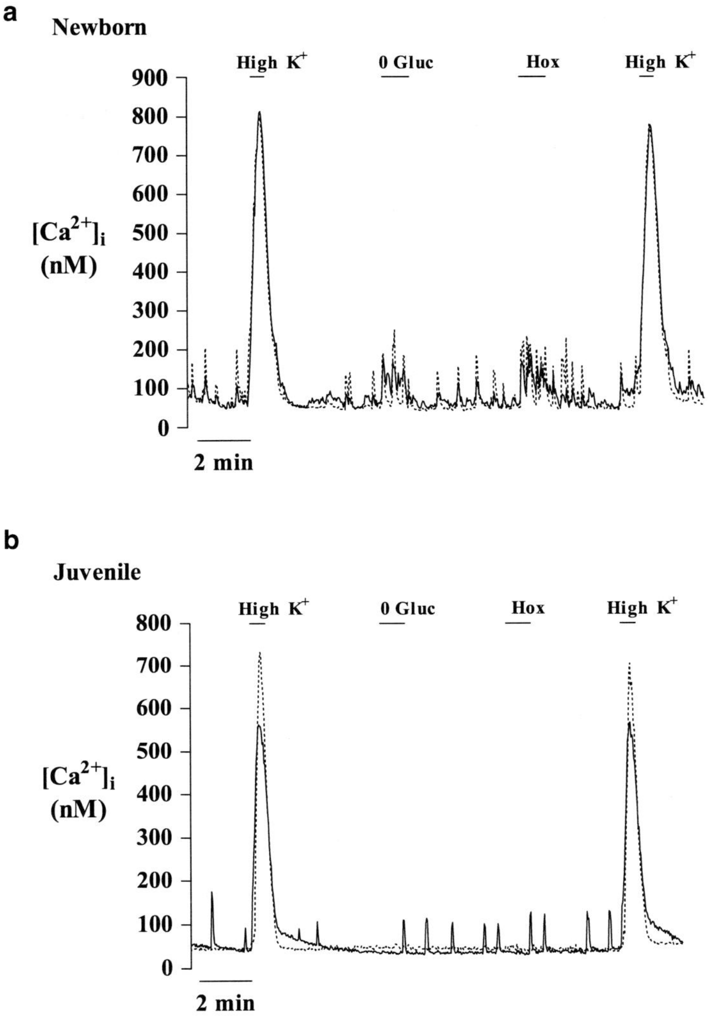

Fig. 27.2 Fura-2 Ca^{2+} recordings from neonatal and juvenile rat AMCs. In (**a**) two newborn AMCs responded to 1 min applications of aglycemia (0 Gluc) and hypoxia (PO_2 ~15 mmHg; Hox) via elevations in basal intracellular Ca^{2+} concentration ($[Ca^{2+}]_i$) and increases in Ca^{2+} spike frequency (e.g. *dotted line*). (**b**) in contrast, significantly fewer juvenile AMCs responded to 0 Gluc and Hox with increases in $[Ca^{2+}]_i$ (15% and 10%, respectively). Note that the presence of spontaneous activity (e.g. *solid line* in **b**) was not correlated with responsiveness

93.4±28 fC/min; 0 mM, 449±103.1 fC/min; $p<0.05$, n=3 cells). Occasionally, a rise in baseline current was measured during aglycemia even though quantal catecholamine secretion was absent suggesting that other cells in the slice, rather the cell under study, were causing a rise in bath catecholamines (i.e. were glucose-sensitive) (Fig. 27.1b; n=7 cells). These responses were notably absent from AMCs in juvenile slices, where no quantal catecholamine secretion or rise in baseline was observed when cells were exposed to aglycemia (Fig. 27.1c; n=10 cells).

27.3.2 Comparison of Aglycemia- and Hypoxia-Evoked Changes in $[Ca^{2+}]_i$ in Neonatal and Juvenile AMCs

Because $[Ca^{2+}]_i$ is required for catecholamine secretion, and aglycemia produces robust increases in $[Ca^{2+}]_i$ in neonatal AMCs, we compared aglycemia-evoked responses from neonatal and juvenile chromaffin cells (Livermore et al. 2011). As illustrated in Fig. 27.2a, some neonatal AMCs produced robust increases in $[Ca^{2+}]_i$ when exposed to aglycemia and hypoxia. Indeed approximately 40%

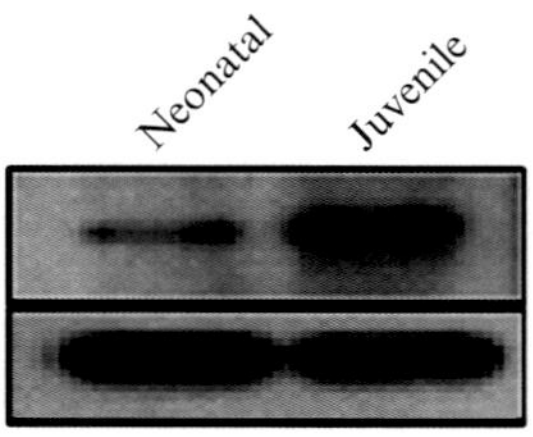

Fig. 27.3 Differential expression of the K_{ATP} channel subunit $K_{ir}6.2$ in juvenile and neonatal adrenal medulla. In Western blots, juvenile adrenal medulla showed a relative increase in $K_{ir}6.2$ protein expression compared to neonatal adrenal medulla. β-actin was used as a loading control

of newborn AMCs responded to hypoxia and aglycemia with an increase in $[Ca^{2+}]_i$ (n=440 cells). In contrast, this response was much less frequent in juvenile AMCs (Fig. 27.2b); hypoxia and aglycemia produced rises in $[Ca^{2+}]_i$ in only ~10% and ~15% of cells , respectively (n=474 cells). Moreover the average aglycemia-evoked $\Delta[Ca^{2+}]_i$ was markedly reduced from 21.7±1.6 nM in neonatal AMCs to 7.7±0.9 nM in juvenile AMCs (means include data from all cells tested, including non-responsive cells).

27.3.3 Developmental Upregulation of the K_{ATP} Channel

Given that the aglycemia-evoked rise in $[Ca^{2+}]_i$ and catecholamine secretion is blunted in juvenile AMCs, and that aglycemia activates an inhibitory pathway involving the K_{ATP} channel in neonatal AMCs, we wondered whether the developmental changes in the glucosensing observed in juvenile animals were related to changes in K_{ATP} channel expression (Livermore et al. 2011). We carried out western blot analysis of protein extracts from the neonatal and juvenile adrenal medulla for $K_{ir}6.2$, the pore-forming subunit of the K_{ATP} channel (Fig. 27.3). Our analysis indicates that K_{ATP} channel expression is ~2.9 fold higher in juvenile adrenal medulla compared to neonates (n =3).

27.4 Discussion

The primary findings from this study are that (1) aglycemia evokes catecholamine secretion from neonatal AMCs in thin slices and this property is absent in juvenile AMCs, (2) the aglycemia- and hypoxia-evoked rise in $[Ca^{2+}]_i$ is suppressed in juvenile AMCs, and (3) that these changes occur in parallel with the upregulation of the K_{ATP} channel in juvenile adrenal medulla. Our amperometric studies confirm and extend our findings from AMCs *in vitro*, that aglycemia evokes catecholamine secretion from neonatal AMCs but not in juvenile AMCs. Indeed, the magnitude of the secretory response is similar from neonatal AMCs in slices and *in vitro* (Livermore et al. 2011). Moreover, these secretory responses are qualitatively similar to responses observed from neonatal thin slices in response to hypoxia and hypercapnia (García-Fernández et al. 2007; Munoz-Cabello et al. 2005). Ratiometric Ca^{2+} imaging revealed that the glucosensitivity of juvenile AMCs was severely blunted. Indeed, both the magnitude of the $[Ca^{2+}]_i$ rise and the proportion of responsive cells were lower in juvenile AMCs, mirroring the developmental changes in hypoxia- and hypercapnia-sensing from AMCs (García-Fernández et al. 2007; Munoz-Cabello et al. 2005; Thompson et al. 1997, 2002). However, the observation that a small population of juvenile AMCs (i.e. <15%) retains its glucosensitivity may explain why splanchnic denervated, but not adrenalectomized, adult rats can still produce a rise in adrenaline in response to hypoglycemia (Khalil et al. 1986). Interestingly, these physiological changes in glucosensing coincide with a ~2.9 fold upregulation of the K_{ATP} channel in juvenile AMCs. Previously, we found that mimicking splanchnic innervation *in utero* via chronic nicotinic receptor stimulation abolished hypoxia-sensing in neonatal AMCs, and that this was paralleled by an ~1.7 fold increase in K_{ATP} channel expression (Buttigieg et al. 2009).

Further experiments must be performed to test the hypothesis that upregulation of the K_{ATP} channel blunts glucosensing in juvenile AMCs. First, there should be a functional increase in the K_{ATP} current in juvenile AMCs exposed to aglycemia. Moreover, in current-clamp recordings from juvenile AMCs, the K_{ATP}-blocker glibenclamide should render some glucose-insensitive cells glucose-sensitive. Indeed, we found that in neonatal AMCs this protocol revealed that K_{ATP} channel activation during aglycemia prevented membrane depolarization and action potential firing in some glucose-insensitive cells (Livermore et al. 2011). Second, mimicking splanchnic innervation *in vitro* or *in utero* via chronic nicotine exposure (which upregulates the K_{ATP} channel) should render neonatal AMCs glucose-insensitive. Third, denervation of the adrenal medulla *in vivo* in juvenile rats, which restores hypoxia-sensitivity to juvenile AMCs, should restore glucose-sensing in juvenile AMCs concomitant with a reduction in K_{ATP} channel expression (Levitsky and Lopez-Barneo 2009). In conclusion, our work helps clarify the developmental regulation of glucosensing in AMCs, which is critical for perinatal development and metabolic regulation.

Acknowledgements This work was supported by a Discovery grant from the Natural Sciences and Engineering Research Council of Canada (NSERC) and an operating grant from the Canadian Institutes of Health Research (CIHR; grant #12037) to CAN. SL and NAP were supported by an Ontario Graduate Scholarship and an NSERC Vanier Canada Graduate Scholarship, respectively.

References

Barbara J-G, Christophe Poncer J, Anne McKinney R, Takeda K (1998) An adrenal slice preparation for the study of chromaffin cells and their cholinergic innervation. J Neurosci Meth 80:181–189

Bournaud R, Hidalgo J, Yu H, Girard E, Shimahara T (2007) Catecholamine secretion from rat foetal adrenal chromaffin cells and hypoxia sensitivity. Pflügers Archiv Euro J Physiol 454:83–92

Buttigieg J, Brown S, Zhang M, Lowe M, Holloway AC, Nurse CA (2008) Chronic nicotine in utero selectively suppresses hypoxic sensitivity in neonatal rat adrenal chromaffin cells. FASEB J 22:1317–1326

Buttigieg J, Brown S, Holloway AC, Nurse CA (2009) Chronic nicotine blunts hypoxic sensitivity in perinatal rat adrenal chromaffin cells via upregulation of KATP channels: role of {alpha}7 Nicotinic acetylcholine receptor and hypoxia-inducible Factor-2{alpha}. J Neurosci 29:7137–7147

Cuezva JM, Burkett ES, Kerr DS, Rodman HM, Patel MS (1982) The newborn of diabetic rat. I. Hormonal and metabolic changes in the postnatal period. Pediatr Res 16:632–637

de Araujo E, de Souza K, Garcia R, de Freitas Mathias P (1999) Glucose does not affect catecholamine stimulus-secretion coupling in rat adrenal medulla: relationship to low changes in osmolarity and to insulin. Res Commun Mol Pathol Pharmacol 105:200–212

García-Fernández M, Mejías R, López-Barneo J (2007) Developmental changes of chromaffin cell secretory response to hypoxia studied in thin adrenal slices. Pflügers Archiv Euro J Physiol 454:93–100

Girard JR, Cuendet GS, Marliss EB, Kervran A, Rieutort M, Assan R (1973) Fuels, hormones, and liver metabolism at term and during the early postnatal period in the rat. J Clin Invest 52:3190–3200

Khalil Z, Marley PD, Livett BG (1986) Elevation in plasma catecholamines in response to insulin stress is under both neuronal and nonneuronal control. Endocrinology 119:159–167

Lagercrantz H, Slotkin TA (1986) The “stress” of being born. Sci Am 254:100–107

Levitsky KL, Lopez-Barneo J (2009) Developmental change of T-type Ca2+ channel expression and its role in rat chromaffin cell responsiveness to acute hypoxia. J Physiol 587:1917–1929

Livermore S, Piskuric NA, Buttigieg J, Zhang M, Nurse C (2011) Low glucose sensitivity and polymodal chemosensing in neonatal rat adrenomedullary chromaffin cells. Am J of Physiol Cell Physiol 301(5):C1104–15

Mochizuki-Oda N, Takeuchi Y, Matsumura K, Oosawa Y, Watanabe Y (1997) Hypoxia-induced catecholamine release and intracellular Ca2+ increase via suppression of K+channels in cultured rat adrenal chromaffin cells. J Neurochem 69:377–387

Munoz-Cabello AM, Toledo-Aral JJ, Lopez-Barneo J, Echevarria M (2005) Rat adrenal chromaffin cells are neonatal CO2 sensors. J Neurosci 25:6631–6640
Piskuric NA, Brown ST, Zhang M, Nurse CA (2008) Glucosensing in an immortalized adrenomedullary chromaffin cell line: role of ATP-sensitive K+channels. Neurosci Lett 445:94–98
Seidler FJ, Slotkin TA (1986) Ontogeny of adrenomedullary responses to hypoxia and hypoglycemia: role of splanchnic innervation. Brain Res Bull 16:11–14
Thompson RJ, Jackson A, Nurse CA (1997) Developmental loss of hypoxic chemosensitivity in rat adrenomedullary chromaffin cells. J Physiol 498:503–510
Thompson R, Farragher S, Cutz E, Nurse C (2002) Developmental regulation of O(2) sensing in neonatal adrenal chromaffin cells from wild-type and NADPH-oxidase-deficient mice. Pflügers Archiv Euro J Physiol 444:539–548

Chapter 28
Contribution of Inflammation on Carotid Body Chemosensory Potentiation Induced by Intermittent Hypoxia

Rodrigo Del Rio, Esteban A. Moya, and Rodrigo Iturriaga

Abstract Exposure to chronic intermittent hypoxia (CIH) produces hypertension. A critical process involved in the CIH-induced hypertension is the potentiation of the carotid body (CB) chemosensory responses to acute hypoxia. The CIH-induced changes in the CB chemosensory process have been related to an enhanced reactive oxygen species (ROS) production. However, it is still a matter of debate where ROS could directly modify the CB chemosensory discharge. Recently, we found that CIH-induced increase expression of TNF-α and IL-1β within the CB. Thus, we studied the contribution of these pro-inflammatory cytokines on the enhanced CB chemosensory response to acute hypoxia in rats exposed to CIH. To study the role of TNF-α and IL-1β, male Sprague-Dawley rats were submitted to CIH (5% O_2, 12 times/hr for 8 hr/day) and received chronic ibuprofen treatment (40 mg/kg). Following 21 days of CIH, rats were anaesthetized and the CB chemosensory discharge was recorded in response to several levels FiO_2 (5-100%). Exposure to CIH significantly increases the immunorreactive levels of TNF-α and IL-1β in the CB, along with an increase accumulation of the p65 NF-κβ subunit. Treating rats with ibuprofen significantly prevents the CIH-induced increases in TNF-α and IL-1β in the CB chemoreceptor cells but failed to decrease the enhanced CB chemosensory reactivity to hypoxia. Our results suggest that the mechanisms underlying the potentiation of the CB chemosensory response to acute hypoxia are not linked to the increased expression of TNF-α and IL-1β within the CBs of CIH-exposed rats.

Keywords Inflammation • Cytokines • TNF-α • IL-1β • Carotid body • Chemosensory discharge • Intermittent hypoxia • Obstructive sleep apnea

28.1 Introduction

Chronic intermittent hypoxia (CIH), a main feature of obstructive sleep apnea (Garvey et al. 2009; Somers et al. 2008), enhances carotid body (CB) chemosensory responses to acute hypoxia (Peng et al. 2003; Rey et al. 2004; Del Rio et al. 2010). The CIH-induced CB chemosensory potentiation has been attributed to increased levels of reactive oxygen species (ROS) in the CB (Pawar et al. 2009; Iturriaga

R. Del Rio • E.A. Moya • R. Iturriaga (✉)
Laboratorio de Neurobiología, Facultad de Ciencias Biológicas, Pontificia Universidad Católica de Chile, Alameda 340, Santiago, Region Metropolitana, Chile
e-mail: riturriaga@bio.puc.cl

C.A. Nurse et al. (eds.), *Arterial Chemoreception: From Molecules to Systems*, Advances in Experimental Medicine and Biology 758, DOI 10.1007/978-94-007-4584-1_28,

et al. 2009), but it is matter of debate if ROS *per se* may increase the CB chemosensory discharges (Gonzalez et al. 2007). Thus, it is likely that molecules downstream of the ROS signals may mediate the CIH-induced effects of ROS on CB chemoreception. Among other molecules upregulated in the CB by CIH, such as ET-1 and iNOS (Iturriaga et al. 2009; Del Rio et al. 2010, 2011), pro-inflammatory cytokines has been proposed as mediators of the CB chemosensory potentiation induced by CIH (Iturriaga et al. 2009; Del Rio et al. 2011). Recently, we found that CIH induced a ROS-dependent increases of TNF-α and IL-1β within the CB, suggesting that these pro-inflammatory cytokines may mediate the ROS-induced potentiation (Del Rio et al. 2011). To test this hypothesis, we study the effects of the anti-inflammatory drug ibuprofen on the increased immunoreactive levels of TNF-α and IL-1β in the rat CB, and the potentiation of the carotid chemosensory responses to hypoxia.

28.2 Methods

28.2.1 Exposure to Intermittent Hypoxia and Subcutaneous Ibuprofen Therapy

Unrestrained, freely moving rats housed in individual chambers were exposed to hypoxic cycles of 5% O_2 for 20 s, followed by room air for 280 s, applied 12 times/h, 8 h/day during 21 days (Del Rio et al. 2010). In the Sham condition, the hypoxic exposure was replaced by means of flushing equal flow of compressed air into the chambers. The room temperature was kept at 23–25°C. The experimental procedures were approved by the Bio-Ethical Committee of the Biological Sciences Faculty, P. Universidad Católica de Chile, and were performed according to the NIH Guide for the Care and Use of Laboratory Animals. Two days before the onset of CIH or sham exposures, animals were anesthetized with isofluorane 3% in oxygen and an osmotic minipump (Alzet Scientific Products, USA) implanted subcutaneously in the back of each animal. The pumps delivered 2.5 μl/h and were filled with 400 mg ibuprofen (40 mg/kg/day) in 2 ml NaCl 0.9% or NaCl 0.9% alone.

28.2.2 Recording of Carotid Body Chemosensory Discharges

Rats were anesthetized with sodium pentobarbitone (40 mg/kg i.p), placed in supine position and the body temperature was maintained at 38.0 ± 0.5 C. One carotid sinus nerve was dissected, placed on a pair of platinum electrodes and covered with warm mineral oil. The neural signal was pre-amplified, filtered and fed to an electronic spike-amplitude discriminator, allowing the selection of action potentials of given amplitude above the noise to be counted with a frequency meter for measuring the frequency of carotid chemosensory discharge expressed in Hz. Barosensory fibers were eliminated by crushing the common carotid artery wall between the carotid sinus and the carotid body. The other carotid sinus nerve was cut to prevent cardioventilatory reflexes evoked by the activation of the CB. The chemosensory discharge was measured at several isocapnic levels of PO_2. At the end of the experiments rats were killed by an overdose of sodium pentobarbitone (100 mg/kg i.p.).

28.2.3 Western Blot for p65 Subunit of the NF-κB Transcription Factor

The CB tissue was rapidly harvested from anesthetized rats, and the CBs were frozen and stored at −80°C until analyzed. The CBs from 4 to 5 rats were pooled and the protein was extracted by sonication in lysing buffer plus 2% protease inhibitor cocktail. Following a centrifugation at 12,000 g for 20 min at 4°C, protein concentration in the supernatant was measured using a Lowry assay kit. Samples were adjusted to the same concentrations of protein using equal volumes of loading buffer containing β-mercaptoethanol. For electrophoresis, 25 μg of protein were fractionated in a 10% polyacrylamide gel. Proteins were transferred onto membranes and treated overnight at 4°C with mouse anti-p65 primary antibody (1:500, Cell Signaling, USA) followed by one hour incubation with the secondary antibody conjugated with horseradish peroxidase (rabbit anti-mouse IgG, 1:2,500, Sigma, USA). Immune complexes on the membrane were visualized using enhanced chemiluminescence and the specific bands according to the expected molecular weight were analyzed with the ImageJ software (NIH, USA).

28.2.4 Immunohistochemical Detection of Cytokines in the Carotid Body

Quantitative immunohistochemistry was used to measure the levels of TNF-α and IL-1β in the rat CB exposed to CIH for 21 days. Anesthetized rats were perfused intracardially with phosphate saline buffer followed by buffered 4% paraformaldehyde, and post-fixed at 4°C for 12 h. Samples were dehydrated in ethanol, included in paraffin, cut in 5 μm sections and mounted on silanized slides. Deparaffinized samples were submitted to microwave based antigen retrieval protocol. Samples were incubated with 0.3% H_2O_2 to inhibit endogenous peroxidase and then in blocking normal horse serum solution. Slides were incubated with specific antibodies overnight at 4°C in humidity chambers for detection of TNF-α (1:20, goat anti- TNF-α, Santa Cruz Biotech., USA) and IL-1β (1:100, rabbit anti- IL-1β, Santa Cruz Biotech., USA). Negative controls were performed by omission of the primary antibody. After rinse in cold PBS, samples were incubated with secondary antibodies conjugated to biotin and revealed at 37°C in a dark chamber with 3,3-diaminobenzidine tetrahydrochloride. Finally, samples were counterstained with Harris Hematoxylin and permanently mounted. Photomicrographs of the CB tissue were taken at 100x with a CCD-camera coupled to a microscope, digitized and analyzed using a color deconvolution algorithm with the ImageJ software (NIH, USA).

28.2.5 Statistical Data Analysis

Data are expressed as means ± SEM. Differences between two groups were assessed by Student T-test comparisons. Differences between more groups were assessed with one or two-way ANOVA tests, followed by appropriated posthoc comparisons.

28.3 Results

28.3.1 Intermittent Hypoxia Increased the Expression of the p65 Subunit of the NF-κB Transcription Factor

Since an enhanced production of ROS induced by hypoxia-reoxygenation increased the expression and the translocation of NF-κB to the nucleus, inducing the expression of IL-1β and TNF-α (Janseen-Heininger et al. 2000), we studied if intermittent hypoxia increased the levels of the p65 subunit of the

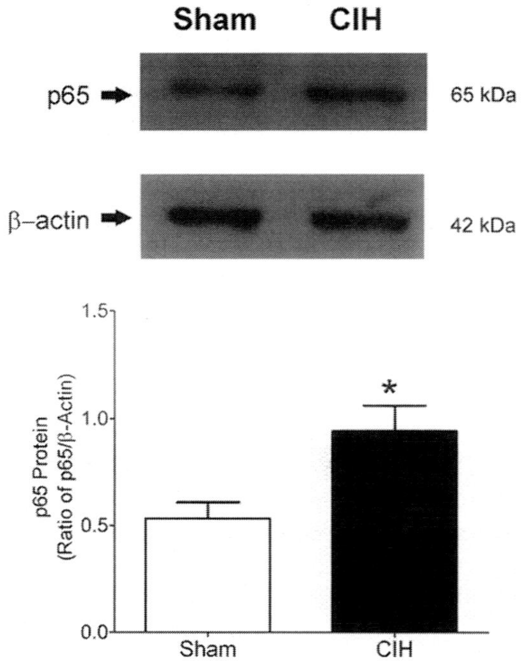

Fig. 28.1 CIH-induced accumulation of NF-κB in the rat CB. *Upper panel*, immunoblot for the p65 subunit of the NF-κB transcription factor. *Lower panel*, band densitometry quantification for sham and CIH rat CBs. *, $p<0.05$ Student T-test, n=4–5 CBs per group

NF-κB transcription factor in the CBs from rats exposed to CIH. Figure 28.1 shows an immunoblot of the p65 subunit of the NF-κB transcription factor. Clearly, CIH produced a significant increase in the p65 levels in the CBs from rats exposed to CIH for 21 days.

28.3.2 *Effect of CIH on TNF-α and IL-1β Immunoreactivity in the Rat CB*

The exposure to CIH for 21 days increased the immunoreactive levels of TNF-α and IL-1β in the rat CB, as is illustrated in Fig. 28.2. Note that TNF-α and IL-1β were mainly confined to the glomus cells, but it is worth noting that not all glomus cells showed positive immunoreactive staining for the cytokines.

28.3.3 *Effects of Ibuprofen on TNF-α and IL-1β Immunoreactivity in the Rat CB*

We found a significant increased of TNF- and IL-1β immunoreactivity in the CB from rats exposed to CIH for 21 days, which was prevent by ibuprofen treatment (see Table 28.1). Ibuprofen prevented the increased TNF-α and IL-1β levels, compared with the optical integrated intensity measured in the CBs of CIH-treated rats.

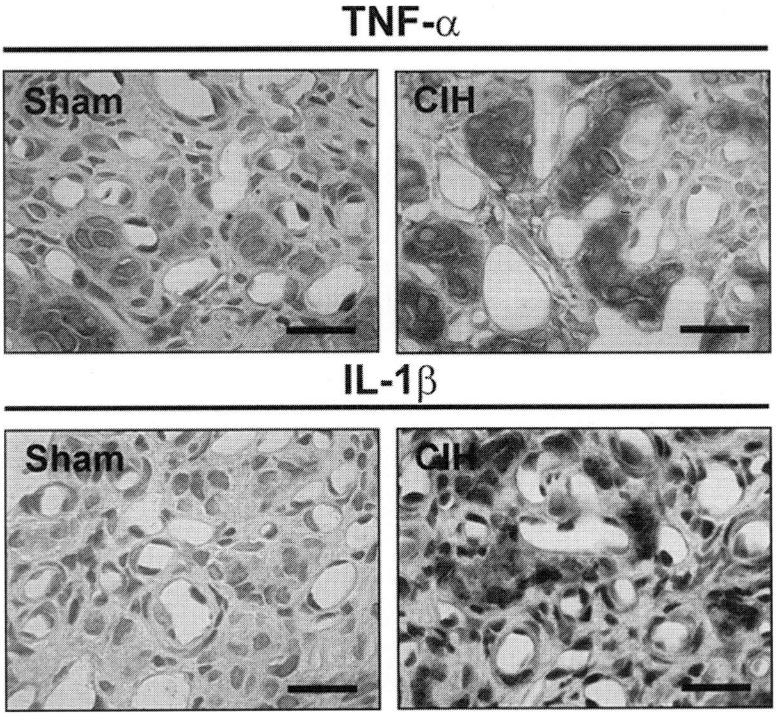

Fig. 28.2 Micrographs showing positive immunoreactivity for TNF-α and IL-1β in the CB from a Sham rat and CIH-treated rat. Note that positive staining is mainly confined to round to ovoid cells sharing histological morphology with CB chemoreceptor cells. Scale bars 20 μm

Table 28.1 Effect of ibuprofen on CIH-induced enhanced expression of TNF-α and IL-1β in the CB

	Sham	CIH	CIH+IB
TNF-α –ir (a.u.)	11.8±2.3	49.3±1.9**	9.8±1.8
IL-1β –ir (a.u.)	16.5±1.5	36.8±2.6**	17.2±1.1

Data are presented as mean±SEM. Sham: control animals; CIH: rats exposed to chronic intermittent hypoxia; CIH+IB: rats treated with ibuprofen during the exposure to chronic intermittent hypoxia; TNF-α –ir: tumor necrosis factor alpha immunoreactivity; IL-1β –ir: interleukin 1 beta immunoreactivity

**p<0.01 compared to Sham

28.3.4 Effects of Ibuprofen on CB Chemosensory Potentiation Induced by CIH

Rats exposed to CIH showed enhanced CB chemosensory responses to acute hypoxia as compared with Sham, and Sham-IB rats (Fig. 28.3). However, the CIH-induced potentiation of the CB chemosensory response to acute hypoxia was not prevented by ibuprofen. The two-way ANOVA analysis showed that the overall curve for CIH and CIH-IB treated groups were not different.

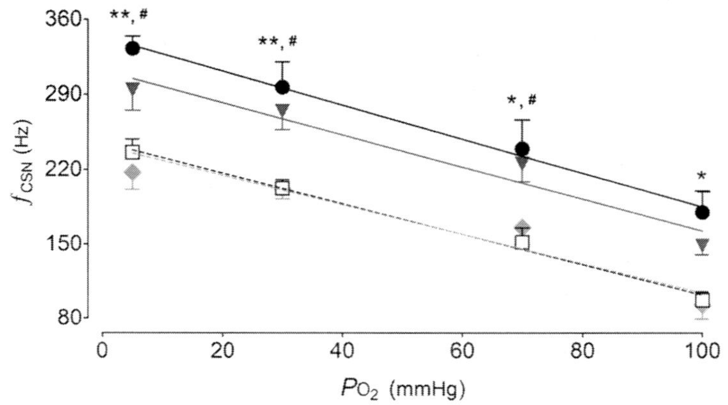

Fig. 28.3 Anti-inflammatory treatment failed to prevent CB chemosensory potentiation induced by CIH. Summary of the CB chemosensory responses induced by several levels of inspired PO_2 in 4 Sham rats (□), 4 CIH-treated rats (•), 4 Sham + IB rats (♦) and 4 CIH + IB (▼). f_{CSN}, frequency of chemosensory discharges in Hz. **, $p<0.01$; *, $p<0.05$ CIH compared to sham; #, $p<0.05$ CIH + IB compared to sham, Bonferroni test after 2-way ANOVA

28.4 Discussion

Present results showed that ibuprofen, which prevented the CIH-increased TNF-α and IL-1β expression in the CB, did not prevent the potentiation of the CB chemosensory responses to acute hypoxia. Thus, the CIH-induced potentiation of the CB chemosensory responses does not depend on the increased TNF-α and IL-1β levels in the CB. The inhibitory effect of ibuprofen on the cytokine accumulation induced by CIH in the CB is consistent with its known anti-inflammatory effect. Although, ibuprofen is considered a non-selective inhibitor of cyclooxygenases 1 and 2, is a well known inhibitor of the nuclear translocation of the transcription factor, NF-κB, which mediates the TNFα and IL-1β production (Jenseen-Heininger et al. 2000).

Some studies have shown that the pro-inflammatory cytokines IL-1β, IL-6, and TNF-α, are excitatory modulators of CB oxygen transduction process in the rat (Lam et al. 2008; Liu et al. 2009; Shu et al. 2007). It is known that the rat glomus cells express these cytokines along with their functional receptors, IL-1R1 and TNF-R1 (Lam et al. 2008; Wang et al. 2002). In addition, Shu et al. (2007) found that the application of IL-1β to glomus cells of the rat CB inhibit the O_2-dependent voltage-gated K^+ currents and the increase of intracellular Ca^{2+}.

Intermittent hypoxia increases the levels of ROS in the rat CB (Pawar et al. 2009; Del Rio et al. 2010). In response to oxidative stress, NF-κB induces the expression of pro-inflammatory cytokines such as IL-1β and TNF-α in the CB, suggesting that chemoreceptor cells can synthesize and release cytokines (Del Rio et al. 2011). Inflammatory processes are also involved in the enhanced CB chemosensory response to hypoxia in rats exposed to sustained hypoxia (Lam et al. 2008; Liu et al. 2009). In fact, Lam et al. (2008) found that sustained hypobaric hypoxia increased the mRNA expression of IL-1β and TNF-α, and the IL-1R1 and TNF-R1 receptors. Besides that, Liu et al. (2009) found that the concurrent administration of ibuprofen and dexamethasone to rats exposed to sustained hypoxia reduced the potentiation of the CB chemosensory response to acute hypoxia and reduced the cytokine mRNA expression. We found that intermittent hypoxia produces a progressive increase of the TNF-α and IL-1β levels in the rat CB (Del Rio et al. 2011). Present results showed that the enhanced CB chemosensory response to hypoxia induced by CIH was not prevented by ibuprofen, while the increased levels of IL-1β and TNF-α in the CB were blocked by the anti-inflammatory treatment. Thus, our results suggest that the mechanisms underlying the hypoxic CB chemosensory potentiation induced by sustained and intermittent hypoxia are different.

In summary, ibuprofen prevented the increased CB levels of TNF-α and IL-1β induced by CIH, but did not prevent the enhanced CB chemosensory responses to acute hypoxia induced by CIH. Thus, our results suggest that IL-1β and TNF-α do not mediate the potentiation of the CB chemosensory response to hypoxia induced by CIH.

Acknowledgements This work was supported by grant 1100405 from the National Fund for Scientific and Technological Development of Chile (FONDECYT).

References

Del Rio R, Moya EA, Iturriaga R (2010) Carotid body and cardiorespiratory alterations in intermittent hypoxia: the oxidative link. Eur Respir J 36:143–150

Del Rio R, Moya EA, Iturriaga R (2011) Differential expression of pro-inflammatory cytokines, endothelin-1 and nitric oxide synthases in the rat carotid body exposed to intermittent hypoxia. Brain Res 1395:74–85

Garvey JF, Taylor CT, McNicholas WT (2009) Cardiovascular disease in obstructive sleep apnoea syndrome: the role of intermittent hypoxia and inflammation. Eur Respir J 33:1195–1205

González C, Agapito MT, Rocher A, González-Martin MC, Vega-Agapito V, Gomez-Niño A, Rigual R, Castañeda J, Obeso A (2007) Chemoreception in the context of the general biology of ROS. Respir Physiol Neurobiol 157:30–44

Iturriaga R, Moya EA, Del Rio R (2009) Carotid body potentiation induced by intermittent hypoxia: implications for cardiorespiratory changes induced by sleep apnoea. Clin Exp Pharmacol Physiol 36:1197–1204

Janssen-Heininger YM, Poynter ME, Baeuerle PA (2000) Recent advances towards understanding redox mechanisms in the activation of nuclear factor kappaB. Free Radic Biol Med 28:1317–1327

Lam SY, Tipoe GL, Liong EC, Fung ML (2008) Chronic hypoxia upregulates the expression and function of proinflammatory cytokines in the rat carotid body. Histochem Cell Biol 130:549–559

Liu X, He L, Stensaas L, Dinger B, Fidone S (2009) Adaptation to chronic hypoxia involves immune cell invasion and increased expression of inflammatory cytokines in rat carotid body. Am J Physiol Lung Cell Mol Physiol 296:L158–L166

Pawar A, Nanduri J, Yuan G, Khan SA, Wang N, Kumar GK, Prabhakar NR (2009) Reactive oxygen species-dependent endothelin signaling is required for augmented hypoxic sensory response of the neonatal carotid body by intermittent hypoxia. Am J Physiol Regul Integr Comp Physiol 296:R735–R742

Peng YJ, Overholt JL, Kline D, Kumar GK, Prabhakar NR (2003) Induction of sensory long-term facilitation in the carotid body by intermittent hypoxia: implications for recurrent apneas. Proc Natl Acad Sci U S A 100:10073–10078

Rey S, Del Rio R, Alcayaga J, Iturriaga R (2004) Chronic intermittent hypoxia enhances cat chemosensory and ventilatory responses to hypoxia. J Physiol 560:577–586

Shu HF, Wang BR, Wang SR, Yao W, Huang HP, Zhou Z, Wang X, Fan J, Wang T, Ju G (2007) IL-1βeta inhibits IK and increases $[Ca^{2+}]i$ in the carotid body glomus cells and increases carotid sinus nerve firings in the rat. Eur J Neurosci 25:3638–3647

Somers VK, White DP, Amin R, Abraham WT, Costa F, Culebras A, Daniels S, Floras JS, Hunt CE, Olson LJ, Pickering TG, Russell R, Woo M, Young T (2008) Sleep apnea and cardiovascular disease. J Am Coll Cardiol 52:686–717

Wang X, Wang BR, Duan XL, Zhang P, Ding YQ, Jia Y, Jiao XY, Ju G (2002) Strong expression of interleukin-1 receptor type I in the rat carotid body. J Histochem Cytochem 50:1677–1684

Chapter 29
Spexin Is Expressed in the Carotid Body and Is Upregulated by Postnatal Hyperoxia Exposure

Andrea Porzionato, Marcin Rucinski, Veronica Macchi, Carla Stecco, Gloria Sarasin, Maria M. Sfriso, Camillo Di Giulio, Ludwik K. Malendowicz, and Raffaele De Caro

Abstract Spexin is a recently identified peptide which is expressed in many different endocrine and nervous tissues. Due to the absence of data regarding spexin expression in the carotid body, the first aim of the present study was to investigate, through immunohistochemistry and Real-Time PCR, the expression and distribution of spexin in the rat and human carotid body. Moreover, the carotid body is known to undergo various structural and functional modifications in response to hyperoxic stimuli during the first postnatal period. Thus, we also evaluated if hyperoxia during the first postnatal weeks may produce changes in the spexin expression. Materials consisted of carotid bodies obtained at autopsy from five human adult subjects and sampled from 10 six-weeks old Sprague–Dawley rats. Five rats were maintained in normoxia for the first six postnatal weeks; five rats were exposed to 60% hyperoxia for 2 weeks and then maintained in normoxia for other 4 weeks. Diffuse anti-spexin immunoreactivity was found in type I cells of both humans and rats. No spexin immunoreactivity was visible in the type II cells. Hyperoxia exposure during the first 2 weeks of postnatal life caused a reduction of volume in the carotid body still apparent after 4 weeks of normoxia. Using real-time PCR, spexin expression was 6–7 times higher in hyperoxia-exposed rats than in normoxia-exposed ones. The expression of spexin in type I cells suggests a possible modulator role in peripheral chemoreception. Moreover, the ascertained role of spexin in the regulation of cell proliferation in other tissues (e.g., adrenal gland cortex) suggests a possible role of spexin also in the hyperoxia-induced plasticity of the carotid body.

A. Porzionato (✉)
Department of Human Anatomy and Physiology, University of Padua, Via A. Gabelli 65, 35121 Padova, Italy
e-mail: andrea.porzionato@unipd.it

Section of Anatomy, Department of Human Anatomy and Physiology, University of Padua, Via A. Gabelli 65, 35121 Padova, Italy
e-mail: andrea.porzionato@unipd.it

M. Rucinski • L.K. Malendowicz
Department of Histology and Embryology, Poznan University of Medical Sciences, Poznan, Poland

V. Macchi • C. Stecco • G. Sarasin • M.M. Sfriso • R. De Caro
Department of Human Anatomy and Physiology, University of Padua, Via A. Gabelli 65, 35121 Padova, Italy

C. Di Giulio
Department of Neurosciences and Imaging, University of Chieti, Via Dei Vestini, 31, Chieti, CH 66100 , Italy

C.A. Nurse et al. (eds.), *Arterial Chemoreception: From Molecules to Systems*, Advances in Experimental Medicine and Biology 758, DOI 10.1007/978-94-007-4584-1_29,

Keywords Peripheral Arterial Chemoreceptors • Carotid body • Hyperoxia • Spexin • Real Time PCR • Immunohistochemistry • Human • Rat

29.1 Introduction

Spexin (also called NPQ) is a peptide which was recently identified through bioinformatics search strategies (Mirabeau et al. 2007; Sonmez et al. 2009) and following biochemical characterization (Mirabeau et al. 2007). The gene produces a preprohormone containing a signal peptide sequence and putative dibasic prohormone cleavage sites. Spexin is a small amino acid region between dibasic cleavage sites which is highly conserved and is considered the most likely active neuropeptide (Mirabeau et al. 2007; Sonmez et al. 2009), although other three short potential peptides have also been suggested to have biological activity (Sonmez et al. 2009). Spexin has been found to be expressed in many different tissues by RT-PCR, Northern blot analysis, in situ hybridization and immunohistochemistry (Mirabeau et al. 2007; Porzionato et al. 2010; Sonmez et al. 2009). In particular, it has been found in neuronal and endocrine cell types. Conversely, there are no data about its possible expression in the carotid body. Thus, the first aim of the present work was to verify its possible expression in the carotid body. As a consequence of postnatal hyperoxia exposure it is known that the carotid body undergoes a series of biochemical, morphological and functional changes which may persist after return to normoxia (e.g., Bisgard et al. 2003; Carroll et al. 2009). Spexin has been found to modulate proliferation of adrenocortical cells and its expression has been found to be changed in adrenocortical regeneration (Rucinski et al. 2010) so that further aim of the study was to analyse possible changes in its expression following postnatal hyperoxia and a consequent period of recovery.

29.2 Materials and Methods

29.2.1 Materials

Female Sprague–Dawley rats and their offspring were housed and handled in accordance with the guidelines of Helsinki Declaration and recommendations of the Public Health Office. Mothers and litters were placed in clear polished acrylic chambers (BioSpherix, OxyCycler A84XOV, Redfield, NY) provided with software enabling modification and continuous monitoring of O_2 and CO_2. Twelve-hour light/dark cycles, humidity level maintained at 50%, and ambient temperature at 24°C were the set conditions.

After term gestation, the newborn rats were randomly distributed between two experimental groups: rats of the control group (n=5) were raised in ambient air for 6 weeks; rats of the hyperoxia

Table 29.1 RT-PCR and QPCR primers of spexin and PBGD (porphobilinogen deaminase). Oligonucleotide sequences for sense (S) and antisense (A) primers are shown

cDNA	Genbank Accession number	Primer	Primer sequence (5′–3′)	Position	PCR product size (bp)
Spexin	XM 001076968	S	CTGGTGCTGTCTGTTCTG	46–63	178
		A	TTGGGTTTCGTCTTTCTGG	205–223	
PBGD	NM 013168	S	GAAAGACCCTGGAAACCTTG	397–416	148
		A	TGCTCATCCAGCTTCCGTA	526–544	

group (n = 5) were exposed to 60% oxygen for the first two postnatal weeks and then maintained in ambient air for other 4 weeks. At the experimental endpoint, animals were euthanized and carotid bodies were collected. For real-time PCR, left carotid bifurcations of rats were dissected and carotid bodies were cleaned from the surrounding tissues under a dissecting microscope, frozen immediately on dry ice and stored at −80°C. For immunohistochemistry, right carotid bifurcations were fixed in Bouin solution, embedded in paraffin wax and serially cut in 5 μm sections.

Human autoptic carotid bodies were also obtained from five adult human subjects (three male, two female; mean age 51 years, Standard Deviation (SD) ± 3.6), clinically negative for chronic pulmonary or cardiovascular diseases, under the supervision of the internal Ethical Committee. Autopsies were performed between 24 and 30 h after death.

29.2.2 *Morphometrical Analysis*

Cavalieri's method was applied to evaluate morphometrical effects of hyperoxia. Every fourth section containing the carotid body was stained with haematoxylin and eosin and used for volume determination. The area of the carotid body was assessed in each section with a system for image analysis (VIDS V, AMS, Snowhill, UK) linked to a stereomicroscope (SZ 40 Olympus) by a camera. Carotid body volume was then estimated by the area of the carotid body on each section, section thickness, and the total number of sections containing the carotid body.

29.2.3 *Conventional RT-PCR and QPCR*

Spexin expression was analyzed by PCR according to (Rucinski et al. 2010). The primers used were designed by Primer 3 software (Whitehead Institute for Biomedical Research, Cambridge, MA) and the reaction amplified a fragments spanning introns (Table 1). In control reactions water (instead of cDNA) and omission of reverse transcriptase (−RT) were used. Each of 20 μl reaction mixture contained 4 μl template cDNA (or standards, or control) 0.5 μM each gene-specific primer and a previously determined optimal $MgCl_2$ concentration (3.5 μM for one reaction). LightCycler FastStart DNA Master SYBR Green I mix (ROCHE) was used. Obtained results were normalized with PBGD (porphobilinogen deaminase) reference gene. Amplification of tyrosine hydroxylase mRNA was performed to confirm the correspondence of RNA assayed with carotid body RNA. PCR efficiency was also assessed by a serial dilution method.

29.2.4 *Immunohistochemistry*

Anti-spexin immunohistochemistry was performed according to (Porzionato et al. 2010). Sections were incubated with primary rabbit polyclonal antibody recognising human, rat and mouse spexin [H-023–81, Phoenix Pharmaceuticals Inc., Burlingame, CA, USA] diluted 1:600 in phosphate buffer saline (PBS) for 1 h at room temperature. Negative controls were performed by omission of primary antibody. Spexin immunoreaction detected in human stomach was used as positive control. In order to verify the immunohistochemical specificity of the reaction, absorption tests with spexin (Phoenix Pharmaceuticals, No. 023–81) were also performed.

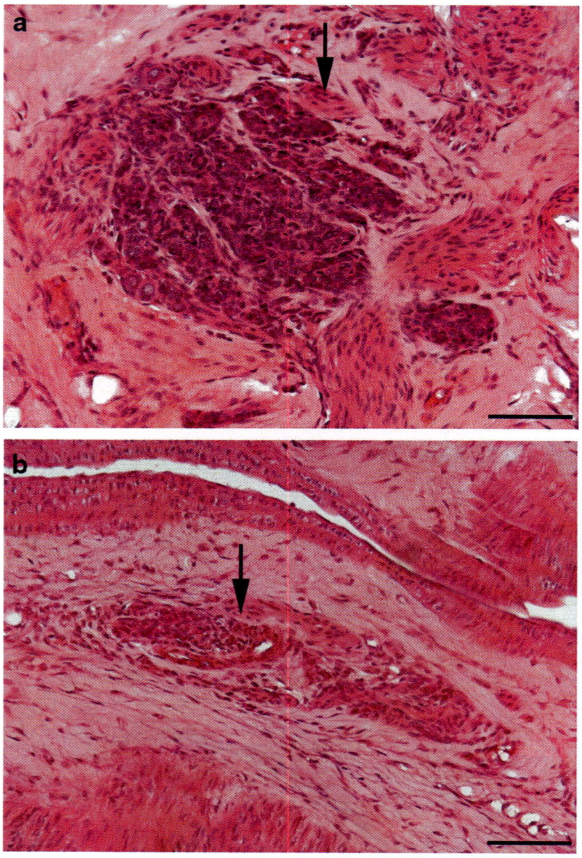

Fig. 29.1 Representative sections of carotid bodies (*arrows*) sampled from rats exposed to normoxia in the first six postnatal weeks (Normoxia group, **a**) or hyperoxia for the first two postnatal weeks and then to normoxia for the following 4 weeks (Hyperoxia group, **b**). Note the decreased section area and smaller cells in the hyperoxic carotid body. Scale bars = 75 μm

29.2.5 *Statistical Analysis*

Groups were compared with Mann–Whitney test and a p value of 0.05 was considered significant. Statistical calculations were performed using Statigraphic 4.0 software (STSC Inc., USA).

29.3 Results

A significant reduction of volume of the carotid body was found in rats exposed to hyperoxia during the first two postnatal weeks (10.3 (± 0.3) × 10^6 μm^3 versus 4.5 (± 0.3) × 10^6 μm^3, $p < 0.01$) (Fig. 29.1a, b).

PCR amplification of tyrosine hydroxylase mRNA confirmed the correspondence of RNA assayed with carotid body RNA. Classic RT-PCR method demonstrated the presence of spexin mRNA in the carotid bodies of rats of both the normoxic and hyperoxic groups. Specific primer RNA amplification by means of RT-PCR revealed the presence of reaction products with expected size. There were no PCR products in negative control reactions. Quantitative PCR revealed approximatively 6–7 times higher relative expression for spexin gene in carotid bodies of rats exposed to 60% hyperoxia during the first two postnatal weeks ($P < 0.001$) (Fig. 29.2). Conversely, differences were not found in the relative expressions of other peptides, such as augurin, beacon, ghrelin and neuromedin S, between the two experimental groups (unpublished observations).

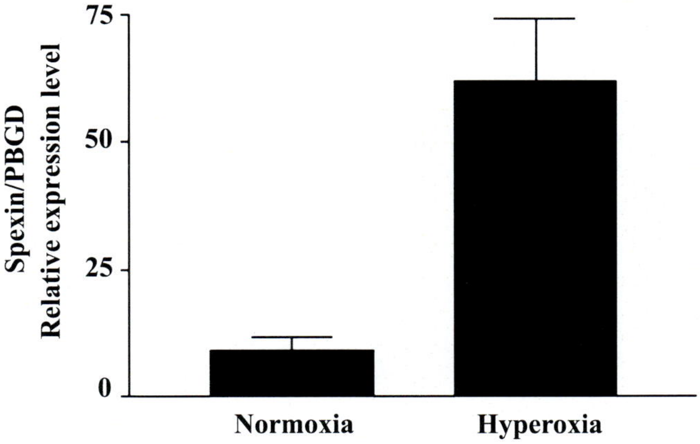

Fig. 29.2 QPCR analyses of spexin gene expression in rat carotid bodies exposed to normoxia in the first six postnatal weeks (Normoxia group) or hyperoxia for the first two postnatal weeks and then to normoxia for the following 4 weeks (Hyperoxia group). All samples were amplified in triplicate. Means ± SEM are shown

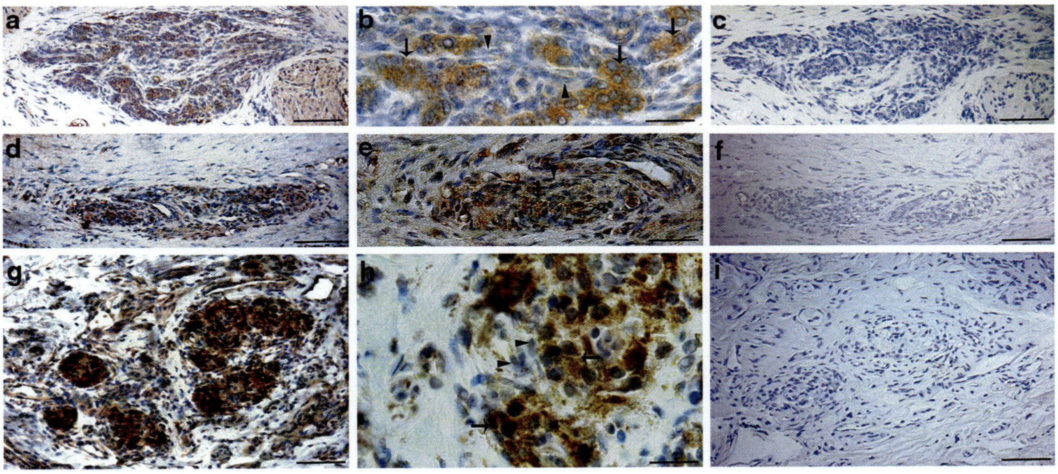

Fig. 29.3 Anti-spexin immunohistochemistry in rat (a-b, normoxic; d-e, hyperoxic) and human (g-h) carotid bodies, showing quite diffuse positivity in clusters of type I cells (*arrows*). Note the very strong reaction in the human sample, the negativity of elongated type II cells (*arrowheads*) surrounding lobules of type I cells and the absence of immunostaining in correspondent negative controls (c,f,i). Scale bars: a,c,d,f,g,i = 75 μm; b,e,h = 24 μm

Immunohistochemical analysis confirmed the expression of spexin both in human and rat type I cells. In rat carotid bodies, cytoplasmic staining for spexin was diffusely positive in most glomic type I cells (Fig. 29.3a, b, d, e). Immunohistochemistry did not permit a clear evaluation of staining differences between normoxic and hyperoxic groups. In human carotid bodies, immunostaining was particularly strong and diffuse in clusters of glomic type I cells (Fig. 29.3g, h). No immunoreactivity for spexin was observed in type II cells in both species. Nerve endings immunostained for spexin were not clearly visible. Glomic cell immunostainings were eliminated when preabsorbed antiserum were used or primary antibodies were omitted (Fig. 29.3c, f, i).

29.4 Discussion

Spexin has been found to be expressed in many different epithelial, endocrine and neuronal cell types through RT-PCR and immunohistochemistry (Porzionato et al. 2010; Rucinski et al. 2010). In the central nervous system, spexin shows different expressions in the various nuclei (Porzionato et al. 2010; Sonmez et al. 2009). In the peripheral nervous system, spexin expression has been observed in sensory and sympathetic ganglia (Porzionato et al. 2010) and *in situ* hybridization identified spexin mRNA in the submucosal layer of esophagus and stomach fundus, probably in the submucous plexus of the enteric nervous system (Mirabeau et al. 2007). The present work is the first in the literature to demonstrate spexin expression in type I cells of human and rat carotid bodies. Little is known about spexin functions. Spexin peptide has originally been found to dose-dependently induce muscle contraction in a stomach explant contractility assay (Mirabeau et al. 2007) but the widespread distribution of spexin in nervous and endocrine tissues, together with its capability to be secreted (Mirabeau et al. 2007), also suggested a role of this peptide in modulation of nervous and endocrine functions (Porzionato et al. 2010). For instance, spexin is highly expressed in the adrenal cortex and spexin exposure has also been demonstrated to increase basal aldosterone secretion in isolated zona glomerularis cells and corticosterone secretion in primary cultures of rat adrenocortical cells (Rucinski et al. 2010). Thus, the expression of spexin in type I cells, which represent the true chemoreceptive elements of the carotid body, suggests a possible modulator role in peripheral chemoreception, although functional experimental data will be necessary.

In response to environmental stimuli the carotid body is known to undergo a series of morphological and cellular changes which are mediated by many different trophic factors (Porzionato et al. 2008a; Porzionato et al. 2008b). In particular, the effects of postnatal hyperoxia on carotid body have been widely studied in rats (Bavis et al. 2008; Donnelly 2000; Gauda and Lawson 2000; Wang and Bisgard 2005). Postnatal hyperoxia reduces the carotid body response to acute hypoxia (e.g., Carroll et al. 2009; Erickson et al. 1998) reported carotid body hypoplasia after one postnatal week of 30 or 60% hyperoxia. Di Giulio et al. (1998) also reported reduction of clear-cored vesicles in the nerve ending synapses and reduction of mean glomus cells cytoplasmic and nuclear areas in type I cells of carotid bodies of adult rats exposed to 98–100% hyperoxia for 60–65 h. Moreover, postnatal hyperoxia exposure is known to produce long-lasting impairment of the ventilatory response to hypoxia still present in the adulthood (e.g., Carroll et al. 2009) and reduced volume of the adult carotid body has been reported in the literature after one (Bisgard et al. 2003), two (Bavis et al. 2008; Wenninger et al. 2006), or four (Fuller et al. 2001, 2002; Prieto-Lloret et al. 2004) weeks of postnatal 60% hyperoxia. Wang and Bisgard (2005) found that 60% hyperoxia produced decreases in proliferating and total type I cells with decreases in tyrosine hydroxylase and synaptophysin contents. In the present study, we confirmed that hyperoxia exposure during the first postnatal period causes a reduction in the carotid body volume which persists also in the following weeks. Carotid body hypoplasia due to hyperoxia has been ascribed to oxidative injuries, functional inactivity and/or changes in the expression of trophic factors (Dmitrieff et al. 2011). Direct oxidative injury on glomic cells has been considered less important by some authors (Wang and Bisgard 2005) who did not find apoptosis in carotid bodies of 1-week-old, normoxic and 60 % hyperoxic, rat pups. Conversely, it has also been suggested that the release of trophic factors by the carotid body may be modulated in an activity-dependent manner as hyperoxia has been found to upregulate or downregulate specific trophic factors (Bavis 2005; Bavis et al. 2008; Dmitrieff et al. 2011). In particular, hyperoxia exposure during the first 3 postnatal days has been found to upregulate transcript levels of cerebellin 1 precursor (Dmitrieff et al. 2011). In the present work we found upregulation of spexin expression in the carotid bodies of rats exposed to hyperoxia during the first two postnatal weeks and then maintained in normoxia for other four weeks. Little is know about trophic effects of spexin but in primary cultures of rat adrenocortical cells spexin has been found to inhibit BrdU incorporation (Rucinski et al. 2010). Thus, the upregulation of spexin expression in the carotid bodies is suggestive of a possible modulator role in hyperoxia-induced plasticity.

Acknowledgements The authors are grateful to Anna Rambaldo for skillful technical assistance.

References

Bavis RW (2005) Developmental plasticity of the hypoxic ventilatory response after perinatal hyperoxia and hypoxia. Respir Physiol Neurobiol 149:287–299

Bavis RW, Wenninger JM, Miller BM, Dmitrieff EF, Olson EB Jr, Mitchell GS, Bisgard GE (2008) Respiratory plasticity after perinatal hyperoxia is not prevented by antioxidant supplementation. Respir Physiol Neurobiol 160:301–312

Bisgard GE, Olson EB Jr, Wang ZY, Bavis RW, Fuller DD, Mitchell GS (2003) Adult carotid chemoafferent responses to hypoxia after 1, 2, and 4 wk of postnatal hyperoxia. J Appl Physiol 95:946–952

Carroll JL, Kim I, Dbouk H, Yang DJ, Bavis RW, Donnelly DF (2009) Time-dependence of hyperoxia-induced impairment in peripheral chemoreceptor activity and glomus cell calcium response. Adv Exp Med Biol 648:299–306

Di Giulio C, Di Muzio M, Sabatino G, Spoletini L, Amicarelli F, Di Ilio C, Modesti A (1998) Effect of chronic hyperoxia on young and old rat carotid body ultrastructure. Exp Gerontol 33:319–329

Dmitrieff EF, Wilson JT, Dunmire KB, Bavis RW (2011) Chronic hyperoxia alters the expression of neurotrophic factors in the carotid body of neonatal rats. Respir Physiol Neurobiol 175:220–227

Donnelly DF (2000) Developmental aspects of oxygen sensing by the carotid body. J Appl Physiol 88:2296–2301

Erickson JT, Mayer C, Jawa A, Ling L, Olson EB Jr, Vidruk EH, Mitchell GS, Katz DM (1998) Chemoafferent degeneration and carotid body hypoplasia following chronic hyperoxia in newborn rats. J Physiol 509:519–526

Fuller DD, Wang ZY, Ling L, Olson EB, Bisgard GE, Mitchell GS (2001) Induced recovery of hypoxic phrenic responses in adult rats exposed to hyperoxia for the first month of life. J Physiol 536:917–926

Fuller DD, Bavis RW, Vidruk EH, Wang ZY, Olson EB Jr, Bisgard GE, Mitchell GS (2002) Life-long impairment of hypoxic phrenic responses in rats following 1 month of developmental hyperoxia. J Physiol 538:947–955

Gauda EB, Lawson EE (2000) Developmental influences on carotid body responses to hypoxia. Respir Physiol 121:199–208

Mirabeau O, Perlas E, Severini C, Audero E, Gascuel O, Possenti R, Birney E, Rosenthal N, Gross C (2007) Identification of novel peptide hormones in the human proteome by hidden Markov model screening. Genome Res 17:320–327

Porzionato A, Macchi V, Parenti A, De Caro R (2008a) Trophic factors in the carotid body. Int Rev Cell Mol Biol 269:1–58

Porzionato A, Macchi V, Parenti A, Matturri L, De Caro R (2008b) Peripheral chemoreceptors: postnatal development and cytochemical findings in Sudden Infant Death Syndrome. Histol Histopathol 23:351–365

Porzionato A, Rucinski M, Macchi V, Stecco C, Malendowicz LK, De Caro R (2010) Spexin expression in normal rat tissues. J Histochem Cytochem 58:825–837

Prieto-Lloret J, Caceres AI, Obeso A, Rocher A, Rigual R, Agapito MT, Bustamante R, Castañeda J, Perez-Garcia MT, Lopez-Lopez JR, Gonzalez C (2004) Ventilatory responses and carotid body function in adult rats perinatally exposed to hyperoxia. J Physiol 554:126–144

Rucinski M, Porzionato A, Ziolkowska A, Szyszka M, Macchi V, De Caro R, Malendowicz LK (2010) Expression of the spexin gene in the rat adrenal gland and evidences suggesting that spexin inhibits adrenocortical cell proliferation. Peptides 31:676–682

Sonmez K, Zaveri N, Kerman I, Burke S, Neal CR, Xie X, Watson SJ, Toll L (2009) Evolutionary sequence modeling for discovery of peptide hormones. PLoS Comput Biol 5:e1000258

Wang ZY, Bisgard GE (2005) Postnatal growth of the carotid body. Respir Physiol Neurobiol 149:181–190

Wenninger JM, Olson EB, Wang Z, Keith IM, Mitchell GS, Bisgard GE (2006) Carotid sinus nerve responses and ventilatory acclimatization to hypoxia in adult rats following 2 weeks of postnatal hyperoxia. Respir Physiol Neurobiol 150:155–164

Chapter 30
Cyclic AMP and Epac Contribute to the Genesis of the Positive Interaction Between Hypoxia and Hypercapnia in the Carotid Body

Maria Ramirez, Laura Almaraz, Constancio Gonzalez, and Asuncion Rocher

Abstract Carotid body chemoreceptor cells in response to hypoxic and hypercapnic stimulus increase their resting rate of release of neurotransmitters and their action potential frequency in the carotid sinus sensory nerve. When chemoreceptor activity is assessed at the level of the carotid sinus nerve and on ventilation, there exists an interaction between hypoxic and hypercapnic stimulus so that the response to both stimuli combined is additive or more than additive, over a wide range of stimulation. It is not clear if this interaction occurs at chemoreceptor cell or directly acting on the sensory nerve. In the present work we demonstrate for the first time the existence of a positive interaction between hypoxic and hypercapnic-acidotic stimuli at the level of both, membrane potential depolarization and neurotransmitter release in rat and rabbit carotid body. Inhibition of adenylate cyclase (SQ-22536) abolished the positive interaction between stimuli and the Epac (exchange proteins activated by cAMP) activator 8-pCPT-2′-O-Me-cAMP reversed the effect of adenylate cyclase inhibition. These results suggest that this interaction between the two natural stimuli is mediated by cAMP via an Epac-dependent pathway, at least at the level of neurotransmitter release.

Keywords cAMP • Protein kinase A • Epac • Membrane potential • Catecholamine release

30.1 Introduction

Carotid bodies (CBs) are responsible for the entire hyperventilation that occurs in hypoxic hypoxia (Lahiri 1976) and for approximately 25–40% of the compensatory hyperventilation that occurs in acidosis (Nattie 1999). These reflex ventilatory responses are initiated at the level of the chemoreceptor cell which represents the PO_2 and H^+ sensing elements in this chemoreceptor organ (Gonzalez et al. 1994). Chemoreceptor cells in response to these stimuli increase their resting rate of release of

M. Ramirez • C. Gonzalez • A. Rocher (✉)
Departamento de Bioquímica y Biología Molecular y Fisiología. IBGM. Facultad de Medicina, CIBERES, Instituto de Salud Carlos III, Universidad de Valladolid-CSIC, Valladolid, 47005, Spain
e-mail: constanc@ibgm.uva.es; rocher@ibgm.uva.es

L. Almaraz
Instituto de Neurociencias, Universidad Miguel Hernández-CSIC, San Juan de Alicante, 03550, Spain

C.A. Nurse et al. (eds.), *Arterial Chemoreception: From Molecules to Systems*, Advances in Experimental Medicine and Biology 758, DOI 10.1007/978-94-007-4584-1_30,

neurotransmitters which in turn augment the action potential frequency in the carotid sinus sensory nerve (CSN), although acidic stimuli might also stimulate directly the sensory nerve endings of the CSN contributing to increase the CSN frequency in situation of acidosis (Rigual et al. 1984). When chemoreceptor activity is assessed at the level of the CSN, or ventilation, there exists an interaction between hypoxic and hypercapnic stimulus so that the response to both stimuli simultaneously applied is additive or more than additive over a wide range of the hypoxic level of stimulation, although severe hypoxias elicit maximal responses independently of the intensity of acidic stimulation (Fitzgerald and Parks 1971; Lahiri and Delaney 1975; Fitzgerald 1976; Pepper et al. 1995). Since in phaeochromocytoma PC12 cells (Taylor et al. 1999) and in the rat neonatal adrenal medulla (Rico et al. 2005) acidic and hypoxic stimulation were synergistic in eliciting secretion of catecholamine, it would appear that in CB the interaction would also occur at the level of chemoreceptor cell. In fact, in isolated chemoreceptor cells hypoxia and acidic stimulus also were synergistic in eliciting an increase in intracellular Ca^{2+} levels (Dasso et al. 2000). Yet, to date, it is not known if the interaction between both physiological stimuli also occurs at the level of neurotransmitter release or at very early steps in the transduction cascade, and if so, what are the molecular pathways mediating it. The purpose of this work has been to investigate if hypoxia and hypercapnic acidosis interact at level of carotid body chemoreceptor cells by analyzing their effects on membrane potential and neurotransmitter release, and to define if Epac dependent pathways modify such putative interaction. We have performed electrophysiological and neurochemical experiments aimed to provide answers to these questions.

30.2 Methods

30.2.1 Animals and Surgery

Experiments were performed with CB from adult New Zealand White rabbits and from adult Wistar rats, anaesthetized with sodium pentobarbital (40 mg/Kg; i.v. lateral vein of the rabbit ear; 60 mg/Kg body wt; i.p. rat). Carotid bifurcations were excised and the CB cleaned of surrounding tissue under a dissecting microscope in a Lucite chamber filled with ice-cold Tyrode solution (in mM: 140 NaCl, 5 KCl, 2 $CaCl_2$, 1.1 $MgCl_2$, 5 glucose and 10 HEPES; pH 7.4). All protocols were approved by the Institutional Animal Care and Use Committee of the University of Valladolid.

30.2.2 Measurement of the Release of CA

CBs endogenous deposits of catecholamines (CA) were isotopicaly labeled by incubating the organs during 2 h at 37°C in a medium containing 30 μM 3,5-[^{3}H]tyrosine (48 Ci/mmol) as described previously (Vicario et al. 2000; Rocher et al. 2005). After labeling, individual CBs were washed in precursor-free Tyrode bicarbonate buffered solution (in mM: NaCl, 116; KCl, 5; $CaCl_2$, 2; $MgCl_2$, 1.1; glucose 5.5; HEPES, 10; $NaHCO_3$, 24), continuously bubbled with 20% O_2/5% CO_2/rest N_2 at 37°C. After a washing period, superfusion solutions (control, hypoxic or acidic/hypercapnic/hypoxic) were renewed every 10 min, collected and analyzed for their ^{3}H-CA content by adsorption on alumina and counting of the eluates in a scintillation spectrometer, as described (Vicario et al. 2000). Control and experimental CBs were stimulated twice (stimulus 1 and 2; S1 and S2) with any given stimulus, but in the experimental CBs the second stimulus included the presence of a drug. Test solution at pH 6.6 was attained by lowering HCO_3^- to 14 mM and bubbling with 20% CO_2/20% O_2, rest N_2. To study the interactions between hypoxic and acidic/hypercapnic stimulus both stimuli were combined correspondingly by adjusting and O_2/CO_2 mixture in the superfusing solution during second HCO_3^- stimulus challenge (see Results). The evaluation of the drug effect or hypercapnic solution effect was assessed by comparing the ratios of the amplitude of the release responses (as the [^{3}H]-CA released (dpm) above basal conditions) in experimental CBs (S1/S2) with that obtained in control organs. Data are presented as mean ± SEM and statistical significance of the observed differences was assessed by the Student's *t* test for unpaired data.

30.2.3 Cell Preparation and Electrophysiological Recordings

Chemoreceptor cells were obtained from rabbit CB as described (Rocher et al. 2005). Briefly, CB were enzymatically dispersed, plated in poly-lysine coated coverslips and cultured up to 72 h in growth medium at 37°C and 5% CO_2/air atmosphere. Recordings of membrane potential were performed using the perforated-patch configuration at 35–36°C. Bath perfusing solutions for electrophysiological recordings were bicarbonate-buffered and identical to those used in the ^{3}H-CA release experiments (see above). Test solutions at different gas composition and pH were attained as described in previous paragraph. Pipette solution was (in mM): KCl, 35; potassium gluconate, 95; $MgCl_2$, 3; EGTA, 5; HEPES, 10; nystatin 100 μg/ml; pH adjusted at 7.2. Resistance of pipettes filled with internal solution was 2.0–3.5 MΩ. Voltage signals were recorded with an EPC-7 amplifier. Pulse generation, data acquisition and analysis were made through a Digidata 1322A (Axon Instruments, California, USA) and CLAMPEX 9.2 software. Voltage recordings were filtered at 2 KHz and sampled at 10–16 KHz.

30.2.4 Western Blot Analysis

For analysis of Epac-1 and Epac-2 expression, frozen CBs from normoxic and chronically hypoxic rats (7 days, 11%O_2 atmosphere) were homogenized in lysis buffer and solubilized proteins were electrophoretically fractioned on 10% SDS-polyacrylamide gels and protein bands transferred to PVDF membranes by standard procedures. Membranes were blocked in 5% non-fat dry milk and then incubated overnight at 4°C with primary monoclonal anti-Epac-1 and anti-Epac-2 (dilution 1:100; Cell Signalling) and anti β-actin (dilution 1:5000; Abcam). Membranes were reincubated with secondary HPR-conjugated anti-mouse IgG (1:1000–1:10000; Abcam). Proteins were visualized by enhanced chemiluminescence and quantified by a pdi Kodak scanner and Image J Analysis software. Optical densities of Epac signals were normalized to β-actin signals to assess equal protein loading. Normalized values were then averaged for all the replicated gels and used to calculate the relative ratio of hypoxic to normoxic values of the same gel.

30.3 Results

30.3.1 Synergism Between Acidosis and Hypoxia at the Chemoreceptor Cell Level in the Rabbit CB

This group of experiments was directed to explore the possible interaction between hypoxic and hypercapnic-acidotic stimulus on the exocytosis of neurotransmitters and on electrical behaviour of chemoreceptor cells. CBs were superfused with control solutions (5% CO_2/20% O_2, rest N_2) to measure

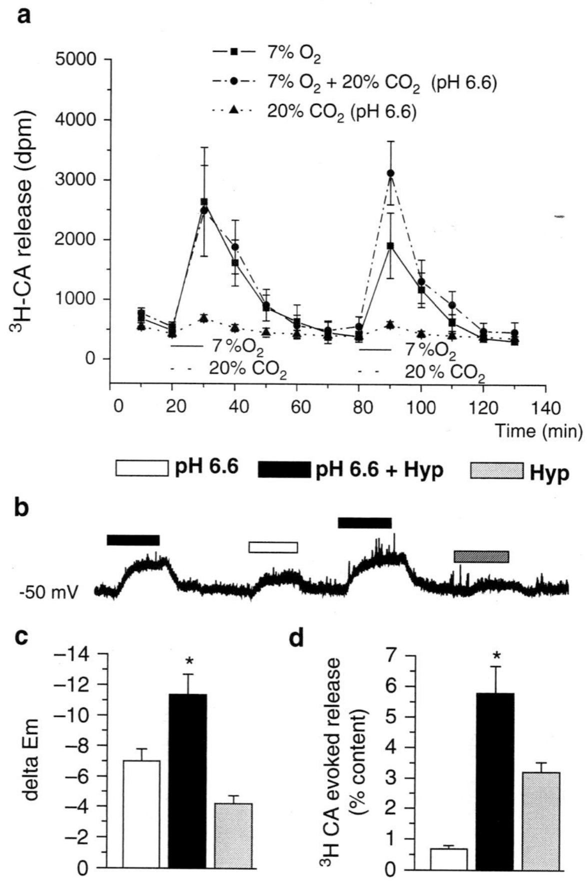

Fig. 30.1 Synergism between acidosis and hypoxia at the chemoreceptor cell level. (**a**) ^{3}H-CA release from three groups of CBs. One group was stimulated twice with hypoxia (incubating solution equilibrated with 7% O_2, pO_2 46 mmHg); a second group was stimulated twice with hypercapnia (incubating solution equilibrated with 20% CO_2 pH 6.6) and the third group was stimulated initially with hypoxia and the second stimulus was hypoxia + hypercapnia, (**b**) Changes in the membrane potential of a cultured chemoreceptor cell during exposure to hypercapnic-acidic (20% CO_2, pH 6.6), hypoxic (Hyp = 25–30 mmHg), and acidic/hypoxic perfusing solutions (pH 6.6+Hyp). (**c**) Mean ± SEM pO_2 of membrane potential changes from six cells exposed to above referred stimuli. (**d**) Evoked release expressed as percentage of the ^{3}H-CA tissue content present in the CB before stimulus application as in (**a**) (mean ± SEM, n = 12; *p < 0.01)

basal release or with stimuli solutions (7%O_2 or 20% CO_2/20% O_2, rest N_2) to measure stimulus evoked secretion. Figure 30.1a shows the time course of the release response elicited by individual stimuli in three CB groups in a first presentation of the stimulus; in the second stimulation, two out of the three groups received again individualized stimulus, but the third group received the hypoxic and hypercapnic stimuli combined (dashed line); note the more than additive response. A precise quantification of the interaction is shown in Fig. 30.1d: acidic hypercapnic stimuli released 0.7% ± 0.1% of the ^{3}H-CA content in the CB, the hypoxic stimuli 3.2% ± 0.3%, and when combined they released 5.8% ± 0.8%. Figure 30.1b shows the changes in the membrane potential of a cultured chemoreceptor cell during exposure to single acidic (20%CO_2, pH 6.6) or hypoxic stimulus (Hyp; $pO_2 \approx 25$ mmHg) and acidic/hypoxic combined stimuli. Figure 30.1c shows mean values of the Δ membrane potential of six chemoreceptor cells during superfusion with acidic, hypoxic and acidic/hypoxic media. As it can be seen, a slightly higher than additive effect is also observed at membrane potential level. The different ability of individual stimuli to depolarize chemoreceptor cells and to elicit a release response has been discussed in detail in a previous publication (Rocher et al. 2009).

30.3.2 Participation of cAMP and Epac Pathway in Acidic and Hypoxic Stimulus Potentiation in the Rat CB

Two decades ago cyclic AMP was implicated in O_2-sensing/transduction mechanism (Perez-Garcia et al. 1991) and recently we have described that a cAMP binding protein (Epac) activated pathway is involved in rabbit CB response to acute hypoxia (Rocher et al. 2009). Here we have tested the hypothesis that the cAMP/Epac mediated pathway would participate in the synergic response of CB to combined hypoxic and acidic stimulus.

Figure 30.2a shows that the positive interaction between hypoxia and hypercapnic acidosis is mediated by cAMP because the adenylate cyclase inhibitor SQ-22536 (100 μM) eliminated the potentiation of the ^{3}H-CA release when it was applied 10 min previous and during second stimulation period. Figure 30.2b summarizes data on this group of experiments: First column shows the release induced by hypoxia, second the release induced by combining hypoxia and hypercapnia, third the release of combined stimuli in the presence of the cyclase inhibitor and the fourth column as the third + the Epac activator 8-pCPT-2′-O-Me-cAMP pCPT-2-O-Me-cAMP. Finally, Fig. 30.2c shows data obtained in control and chronically hypoxic CBs. Notice the lack of effect of SQ-22536 on basal release of CA (5% CO_2) and on the hypercapnic-acidic (20% CO_2 pH 6.6) elicited response in control CB. By the contrary, in chronic hypoxic CB, SQ-22536 inhibits by nearly 40% the secretory release elicited by 20%CO_2 (pH 6.6), i.e., it appears that in chronically hypoxic (CH) sensitized CB the CO_2/acidic transduction occurs by mechanisms involving the cAMP-Epac pathway.

30.3.3 Western Blot Analysis of Epac-1 and Epac-2

In Fig. 30.3 we present Epac-1 and Epac-2 protein expression in rat CB. Figure 30.3a shows a representative immunoblot for both proteins obtained from normoxic and chronically hypoxic rat CBs. We observe bands of the expected molecular weight for Epac-1 and Epac-2 (≈100 kDa) in both, normoxic and hypoxic CB. Superior cervical ganglion (SCG) was used as positive control tissue. In the same conditions, Epac-1 was the most abundant isoform and Epac-2 is represented by a double band (≈100 kDa) barely detected by immunoblot. To assess equal protein loading, densities of β-actin bands were compared and represented in Fig. 30.3b. Whereas no significant change was observed in Epac-1 expression, Epac-2 lower band up regulated (40%) after chronic hypoxic treatment.

30.4 Discussion

The known interactions between hypoxia and hypercapnia on intracellular Ca^{2+} levels and CSN discharges (see Introduction), rendered of special interest to define the interaction between hypoxic and hypercapnic-acidotic stimulus on the exocytosis of neurotransmitters. Our present findings are consistent with the notion that the interaction between hypoxic and acidic hypercapnic stimulus is generated at the level of chemoreceptor cells. First, we demonstrate a positive interaction between hypoxic and

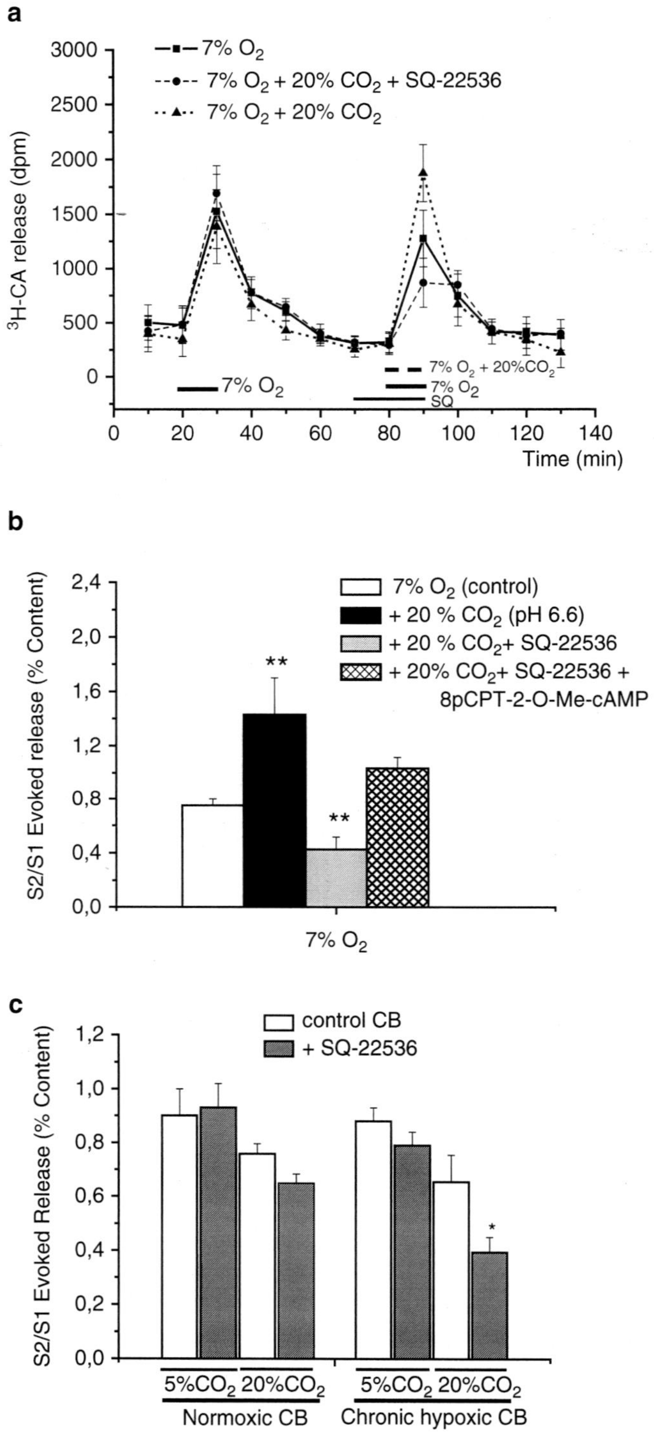

Fig. 30.2 The potentiating effect of acidotic hypercapnia/hypoxia on rat CB secretory response is mediated by cAMP/Epac. (**a**) Time course of ^{3}H-CA release elicited by two consecutive stimuli in three groups of CBs: first stimulus in the three groups was hypoxia (7%O_2) and second stimulus was hypoxia (*continuous line*; n=6), hypoxia + hypercapnia (7%O_2+20%CO_2, pH 6.6; *dotted line*; n=6), and hypoxia + hypercapnia acidotic in presence of SQ-22536 (*dashed line*; n=6). (**b**) Effect of SQ-22536 on potentiating effect of combined stimuli and its reversion by 8pCPT-2-O-Me-cAMP expressed as S2/S1 evoked release of ^{3}H-CA (n=8; **p<0.01). (**c**) Effect of SQ 22536 on 20% CO_2 (pH 6.6) elicited secretory response from normoxic and chronically hypoxic rat CB

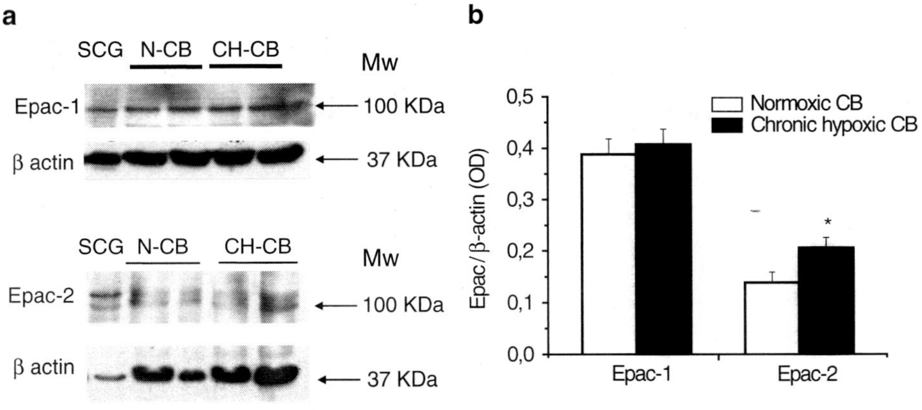

Fig. 30.3 Expression of Epac-1 and Epac-2 in rat CB by Western blot. (**a**) CB proteins fractionated on 10% SDS-PAGE and incubated with monoclonal anti-Epac-1 or anti-Epac-2 and β-actin. *N-CB* normoxic CB, *CH-CB* chronically hypoxic CB, *SCG* superior cervical ganglion. (**b**) Ratio Epac/β-actin from normoxic and hypoxic CB. Mean ± SEM ($n=8$; $*p<0.05$)

hypercapnic-acidotic stimuli at the level of membrane depolarization and second, at the level of secretory response in rat and rabbit CB chemoreceptor cells. This last interaction is mediated by a cAMP dependent pathway because inhibition of adenylate cyclase by the specific blocker SQ-22536 abolishes the potentiating response to both stimuli. Although more experimental data are required to envision the mechanisms involved in cAMP potentiation of the hypoxic /hypercapnic acidotic release, our preliminary findings that a permeable activator of Epac, 8-pCPT-2′-O-Me-cAMP, reverses the adenylate cyclase inhibition suggests that it can be mediated by these exchange proteins directly activated by cAMP (Epacs) which are known to have many cell targets including the exocytotic machinery, Ca^{2+} and K^{+} channels. In this context we also demonstrate, for the first time, that Epac-1 and Epac-2 are expressed in rat CB and one of them or both proteins could mediate the described potentiating effect. It has been reported that in the rat, Epac1 mRNA is widely expressed whereas Epac-2 mRNA is more restricted to neural tissues. Epac protein seems to be developmentally regulated in the rat central nervous system, being Epac-1 the dominant isoform in embryonic tissues, but down-regulated in adult tissues. Conversely, Epac-2 is upregulated after birth (Peace and Shewan 2011). In rat CB we found that Epac-1 is the predominant isoform relative to β-actin expression, being Epac-2 a hypoxic inducible isoform. Although it is unclear whether this dynamic expression pattern has any functional significance, it would suggest distinct roles for Epac-1 and Epac-2. The additional observation that the cAMP independent secretory response elicited by hypercapnic acidosis acting as single stimulus in normoxic CB became cAMP dependent after chronic hypoxic treatment (see Fig. 30.2c) and that Epac-2 is up-regulated after chronic hypoxia suggest a more pleiotropic role of Epac pathway in CB chemoreception. Our experiments in acidic transduction after chronic hypoxic treatment do not permit to discard the participation of a signal transduction pathway activated by PKA because Epac and PKA can independently mediate cAMP-dependent cellular functions and here we haven't explore this alternative.

Our data using the release of neurotransmitters as an index of Ca^{2+} dynamics in CB chemoreceptor cells are compatible with the findings of Dasso et al. (2000) showing that in isolated chemoreceptor cells hypoxia and acidic stimulus are synergistic in eliciting an increase in intracellular Ca^{2+} levels. The finding that an endogenous cAMP-dependent mechanism is enhancing the coupling of the hypoxic and hypercapnic acidic stimulus to exocytosis would suggest that the Epac pathway is activating Ca^{2+} entry and subsequent Ca^{2+}-dependent neurotransmitter release. Although Ca^{2+} channels have been described as potential targets for Epac (Seino and Shibasaki 2005) this possibility could be excluded in the case of chemoreceptor cells because cAMP analogs do not modify Ca^{2+} currents in chemoreceptor cells (Lopez-Lopez et al. 1993).

Exocytotic machinery and K^+ channels would be alternative targets of the Epac-dependent regulation mediated by cAMP during hypoxic /hypercapnic acidic stimulation in accord to the above mentioned potentiation of the exocytotic release and membrane potential depolarization induced by both stimuli. Recently, using adenylate cyclase and PKA and Epac specific inhibitors we have shown in rabbit CB that the high gain of the hypoxic stimulus on the exocytosis is mediated by cAMP, PKA-independent, Epac-mediated mechanism (Rocher et al. 2009). It has been described that the facilitation of exocytosis is mediated by interactions of Epac with Rab-3, a small G-protein which is a key modulator of exocytotic machinery (Sudhof 2004) increasing the release probability of secretory granules already in the readily releasable pool and accelerates the refilling of this pool (Renstrom et al. 1997). Finally, the additional observation that cAMP analogs mimic hypoxia in inhibiting K^+ currents in isolated rabbit chemoreceptor cells (Lopez-Lopez et al. 1993) suggests that K^+ channels may also be targets of the Epac dependent regulation mediated by cAMP during hypoxic/acidic combined stimulation and could explain the synergy on membrane depolarization. Several K^+ channel candidates would be targets for Epac pathways (Gonzalez et al. 2009) but to date we do not have any evidence and more experiments will be required to clarify the effect.

Acknowledgements This work was supported by grants BFU2007-61848 (DGICYT), CIBER CB06/06/0050 (FISS-ICIII).

References

Dasso LL, Buckler KJ, Vaughan-Jones RD (2000) Interactions between hypoxia and hypercapnic acidosis on calcium signalling in carotid body type I cells. Am J Physiol Lung Cell Mol Physiol 279:L36–L42

Fitzgerald RS (1976) Single fiber chemoreceptor responses of carotid and aortic bodies. In: Paintal AS (ed) Morphology and mechanisms of chemoreceptors. Vallabhbhai Patel Chest Institute, Delhi, pp 27–35

Fitzgerald RS, Parks DC (1971) Effect of hypoxia on carotid chemoreceptor response to carbon dioxide in cats. Respir Physiol 12:218–229

Gonzalez C, Almaraz L, Obeso A, Rigual R (1994) Carotid body chemoreceptors: from natural stimuli to sensory discharges. Physiol Rev 74:829–898

Gonzalez C, Vaquero LM, López-López JR, Pérez-García MT (2009) Oxygen-sensitive potassium channels in chemoreceptor cell physiology: making a virtue of necessity. Ann N Y Acad Sci 1177:82–88

Lahiri S (1976) Depressent effect of acute and chronic hypoxia on ventilation. In: Paintal AS (ed) Morphology and mechanisms of chemoreceptors. Vallabhbhai Patel Chest Institute, Delhi, pp 138–146

Lahiri S, Delaney RG (1975) Stimulus interaction in the *responses* of carotid body chemoreceptor single afferent fibers. Respir Physiol 24:249–266

Lopez-Lopez JR, De Luis A, Gonzalez C (1993) Properties of a transient K^+ current in chemoreceptor cells of rabbit carotid body. J Physiol 460:15–32

Nattie E (1999) CO_2, brainstem chemoreceptors and breathing. Prog Neurobiol 59:299–331

Peace AG, Shewan DA (2011) New perspectives in cyclic AMP-mediated axon growth and guidance: the emerging epoch of Epac. Brain Res Bull 84:280–288

Pepper DR, Landauer RC, Kumar P (1995) Postnatal development of CO2-O2 interactionin the rat carotid body in vitro. J Physiol 485:531–541

Perez-Garcia MT, Almaraz L, Gonzalez C (1991) Cyclic AMP modulates differentially the release of dopamine induced by hypoxia and other stimuli and increases dopamine synthesis in the rabbit carotid body. J Neurochem 57:1992–2000

Renstrom E, Eliasson L, Rorsman P (1997) Protein kinase A-dependent and -independent stimulation of exocytosis by cAMP in mouse pancreatic B-cells. J Physiol 502:105–118

Rico AJ, Prieto-Lloret J, Gonzalez C, Rigual R (2005) Hypoxia and acidosis increase the secretion of catecholamines in the neonatal rat adrenal medulla: an in vitro study. Am J Physiol Cell Physiol 289:C1417–C1425

Rigual R, Gonzalez E, Fidone S, Gonzalez C (1984) Effects of low pH on synthesis and release of catecholamines in the cat carotid body in vitro. Brain Res 309:178–181

Rocher A, Geijo-Barrientos E, Caceres AI, Rigual R, Gonzalez AL (2005) Role of voltage-dependent calcium channels in stimulus-secretion coupling in rabbit carotid body chemoreceptor cells. J Physiol 562:407–420

Rocher A, Caceres A, Almaraz L, Gonzalez C (2009) Epac signalling pathways are involved in low PO2 chemoreception in CB chemoreceptor cells. J Physiol 587:4015–4027

Seino S, Shibasaki T (2005) PKA-dependent and PKA-independent pathways for cAMP-regulated exocytosis. Physiol Rev 85:1303–1342

Sudhof TC (2004) The synaptic vesicle cycle. Annu Rev Neurosci 27:509–547

Taylor SC, Roberts ML, Peers C (1999) Acid-evoked quantal catecholamine secretion from rat phaeochromocytoma cells and its interaction with hypoxia-evoked secretion. J Physiol 519:765–774

Vicario I, Rigual R, Obeso A, Gonzalez C (2000) Characterization of the synthesis and release of catecholamine in the rat carotid body in vitro. Am J Physiol Cell Physiol 278:C490–C499

Chapter 31
Interactions Between Postnatal Sustained Hypoxia and Intermittent Hypoxia in the Adulthood to Alter Brainstem Structures and Respiratory Function

Elena Olea, Susana P. Gaytan, Ana Obeso, Constancio Gonzalez, and Rosario Pasaro

Abstract Neural plasticity is defined as a persistent change in the morphology and/or function based on prior experiences. Plasticity is well evident when the triggering experience occurs early in life, but in the case of respiratory control plasticity, it also can be triggered in adult life. We have combined a 10 days postnatal hypoxic (PH) (0–10 days of age;11% O_2) and a 15 days intermittent hypoxia (IH) exposures in the adulthood (90–105 days of age; 5% O_2, 40 s/20% O_2, 80 s; 8 h/day) to test if early PH interacts with IH of the adulthood to generate detrimental plastic changes. After recording of ventilatory parameters, the brains were studied immunocytochemically for localization of the organization pattern of non-phosphorylated subunit of neurofilament H (NFH) and tyrosine hydroxylase (TH) expression in the nucleus tractus solitarius (Sol) and caudal (CVL) and rostral ventrolateral reticular (RVL) nuclei, areas related to central cardio-respiratory regulation. In comparison to control, PH male rats (but not females) at 1 month of age hyperventilated at rest, in response to moderate hypoxia (12% O_2) and 5% CO_2, the effect being due to increased tidal volume. At 3.5 months sex differences in ventilation disappeared and it was indistinguishable between control and PH. IH tended to decrease ventilation in both control (C) and PH animals. PH augmented PENH values in air and in hypoxic conditions when compared with C group. IH in both groups, tended to decrease the PENH value, being statistically different in PH+IH. Results also show an increment of disorganization of NFH-positive labeled structures at the level of Sol and CVL/RVL nuclei in PH, IH and HP+HI groups. PH rats showed differences in the number of TH-positive neurons at the level of CVL/RVL nuclei, which was increased in the PH and PH+IH groups with respect to C one. In conclusion, PH alters the central morpho-physiological organization and the catecholaminergic components of cardio-respiratory nuclei, whose effects were enhanced after a period of IH in the adulthood.

Keywords Neural plasticity • Nucleus tractus solitarius • Caudal ventrolateral reticular nucleus • Rostral ventrolateral reticular nucleus • Respiratory control plasticity • Carotid body chemoreflex

E. Olea • A. Obeso • C. Gonzalez
Department of Biochemistry and Molecular Biology and Physiology and IBGM, CIBERES-Institute Carlos III, University of Valladolid School of Medicine and CSIC, Ramon y Cajal 7, Valladolid, 47005, Spain
e-mail: aobeso@ibgm.uva.es; constanc@ibgm.uva.es

S.P. Gaytan • R. Pasaro (✉)
Department of Physiology and Zoology, University of Seville, Avda Reina Mercedes, 6, Seville, 41012, Spain
e-mail: sgaytan@us.es; mrpasaro@us.es

C.A. Nurse et al. (eds.), *Arterial Chemoreception: From Molecules to Systems*, Advances in Experimental Medicine and Biology 758, DOI 10.1007/978-94-007-4584-1_31,

31.1 Introduction

Under chronic conditions, hypoxia (both continuous and intermittent) appears to induce central nervous system plasticity of respiratory and sympathetic functions, leading to cardiovascular abnormalities, such as hypertension (Prabhakar et al. 2007; Zoccal et al. 2009). Perinatal hypoxia in the neonate originates numerous functional deficits, such as impaired resting ventilation and ventilatory response to hypoxia, as well as alterations in the catecholaminergic components of the chemoafferent pathway (Braga et al. 2006). Since intermittent hypoxia (IH) is a prominent feature of sleep-disordered breathing, neuroplasticity may compensate for factors that predispose to sleep-disordered breathing, particularly during obstructive sleep apnea (OSA) (Mahamed and Mitchell 2007). These patients suffer repetitive episodes of obstruction of upper airways occurring during sleep (>30 episodes/ sleep hour in some patients). Each obstruction produces hypoxic challenge that trigger cardio-respiratory reflexes, most significantly hyperventilation, aimed to normalize arterial blood gases, taking into account the homeostatic role of the carotid body (CB) function (Gonzalez et al. 1994). Frequently, patients suffering OSA show a number of associated pathologies, mostly cardiovascular (hypertension, augmented acute vascular events). It has been proposed that the CB chemoreflex plays a key pathogenic role in the cardiovascular pathology in OSA: repeated stimulation during the apneic episodes would cause a sensitization of the CB chemoreflex that would ultimately produce a sustained sympathetic drive and increase in circulating catecholamine (Bao et al. 1997; Kumar et al. 2006).

Anatomical and physiological experiments have shown that the dorsomedial part of the nucleus tractus solitarius (Sol) is the primary termination site of glossopharyngeal and vagal baroreceptors, integrating the baroreceptor afferents, while the midline area caudal to the calamus scriptorius has been identified as a primary central termination site for CB afferents (Guyenet 2006). Baroreceptor input is integrated in the Sol and conveyed to the caudal (CVL) and rostral ventrolateral reticular (RVL) medullary nuclei to modulate vagal and sympathetic outflow to the heart and peripheral vasculature (Guyenet 2006; Kline et al. 2010). Both regions also contain the noradrenergic/adrenergic cell groups, A2/C2 and A1/C1, respectively, which are adjacent to or intermingled with the respiratory neurons (Pilowsky et al. 1990). The Sol neurons are stimulated by hypoxia or hypercapnia, and most profoundly by a combination of both, and induce a chemoreceptor reflex response that includes hyperpnea, bradycardia and a sympathetically mediated vasoconstriction (Kline et al. 2010), and the long-term acclimatization to hypoxia has been associated with tyrosine hydroxylase (TH) upregulation (Roux et al. 2000).

In the present work we explore the influence of postnatal hypoxia (PH) and/or adult chronic IH to test if early PH interacts with IH in the adulthood to generate detrimental plastic changes that could impinge the chemoreceptor reflex response and the corresponding medullary cardio-respiratory neuronal network. We used immunohistochemical methods to study the TH expression and the organization pattern of non-phosphorylated subunit of Neurofilament H (heavy)(NFH) matrix, which defines the integrity of neuronal cell bodies, dendrites and some thick axons in the medulla (Perrot et al. 2008). The importance of NFH is rooted in the fact that its appearance occurs after synaptogenesis (Carden et al. 1987). Immunohistochemistry against NFH has been also used to delimitate different subnuclei in the mammalian brain (Koutcherov et al. 2000), as well as for determining the neuropathological process in several nervous tissue diseases (Perrot et al. 2008).

31.2 Methods

Experiments were performed in 40 adult Wistar rats following the European Community Council directive (86/609/EEC) for the Care and Use of Laboratory Animals. Animals were initially divided in two 20 animal groups: control (C) and postnatal hypoxic (exposed to a sustained hypoxic atmosphere, 11% O_2, from days 0–10 of age, PH). When they reached 3 months of age, each group was subdivided in two, with half of the animals in each group being exposed to IH for 15 days, so that by the age of 3.5 months we have four groups of 10 animals: C, PH, C+IH, and PH+IH. Intermittent hypoxia protocol was as follows: 5% O_2, 40 s/20% O_2, 80 s; 8 h/Day; 15 days.

We have measured, in all the animals, respiratory parameters and airway resistances (PENH values) at 1 and 3.5 months of age: breathing frequency (Bf), tidal volume (TV), minute ventilation/Kg of body weight (MV/Kg), and PENH values while the animals were breathing air, 12, 10, and 7% O_2 and 5% CO_2 in air. Ventilation was measured in conscious freely moving rats by whole body plethysmography as previously described (Gonzalez-Martín et al. 2011) (Emka Technologies, Paris, France; 0.5 or 2 L/min for rats at 1 or 3.5 months of age, respectively). Temperature was maintained in the chamber within the thermo-neutral range (22–24°C). Data were presented as means ± SEM. Statistical significance of differences was assessed using a two-tailed t-test for unpaired data and for comparisons of more than two groups we have used a two-way ANOVA followed by Bonferroni multicomparison test.

After all the recording parameters, all the animals were deeply anesthetized with sodium pentobarbital (Nembutal, 150 mg/Kg, i.p.) and perfused transcardially with saline followed by 4% formaldehyde in 0.1 M phosphate-buffer (PB). The brainstem was removed, postfixed overnight, and stored in a cryo-protecting solution (sucrose 30%). Coronal sections (40 μm thick) were cut on a cryostat (CM1850, Leica Microsystems, Wetzlar, Germany) serially collected in 0.1 M phosphate-buffered saline (PBS), and processed for localization of TH expression and the organization pattern of NFH. First, following the routine procedure (Pasaro et al. 2009), 50% of sections were incubated (72 h, 4°C) with a rabbit anti-NFH (SMI 32) monoclonal antibody (1:500; Covance, Emeryville, USA) followed by incubation with a biotinylated goat anti-mouse IgG (1:200; Jackson-ImmunoResearch Labs, overnight, 4°C) and the standard revealing process (Pasaro et al. 2009). The other 50% of sections were incubated (72 h, 4°C) with a rabbit anti-TH antibody (1:1000; Santa Cruz Biotechnology, Santa Cruz, USA) followed by incubation with a biotinylated donkey anti-rabbit IgG (1:500; Jackson-ImmunoResearch Labs., West Grove, USA, overnight, 4°C) and the same protocol for revealing the peroxidase activity. After washing in PBS, the sections were mounted on gelatin-coated slides, air-dried, dehydrated, cleared with xylene, and cover-slipped with DPX mounting media (VWR Inter. Ltd., Poole, England). The sections were analyzed using a light microscope (Olympus BX61). Counting of TH-positive cells and NFH-positive cells and fibers was performed in one out of three sections from −6.2 to −2.8 mm interaural, with the help of the Paxinos and Watson Atlas (1997). The results were submitted to statistical analysis using the SSPS 15.0 software, using a t-test for independent groups for equal variances tested by "Levene's Test for the homogeneity of variance. The basic criterion for statistical significance is a "2-tailed significance" <0.05.

31.3 Results

The pups after 10 days of PH showed a significant ($p<0.001$) decrease in the body weight at 1 month of age (50.4 ± 1.40 g) with respect to control (57.3 ± 1.34 g), but not at 3.5 months of age (control: 255.7 ± 9.2 g vs. PH pups 241.5 ± 6.9 g, $p<0.05$).

In comparison to control, PH male rats (but not females) at 1 month of age hyperventilated at rest and in response to moderate hypoxia (12% O_2) and 5% CO_2, the effect being due to increased TV (Fig. 31.1). At 3.5 months of age, sex related differences disappeared and, additionally, ventilation was indistinguishable between the C and PH groups. IH tended to decrease ventilation in both control and PH animals (Fig. 31.2a). PENH values at 1 month were not different between C and PH groups and there were no sex related differences. At 3.5 months PENH was equal in both sexes. In PH group PENH value tended to augment compared to C group, being statistically different when rats breathe

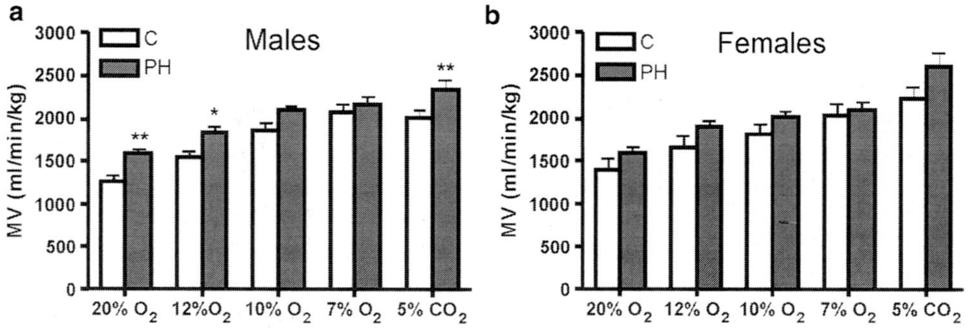

Fig. 31.1 Minute ventilation (MV) of control (C) and postnatal hypoxic (PH) rats. PH male rats (**a**) but not females (**b**) at 1 month of age hyperventilated at rest, in response to moderate hypoxia (12% O_2) and 5% CO_2, the effect being due to increased tidal volume (TV). At 3.5 months no changes were observed

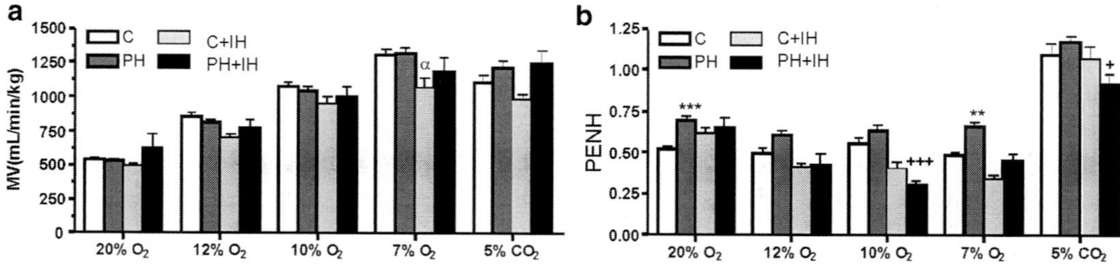

Fig. 31.2 (**a**) Minute ventilation (MV) in control (C) and postnatal hypoxic (PH)(female + male) rats, before and after intermittent hypoxia (IH) at 3.5 months, while breathing air, 12% O_2, 10% O_2, 7% O_2 and 5% CO_2. There are not differences between control and PH groups. IH tended to decrease MV in both groups, C+IH and PH+IH, but mainly in C+IH group being statistically different at 7% of O_2 (α $p<0.05$ C+IH vs. C group). (**b**) Airways resistances (PENH) values in C and PH (female + male) rats, before and after IH at 3.5 months of age while breathing air, 12% O2, 10% O2, 7% O2 and 5% CO2. PH augmented PENH values in air and 7% O_2 (*** $p<0.001$ and ** $p<0.01$, respectively) when compared with C group. IH in both groups, C+IH and PH+IH, tended to decrease the PENH value, being statistically different in PH+IH when the animals breathe 10% O2 +++ $p<0.001$ and 5% CO_2 + $p<0.05$ vs. PH group

air and 7% O_2 ($p<0.001$ and $p<0.01$, respectively). IH tended to decrease PENH value in both groups (C+IH and PH+IH), mainly when breathing in hypoxic and hypercapnic atmospheres (12%, 10%, and 10% O_2, and 5% CO_2 in air) (Fig. 31.2b).

On the other hand, a strong difference in the organization pattern of somatic and dendritic trees of the neurons located in the medulla was found. This was more evident in the Sol and CVL/RVL nuclei, comparing the C group with respect to the HP, C+IH and HP+HI, where there was a pattern of the NFH-positive structures more loosely and randomly distributed through the medulla in the latter one (Fig. 31.3a–d and i–l). TH-positive neurons were found along the A2/C2 and A1/C1 groups in all of the experimental groups, showing the characteristic somatic and dendritic pattern of catecholaminergic neurons (Fig. 31.3e–g, m and n), as well as dendritic varicosities around several medullary nuclei motoneurons (Fig. 31.3h). The quantitative results did not show significant statistical differences with respect to the number of cells in the Sol nucleus among all experimental groups. However, in the area around −5.8 mm interaural (around the middle of the Sol nucleus), it was shown a moderate statistically difference between the number of TH-positive neurons in the PH group with respect to the C one, this difference was even greater comparing the C and the PH+IH group. There was a statistically significant difference in the number of TH-positive neurons in the CVL/RVL nuclei when comparing the PH group with respect to the C one; the difference was even higher when comparing the C with respect to the PH+IH group (Fig. 31.3o).

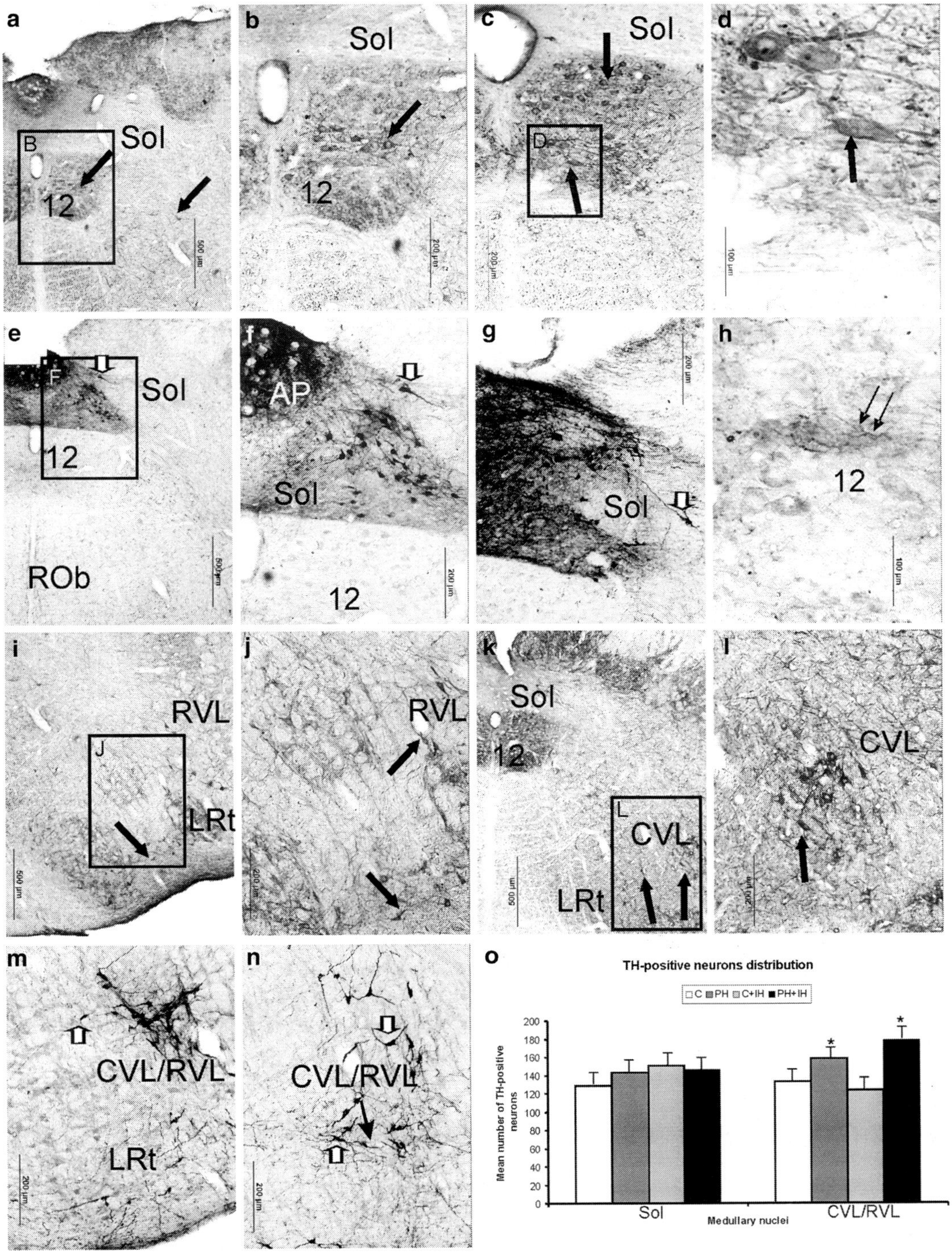

Fig. 31.3 Microphotographs of coronal sections of the rat medulla showing the distribution of NFH-positive cells (*black arrows*) and fibers (**a–d** and **i–l**) and TH-positive neurons (*open black arrows*) (**e–g**, **m** and **n**) and fibers (*double black arrows*)(**h**). (**a–d**) NFH-positive cells and fibers in the solitary tract nucleus (Sol) of control (**a**, **b**) and perinatal hypoxic (PH) (**c**, **d**) rats. (**e–g**) TH-positive neurons in the caudal Sol of control (**e**, **f**) and PH (**g**, **h**) rat. (**h**) Higher magnification of TH-positive fibers surrounding Neutral Red stained genioglossus motoneurons. (**i–l**): NFH-positive cells and fibers in the caudal (CVL) and rostral ventrolateral medulla (RVL) of control (**i**, **j**) and PH (**k**, **l**) rats. (**m, n**): TH-positive neurons in the CVL and RVL of control (M) and PH (N) rats. (**o**) Diagrams showing the mean number of TH-positive neurons in the Sol and CVL/RVL nuclei of the medulla. Abbreviations: *12* hypoglossal nucleus, *IO* inferior olive, *LRt* lateral reticular nucleus, *ROb* raphe obscurus nucleus. Note that (**b**, **d**, **f**, **j** and **l**) show higher magnifications of the microphotographs in (**a**, **c**, **e**, **g**, **i** and **k**)

31.4 Discussion

A recent study has found that IH rats exhibited marked hypoventilation in all studied conditions and an increased turnover of norepinephrine in sympathetic endings, indicating that IH produced a bias in the integration of the input arising from the CB with a diminished drive of ventilation and an exaggerated activation of brainstem sympathetic neurons (Gonzalez-Martín et al. 2011). These facts imply profound changes in the brainstem integrating respiratory centers (Kline 2010) which are accompanied by marked changes in the dynamics of norepinephrine in vascular sympathetic endings. Present results, using a more intense IH paradigm, show only a tendency in the MV/Kg to decrease. Additionally, we found that the number of TH-positive neurons increase in PH and particularly in PH+IH animals, implying plastic changes in brain stem organization. Finally we observed that airways resistance, as evidenced by PEHN values are increased in PH and PH+IH groups.

The caudal Sol neurons receive inputs from CB chemoreceptors, a great majority of them projected to the CVL/RVL nuclei (Guyenet 2006). The medial nucleus of the Sol, a brain region containing catecholaminergic neurons, plays an important role in cardiovascular and respiratory control mechanisms (Guyenet 2006). In the present study, it has been found in this area related to cardiovascular afferent integration, the strongest difference in the number of TH-positive neurons along the Sol nucleus. Previous studies have concluded that different types of hypoxia such a acute or chronic and long term IH elicits different changes in TH activity, suggesting also a region specific modulation of catecholamine response (Adams et al. 2001). Additionally, IH, which represents the main trait of OSA and an independent risk factor for hypertension, also induces cellular changes, architectural disorganization, and apoptosis in several brain regions (Gozal et al. 2005). In the present report, it was found an increment in the number of TH-positive neurons, indicating a higher number of cells expressing the TH enzyme, as well as a disorganization of the NFH-positive elements, that could be related to a deterioration of the neurons at the level of the Sol and CVL/RVL nuclei.

Acknowledgements Supported by Grants BFU2007-61848 (DGICYT), CIBER CB06/06/0050 (ISCiii), and P08-CVI-03934 and P09-CVI-4617 (Junta de Andalucía). Thanks to the technical assistance of Mª Angeles Bueso and Elena Gonzalez.

References

Adams MB, Brown RE, Gibson C, Coulter CL, McMillen IC (2001) Tyrosine hydroxylase protein content in the medulla oblongata of the foetal sheep brain increases in response to acute but not chronic hypoxia. Neurosci Lett 316(2):63–66

Bao G, Metreveli N, Li R, Taylor A, Fletcher EC (1997) Blood pressure response to chronic episodic hypoxia: role of the sympathetic nervous system. J Appl Physiol 83:95–101

Braga VA, Soriano RN, Machado BH (2006) Sympathoexcitatory response to peripheral chemoreflex activation is enhanced in juvenile rats exposed to chronic intermittent hypoxia. Exp Physiol 91:1025–1031

Carden MJ, Trojanowski JQ, Schlaepfer WW, Lee VM (1987) Two-stage expression of neurofilament polypeptides during rat neurogenesis with early establishment of adult phosphorylation patterns. J Neurosci 7(11):3489–3504

Gonzalez C, Almaraz L, Obeso A, Rigual R (1994) Carotid body chemoreceptors: from natural stimuli to sensory discharges. Physiol Rev 74(4):829–898

Gonzalez-Martín MC, Vega-Agapito MV, Conde S, Castañeda J, Bustamante R, Olea E, Perez-Vizcaino F, Gonzalez C, Obeso A (2011) Carotid body function and ventilatory responses in intermittent hypoxia. Evidence for anomalous brainstem integration of arterial chemoreceptor input. J Cell Physiol 226(8):1961–1969

Gozal E, Sachleben LR Jr, Rane MJ, Vega C, Gozal D (2005) Mild sustained and intermittent hypoxia induce apoptosis in PC-12 cells via different mechanisms. Am J Physiol Cell Physiol 288(3):C535–C542

Guyenet PG (2006) The sympathetic control of blood pressure. Nat Rev Neurosci 7(5):335–346

Kline DD (2010) Chronic intermittent hypoxia affects integration of sensory input by neurons in the nucleus tractus solitarii. Respir Physiol Neurobiol 174:29–36

Kline DD, King TL, Austgen JR, Heesch CM, Hasser EM (2010) Sensory afferent and hypoxia-mediated activation of nucleus tractus solitarius neurons that project to the rostral ventrolateral medulla. Neuroscience 167(2):510–527

Koutcherov Y, Mai JK, Ashwell KW, Paxinos G (2000) Organization of the human paraventricular hypothalamic nucleus. J Comp Neurol 423(2):299–318

Kumar GK, Rai V, Sharma SD, Ramakrishnan DP, Peng YJ, Souvannakitti D, Prabhakar NR (2006) Chronic intermittent hypoxia induces hypoxia-evoked catecholamine efflux in adult rat adrenal medulla via oxidative stress. J Physiol 575:229–239

Mahamed S, Mitchell GS (2007) Is there a link between intermittent hypoxia-induced respiratory plasticity and obstructive sleep apnoea? Exp Physiol 92(1):27–37

Pasaro R, Ribas-Salgueiro JL, Matarredona ER, Sarmiento M, Ribas J (2009) Systemic inhibition of the Na(+)/H(+) exchanger type 3 in intact rats activates brainstem respiratory regions. Adv Exp Med Biol 648:395–401

Paxinos G, Watson C (1997) The rat brain in stereotaxic coordinates, 3rd edn. Academic, Sydney

Perrot R, Berges R, Bocquet A, Eyer J (2008) Review of the multiple aspects of neurofilament functions, and their possible contribution to neurodegeneration. Mol Neurobiol 38(1):27–65

Pilowsky PM, Jiang C, Lipski J (1990) An intracellular study of respiratory neurons in the rostral ventrolateral medulla of the rat and their relationship to catecholamine-containing neurons. J Comp Neurol 301:604–617

Prabhakar NR, Dick TE, Nanduri J, Kumar GK (2007) Systemic, cellular and molecular analysis of chemoreflex-mediated sympathoexcitation by chronic intermittent hypoxia. Exp Physiol 92(1):39–44

Roux JC, Pequignot JM, Dumas S, Pascual O, Ghilini G, Pequignot J, Mallet J, Denavit-Saubié M (2000) O2-sensing after carotid chemodenervation: hypoxic ventilatory responsiveness and upregulation of tyrosine hydroxylase mRNA in brainstem catecholaminergic cells. Eur J Neurosci 12(9):3181–3190

Zoccal DB, Paton JF, Machado BH (2009) Do changes in the coupling between respiratory and sympathetic activities contribute to neurogenic hypertension? Clin Exp Pharmacol Physiol 36(12):1188–1196

Chapter 32
Brain-Derived Neurotrophic Factor in the Nucleus Tractus Solitarii Modulates Glucose Homeostasis After Carotid Chemoreceptor Stimulation in Rats

Sergio Montero, Ricardo Cuéllar, Mónica Lemus, Reyes Ávalos, Gladys Ramírez, and Elena Roces de Álvarez-Buylla

Abstract Neuronal systems, which regulate energy intake, energy expenditure and endogenous glucose production, sense and respond to input from hormonal related signals that convey information from body energy availability. Carotid chemoreceptors (CChr) function as sensors for circulating glucose levels and contribute to glycemic counterregulatory responses. Brain-derived neurotrophic factor (BDNF) that plays an important role in the endocrine system to regulate glucose metabolism could play a role in hyperglycemic glucose reflex with brain glucose retention (BGR) evoked by anoxic CChr stimulation. Infusing BDNF into the nucleus tractus solitarii (NTS) before CChr stimulation, showed that this neurotrophin increased arterial glucose and BGR. In contrast, BDNF receptor (TrkB) antagonist (K252a) infusions in NTS resulted in a decrease in both glucose variables.

Keywords BDNF • NTS • Brain glucose retention • Carotid chemoreceptor stimulation

32.1 Introduction

BDNF together with its receptor TrkB promotes trophic support to the neurons in the central and peripheral nervous system. BDNF also participates in neuronal development and glucose homeostasis (Krabbe et al. 2007). Observations *in vitro* provide evidence that BDNF plays a determinant role as a stimulator of glucose utilization in cortical neurons in response to the augmented energy demand (Burkhalter et al. 2003). Peripheral injections of BDNF exhibit hypoglycemic effects in obese diabetic animals such *db/db* mice (Nonomura et al. 2001). BDNF and glial cell line-derived neurotrophic factor (GDNF) are critical for carotid body development (Katz 2005).

S. Montero
Centro Universitario de Investigaciones Biomédicas, Universidad de Colima,
Ave. 25 de Julio s/n, Col. Villas de San Sebastián, Colima, Col C.P. 28045, Mexico

Facultad de Medicina, Universidad de Colima, Colima, Mexico

R. Cuéllar • M. Lemus • E.R. de Álvarez-Buylla (✉)
Centro Universitario de Investigaciones Biomédicas, Universidad de Colima,
Ave. 25 de Julio s/n, Col. Villas de San Sebastián, Colima, Col C.P. 28045, Mexico
e-mail: rab@ucol.mx

R. Ávalos • G. Ramírez
Facultad de Medicina, Universidad de Colima, Colima, Mexico

C.A. Nurse et al. (eds.), *Arterial Chemoreception: From Molecules to Systems*,
Advances in Experimental Medicine and Biology 758, DOI 10.1007/978-94-007-4584-1_32,

As NTS is a major interface between sensory visceral afferents and the CNS (Appleyard et al. 2007), NTS neurons are ideally suited to coordinate complex responses as occurs in energy metabolism. We and others previously identified that nitrergic pathways between CChr-baroreceptors and NTS play a role in glucose homeostasis (Lemus et al. 2009; Hsieh et al. 2010). All these facts led us to study whether BDNF in the NTS regulates BGR elicited by CChr stimulation with cyanide (NaCN) infused into the isolated carotid sinus (Alvarez-Buylla and Alvarez-Buylla 1988).

32.2 Methods

32.2.1 Animals and Surgical Procedures

Male Wistar rats (280–300 g) were used. Procedures were in accord to the Guide for the Care and Use of Laboratory Animals from National Institutes of Health, USA. Rats were housed at 22–23°C under a 12:12 light–dark cycle. Food was removed 12 h before surgery, but animals had free access to water containing 10% glucose. Rats were anesthetized with sodium pentobarbital (3 mg/100 g i.p.) supplemented by a continuous i.p. infusion of the same anesthetic (0.063 mg/min). Buprenorphine (0.03 mg/kg s.c.) was used 5 min before surgical procedures to minimize suffering. Body temperature was kept at 37°C ± 1°C. Animals were artificially ventilated through a tracheal cannula. Respiratory rate and tidal volume were based on pH, pO_2 and pCO_2 values in arterial blood obtained during experimental procedures, 10 min before and at the end of experiment. Permanent silastic catheters filled with heparin (1,000 U/mL) were inserted into the abdominal aorta and jugular sinus without interrupting circulation in these vessels (Alvarez-Buylla and Alvarez-Buylla 1988). The correct placement of catheters was verified at the end of each experiment.

32.2.2 Drugs

The drugs used were: (a) sodium cyanide (NaCN, Fluka Biochemika), 5 μg/100 g diluted in 100 nL of freshly prepared sterile saline-sal; (b) artificial cerebrospinal fluid (aCSF – 100 nL, containing NaCl 145 mM, KCl 2.7 mM, MgCl 1.0 mM, $CaCl_2$ 1.2 mM, ascorbate 2.0 mM, NaH_2PO_4 2 mM, pH 7.3–7.4); (c) BDNF (1 ng diluted in 100 nL of aCSF) (Balkowiec and Katz 2000); (d) BDNF receptor antagonist (K252a) (1 ng/100 nL of aCSF) (Shen et al. 2006). In control experiments, the same volume of aCSF was injected instead of BDNF drugs.

32.2.3 CChr Stimulation

CChr stimulation with NaCN was performed as described (Alvarez-Buylla and Alvarez-Buylla 1988). Briefly, 5 μg/100 g NaCN/100 μL sal-/2 s was injected into the local circulation of the left carotid sinus, avoiding baroreceptor stimulation. To locally stimulate just one carotid sinus of the rat with a micro-dose of NaCN, the left carotid sinus was temporarily isolated from the cephalic circulation, and the right carotid sinus was denervated. Only the left carotid sinus was exposed to NaCN and within 15–16 s, NaCN is cleared into a washing cannula (Fig. 32.1a). NaCN administered as described above, but after carotid nerve section did not have effect (Alvarez-Buylla and Alvarez-Buylla 1988) indicating that the observed responses were due to local CChr stimulation.

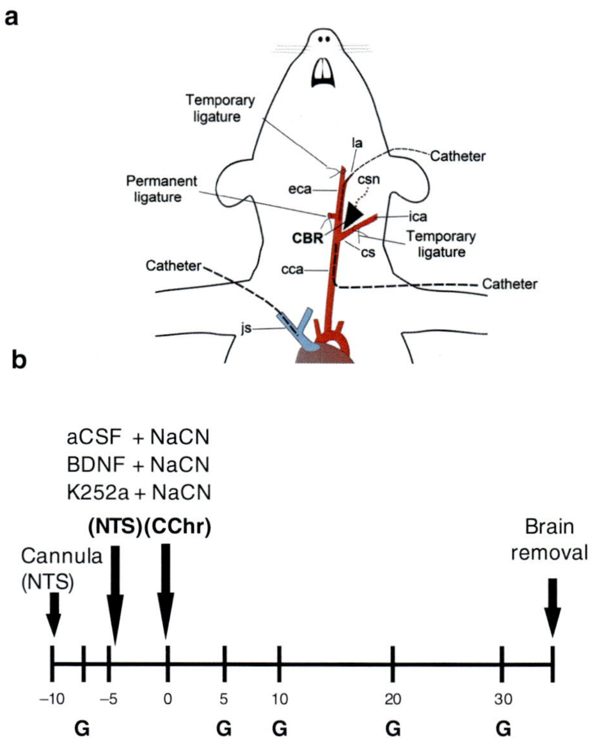

Fig. 32.1 (**a**) Placement of catheters and ligatures to locally perfuse *left* carotid sinus to stimulate carotid body chemoreceptors (CChr) with cyanide (NaCN). (**b**) Experimental protocol. *aCSF* artificial cerebrospinal fluid, *BDNF* brain-derived neurotrophic factor, *CBR* carotid body receptors, *cc* common carotid artery, *csn* carotid sinus nerve, *cs* circulatory isolated carotid sinus, *eca* external carotid artery, *G* glucose determination, *ica* internal carotid artery, *js* jugular sinus, *la* lingual artery, *K252a* BDNF receptor antagonist, *NaCN* sodium cyanide, *ph* pharyngeal artery

32.2.4 Microinjection of Drugs into the NTS

NTS injections were done in a stereotaxic frame (Stoelting Co.) and the following coordinates from bregma: AP = −12.7 mm, L = 1.45 mm, V = 7.7 mm; incisor bar 3.3 mm above zero point (Paxinos and Watson 1986). Once the brain surface was reached, a glass micropipette (50–60 μm external tip diameter, Microcap capillaries) filled with the solution to be injected, was inserted into the left NTS. The micropipette was connected to a 0.5 mL Hamilton microsyringe with polyethylene tubing (PE 20) for injections. BDNF drugs were delivered in 100 nL of aCSF during 3 s approx. The volume of each injection was determined by measuring the movement of the fluid meniscus within the microinjector pipette. In control experiments the micropipette was directed to the gigantocellular reticular nucleus (Gi), the coordinates in this case were AP = −12.7 mm, L = 1.45 mm, V = 9.2 mm (Paxinos and Watson 1986). Once the last blood sample was drawn, the correct positioning of the micropipette tip site was corroborated by injecting 50–100 nL of methylene blue (10%) through it. Anesthetized rats were decapitated; the brains removed, immediately frozen, and sectioned at 40 μm in a cryostat (CM-1800, Leica Microsystems). Sections were stained with cresyl violet for histological verification of the microinjection site and tissue damage.

32.2.5 Blood Sampling and Measurements

Blood samples (arterial and venous) were taken from the catheters as follows: two basal samples at t = −10 min and t = −5 min (averaged to obtain a basal level in t = −7.5 min). BDNF drugs or aCSF, as control, were injected into the NTS at t = −4 min, while CChr stimulation (NaCN injection) was done into the local circulation of the isolated carotid sinus at t=0 min. Four samples were then collected at t=5 min, t=10 min, t=20 min and t=30. After each sample was taken (0.15 mL arterial blood and 0.15 mL of venous blood), rats received 0.3 mL to compensate for fluid loss. Blood volume taken amounted to 1.8 mL, less than 8% of total circulating volume (Fig. 32.1b). Blood was centrifuged and plasma was kept chilled until assayed. Plasma glucose concentration in μmol/mL was determined by the glucose-oxidase method with a glucose analyzer (Beckman Autoanalyzer). BGR was determined by arterial-venous (A-V) glucose differences in μmol/mL. Blood flow was not considered for glucose retention estimation as it has been shown that it does not change significantly after local CChr stimulation (Alvarez-Buylla et al. 1997). To discard any possible change in blood pressure after NaCN injection into the local carotid sinus circulation, arterial blood pressure was measured in the femoral artery in two control experiments with a pressure transducer (Ohmeda Pte Ltd) connected to a Grass polygraph (Grass Instruments Co.); the results obtained in these experiments did not show significant changes. Glucose, PO_2, PCO_2 and pH values were within the range of standard curves at all times.

32.2.6 Experimental Protocol

Animals were allowed to stabilize for 30 min after surgery, when they were randomized into the following groups: **(a)** aCSF+NaCN (100 nL of aCSF into the NTS followed by NaCN, 5 μg/100 g/100 μL sal. infused into the isolated carotid sinus), n=9; **(b)** BDNF+NaCN (1 ng of BDNF/100 nL aCSF into de NTS followed by NaCN as in **a**), n=9; **(c)** K252a+NaCN (1 ng/nL of K252a in aCSF into the NTS followed by NaCN as in **a**), n=9.

32.2.7 Data Analysis

Values expressed as means ± SEM were analyzed using the SPSS 12.0 software on a PC computer. For multiple comparisons between groups, ANOVA one way and Scheffe tests were applied. To compare the results between basal and experimental values in a particular group, Student *t*'test was done. Significance was set at $p<0.05$. An asterisk (*) compares basal value of arterial glucose or BGR *vs* experimental values; the plus sign (+) compares aCSF *vs* corresponding BDNF or K252a groups. The pound sign (#) compares BDNF *vs* K252a groups.

32.3 Results

In our experiments the actual basal arterial blood glucose values were 6.33±0.20 μmol/mL (in aCSF rats), 6.70±0.17 (in BDNF rats) and 6.06±0.25 (in K252a rats); these were not statistically different ($p=0.12$).

Rats injected with aCSF (100 nL) into the NTS 4 min before CChr stimulation with NaCN (5 μg/100 g/100 μL sal), increased arterial glucose concentration at all the times studied, as was shown (Alvarez-Buylla and Alvarez-Buylla 1988); values rose from 6.33±0.20 up to 8.46±0.47 μmol/mL at t=30 min ($p<0.01$) (Fig. 32.2). BGR levels also showed a prompt and significant increase, values rose from 0.65±0.08 μmol/mL to 1.42±0.23 μmol/mL ($p<0.01$) (Fig. 32.3). Rats that received a BDNF injection into the NTS before CChr stimulation showed significant increases in glucose concentration levels when compared to aCSF group, with the higher values obtained at t=10 min and t=20 min ($p<0.01$ for both), values rose from 6.69±0.17 to 10.46 μmol/mL at t=20 min ($p<0.01$) (Fig. 32.2). Similar results were observed in BGR values; they increased from 0.78±0.06 μmol/mL to 2.20±0.24 μmol/mL a t=20 min ($p<0.01$) (Fig. 32.3). However, rats injected with K252a into

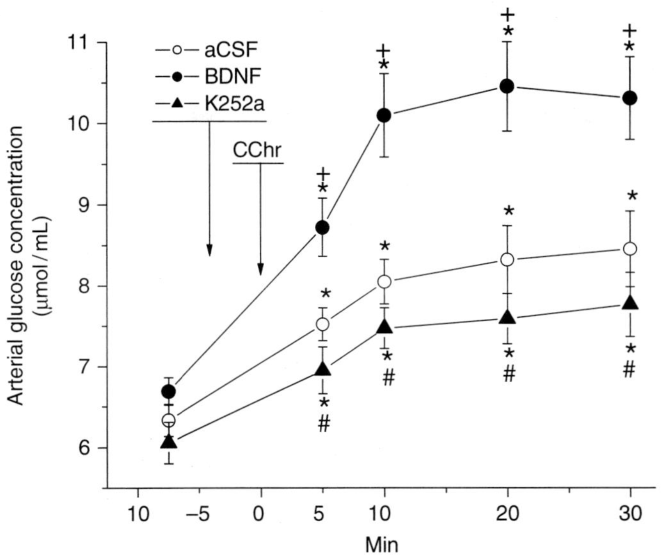

Fig. 32.2 Local CChr stimulation with NaCN (5 μg/100 g) 4 min after aCSF (100 nL) infusion (n=9), or BDNF (1 ng, diluted in 100 nL of freshly prepared aCSF) infusion (n=9), or K252a (1 ng diluted in 100 nL of freshly prepared aCSF) infusion (n=9) in the NTS of anesthetized rats. *aCSF* artificial cerebrospinal fluid, *BDNF* brain-derived neurotrophic factor, *CChr* carotid chemoreceptor, *K252a* TrkB receptor antagonist. The values are means ± SE; $*p<0.05$ compared with its own basal, Student *t*-test; $+p<0.05$ compared with aCSF group; $\#p<0.5$ compares BDNF group with K252a group (ANOVA one way)

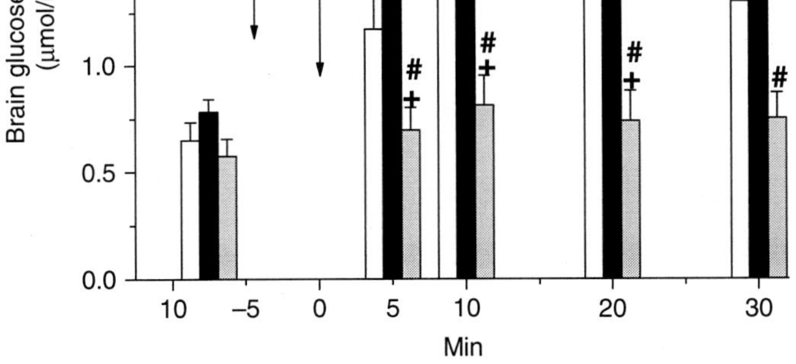

Fig. 32.3 Local CChr stimulation with NaCN (5 μg/100 g) 4 min after aCSF (100 nL) infusion (n=9), or BDNF (1 ng, diluted in 100 nL of freshly prepared aCSF) infusion (n=9), or aK252a (1 ng diluted in 100 nL of freshly prepared aCSF) infusion (n=9) in the NTS of anesthetized rats. *aCSF* artificial cerebrospinal fluid, *BDNF* brain-derived neurotrophic factor, *CChr* carotid chemoreceptor, *K252a* TrkB receptor antagonist. The values are means ± SE; $*p<0.05$ compared with their own basal, Student *t*-test; $+p<0.05$ compares aCSF group with both other groups ; $\#p<0.5$ compares BDNF group with K252a group (ANOVA one way)

NTS 4 min before CChr stimulation, arterial glucose showed lower increases all over the experiment ($p<0.01$, ANOVA), while BGR did not change during the experiment (Figs. 32.2 and 32.3). The comparison between BDNF *vs* K252a values for BGR was significant ($p<0.01$, ANOVA).

32.4 Discussion

Our findings suggest that BDNF is involved in glucose variable effects induced by CChr stimulation. A single unilateral aCSF injection made into the NTS of normal rats 4 min before CChr local anoxic stimulation, increased arterial glucose concentration and BGR levels. When BDNF was infused into the NTS 4 min before CChr stimulation, higher increases in glucose variables were obtained when compared to aCSF group. By contrast, K252a infusion into the NTS before CChr stimulation decreased both glucose variables studied (Fig. 32.2).

Intracerebroventricular administration of BDNF reduces blood glucose in obese *db/db* mice. BDNF also enhances norepinephrine turnover, and suggests that this neurotrophin activates the sympathetic nervous system through the CNS to regulate energy expenditure in these animals (Nonomura et al. 2001; Tsuchida et al. 2001). Sympathetic activation stimulates hepatic glycogenolysis and glucagon secretion, systems designed to provide as much glucose as possible for brain use in a stressful situation (Hoffman 2007). The hyperglycemic reflex with BGR observed here after BDNF infusion into NTS preceding CChr anoxic local stimulation, can be considered as a local stressor stimulus to CChrs, and could partly be due both, to a sympathetic activity, and to a BDNF-induced increase in glucose utilization and glucose transporter GLUT3 expression in cortical neurons (Burkhalter et al. 2003). Anoxic stimulus in CChr activates visceral afferent neurons in the nodose/petrosal/NTS sensory ganglion complex (Erickson et al. 1996) to elicit endogenous BDNF, sympathetic activity and hyperglycemia by binding to the TrkB receptor (Nonomura et al. 2001). The data presented in Fig. 32.3 show that BGR induced by BDNF was induced through TrkB receptor activation, as tyrosine kinase inhibitor K252a abolished this effect.

There are anatomical and functional evidences in relation to peripheral chemoreceptor afferents that make their first synapses in the NTS, mainly in the commissural NTS (cNTS) (Brady et al. 1999; Paton et al. 2001). Recent studies in awake rats show that glutamatergic and purinergic mechanisms are part of the neurotransmission system of the chemoreflex at cNTS level (Braga et al. 2007). We hypothesized a possible involvement of ionotropic glutamate receptors in the effects observed in this paper after CChr stimulation and BDNF infusion in the NTS. In summary, our data, and the facts mentioned above, suggest that this neurotrophin in the NTS could play an important role in central glucose homeostasis.

Acknowledgements Supporte by FRABA 330/205 and CONACYT P49376-Q grants.

References

Alvarez-Buylla R, Alvarez-Buylla E (1988) Carotid sinus receptors participate in glucose homeostasis. Respir Physiol 72:347–360

Alvarez-Buylla R, de Alvarez-Buylla ER, Mendoza H, Montero SA, Alvarez-Buylla A (1997) Pituitary and adrenals are required for hyperglycemic reflex initiated by stimulation of CBR with cyanide. Am J Physiol 272:R392–R399

Appleyard SM, Marks D, Kobayashi K, Okano H, Low MJ, Andresen MC (2007) Visceral afferents directly activate catecholamine neurons in the solitary tract nucleus. J Neurosci 28:13292–13302

Balkowiec A, Katz DM (2000) Activity-dependent release of endogenous brain-derived neurotrophic factor from primary sensory neurons detected by ELISA in situ. J Neurosci 20:7417–7423

Brady R, Zaidi SI, Mayer C, Katz DMJ (1999) BDNF is a target-derived survival factor for arterial baroreceptor and chemoafferent primary sensory neurons. J Neurosci 19:2131–2142

Braga VA, Soriano RN, Braccialli AL, de Paula PM, Bonagamba LGH, Paton JFR, Machado BH (2007) Involvement of L-glutamate and ATP in the neurotransmission of the sympathoexcitatory component of the chemoreflex in the commissural nucleus tractus solitarii of awake rats and in the working heart-brainstem preparation. J Physiol 581:1129–1145

Burkhalter J, Fiumelli H, Allaman I, Chatton J-Y, Martin J-L (2003) Brain-derived neurotrophic factor stimulates energy metabolism in developing cortical neurons. J Neurosci 23:8212–8220

Erickson JT, Connover JC, Borday V, Champagnat J, Barbacid M, Yancopoulos G, Katz DM (1996) Mice lacking brain-derived neurotrophic factor exhibit visceral sensory neuron losses distinct from mice lacking NT4 and display a severe developmental deficit in control of breathing. J Neurosci 16:5361–5371

Hoffman RP (2007) Sympathetic mechanisms of hypoglycemic counterregulation. Curr Diabetes Rev 3:185–193

Hsieh HY, Robertson CL, Vermehren-Schmaedick A, Balkowiec A (2010) Nitric oxide regulates BDNF release from nodose ganglion neurons in a pattern-dependent and cGMP-independent manner. J Neurosci Res 88:1285–1297

Katz DM (2005) Regulation of respiratory neuron development by neurotrophic and transcriptional signaling mechanisms. Respir Physiol Neurobiol 149:99–109

Krabbe KS, Nielsen AR, Krogh-Madsen R, Plomgaard P, Rasmussen P, Erikstrup C, Fischer CP, Lindegaard B, Petersen AMW, Taudorf S, Secher NH, Pilegaard H, Bruunsgard H, Pedersen BK (2007) Brain-derived neurotrophic factor (BDNF) and type 2 diabetes. Diabetologia 50:431–438

Lemus M, Montero S, Luquin S, Garcia J, de Alvarez-Buylla ER (2009) Nitric oxide in the solitary tract nucleus (STn) modulates glucose hemostasis and FOS-ir expression after carotid chemoreceptor stimulation. Adv Exp Med Biol 648:403–410

Nonomura T, Tsuchida A, Ono-Kishino M, Nakagawa T, Taiji M, Noguchi H (2001) Brain-derived neurotrophic factor regulates energy expenditure through the central nervous system in obese diabetic mice. Int J Exp Diabetes Res 2:201–209

Paton JFR, Deuchars J, Li Y-W, Kasparov S (2001) Properties of solitary tract neurons responding to peripheral arterial chemoreceptors. Neuroscience 105:231–248

Paxinos G, Watson C (1986) The rat brain in stereotaxic coordinates. Academic, New York

Shen F, Meredith GE, Napier C (2006) Amphetamine-induced place preference and conditioned motor sensitization requires activation of tyrosine kinase receptors in the hippocampus. J Neurosci 26:11041–11051

Tsuchida A, Nonomura T, Ono-Kishino M, Nakagawa T, Taiji M, Noguchi H (2001) Acute effects of brain-derived neurotrophic factor on energy expenditure in obese diabetic mice. Int J Obes 25:1286–1293

Chapter 33
Hydrogen Sulfide Acting at the Carotid Body and Elsewhere in the Organism

Robert S. Fitzgerald, Machiko Shirahata, Irene Chang, Eric W. Kostuk, and Samara Kiihl

Keywords Carotid Body • Smooth Muscle • Hydrogen Sulfide • Neuroprotection • Data Disparity Causes • Neural Tissue

33.1 Introduction

33.1.1 Hydrogen Sulfide (H_2S)

H_2S, the most recently explored gasotransmitter, has been found to have actions in at least three types of tissues.

1. Vascular Smooth Muscle. In vascular smooth muscle it has been reported to relax cells by opening K_{ATP} channels, thus hyperpolarizing them; the smooth muscle relaxes (Kimura 2009; Zhao et al. 2001; Zhao and Wang 2002). However, other studies report constriction (Lim et al. 2008), and in rat pulmonary artery smooth muscle there is a multiphasic response (Olson et al. 2006).

R.S. Fitzgerald (✉)
Departments of Environmental Health Sciences, of Physiology, and of Medicine,
The Johns Hopkins Medical Institutions, Baltimore, MD 21205, USA
e-mail: rfitzger@jhsph.edu

M. Shirahata
Departments of Environmental Health Sciences, and of Anesthesiology/Critical Care Medicine,
The Johns Hopkins Medical Institutions, Baltimore, MD 21205, USA
e-mail: mshiraha@jhsph.edu

E.W. Kostuk • I. Chang
Department of Environmental Health Sciences, Bloomberg School of Public Health,
The Johns Hopkins University, Baltimore, MD 21205, USA
e-mail: ekostuk@jhsph.edu

S. Kiihl
Department of Biostatistics, Bloomberg School of Public Health, The Johns Hopkins University,
Baltimore, MD 21205, USA

C.A. Nurse et al. (eds.), *Arterial Chemoreception: From Molecules to Systems*,
Advances in Experimental Medicine and Biology 758, DOI 10.1007/978-94-007-4584-1_33,

2. Neural Tissue. In neural tissue H_2S plays multiple roles such as both suppressing EPSPs in high concentrations (Kimura 2002) and facilitating synaptic transmission (Kimura 2002, 2005; Kimura et al. 2005), generating Long Term Potentiation (Kimura 2002), neuroprotecting by increasing glutathione levels (Kimura et al. 2005).
3. Airway Tissues. In airway tissues H_2S attenuates inflammation (Chen et al. 2009) and induces relaxation of smooth muscle (Fitzgerald et al. 2010).

33.1.2 *Carotid Body Studies*

Therefore, it is not surprising that H_2S has also been reported to have an effect on what is arguably the most important interoreceptor in the organism, the carotid bodies (CBs). When stimulated, these bilaterally located structures precipitate an impressive array of reflex responses in the organism in an effort to reestablish and maintain homeostasis, particularly in the respiratory and cardiovascular systems.

1. One group of studies presents evidence that H_2S stimulates an increase in neural output from the CBs (Li et al. 2010; Peng et al. 2010).
2. While a second study presents evidence suggesting that H_2S might attenuate CB neural output (Fitzgerald et al. 2011).

33.2 Results from CB Studies Only

33.2.1 *Peng et al*

This group (Peng et al. 2010) reported the presence of the H_2S synthesizing enzyme cystathionine-γ-lyase (CSE) in the glomus cells of mice CBs. With CSE present neural output increased in response to hypoxia. Whereas in the CSE-knockout mice no such response to hypoxia was observed. Further, with CSE present hypoxia increased the amount of H_2S, but in the knockouts hypoxia did not increase H_2S. In the rat preparation again CSE was located in the glomus cells, and propargylglycine (a CSE blocker) reduced the hypoxia-induced increase in neural output and H_2S genesis. Finally, the application of 30, 50, and 100 μM amounts of NaHS, an H_2S precursor, generated a linear dose/response (neural output) curve. Their methods and complete results have been published (Peng et al. 2010).

33.2.2 *Cat as the Model*

In a second published study (Fitzgerald et al. 2011) cat was the animal model.

1. CBs were first removed for incubation in a physiological salt solution at a constant temperature under normoxic or hypoxic conditions. The CBs were observed to respond as usual to hypoxia in this set-up, namely with an increase in the release of two excitatory neurotransmitters, acetylcholine (ACh) and ATP. Because of the absence of data regarding appropriate concentrations of exogenously applied H_2S or a precursor several different concentrations of Na_2S, a precursor of H_2S, were included in the incubation medium.
2. The initial analysis of the effect included all concentrations. The release of ACh and ATP was less than the release under control conditions (Fig. 33.1).

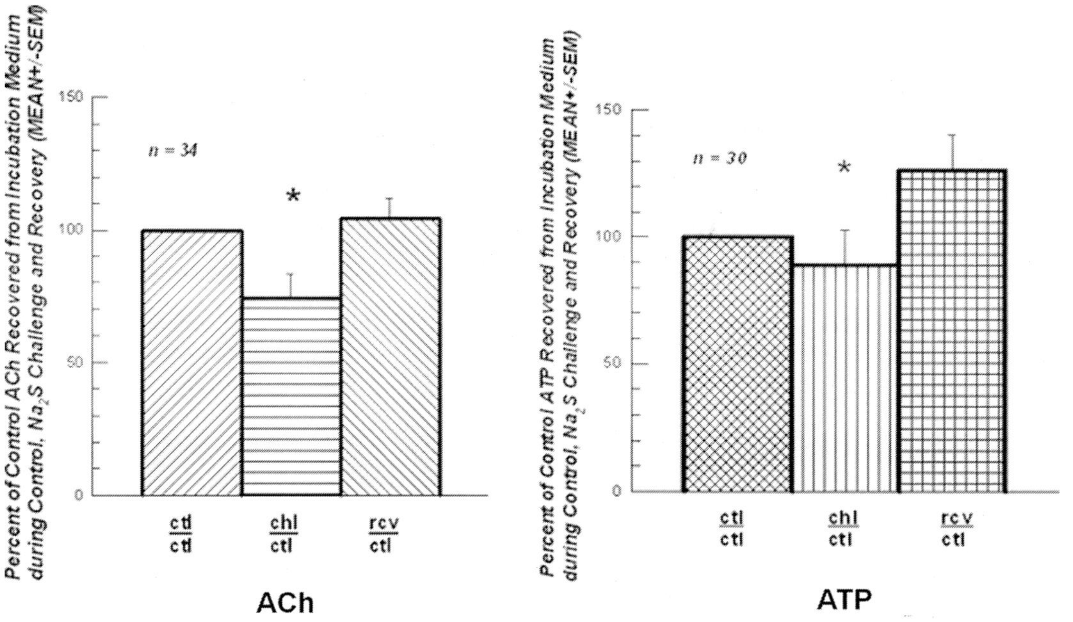

Fig. 33.1 ACh: Shown are 34 trials of various concentrations of Na_2S (nM to μM). Values have been normalized to the control for before (ctl/ctl), during the Na_2S challenge (chl/ctl), and recovery from the challenge (rcv/ctl). *: (chl/ctl) differs significantly (P=<0.001) from (ctl/ctl), and from (rcv/ctl); P=0.014. (ctl/ctl) does not differ from (rcv/ctl). ATP: Shown are 30 trials of the same preparations. *: (chl/ctl) differs significantly (P=0.046) from (ctl/ctl), and from (rcv/ctl); P=0.030. (ctl/ctl) does not differ from (rcv/ctl)

3. Frequently the data displayed a disappointing variability. But this could have been due to the fact that H_2S also increases intracellular concentrations of Ca^{2+} largely by inducing Ca^{2+} influx, and to a lesser extent through the release from intracellular Ca^{2+} stores (Kimura et al. 2005). The data did not lend themselves to the classical linear dose/response curve, nor to the piecemeal linear response.
4. But subsequent attempts to structure a dose/response curve revealed that the ATP data fit the Natural Cubic Spline Regression Model significantly (P=0.036) (Fitzgerald et al. 2011).
5. Another analysis of the data using BoxPlot graphs (Fig. 33.2) demonstrated that as concentrations of Na_2S increased from nM to μM, the reduction in the release of ATP from the control value increased from 35.4% (10 nM) to 50.0% (20 nM) and then began to decrease at 5 μM (5000 nM) to 36.1% and to 22.4% at 100 μM (100,000 nM).
6. Immunohistochemistry studies (Fitzgerald et al. 2011) showed the presence of a second H_2S synthesizing enzyme, cystathionine-β-synthase (CBS) and a subunit of the K_{ATP} channels (Kir6.1) in the glomus cells of the cat CB.

33.3 Discussion

33.3.1 H_2S and Other Tissues

Pertinent to the CB studies reported above suggesting a conflict in the data, it would be worthwhile to review the reported actions of H_2S on other tissues.

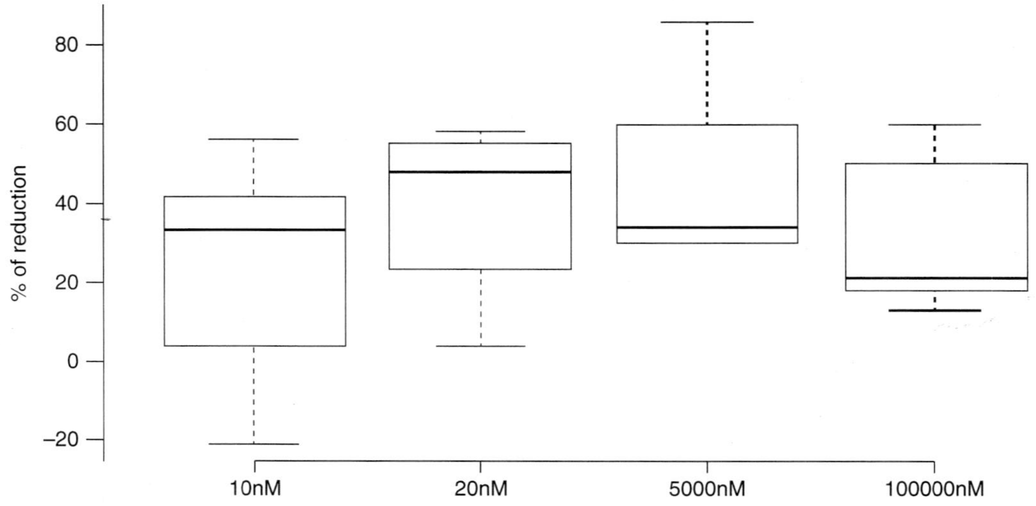

Fig. 33.2 Boxplot of the ATP results from all concentrations of Na_2S. This plot shows in greater detail the effect of the lowest concentrations used. Clearly nM concentrations reduced the release of ATP (% reduction). As the concentrations of Na_2S increased, the reductions began to decrease. The data fitted to a Natural Cubic Spline Regression Model revealed a statistically significant relationship between Na_2S concentration and the reduction in the release of ATP (Fitzgerald et al. 2011). Abscissa shows all concentrations in nM. Hence 5 μM is 5,000 nM; 100 μM is 100,000 nM

1. Vascular Smooth Muscle. Several studies have reported the relaxing effect of H_2S on vascular smooth muscle (Kimura 2009; Kimura et al. 2005; Zhao et al. 2001; Zhao and Wang 2002). However, Lim and colleagues report a vasoconstrictive effect of H_2S (Lim et al. 2008). And Olson and his colleagues (Olson et al. 2006) report a multiphasic action of H_2S on rat pulmonary artery, mimicking the effect of hypoxia in that structure.
2. Neural Tissue.

   (a) Neuroprotection. In neural tissue H_2S has been thought to function as a neuroprotective agent when neurons are vulnerable to oxidative stress. H_2S under such stress generates an increase in the cystine/glutamate antiporter. This action promotes an increase in intracellular cystine, generating an increase in intracellular cysteine as well as γ-GCS activity. This action raises glutathione levels. GSH is a major intracellular antioxidant. This offsets the oxidative stress. Such an action suggests a protective role for H_2S in the glomus cells of very active CBs during hypoxia, when H_2S is said to rise.
   (b) Neural Traffic. But H_2S can also modulate neural traffic. It can inhibit synaptic transmission by blocking the EPSPs; but it can also enhance NMDA receptor-mediated actions.
   (c) Long Term Potentiation. H_2S can also facilitate long term potentiation in the hippocampus (Kimura 2009). Hence, apparently contradictory effects due to H_2S action can be found in these tissues. But these effects are on the phenomenological level. Searching for mechanisms of action might well resolve such apparent contradictions.

33.3.2 Potential K^+ Channel Mechanisms

The results of Peng et al. (2010) and others (Telezhkin et al. 2009) might be due to H_2S acting on the maxi K_{BK} channel closing it off and depolarizing the glomus cell, with the subsequent release of excitatory neurotransmitters (Li et al. 2010). The findings of Telezhkin et al. (2009) and Li et al. (2010)

would be consistent with this mechanism as would be H_2S opening L-type Ca^{++} channels in the glomus cells (Garcia-Bereguiain et al. 2008). However, Sitdikova and colleagues (Sitdikova et al. 2010) found that H_2S activated some maxi K_{BK} channel

33.3.3 *Ion Channels*

What must be kept in mind is the wide variety of sites targeted by H_2S including ion channels (so important in the behavior of glomus cells): K_{ATP} K_{BK} L-type Ca^{++}, T-type Ca^{++}, $TPRV_1$, $TPRA_1$, Cl^- (Tang et al. 2010).

33.3.4 *Disparity of Results*

Other very important considerations must be kept in mind when trying to account for disparity in results.

1. Frequently investigators will extrapolate too readily from one species to another when the animals' survival/reproductive patterns are significantly different, warranting a difference in how neural traffic is distributed both with respect to speed and amount.
2. Secondly, the components of one tissue might not be exactly the same as those of another tissue due to the function of that tissue in the organism's mandate to maintain homeostasis.
3. Thirdly, the density, orientation, affinity of receptors or channels for an agent in the cells of a tissue must be acknowledged. Perhaps a tissue has more K_{BK} channels than K_{ATP} channels.
4. Fourth, a protein channel or receptor in the lipid membrane might not always present its sensitive structural unit(s) to the circulating extracellular fluid under conditions of control *vs* stress.
5. Fifth, the affinity of the structure could also change depending on the internal environment; e.g., the pH optimum for protein function .
6. Finally, the dose/concentration of a stimulating/attenuating agent is very important since it is not at all uncommon for an agent to have one action at a low concentration, but a different action at higher concentrations.

33.3.5 *Cat vs Mice/Rat*

In a previous study Chou and Shirahata (1996) demonstrated with patch clamp studies that the cat CB did not exhibit responses consonant with the presence of K_{BK} channels. Peng et al. (2010) make no mention of cystathionine-β-synthase (CBS) in their rats or mice. However, we did explore and found CBS in the glomus cells of cats (Fitzgerald et al. 2011). Since the CB is a neural crest derivative, this was not surprising to us. Further, Telezhkin et al. (2009) found CBS in the CBs of rats, while Li et al. (2010) found it in mice. Given the results of our functional studies, we looked for and found the Kir6.1 subunit of the K_{ATP} channel in the cats' glomus cells (Fitzgerald et al. 2011). However, to our knowledge there is yet to appear a study reporting the presence/operation of K_{ATP} channels in rats and mice.

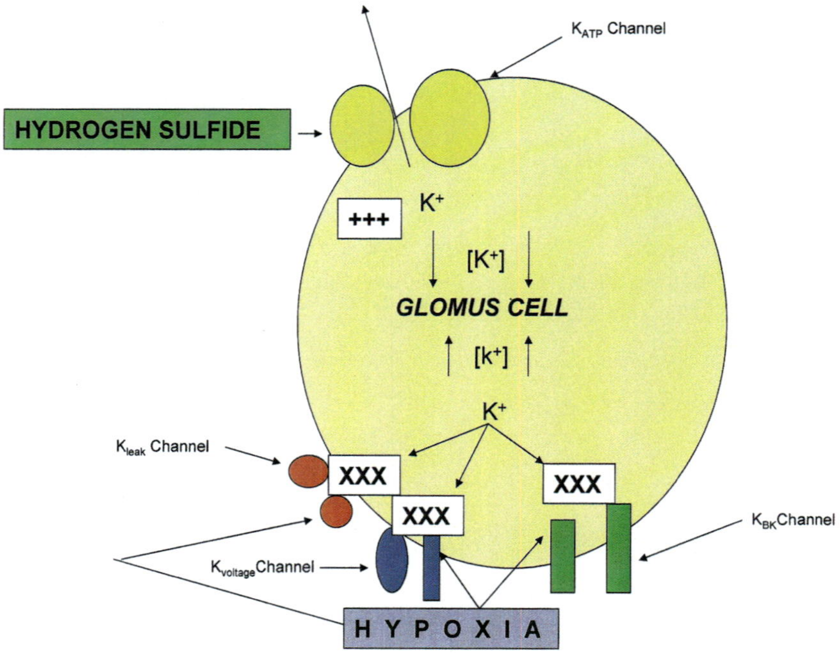

Fig. 33.3 This drawing presents a possible role that H_2S might have in the cat CB. Neuroprotection is one role for H_2S in the CNS, protecting very active neurons from stress. Hypoxia is a strong stimulus to the neurotransmitter-containing glomus cells of the cat CB. H_2S might protect glomus cells from exhausting their supply of excitatory and inhibitory transmitters

33.3.6 Tentative Model of H_2S Action in Cat CB

Given the complexity of the effects of H_2S, one can propose an action of H_2S in the cat CB only with considerable caution. Although the previously mentioned study of Chou and Shirahata (1996) seemed to indicate an absence of K_{BK} channels, we nonetheless postulate that they may have been present but in much lower density than other K channels. We do so since the K_{BK} channel has been demonstrated in CBs of rats and mice.

During an hypoxic challenge when the K channels are blocked and the outward movement of K^+ ions reduced or abolished, the glomus cell undergoes depolarization. A second effect of hypoxia is to increase the levels of H_2S. We propose in Fig. 33.3 that H_2S acts on the K_{ATP} channels in the cat glomus cell allowing the escape of K^+ ions. This modulates the depolarization and prevents the glomus cell from exocytosing the contents of all the transmitter-containing vesicles.

The fact that incubation of the cat CBs with pinacidil, a known K_{ATP} channel opener, had the same effect on ACh/ATP release from the incubated cat CBs as did Na_2S (Fitzgerald et al. 2011) lends support to our proposed role for H_2S in the cat CB. Clearly, much further experimentation is needed in all species to arrive at a more definitive model of H_2S action in the CB.

Acknowledgements The authors gratefully acknowledge the support of the U.S. National Institutes of Health (National Heart Lung Blood Institute) HL-50712, HL-61596, HL-72293.

References

Chen YH, Wu R, Geng B, Qi YF, Wang PP, Yao WZ, Tang CS (2009) Endogenous hydrogen sulfide reduces airway inflammation and remodeling in a rat model of asthma. Cytokine 45:117–123

Chou CL, Shirahata M (1996) Two types of voltage-gated K channels in carotid body cells of adult cats. Brain Res 74:34–42

Fitzgerald RS, Shirahata M, Chang I, Kostuk E (2010) Hypoxia_vs hydrogen sulfide (H_2S) acting at the carotid body (CB) and elsewhere systemically. FASEB J 24:1026.22

Fitzgerald RS, Shirahata M, Chang I, Kostuk E, Kiihl S (2011) The impact of hydrogen sulfide (H_2S) on neurotransmitter release from the cat carotid body. Respir Physiol Neurobiol 176:80–89

Garcia-Bereguiain MA, Samhan-Arias AK, Martin-Romero FJ, Gutierrez-Merino C (2008) Hydrogen sulfide raises cytosolic calcium in neurons through activation of L-type Ca^{2+}channels. Antioxid Redox Signal 10:31–41

Kimura H (2002) Hydrogen sulfide as a neuromodulator. Mol Neurobiol 26:13–19

Kimura H (2005) Hydrogen sulfide as a biological mediator. Antioxid Redox Signal 7:778–780

Kimura H (2009) Hydrogen sulfide: from brain to gut. Antioxid Redox Signal 12:1111–1123

Kimura H, Nagai Y, Umemura K, Kimura Y (2005) Physiological roles of hydrogen sulfide: synaptic modulation, neuroprotection, and smooth muscle relaxation. Antioxid Redox Signal 7:795–803

Li Q, Sun B, Wang X, Jin Z, Zhou Y, Dong L, Jiang L-H, Rong R (2010) A crucial role of hydrogen sulfide in oxygen sensing is modulating large conductance calcium-activated potassium channels. Antioxid Redox Signal 12:1179–1189

Lim JJ, Liu Y-H, Sandar E, Khin W, Bian J-S (2008) Vasoconstrictive effect of hydrogen sulfide involves down regulation of cAMP in vascular smooth muscle cells. Am J Physiol Cell Physiol 295:1261–1270

Olson KR, Dombkowski RA, Russell MJ, Doellman MM, Head SK, Whitfield NL, Madden JA (2006) Hydrogen sulfide as an oxygen sensor/transducer in vertebrate hypoxic vasoconstriction and hypoxic vasodilation. J Exp Biol 209:4011–4023

Peng Y-H, Nanduri J, Raghuraman G, Souvannakitti D, Gadalla MM, Kumar GK, Snyder SH, Prabhakar NR (2010) H_2S mediates O_2 sensing in the carotid body. Proc Natl Acad Sci USA 107:10719–10724

Sitdikova GF, Weiger TM, Hermann A (2010) Hydrogen sulfide increases calcium-activated potassium (BK) channel activity of rat pituitary tumor cells. Pflugers Arch Eur J Physiol 459:389–397

Tang G, Wu L, Wang R (2010) Interaction of hydrogen sulfide with ion channels. Clin Exp Pharmacol Physiol 37:753–763

Telezhkin V, Brazier SP, Cayzac S, Muller CT, Riccardi D, Kemp PJ (2009) Hydrogen sulfide inhibits human BK_{Ca} channels. Adv Exp Med Biol 648:65–72

Zhao W, Wang R (2002) H2S-induced vasorelaxation and underlying cellular and molecular mechanisms. Am J Physiol Heart Circ Physiol 283:H474–H480

Zhao W, Zhang J, Lu Y, Wang R (2001) The vasorelaxant effect of H_2S as a novel endogenous gaseous K_{ATP} channel opener. EMBO J 20:6008–6016

Chapter 34
Purinergic Modulation of Carotid Body Glomus Cell Hypoxia Response During Postnatal Maturation in Rats

John L. Carroll, Amit Agarwal, David F. Donnelly, and Insook Kim

Abstract Carotid body (CB) glomus cells respond to hypoxia by releasing neurotransmitters, such as ATP, which are believed to stimulate excitatory receptors on apposed nerve endings of the carotid sinus nerves as well as bind to autoreceptors on the glomus cell membrane to modulate response magnitude. The CB response to hypoxia is small at birth and increases during postnatal maturation in mammals. As ATP has been shown to inhibit the glomus cell response to hypoxia via an autoreceptor mechanism, we hypothesized that ATP-mediated inhibition may vary with age and play a role in postnatal development of the hypoxia response magnitude. The effects of ATP on CB glomus cell intracellular calcium ($[Ca^{2+}]_i$) responses to hypoxia were studied at two ages, P0-1 and P14-18. The inhibitory effect of ATP or a stable ATP analog on the glomus cell response to hypoxia was greater in newborn rats compared to the more mature age group. Use of selective P2Y receptor agonists and antagonists suggests that the inhibitory effect of ATP on the glomus cell $[Ca^{2+}]_i$ response to hypoxia may be mediated by a P2Y12 receptor. Thus, developmental changes in ATP-mediated glomus cell inhibition may play a role in carotid chemoreceptor postnatal maturation.

Keywords Carotid body • Glomus cells • Hypoxia • ATP • Postnatal • Development • Maturation • Purinergic receptor

34.1 Introduction

The carotid body (CB) chemoreceptors, the main arterial oxygen sensors in mammals, respond minimally to hypoxia immediately after birth and require ~2 weeks time (in rats) to mature functionally to adult levels. Although the mechanisms of O_2 sensing are not fully understood,

J.L. Carroll (✉) A. Agarwal • I. Kim
Division of Pediatric Pulmonary Medicine, Department of Pediatrics, University of Arkansas for Medical Sciences, 1 Children's Way, Little Rock, AR 72202, USA
e-mail: carrolljohnl@uams.edu; kiminsook@uams.edu

D.F. Donnelly
Dept of Pediatrics, Yale University School of Medicine, 333 Cedar St., New Haven, CT, USA
e-mail: David.Donnelly@Yale.edu.

C.A. Nurse et al. (eds.), *Arterial Chemoreception: From Molecules to Systems*, Advances in Experimental Medicine and Biology 758, DOI 10.1007/978-94-007-4584-1_34,

the current view is that the CB glomus cell responds to low PO_2 with cell membrane depolarization and release of one or more neurotransmitters (eg., ATP, acetylcholine) which cause excitation of apposed carotid sinus nerve endings. The released neurotransmitters may also stimulate inhibitory autoreceptors on glomus cells and thus modulate the response magnitude (Bairam and Carroll 2005).

A clear postnatal increase in the CB response to hypoxia has been demonstrated at the level of carotid sinus nerve output, neurotransmitter release, glomus cell $[Ca^{2+}]_i$ response, glomus cell depolarization and the O_2 sensitivity of glomus cell potassium channels such as TASK1 and TASK3, which modulate glomus cell excitability. ATP, dopamine and other neurotransmitters have been shown to inhibit the glomus cell response to hypoxia via autoreceptors (Bairam and Carroll 2005; Benot and Lopez-Barneo 1990; Xu et al. 2005). While the dopamine D2 autoreceptor has been relatively well-studied, the autoinhibitory role of ATP in CB postnatal maturation remains unresolved. Therefore, we sought to understand whether the inhibitory effect of ATP on glomus cell $[Ca^{2+}]_i$ response to hypoxia changes with age and, if so, what subtype of purinergic receptor mediates the effect.

34.2 Methods

34.2.1 Animal Model

The use of animals in this study was approved by the Animal Care and Use Committee of the University of Arkansas for Medical Sciences. Experiments were conducted on Sprague–Dawley rat pups of both sexes. Carotid bodies were harvested from rats aged 0–1 day (P0-1) and 14–18 days (P14-18).

34.2.2 Carotid Body Cell Isolation

Rat pups were anesthetized with isoflurane, decapitated and the heads placed in ice-cold buffered saline solution (BSS). Whole CBs were dissected out and placed in ice-cold low Ca^{2+}/Mg^{2+} BSS. Isolated CBs were dissociated with an enzyme cocktail containing trypsin and collagenase by incubating at 37°C for 20–25 min. Dissociated CB cells were placed on poly-D-lysine coated glass coverslips and incubated at 37°C until use.

34.2.3 Intracellular Calcium Measurement

Intracellular Ca^{2+} concentration was measured using the calcium-sensitive fluorescent probe fura-2. Plated CB cells were loaded with 4 mM of fura-2 acetoxymethyl ester (fura-2 AM; Molecular Probes) for 30 min at 37°C in 21%O2/5%CO2. Fura-2 fluorescent emission was measured at 510 nm in response to alternating excitation at 340 and 380 nm. Images were acquired and stored using a NIKON TE300 microscope and CCD (CoolSNAP HQ2) camera under computer control (MetaFluor: Molecular Devices). For each coverslip, the background light levels were subtracted from each image before measurement of fluorescence intensity. $[Ca^{2+}]_i$ was determined using the 340/380 fluorescence ratio as described previously (Wasicko et al. 1999). Calibration was performed using cell-free solutions.

34.2.4 Experimental Protocol

Dissociated CB glomus cells were superfused (~35°C) in a closed imaging chamber with a bicarbonate-buffered solution containing (mM): 118 NaCl, 23 $NaHCO_3$, 3 KCl, 2 KH_2PO_4, 1.2 $CaCl_2$, 1 $MgCl_2$ and 10 glucose. Single cells and clusters of live dissociated CB cells which were identified as glomus cells by Rhodamine-tagged Peanut Agglutinin (Rhod-PNA) were selected for study (Kim et al. 2009). Glomus cells from newborn rats P0-1 and mature rats P14-18 were exposed, from a normoxia baseline, to BSS with 20 mM KCl, hypoxia (0% O_2 alone), hypoxia (0% O_2) + 2-MeSATP (stable ATP analog) and five different concentrations of 2-methyl-thio-ATP (2-MeSATP) (concentrations: 0.001, 0.01, 0.1, 1 and 10 μM). Dose–response curves were constructed in both age groups. Hypoxia challenges were 2 min, KCl challenge duration was 30 s, and each challenge was followed by 5 min recovery period.

To determine if the effect of 2-MeSATP also occurred with naturally occurring ATP, the dissociated glomus cells from P14-18 cells were exposed to hypoxia+ATP (concentrations: 1, 10 and 100 μM) following same protocol as above. In some experiments glomus cells were exposed to hypoxia+ MRS2365 (P2Y1 receptor specific agonist) as well as hypoxia+100 μM ATP+30 μM reactive blue 2 (a potent nonspecific P2Yreceptor antagonist).

In other experiments, cells from P14-18 rats were exposed to hypoxia+2-MeSATP (1 μM)+MRS2395 (0.1 μM) (P2Y12 receptor specific antagonist) or hypoxia+2-MeSATP (1 μM)+ MRS2179 (0.1 μM) (P2Y1 receptor specific antagonist).

34.2.5 Data Analysis

The glomus cell $[Ca^{2+}]_i$ response to hypoxia was determined from the peak $[Ca^{2+}]_i$ level above the pre-hypoxia baseline $[Ca^{2+}]_i$ level. Because the magnitude of response differs between newborn and mature age groups (Wasicko et al. 1999, 2006), the inhibition of the hypoxia response by ATP (or ATP analog) was expressed as % inhibition compared to the control response to hypoxia (without drug).

34.3 Results

34.3.1 Effect of ATP on Glomus Cell $[Ca^{2+}]_i$ Response to Hypoxia

CB glomus cells from both ages responded with a brisk increase in $[Ca^{2+}]_i$ when exposed to hypoxic superfusate. As previously reported, the $[Ca^{2+}]_i$ response of mature cells is about threefold larger than that of the newborn cells (Wasicko et al. 2006). Preliminary results indicate that ATP inhibited the glomus cell $[Ca^{2+}]_i$ response to hypoxia at both ages (Fig. 34.1). However, at the intermediate ATP concentration (10 μM), inhibition of the glomus cell $[Ca^{2+}]_i$ response to hypoxia was greater in cells from P0-1 compared to cells from P14-18 (Fig. 34.1). Similar results were obtained with 2-MeSATP although the concentration at which inhibition occurred was lower, due to the higher potency of the agonist.

34.3.2 CB Glomus Cell Purinergic Autoreceptor

The inhibitory effect of ATP on the glomus cell $[Ca^{2+}]_i$ response to hypoxia was blocked by reactive blue 2, a non-specific P2Y receptor antagonist. Preliminary results suggested that inhibition of the hypoxic $[Ca^{2+}]_i$ response by ATP was also blocked by MRS2395, a specific P2Y12 receptor antagonist, but not by MRS2179, a P2Y1 receptor antagonist.

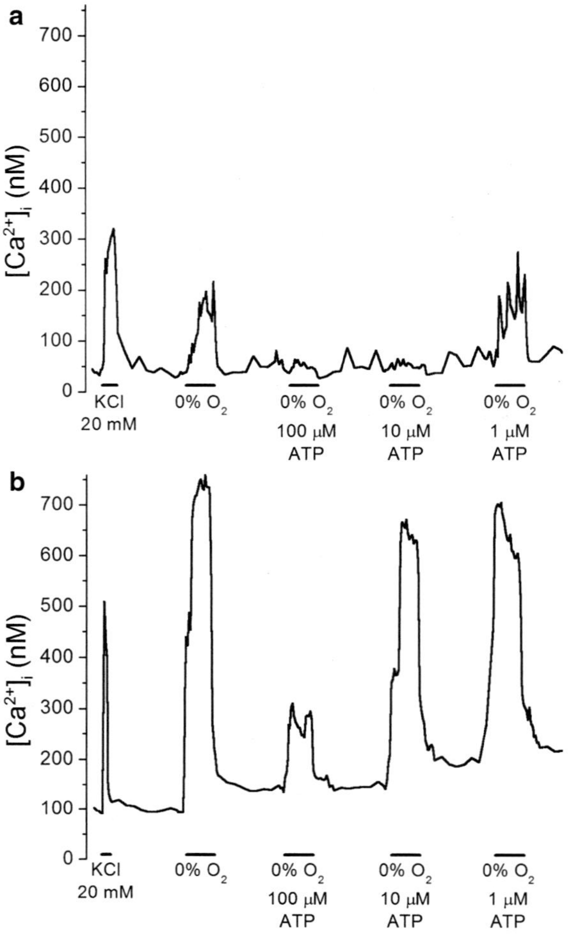

Fig. 34.1 Representative tracings of glomus cell $[Ca^{2+}]_i$ increase in response to 20 mM KCl, hypoxia (0% O_2) and hypoxia+ATP (various concentrations). (**a**) Glomus cells from newborn rats (P0-1) and (**b**) cells from mature rats (P14-18)

34.4 Discussion

Our preliminary results showed that the inhibitory effect of ATP on the CB glomus cell $[Ca^{2+}]_i$ response to hypoxia was stronger in cells dissociated from newborn rats compared to mature glomus cells from P14-18 days old rats. Based on the studies with pharmacologic purinergic agonists and antagonists, this inhibitory effect may be mediated by a P2Y12 purinergic receptor subtype.

The emerging picture of CB glomus cell postnatal maturation is one of multiple mechanisms acting at multiple levels within the carotid body. The identity of the glomus cell oxygen sensor (or sensors) is still a matter of debate and it is not yet known whether O_2 sensing matures after birth at this level. Current popular views suggest that the mitochondria are a main site of O_2 sensing (Wyatt and Buckler 2004), linked via signaling pathways to cell membrane ion channels (Wyatt and Evans 2007), with other O_2 sensors such as heme oxygenase functionally linked to specific ion channels (Kemp 2005; Ortega-Saenz et al. 2006). Whatever the O_2 sensing mechanisms are, exposure to hypoxia leads to glomus cell depolarization and release of several neurotransmitters with potential for auto-modulation of the glomus cell response magnitude via autoreceptors. Dopamine acts via D2 receptors to dampen the response to hypoxia in a developmentally regulated fashion (Tomares et al. 1994). In the present study, the results suggest that ATP may also modulate glomus cell response to hypoxia in a differential manner with respect to maturation.

It is important to note that much remains unknown about the function of glomus cells in situ, where high concentrations of multiple neurotransmitters may have synergistic or antagonistic effects. In addition, adenosine receptors are present within the carotid body (Conde et al. 2006; Gauda et al. 2000) and may be stimulated by adenosine release or following breakdown of ATP by ectoATPases. Further studies will be required to understand basic mechanisms and the relative importance of ATP in CB O_2 sensing and functional maturation of organ function.

Acknowledgements This study was supported by grants from National Institutes of Health (HL-05621 NIH RO1) and Arkansas Biosciences Institute (ABI).

References

Bairam A, Carroll JL (2005) Neurotransmitters in carotid body development. Respir Physiol Neurobiol 149:217–232

Benot AR, Lopez-Barneo J (1990) Feedback inhibition of Ca^{2+} currents by dopamine in glomus cells of the carotid body. Eur J Neurosci 2:809–812

Conde SV, Obeso A, Vicario I, Rigual R, Rocher A, Gonzalez C (2006) Caffeine inhibition of rat carotid body chemoreceptors is mediated by A2A and A2B adenosine receptors. J Neurochem 98:616–628

Gauda EB, Northington FJ, Linden J, Rosin DL (2000) Differential expression of a(2a), A(1)-adenosine and D(2)-dopamine receptor genes in rat peripheral arterial chemoreceptors during postnatal development. Brain Res 872:1–10

Kemp PJ (2005) Hemeoxygenase-2 as an O_2 sensor in K^+ channel-dependent chemotransduction. Biochem Biophys Res Commun 338:648–652

Kim I, Yang DJ, Donnelly DF, Carroll JL (2009) Fluoresceinated peanut agglutinin (PNA) is a marker for live O(2) sensing glomus cells in rat carotid body. Adv Exp Med Biol 648:185–190

Ortega-Saenz P, Pascual A, Gomez-Diaz R, Lopez-Barneo J (2006) Acute oxygen sensing in heme oxygenase-2 null mice. J Gen Physiol 128:405–411

Tomares SM, Bamford OS, Sterni LM, Fitzgerald RS, Carroll JL (1994) Effects of domperidone on neonatal and adult carotid chemoreceptors in the cat. J Appl Physiol 77:1274–1280

Wasicko MJ, Sterni LM, Bamford OS, Montrose MH, Carroll JL (1999) Resetting and postnatal maturation of oxygen chemosensitivity in rat carotid chemoreceptor cells. J Physiol 514(Pt 2):493–503

Wasicko MJ, Breitwieser GE, Kim I, Carroll JL (2006) Postnatal development of carotid body glomus cell response to hypoxia. Respir Physiol Neurobiol 154:356–371

Wyatt CN, Buckler KJ (2004) The effect of mitochondrial inhibitors on membrane currents in isolated neonatal rat carotid body type I cells. J Physiol 556:175–191

Wyatt CN, Evans AM (2007) AMP-activated protein kinase and chemotransduction in the carotid body. Respir Physiol Neurobiol 157:22–29

Xu J, Xu F, Tse FW, Tse A (2005) ATP inhibits the hypoxia response in type I cells of rat carotid bodies. J Neurochem 92:1419–1430

Chapter 35
Serotonin Dynamics and Actions in the Rat Carotid Body: Preliminary Findings

Maria Ramirez, Teresa Gallego-Martin, Elena Olea, Asuncion Rocher, Ana Obeso, and Constancio Gonzalez

Abstract Serotonin or 5-HT is a biogenic amine present in the carotid body (CB) of several species as evidenced in many immunocytochemical studies and in a few biochemical measurements. Early literature on 5-HT actions in the CB in all studied species has lead to the conclusion that it does not participate in the setting of conducted action potentials in the sensory nerve of the CB. However, during the last 10 years very important roles in the cellular physiology of the CB have been proposed for this biogenic amine. These roles include a primary role in setting the excitability of chemoreceptor cells via an autocrine or paracrine action, and thereby, the conducted activity in the carotid sinus nerve, and a critical role in the genesis of long term sensory facilitation observed in CBs of animals exposed to intermittent hypoxia. These facts, along with important discrepancies in the endogenous levels of 5-HT in the CB prompted present study conducted in rat CBs. We measured CB endogenous 5-HT content by HPLC with electrochemical detection and found levels of 5-HT in the range of 15–22 pmole/mg tissue in control and chronically hypoxic animals either sustained or intermittent, with no significant differences among them. 5-HT and the 5-HT2A antagonist ketanserin dose-dependently activated chemoreceptor cells as assessed by their capacity to release catecholamines from freshly isolated CB. In preliminary experiments we have observed that intense hypoxia and high extracellular K^+ promote a small release of 5-HT from CB which is not dependent on the presence of extracellular Ca^{2+}. Further studies are needed to firmly establish the dynamics of 5-HT in the CB of the rat.

Keywords 5-HT • Carotid body • Catecholamine • Ketanserin • Hypoxia • HPLC.

35.1 Introduction

Intracarotid serotonin or 5- hydroxytryptamine (5-HT) in the cat causes a stimulation of respiration and section of the carotid sinus nerve (CSN) eliminated the hyperventilatory effect (Reid and Rand 1951; Ginzel and Kottegoda 1954). In the dog, McCubbin et al. (1956) found that 5-HT increases CSN discharges and ventilation, with both responses disappearing when the carotid sinus region was excluded from the circulation. Salmoiraghi et al. (1956) performed the initial studies in the rat finding

M. Ramirez • T. Gallego-Martin • E. Olea • A. Rocher • A. Obeso • C. Gonzalez (✉)
Department of Biochemistry and Molecular Biology and Physiology, IBGM School of Medicine, University of Valladolid CSIC, CIBERES-Institute Carlos III, Valladolid, 47005, Spain
e-mail: rocher@ibgm.uva.es; aobeso@ibgm.uva.es; constanc@ibgm.uva.es

C.A. Nurse et al. (eds.), *Arterial Chemoreception: From Molecules to Systems*, Advances in Experimental Medicine and Biology 758, DOI 10.1007/978-94-007-4584-1_35,

that 5-HT also produced CSN dependent transient hyperventilation followed in some occasions by apnea and in some other by moderate tachypnea. Scheider and Yonkman (1954) carried out a study in dogs, cats and rabbits and found that 5-HT i.v. always produced hyperventilation, although the appearance of other short lasting responses (e.g., early or late apneas) were species dependent. Therefore from these earlier studies the emerging notion was 5-HT augments ventilation mediated entirely or in its greatest part by CB chemoreceptors.

From more recent studies we want to summarize the findings of Kirby and McQueen (1984) in the cat. These authors used intracarotid 5-HT (1–50 μg) and monitored CSN discharges observing three components in the response: 1) a transient burst of activity during the injection period in 56% cases, 2) a dose-related chemodepression commenced a few seconds after completing the injection, and 3) a delayed longer-lasting chemoexcitation occurred in many experiments, concomitant with a fall in blood pressure. MDL72222, a 5-HT3 antagonist, abolished the initial excitatory effect, augmented the 5-HT dose required to cause depression, and it had no effect on the response evoked by hypoxia. Ketanserin, a 5-HT2A antagonist, did not affect the initial excitatory effect, augmented the 5-HT dose required to cause depression, markedly reduced the delayed chemoexcitation and hypotension, and did not affect the CSN activity elicited by hypoxia. MDL72222 and ketanserin combined eliminated all the effects of 5-HT but did not affect the response elicited by hypoxia. They concluded that 5-HT is not crucial in chemoreception.

Sapru and Krieger (1977) observed that intracarotid 5-HT (2–5 μg/kg) produced an increase in CSN activity and in ventilation, with disappearance of the ventilatory effects after ipsilateral CNS section. The CSN response had a fast onset and faded out in 10–15s. 5-HT action was not affected by methysergide (a wide spectrum blocker of 5-HT receptors; 500 μg/kg of body weight). Yoshioka (1989) reported that 1–100 μg/kg of 5-HT injected into the femoral vein produced a dose dependent ($EC_{50} \approx 10$ μg/kg), rapid onset, and short lived increase in CSN activity, mimicked by 2-methyl-5-HT (a purported 5-HT3 receptor agonist; $EC_{50} \approx 7$ μg/kg), and unaffected by methysergide and ketanserin. The 5-HT3 antagonist GR38032F by itself did not change ongoing or NaCN elicited CSN activity, but right shifted the excitatory action of 5-HT without affecting maximum effect (Gonzalez et al. 1994).

Zhang and Nurse (2000) observed that the spontaneous voltage fluctuations or spontaneous spikes recorded in cells from cultured clusters of rat CB, were augmented or induced by hypoxia or 5-HT (2–10 μM), and inhibited by ketanserin (50–100 μM), concluding that endogenous 5-HT may contribute to spontaneous firing of chemorceptor cells acting in an autocrine/paracrine manner. In 2003, Zhang et al. expanded the initial study to show that: 1) membrane depolarization-spikes of chemoreceptor cells were inhibited by ketanserin (10–50 μM) and by protein kinase C (PKC) inhibitors; the PKC activator 1-oleoyl-2-acetylglycerol (OAG) mimicked 5-HT effects, 2) 5-HT acted like hypoxia causing an inhibition Ca^{2+} dependent K^+ currents, 3) the 5-HT2A receptor blockers ketanserin and ritanserin reversibly inhibited spontaneous and hypoxia induced action potential recorded from clustered chemoreceptor cells, and, in addition, they also inhibited the hypoxic responses in petrosal neurons functionally coupled to co-cultured chemoreceptor cells, 4) RT-PCR and confocal immunofluorescence revealed 5-HT2A receptor expression in rat chemoreceptor cells, 5) the 5-HT3 antagonist MDL72222 did not affect the responses (but see above). They concluded that chemoreceptor cells control their own excitability by releasing 5-HT which acts in an autocrine–paracrine manner to stimulate 5-HT2A receptors that, via PKC, modulate Ca^{2+}dependent K^+ currents.

In a first study Prabhakar's laboratory (Jacono et al. 2005) reported that hypoxia elicited activity in the CSN in the rat was not affected by 5-HT (3 μM) on its onset or magnitude, although it prolonged the time required for the CSN activity to return to basal after the hypoxic challenge. They also reported that ketanserin (40 μM) prevented this 5-HT effect, but had no effect on hypoxic response in the absence of 5-HT. Finally, they measured endogenous 5-HT in the CB and found very small levels (0.2 ± 0.003 pmol/mg of protein, equivalent to ≈ 0.02 pmol/mg of tissue or ≈ 0.001 pmol/CB or 1 fmol/CB) and a proportionally very high release (1.4 ± 0.09 fmol/min/4 CBs) that was not altered by hypoxia. They concluded that 5-HT is not critical for the genesis of the hypoxic response in the rat CB, but it can modulate the dynamics of the hypoxic response via 5-HT2A receptors. Peng et al. (2006) observed that repeated 5-HT application elicited an injection-related chemosensory burst of activity and long term sensory facilitation (LTSF) in isolated CB-CSN preparations of control rat and mice. This LTSF, at 60 min after the last 5-HT injection, resulted a x4 increase in CSN activity when compared with preinjection activity. Ketanserin had no effect on the initial 5-HT induced burst-like sensory excitation but prevented LTSF induced by 5-HT. (This 5-HT-induced LTSF is comparable to that found after acute repetitive hypoxic stimulation in rats exposed to intermittent hypoxia (IH) for 10 days, but absent in control rats; Peng et al. 2003). Additional experiments led them to conclude that 5-HT elicits LTSF by a coupling of 5-HT2A receptors to PKC, which in turn activates NADPH oxidase, being NADPH-derived ROS involved in, or responsible for, the observed LTSF because it was absent in gp91$^{phox-/-}$mice. In 2009, Peng et al. showed three important things: 1)NADPH oxidase involved was NOX2, 2) 5-HT was necessary for IH to induce LTSF induced by repetitive hypoxic stimulation, and 3) in the CB of IH rats acute hypoxia produced a robust 5-HT-release (from 1.6 ± 0.08 in normoxia to 8.1 ± 0.1 femtomoles/min/CB in hypoxia).

As a whole these set of conflicting, but very important observations, drove us to perform present study.

35.2 Material and Methods

35.2.1 Animals and Surgical Procedures

Experiments were performed in Wistar adult rats of both sexes (250–350 g) obtained in the vivarium of the Faculty of Medicine of the University of Valladolid. The Institutional Committee of the University of Valladolid for Animal Care and Use approved the protocols. Rats were anaesthetized with sodium pentobarbital (60 mg/kg i.p.), tracheostomized and the carotid arteries were dissected. A block of tissue including the carotid sinus area was removed and placed in a Lucite chamber. CBs were cleaned free of CSN and nearby connective tissue as previously described (Vicario et al. 2000). Animals were euthanized by intracardiac overdoses of sodium pentobarbital until the beating of the heart ceased.

35.2.2 Exposure to Chronic Sustained and Intermittent Hypoxia

Rats were exposed to chronic sustained hypoxia (CH) in a glass chamber continuously fluxed with a gas mixture (10–11% O_2 in N_2; $PO_2 \approx 80$ mmHg; 15 days). Accumulation of CO_2 was prevented by the continuous flow of the gas mixtures and the presence in the chamber floor of a layer of soda lime. The rats had free access to food and water, and remained in this atmosphere except for 30 min/3 days for routine cleaning and maintenance. At day 15 CBs were removed and cleaned as above.

Exposure to IH was carried out as described by Gonzalez-Martin et al. (2009, 2011) except for the nadir percentage O_2 attained in the exposure chamber was 5%, i.e., the IH protocol was 40s, 5%O_2/80s, air, 8h/day (from 08 to 16h), 15 days.

35.2.3 Labelling of Catecholamines Stores: Release of ^{3}H-CA

Protocols to label catecholamine (CA) stores in chemoreceptor cells and to study the release of labelled CA as well as the analytical protocols involved have been described in detail in several studies of the laboratory (Vicario et al. 2000; Conde et al. 2007). Specific aspects of the experiments would be provided in the Results.

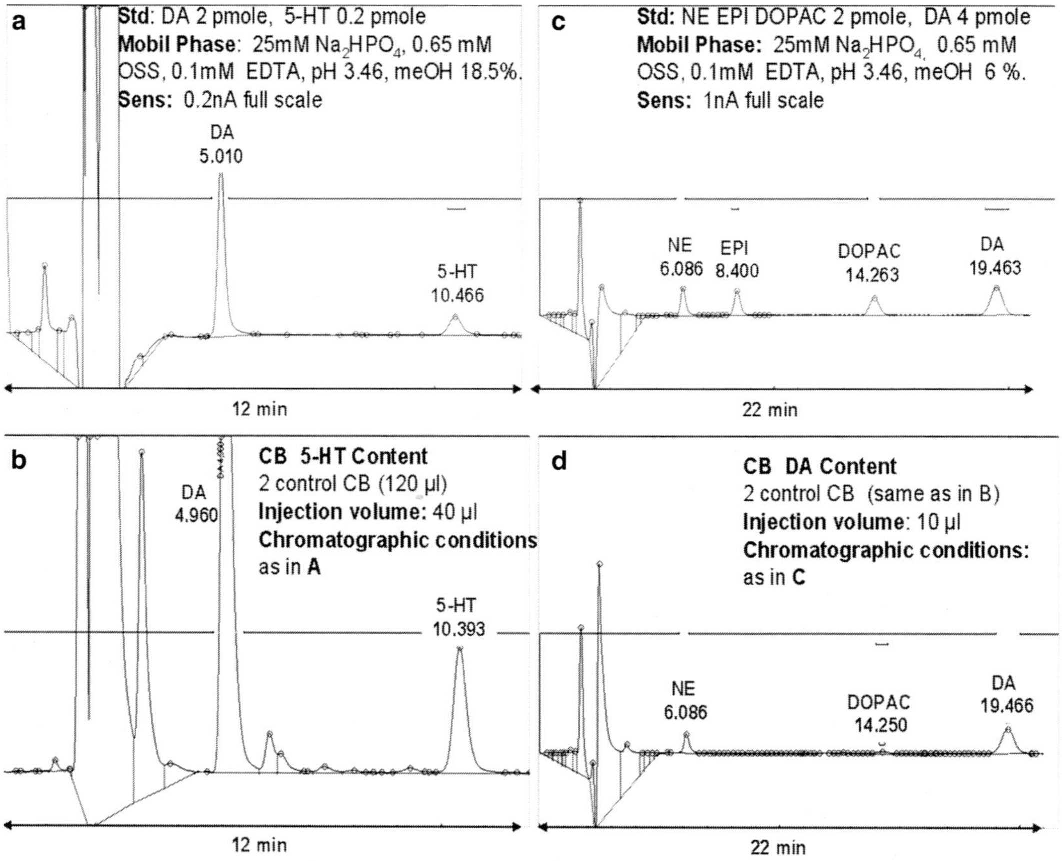

Fig. 35.1 General chromatographic conditions used in present study. See Results section for detailed description

35.2.4 Chromatographic Analysis of Endogenous CA and 5-HT

For the analysis of endogenous unlabelled CA and 5-HT, upon dissection, the two CB from a rat were placed in eppendorf tubes containing 100 µl of ice-cold 0.3 N PCA. After weighing the tissues free of adherent PCA in an electrobalance (Supermicro, Sartorius), tissues were glass to glass homogenized, the homogenizer washed with 20 µl of PCA, centrifuged and aliquots (10–40 µl) of supernatants were directly injected into an HPLC system. The HPLC system was composed of a Waters 600 controller pump, automatic injector Waters 717 plus Autosampler, and BAS LC-4C Amperometric Detector. Mobile phases and sensitivity of the detector varied according to specific interests and are provided in Fig. 35.1. Amine identification was done against external standards. Quantification was made with Peak Sample Data Chromatography System software (Buck Scientific, East Norwalk, CT).

35.2.5 Data Analysis

Data presented are means ± SEM and were evaluated for statistical significance using Graph Pad Prism Software, version 4. The significance of the differences between the means was calculated by One Way Analysis of Variance (ANOVA) with Dunnett's multiple comparison tests. P values of 0.05 or less were considered to represent significant differences.

35.3 Results

35.3.1 Serotonin and Catecholamine Levels in the CB

As mentioned above, Fig. 35.1 shows the chromatographic conditions used to identify catecholamines: dopamine (DA), norepinephrine (NE), epinephrine (E), and dihydroxy phenyl acetic acid (DOPAC) the main catabolite of DA, and 5-HT. Note in the two upper chromatograms (Fig. 35.1a, c) the different concentrations of the organic component of the mobile phase (methanol) to optimize the duration of the chromatographic run and the resolution of the analytes of interest. Note also the different setting of sensitivity in the detector (0.2 nA and 1 nA) needed to detect 5-HT and DA.

Figure 35.1b shows a run at high sensitivity to measure 5-HT in a 40 μl sample taken from a total of 120 μl where 2 CBs were homogenized (injected homogenate corresponds to 2/3 of one CB). Note the well defined peak for 5-HT, the saturation for the DA peak and the lack of resolution of NE that appears in the solvent front. To adequately resolve CA and to obtain measurable peaks it was needed to decrease the organic solvent (to about 1/3), to decrease the sensitivity (to 1/5), and decrease the injected volume (to ¼) (Fig. 35.1d). According to specific results, the injected volumes can be changed from experiment to experiment and increase them up to 100 μl, and sensitivities can be varied within 0.2 nA and 1 nA full scale.

Figure 35.2a shows 5-HT levels in the CBs of rats in control conditions, exposed to IH, and exposed to CH. Mean 5-HT values in the CB of the three groups ranged between 15 and 22 pmole/mg tissue,

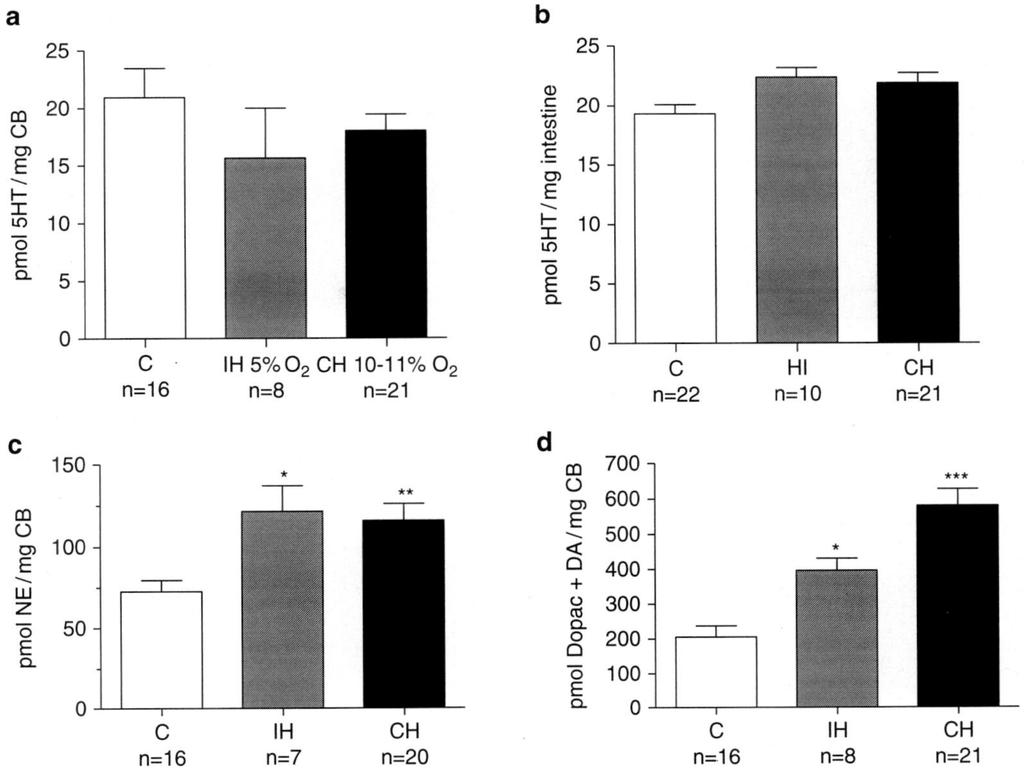

Fig. 35.2 Levels of 5-HT in the CB (**a**) and duodenum (**b**) and of NE (**c**) and DA (**d**) in the CB. C, Control rats; IH, rats subjected to IH, and; CH, rats subjected to sustained hypoxia. Means + SEM of n individual values as shown in the drawing

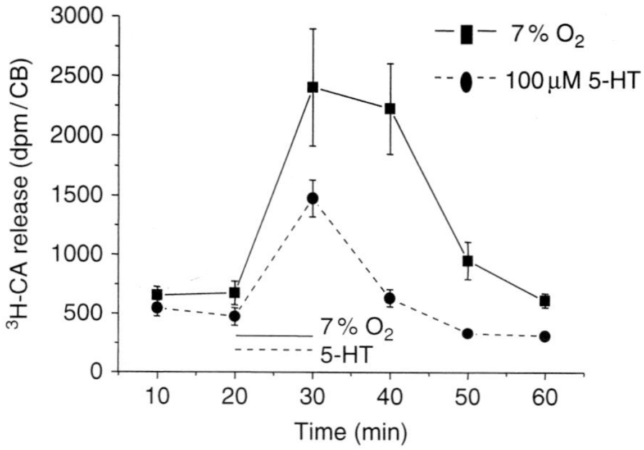

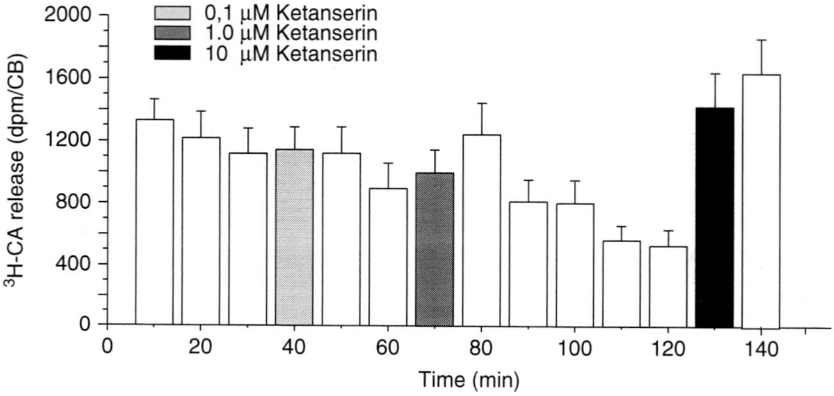

Fig. 35.3 *Top*, time courses of the release of the ^{3}H-CA elicited by hypoxia (10 min superfusion with a hypoxic solution equilibrated with 7% O_2, 5% CO_2, balance N_2) and by 100 μM 5-HT in normoxia (air equilibrated solution). *Bottom*, time course of the release of ^{3}H-CA in CBs superfused with a normoxic solution (5% CO_2 in air) containing ketanserin at concentrations and times as depicted in the drawing

and there were no statistically significant differences among the three groups. Fig. 35.2b shows 5-HT levels in the duodenum of the same groups of animals, and interestingly enough, 5-HT levels are nearly identical to those of the CB with no significant differences among them. Figure 35.2c and d show, respectively levels of NE and DA in the CBs of the same animals. The NE/5-HT ratio was around 3 in Control animals and nearly double in CH and IH due to the increase in NE. The DA/5-HT ratio in Control animals was around 10, reached nearly 20 in IH and 30 in CH.

35.3.2 *Effects of 5-HT and Ketanserin on the Release of CA from Chemoreceptor Cells*

Upper part of Fig. 35.3 shows the time course of the release of CA elicited by 100 μM 5-HT, a concentration producing maximal effect on the CA release rate. For comparison we also show the release response elicited by a moderate hypoxic stimulus. Aside from the obvious smaller magnitude of the release response elicited by 5-HT, it should also be noted that kinetics of the response induced by hypoxia and 5-HT are different, being the 5HT response faster in onset and rapidly subsiding on 5-HT removal. Presumably, the different kinetics reflect different mechanisms. Lower part of the Fig. 35.3 shows the effect of three increasing concentrations of ketanserin on the basal ongoing release of CA. Surprisingly, ketanserin, although a 5-HT antagonist, dose dependently increased the release of CA from chemoreceptor cells, being even more potent that 5-HT itself.

35.3.3 Effects of Hypoxia on 5 HT Release

In preliminary experiments (n=2) we have measured the release of endogenous 5-HT elicited by hypoxia and high eternal K^+ applied during 15 min in each case. Our findings would indicate that both intense hypoxia (2% O_2) and 60 mM K_e^+ elicit a very modest release of 5-HT which in control CBs represents <1/60 of the endogenous content (or 0,30 pmole/mg tiss/15 min); the release of CA from the same CBs was around 22 pmole/mg tissue/15 min or around 1/15 of the endogenous content. The release of 5-HT was independent of extracellular Ca^{2+}while that of CA was inhibited by more than 90% in Ca^{2+} free solutions.

35.4 Discussion

In the present study we have stressed the description of the analytical methodology for 5-HT analysis because there are in the literature very important discrepancies regarding the 5-HT levels in the CB. It is worth noting that Chiocchio et al. (1967, 1971) reported 5-HT levels oscillating between 11 and 39 pmole/mg wet tissue for the cat CB. Hellstrom and Koslow (1975) reported values of 50 pmole/mg wet tissue for the rat CB and Hui et al. (2003) in more recent times reported levels oscillating between *ca.* 8.5 and 11.4 pmole/mg wet tissue for the rat CB. At variance with these reasonable agreements, Jacono et al. (2005) reported 5-HT contents for the rat CB of *ca.* 0.02 pmole/mg wet tissue. These discrepancies advised us to be extremely careful with the analytical procedures and to use a reference tissue, namely rat duodenum (Fig. 35.2b). At the meeting we presented data from 8 different studies spanning between 1972 and 2001, measuring 5-HT in the duodenum of control rats with different assay methods (from colorimetry to HPLC with fluorescent detection) and giving 5-HT concentrations oscillating between 6.8 and 38.6 pmole/mg wet tissue with a mean for all studies of 23.1 pmole/mg wet tissue. As seen in Fig.35.2b our control values were 19.3 pmole/mg wet tissue, therefore giving soundness to our measurements in the CB.

The data of Fig. 35.3 showing the effects of the 5-HT and ketanserin on the release of CA from chemoreceptor cells is perplexing, because it should not be expected that the agonist and the antagonist have comparable effects on any giving parameter. Yet, in this case both augment the activity of chemoreceptor cells as exemplified by their release of CA (Fidone et al. 1982; Gonzalez et al. 1992). Although additional experiments of competition are required to clarify the reasons for this peculiarity, we can assure that the finding is genuine as another 5-HT2A antagonist, ritanserin, has alike effects. On searching for plausible explanations we have found that both antagonists are in fact capable of inhibiting a great variety of 5-HT receptors, adrenergic receptors, and some subtypes of DA receptors (see http://www.iuphar-db.org/DATABASE/ReceptorFamiliesForward?type=GPCR). Then, giving the similarities of effects between the purportedly 5-HT2A receptor antagonists and dopaminergic receptors present in chemoreceptor cells (Dinger et al. 1981; Conde et al. 2008) we would suggest that ketanserin actions result from its targeting of multiple receptors.

Acknowledgements Supported by Grants BFU2007-61848 (DGICYT), CIBER CB06/06/0050 (ISCiii), and Accion Integrada (Micinn, Spain) PT2009-0172.

References

Chiocchio SR, Biscardi AM, Tramezzani JH (1967) 5-Hydroxytryptamine in the carotid body of the cat. Science 158(802):790–791

Chiocchio SR, King MP, Carballo L, Angelakos ET (1971) Monoamines in the carotid body cells of the cat. J Histochem Cytochem 19(10):621–626

Conde SV, Obeso A, Gonzalez C (2007) Low glucose effects on rat carotid body chemoreceptor cells secretory responses and action potential frequency in the carotid sinus nerve. J Physiol 585:721–730

Conde SV, Gonzalez C, Batuca JR, Monteiro EC, Obeso A (2008) An antagonistic interaction between A2B adenosine and D2 dopamine receptors modulates the function of rat carotid body chemoreceptor cells. J Neurochem 107(5):1369–1381

Dinger B, Gonzalez C, Yoshizaki K, Fidone S (1981) [3 H] spiroperidol binding in normal and denervated carotid bodies. Neurosci Lett 21(1):51–55

Fidone S, Gonzalez C, Yoshizaki K (1982) Effects of low oxygen on the release of dopamine from the rabbit carotid body in vitro. J Physiol 333:93–110

Ginzel KH, Kottegoda SR (1954) The action of 5-hydroxytryptamine and tryptamine on aortic and carotid sinus receptors in the cat. J Physiol 123(2):277–288

Gonzalez C, Almaraz L, Obeso A, Rigual R (1992) Oxygen and acid chemoreception in the carotid body chemoreceptors. Trends Neurosci 15(4):146–153

Gonzalez C, Almaraz L, Obeso A, Rigual R (1994) Carotid body chemoreceptors: from natural stimuli to sensory discharges. Physiol Rev 74(4):829–898

Gonzalez-Martín MC, Vega-Agapito V, Prieto-Lloret J, Agapito MT, Castañeda J, Gonzalez C (2009) Effects of intermittent hypoxia on blood gases plasma catecholamine and blood pressure. Adv Exp Med Biol 648:319–328

Gonzalez-Martín MC, Vega-Agapito MV, Conde SV, Castañeda J, Bustamante R, Olea E, Perez-Vizcaino F, Gonzalez C, Obeso A (2011) Carotid body function and ventilatory responses in intermittent hypoxia evidence for anomalous brainstem integration of arterial chemoreceptor input. J Cell Physiol 226(8):1961–1969

Hellström S, Koslow SH (1975) Biogenic amines in carotid body of adult and infant rats–a gas chromatographic-mass spectrometric assay. Acta Physiol Scand 93(4):540–547

Hui AS, Striet JB, Gudelsky G, Soukhova GK, Gozal E, Beitner-Johnson D, Guo SZ, Sachleben LR Jr, Haycock JW, Gozal D, Czyzyk-Krzeska MF (2003) Regulation of catecholamines by sustained and intermittent hypoxia in neuroendocrine cells and sympathetic neurons. Hypertension 42(6):1130–1136

Jacono FJ, Peng YJ, Kumar GK, Prabhakar NR (2005) Modulation of the hypoxic sensory response of the carotid body by 5-hydroxytryptamine: role of the 5-HT2 receptor. Respir Physiol Neurobiol 145(2–3):135–142

Kirby GC, McQueen DS (1984) Effects of the antagonists MDL 72222 and ketanserin on responses of cat carotid body chemoreceptors to 5-hydroxytryptamine. Br J Pharmacol 83(1):259–269

Mccubbin JW, Green JH, Salmoiraghi GC, Page IH (1956) The chemoceptor stimulant action of serotonin in dogs. J Pharmacol Exp Ther 116(2):191–197

Peng YJ, Overholt JL, Kline D, Kumar GK, Prabhakar NR (2003) Induction of sensory long-term facilitation in the carotid body by intermittent hypoxia: implications for recurrent apneas. Proc Natl Acad Sci U S A 100(17): 10073–10078

Peng YJ, Yuan G, Jacono FJ, Kumar GK, Prabhakar NR (2006) 5-HT evokes sensory long-term facilitation of rodent carotid body via activation of NADPH oxidase. J Physiol 576(Pt 1):289–295

Peng YJ, Nanduri J, Yuan G, Wang N, Deneris E, Pendyala S, Natarajan V, Kumar GK, Prabhakar NR (2009) NADPH oxidase is required for the sensory plasticity of the carotid body by chronic intermittent hypoxia. J Neurosci 29(15):4903–4910

Reid G, Rand M (1951) Physiological actions of the partially purified serum vasoconstrictor (serotonin). Aust J Exp Biol Med Sci 29(6):401–415

Salmoiraghi GC, Page IH, McCubbin JW (1956) Cardiovascular and respiratory response to intravenous serotonin in rats. J Pharmacol Exp Ther 118(4):477–481

Sapru HN, Krieger AJ (1977) Effect of 5-hydroxytryptamine on the peripheral chemoreceptors in the rat. Res Commun Chem Pathol Pharmacol 16(2):245–250

Schneider JA, Yonkman FF (1954) Species differences in the respiratory and cardiovascular response to serotonin (5-hydroxytryptamine). J Pharmacol Exp Ther 111(1):84–98

Vicario I, Rigual R, Obeso A, Gonzalez C (2000) Characterization of the synthesis and release of catecholamine in the rat carotid body in vitro. Am J Physiol Cell Physiol 278(3):C490–C499

Yoshioka M (1989) Effect of a novel 5-hydroxytryptamine3-antagonist, GR38032F, on the 5-hydroxytryptamine-induced increase in carotid sinus nerve activity in rats. J Pharmacol Exp Ther 250(2):637–641

Zhang M, Nurse CA (2000) Does endogenous 5-HT mediate spontaneous rhythmic activity in chemoreceptor clusters of rat carotid body? Brain Res 872(1–2):199–203

Zhang M, Fearon IM, Zhong H, Nurse CA (2003) Presynaptic modulation of rat arterial chemoreceptor function by 5-HT: role of K+channel inhibition via protein kinase C. J Physiol 551:825–842

Chapter 36
Human Carotid Body HIF and NGB Expression During Human Development and Aging

Camillo Di Giulio, S. Zara, A. Cataldi, Andrea Porzionato, Mieczyslaw Pokorski, and Raffaele De Caro

Abstract Hypoxia inducible factor 1(HIF-1α) is the regulator of oxygen homeostasis in tissue correlated with neuroglobin (NGB) a member of the family of globins in vertebrates. The present study investigates, the expression and the location of NGB, HIF-1α in human carotid bodies, sampled at autopsy from children (mean age: 2 year ±), young (mean age: 27.5) and 4 old subjects (mean age: 73.5). The percentage of NGB positive area was higher in the old subjects (4.4 ±2.8%), as compared with the young ones (2.4 ±1.8%) and children (1.0 ±1.8%). Positive HIF-1α nuclei were detected in young and old subjects (1.0 ±0.14% *vs* 3.0 ±0.28%, respectively), whereas CB tissues from children did not show any HIF-1α reaction. The increase of NGB and HIF-1α expression suggests a possible role of the two oxygen sensors in the aging processes. Even though the physiological role of NGB is not well understood, it could be suggested that is act as a respiratory protein connected with HIF.

Keywords Aging and development • Human carotid body • HIF • Neuroglobin

36.1 Introduction

Carotid body (CB) undergoes several morphological, physiological and biochemical changes during development and aging (Di Giulio et al. 2003), it releases several substances such as dopamine, acetylcholine, norepinephrine, erythropoietin, substances P and detects pO_2 and pCO_2 levels in the blood, regulates ventilation according to homeostasis for the oxygen needs and body requirements (Prabhakar et al. 2001). During development and aging there is a variation of nerve conductivity, the

C. Di Giulio (✉)
Department of Neurosciences and Imaging, University of Chieti, Via Dei Vestini, 31, Chieti, CH 66100, Italy
e-mail: digiulio@unich.it

S. Zara
Department of Drug Sciences, University of Chieti, Via Dei Vestini, 31, Chieti, CH 66100, Italy

A. Cataldi
Department of Medicine and Aging Sciences, University of Chieti, Via Dei Vestini 31, Chieti, CH 66100, Italy

A. Porzionato • R. De Caro
Department of Human Anatomy and Physiology, University of Padova, Warsaw, Poland

M. Pokorski
Medical Research Center, Polish Academy of Sciences, Warsaw, Poland
e-mail: m_pokorski@hotmail.com

C.A. Nurse et al. (eds.), *Arterial Chemoreception: From Molecules to Systems*, Advances in Experimental Medicine and Biology 758, DOI 10.1007/978-94-007-4584-1_36,

maximum number of impulses per minute in nerves generally decrease with old age and became less sensitivity of peripheral receptor, leading to a reduction of the homeostatic capacity with higher latency in the adaptation responses (Rivner et al. 2001). In response to hypoxia, HIF-1 α (Hypoxic Inducible Factor –1 α) mediates transcriptional activation of a network of genes encoding VEGF (Vascular Endothelial Growth Factor), NOS (Nitric Oxide Synthase) (Bredt et al. 1990), erythropoietin and several glycolytic enzymes (Iyer et al. 1998). Hypoxia inducible factor 1(HIF-1α) is the regulator of oxygen homeostasis in tissue (Semenza 1999). Neuroglobin (NGB), a 151-amino-acid protein with a hexa-coordinate heme-Fe atoms, which displays O_2 affinities and binds CO (Burmester and Hankeln 2009), is a member of the family of globins in vertebrates and is concentrated in the mitochondria-containing areas of neurons, and its distribution is correlated with oxygen consumption rates. We have previously shown that NGB is expressed in rat carotid body (CB) tissues (Verratti et al. 2009). During development and aging the antioxidant capacity is modified and the CO_2-O_2 sensitivity is not the same. Generally the growth, the differentiation, aging and the death of cells are related to a series of factors including oxygen consumption, intracellular pH, free radicals (Sohal et al. 1986; Finkel and Holbrook 2000; Wickens 2001). To test if oxygen-sensitive mechanisms in human carotid bodies are affected in an age-dependent fashion, we studied NGB, HIF-1 α, VEGF and NOS and structural changes in children, young and old subjects.

36.2 Materials and Methods

The present study investigates, by immunohistochemical analysis, the expression and the location of NGB, HIF-1α in human carotid bodies, sampled at autopsy from three children (mean age: 2 year ± 66SD days), 4 young (mean age: 27.5 ±3.4 years) and 4 old subjects (mean age: 73.5 ±3.4 years). All subjects died of trauma and were all clinically without chronic pulmonary or cardiovascular diseases. Autopsies were performed between 18 and 78h after death. Specimens were taken of the right carotid bifurcation, including 20 mm of the common carotid and 20 mm of the internal and external carotid arteries. The study was approved by the local Ethical Committee and was performed according to the Italian laws on autopsied human tissues (Porzionato et al. 2005).

Light microscopy analysis and immunohistochemistry Tissues have been fixed in 10% phosphate-buffered formalin for 72 h and dehydrated through ascending alcohols and xylene and then paraffin embedded. Samples have been then de-waxed (xylene and alcohol progressively lower concentrations) and the slides, 5 μm thick, have been processed for Trichrome Mallory staining (Tricromica kit) (Bio Optica, Milano, Italy), as suggested by the data sheet, to distinguish connective compartment from parenchyma, and for immunohistochemical analysis.

In order to detect Ngb and HIF-1α proteins immunohistochemistry has been performed by means of Ultravision LP Detection System HRP Polymer & DAB Plus Chromogen (Lab Vision Thermo, CA, USA). Slides have been incubated in the presence of mouse anti-Ngb (Biovendor, Heidelberg, Germany), and mouse anti-HIF-1α, (Santa Cruz Biotechnology, CA, USA) and then in the presence of specific HRP-conjugated secondary antibodies. Peroxidase has been developed using diaminobenzidine chromogen (DAB) and nuclei have been hematoxylin counterstained. Negative controls have been performed by omitting the primary antibody.

Samples have been observed by means of light microscopy Leica DM 4000 (Leica Cambridge Ltd, Cambridge, UK) equipped with a Leica DFC 320 camera (Leica Cambridge Ltd, Cambridge, UK) for computerized images.

Computerized Morphometry measurements and Image Analysis After digitizing the images deriving from immunohistochemical stained sections, QWin Plus 3.5 software (Leica Cambridge Ltd, Cambridge, UK) has been used to evaluate NGB and HIF-1α expression. Image analysis of protein expression has been performed through the quantification of threshold area for immunohistochemical brown color per ten fields of light microscope observation. QWin Plus 3.5 assessments have been logged to Microsoft Excel and processed for Standard Deviations and Histograms. The statistical significance of the results has been evaluated using the *T*-Test and the Linear Regression Test, with $p = 0.05$.

36.3 Results

Immunostaining was found in the carotid body. NGB positive area showed a progressive increase of expression from children to adult and from adult to old subjects, so was higher in the old subjects (4.4 ±2.8%), as compared with the young ones (2.4 ±1.8%) and children (1.0 ±1.8%) (Fig. 36.1). Positive HIF-1α nuclei were detected in young and old subjects (1.0 ±0.14% *vs* 3.0 ±0.28%, respectively), whereas CB tissues from children did not show any HIF-1α reaction (Fig. 36.2). Along with NGB and

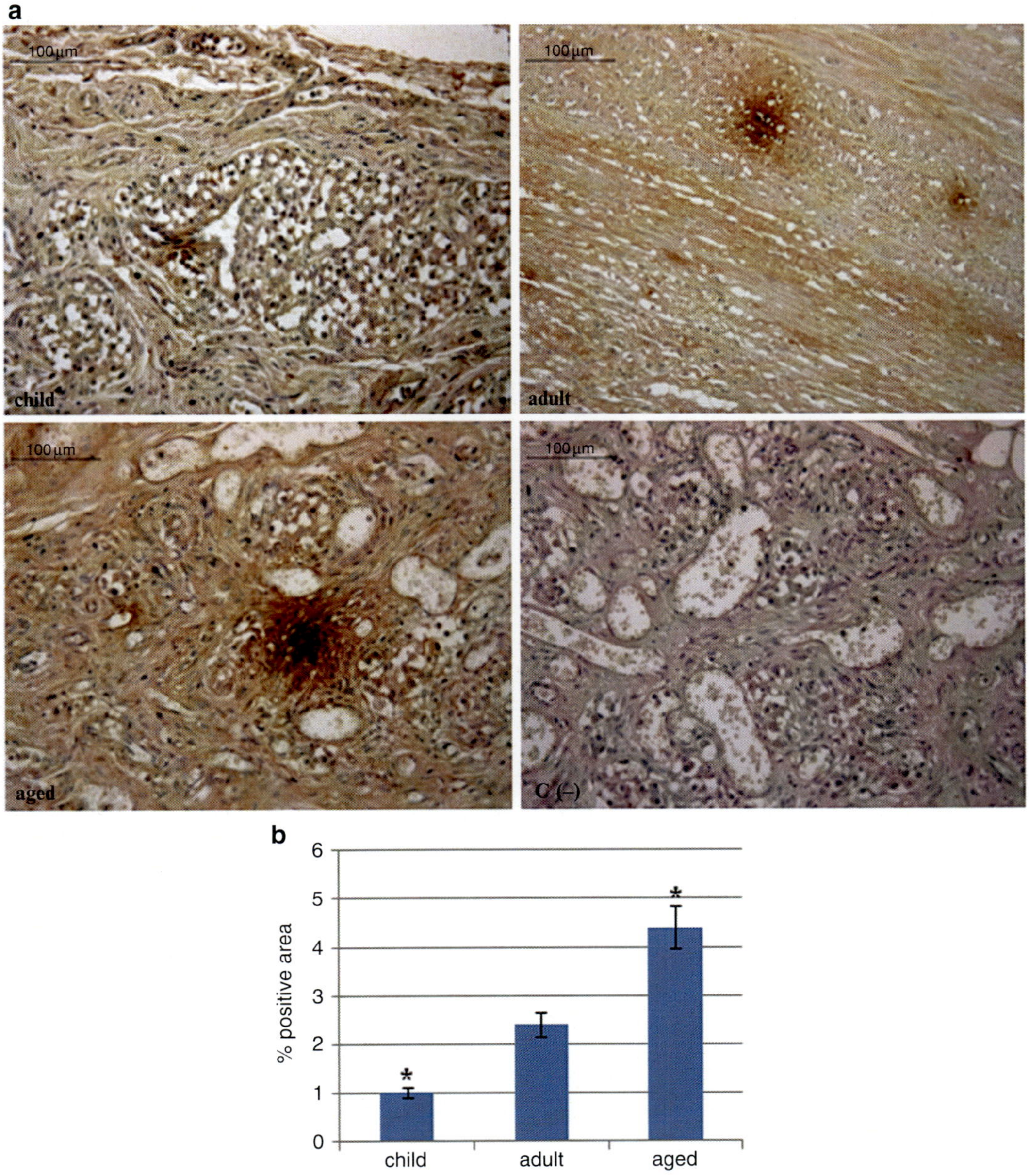

Fig. 36.1 (**a**) Immunohistochemical detection of NGB expression in human carotid bodies of child, adult and aged subjects; C(–) negative control; (**b**) Graphic representation of NGB % positive area (± SD) densitometric analysis determined by direct visual counting of ten fields (mean values) for each of three slides per sample at 40x magnification (* aged NGB *vs* child NGB $p<0.05$)

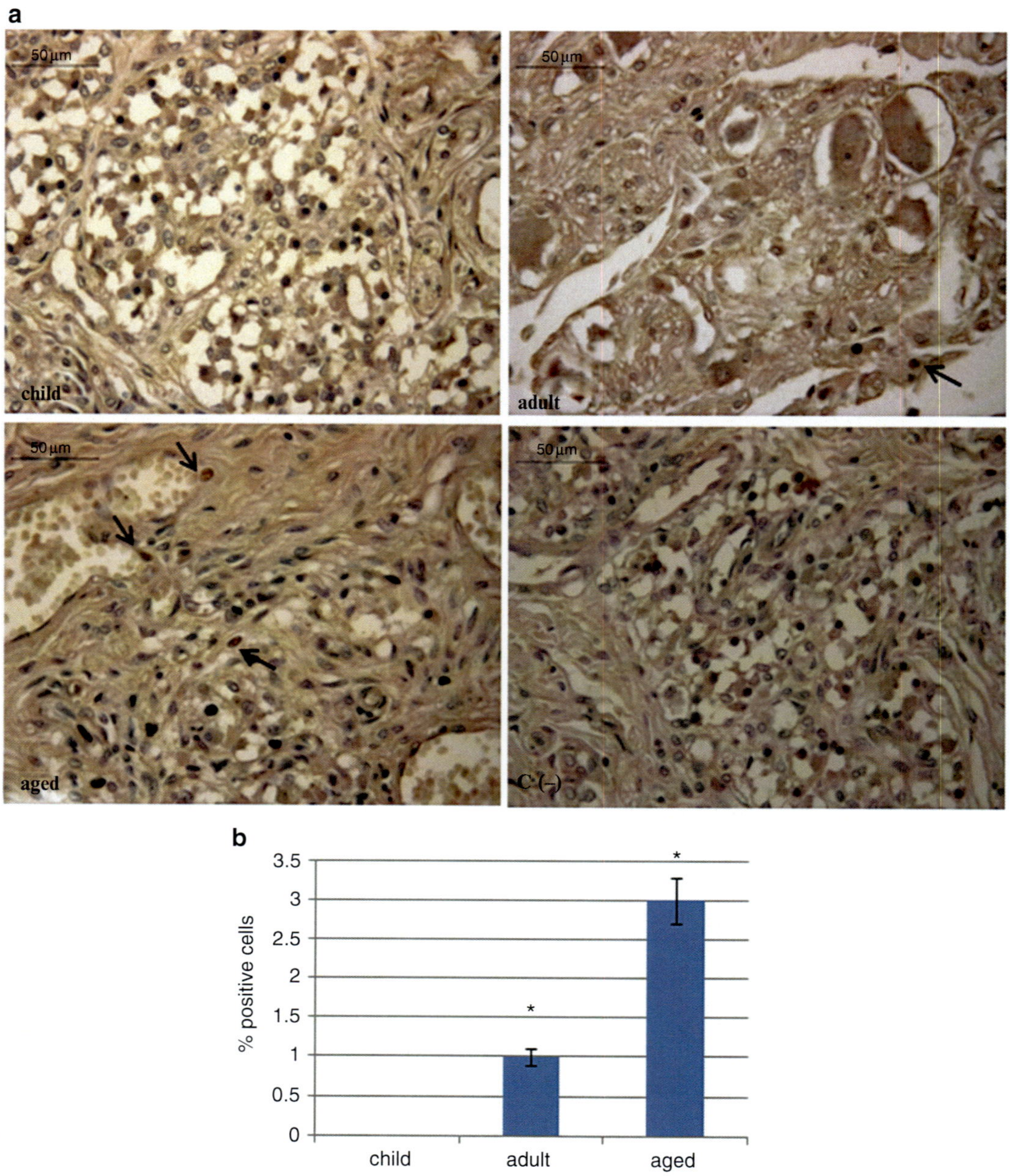

Fig. 36.2 (**a**) Immunohistochemical detection of HIF-1α expression in human carotid bodies of child, adult and aged subjects; C(−) negative control; arrows indicate HIF-1α positive nuclei; (**b**) Graphic representation of HIF-1α positive nuclei % (± SD) densitometric analysis determined by direct visual counting of ten fields (mean values) for each of three slides per sample at 40x magnification (* aged HIF-1α *vs* adult HIF-1α $p<0.05$)

HIF-1α expression evaluation, TUNEL analysis has been performed to estimate apoptosis event occurrence. A significant increase of TUNEL positive nuclei is evidenced in children carotid bodies compared to aged subject ones (14% and 8%, respectively), while adult subjects show an intermediate rate of apoptotic events (10%) (Fig. 36.3).

With Tricromic Mallory staining we found a decrease of CB parenchyma area from child to old subject and a significant increase in intralobular connective tissue was observed in aged subjects respect to child (Fig. 36.4).

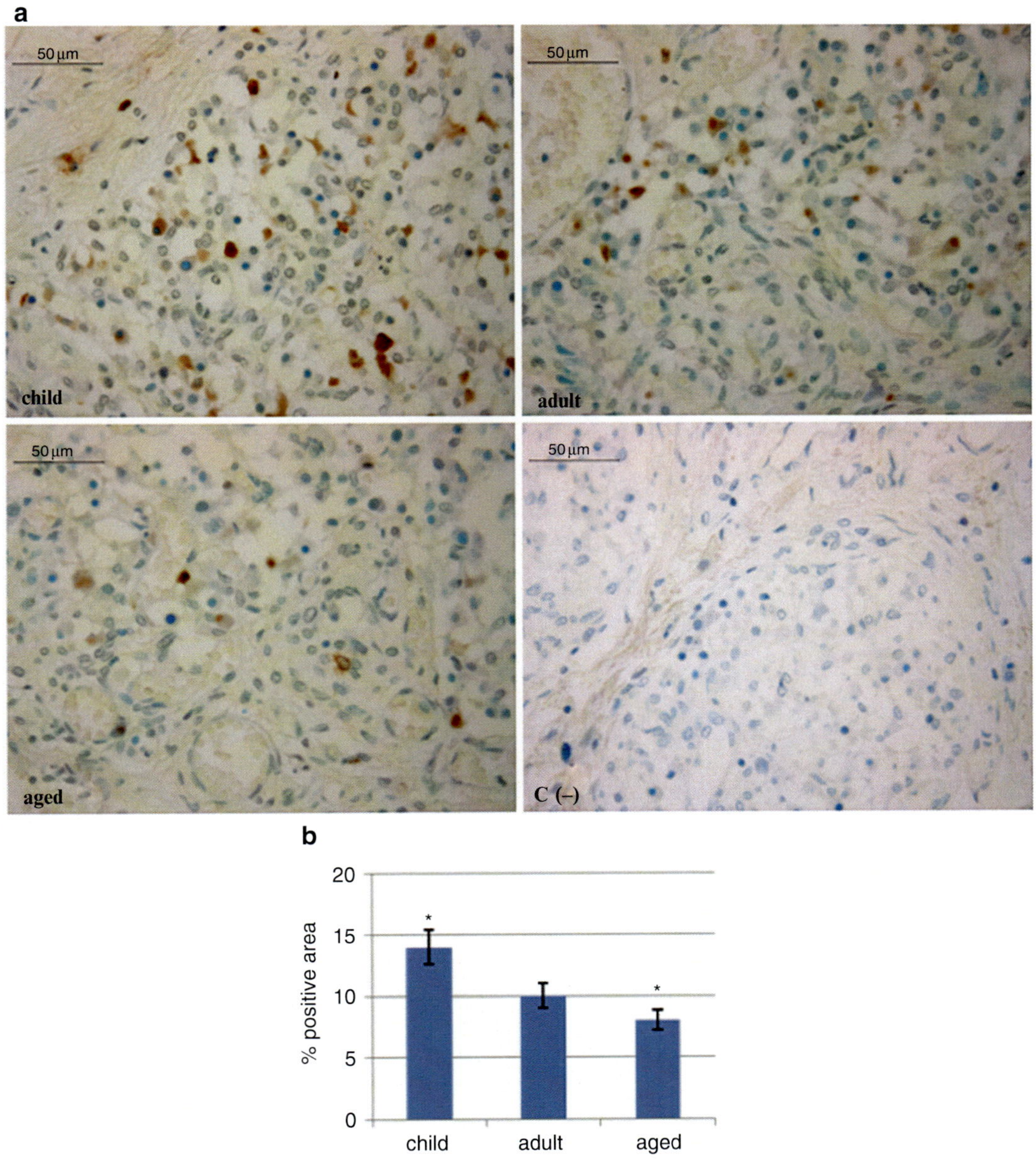

Fig. 36.3 (**a**) TUNEL detection of apoptotic nuclei in human carotid bodies of child, adult and aged subjects. (**b**) Graphic representation of TUNEL positive nuclei % (± SD) densitometer analysis determined by direct visual counting of ten fields (mean values) for each of three slides per sample; (* aged % positive cells *vs* child % positive cells $p<0.05$)

36.4 Conclusion

The increase of NGB and HIF-1α expression suggests a possible role of the two oxygen sensors in the aging processes. Even though the physiological role of NGB is not well understood, it could be suggested that it acts as a respiratory protein connected with HIF in order to modulate the oxygen sensitive mechanisms undergoing alterations during the aging processes (Leiser and Kaeberlein 2010).

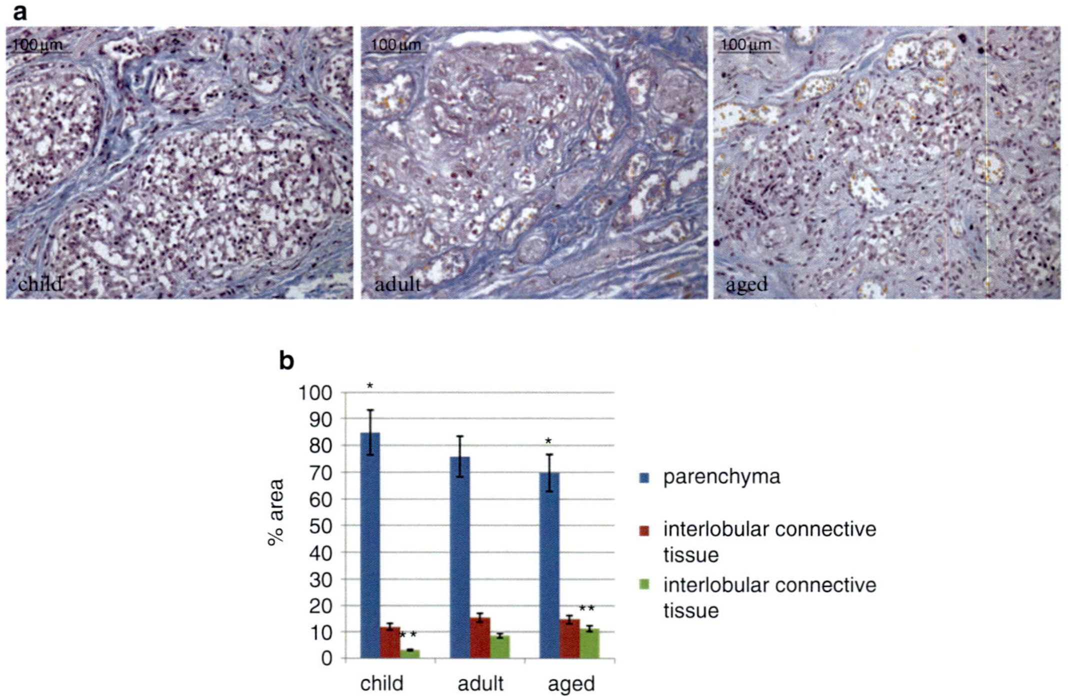

Fig. 36.4 (**a**) Morphometric analysis of human carotid bodies performed on Trichrome Mallory stained slides of child, adult and aged subjects. (**b**) Parenchyma, connective interlobular and intralobular compartment measurements expressed as % area mean (± SD) assessed by direct visual counting of ten fields for each of three slides per each of 5 samples (* Aged parenchyma % area *vs* child parenchyma % area $p<0.05$) (** Aged intralobular connective tissue % area *vs* child intralobular connective tissue % area $p<0.05$)

NGB is thought to participate in processes such as oxygen transport, oxygen storage, and nitric oxide detoxification and similar to hemoglobin may act as a respiratory protein by reversibly binding gaseous ligands (NO and O_2) via the Fe-containing porphyrin ring (Burmester and Hankeln 2009).

The increase of HIF along the life axis suggests a possible involvement of HIF as an oxygen sensor for the aging and development processes (Porzionato et al. 2008). The PO_2 fluctuations would be a key regulator of gene expression during aging and development. Moreover hypoxia and aging (Gelfi et al. 2006, 1990) are considered as "stresses" for the CB, and the increase in HIF and NGB would be connected with free radical production during aging and hypoxia (Jamieson 1989; Finkel and Holbrook 2000) interfering with the oxygen sensitive mechanism. The stimulatory effect of HIF-NGB help to postulate the hypothesis that there is a sharing common mechanism pathway for enzymatic activation during hypoxia and aging. Development and aging interplaying with of all these factors induce adaptive processes in structure and function. Decreased cellular PO_2 induces angiogenesis through NOS release HIF, VEGF accumulation and with HIF playing a physiological role in aged tissues. During aging the increase of NGB expression and HIF-1α positive nuclei along with the increase of intralobular connective tissue and decrease of parenchyma leads to the following hypothesis.

Prevention of oxygen desaturation, reducing all causes of hypoxemia from neonatal life to aging would decrease the incidence of diseases in the elderly population with a lifespan extension. Our conclusion is that CB is a model to study the aging process correlating the high blood flow and metabolism, and is a tool to modulate the aging and developmental processes interfering with cells metabolism.

Acknowledgment This paper is dedicated to the memory of Prof. Sukhamay Lahiri, for his example in science and in life.

References

Bredt DS, Hwang PM, Snyder SH (1990) Localization of nitric oxide synthase indicating a neuronal role for nitric oxide. Nature 347:768

Burmester T, Hankeln T (2009) What is the function of neuroglobin. Exp Biol 212:1423–1428

Di Giulio C, Cacchio M, Bianchi G, Rapino C, Di Ilio C (2003) Carotid body as a model for aging studies: is there a link between oxygen and aging ? J Appl Physiol 95:1755–1758

Finkel T, Holbrook NJ (2000) Oxidants, oxidative stress and the biology of ageing. Nature 408:239–247

Gelfi C, Vigano A, Ripamonti M, Pontoglio A, Begum S, Pellegrino MA, Grassi B, Bottinelli R, Wait R, Cerretelli P (2006) The human muscle proteome in aging. J Proteome Res 5:1344–1353

Iyer NV, Kotch LE, Agani F, Leung SW, Laughner E, Wenger RH, Gassmann M, Gearhart JD, Lawer AM, Yu AY, Semenza GL (1998) Cellular and developmental control of O_2 homeostasis by hypoxia-inducible factor 1-$\propto$. Genes Dev 12:149–162

Jamieson D (1989) Oxygen toxicity and reactive oxygen metabolites in mammals. Free Radic Biol Med 7:87–108

Leiser SF, Kaeberlein M (2010) The hypoxia-inducible factor HIF-1 functions as both a positive and negative modulator of aging. Biol Chem 391:1131–1137

Martinelli M, Winterhalder R, Cerretelli P, Howald H, Hoppeler H (1990) Muscle lipofuscin content and satellite cell volume is increased after high altitude exposure in humans. Experientia 46:672–676

Porzionato A, Macchi V, Guidolin D, Parenti A, Ferrara SD, De Caro R (2005) Histopathology of carotid body in heroin addiction possible chemosensitive impairment. Histopathology 46:296–306

Porzionato A, Macchi V, Parenti A, Matturri L, De Caro R (2008) Peripheral chemoreceptors: postnatal development and cytochemical findings in sudden infant death syndrome. Histol Histopathol 23:351–365

Prabhakar NR, Fields RD, Baker T, Fletcher EC (2001) Intermittent hypoxia: cell to system. Am J Physiol Lung Cell Mol Physiol 281:524–528

Rivner MH, Swift TR, Malik K (2001) Influence of age and height on nerve conduction. Muscle Nerve 24:1134–1141

Semenza GL (1999) Regulation of mammalian O_2 homeostasis by hypoxia-inducible factor 1. Annu Rev Cell Dev Biol 15:551–578

Sohal RS, Toy PL, Allen RG (1986) Relationship between life expectancy, endogenous antioxidants and products of oxygen free radical reactions in the housefly, musca domestica. Mech Aging Dev 36:71–77

Verratti V, Di Giulio C, Bianchi G, Cacchio M, Petruccelli G, Artese L, Lahiri S, Iturriaga R (2009) Neuroglobin in aging carotid bodies. Adv Exper Med Biol 648:191–195

Wickens AP (2001) Ageing and the free radical theory. Respir Physiol 128:379–391

Chapter 37
Propranolol Does Not Affect the Hindlimb Vasodilatation Elicited by Stimulation of Superior Laryngeal Nerve Paraganglia

Edward T. O'Connor, Ken D. O'Halloran, and James F.X. Jones

Keywords Superior laryngeal nerve • Paraganglia • Carotid body • Vagus nerve • Chemoreceptor • Propranolol

37.1 Introduction

The carotid body (CB) consists of a small, highly vascularised, cluster of cells known as chemoreceptors located at the bifurcation of the common carotid artery. These chemoreceptors "taste" chemical stimuli in the blood and convey impulses to the cardiorespiratory centres of the medulla which then cause appropriate alterations in heart rate, blood pressure and respiration. This information is relayed to the medulla via the glossopharyngeal (IXth cranial) nerve. We are interested in another group of chemoreceptors, referred to as vagal paraganglia, and in particular those found at the bifurcation of the superior laryngeal nerve (SLN), which is the largest paraganglion of the vagus (Xth cranial) nerve in the rat (Hughes et al. 2003).

Reflex studies concerned with sensors of low blood oxygen (hypoxia) have been dominated by experiments performed on the CB of dog and cat. Stimulation of the CB results in increased ventilation. However, humans and rats exhibit numerous "paraganglia" scattered along the distribution of the trunk and branches of the vagus nerve. The function of these structures is obscure. Previous work by O'Leary et al. (2004) showed that these paraganglia in the rat are chemosensitive to both hypoxia and cyanide (inhibitors of mitochondrial respiration), with the SLN greatly increasing its discharge rate in response to both of these stimuli, demonstrating that the chemosensitive cells of these paraganglia are not only structurally similar to those of the CB (McDonald and Blewett 1981) but are also excited by similar stimuli.

E.T. O'Connor (✉) • K.D. O'Halloran • J.F.X. Jones
School of Medicine and Medical Science, University College Dublin, Belfield, Dublin 4, Ireland
e-mail: edward.oconnor@ucd.ie; ken.ohalloran@ucd.ie; james.jones@ucd.ie

C.A. Nurse et al. (eds.), *Arterial Chemoreception: From Molecules to Systems*, Advances in Experimental Medicine and Biology 758, DOI 10.1007/978-94-007-4584-1_37,

Studies involving the aortic bodies (AB), a group of vagal paraganglia innervated by the aortic nerve of the vagus, have shown that stimulation of these chemoreceptors results in little or no change in respiration, but does cause cardiovascular effects. Daly and Ungar (1966) showed in the dog that the respiratory effects of the CB were much greater than that of the AB, but the reflex hindlimb vasoconstrictions were equal. Similarly, Daly and Jones (1998) stimulated the AB in the cat, which proved to have little respiratory effect but caused a potent constriction of the femoral vasculature. These data suggested that stimulation of the vagal paraganglia of the SLN may have similar effects to those of the AB, and we hypothesised that the reflex activation of these chemoreceptors causes little respiratory change but influences the cardiovascular system of the rat. Recent work in our lab (O'Connor et al. 2010) has provided evidence to support this hypothesis; when the SLN paraganglia were selectively stimulated by sodium cyanide (NaCN), in the urethane anaesthetised rat, we observed a significant drop in mean arterial pressure (MAP) and an increase in hindlimb conductance (HLC), indicating a vasodilatation, whilst respiratory activity was virtually unaffected. The cardiovascular changes caused by SLN stimulation with NaCN resemble those of a defence like response in the rat, which may be as a result of adrenaline release; therefore, we examined the effect of the β-adrenoceptor antagonist propranolol on this response.

37.2 Methods

All experiments were conducted in accordance with protocols approved by the University animal research ethics committee and licensed by the Irish Department of Health and Children. Wistar rats of both sexes ($n = 11$, body mass 265 ± 6 g) were anaesthetised with 20% urethane (1.5 $g.kg^{-1}$ i.p.). The level of surgical anaesthesia was assessed by the absence of a withdrawal reflex and cardiovascular response to paw pinch. Core body temperature was measured with a rectal thermometer and prevented from dropping below 37 °C using a homeothermic blanket system (Harvard). The left femoral vein was cannulated for administering supplemental anaesthesia, fluids and drugs, and the right carotid artery cannulated for recording arterial blood pressure (BP). A flow probe (0.5 mm, Product no: MA0.5PSB, Transonic Systems Inc.) was placed on the left femoral artery and covered with surgical lubrication gel (Fougena Surgilube) to minimize acoustic echoes. The accuracy of the flow values were tested *ex vivo* using either the carotid artery or special calibration tubing and timed collections of water into a graduated cylinder. A cervical tracheotomy was performed and the trachea cannulated just below the larynx. Both SLNs were dissected free and cut close to the larynx, preserving the region of bifurcation, an area that reliably contains glomus tissue, with the free ends placed in 20 μl tissue baths containing HEPES buffered Tyrode's solution (Sigma, product no: T2145; containing in mM: NaCl 137.0; KCl 2.7; $MgCl_2$ 1.0; $CaCl_2$ 1.36; Na_2HPO_4 0.35; (D)-Glucose 5.5; HEPES 10.0; pH 7.4.). The baths were then surrounded by Vaseline to prevent NaCN leakage and desiccation of the underlying tissue.

The SLN paraganglia were selectively stimulated by adding 10 μl NaCN (0.1 $mg.ml^{-1}$) directly to the baths for a period of 100 s, after which the NaCN solution was replaced by normal Tyrode's. Two of these NaCN challenges were performed, followed by administration of the β-adrenoceptor antagonist propranolol (1 $mg.kg^{-1}$ i.v.), after which two more tests were carried out. To further investigate the role of adrenaline in this response, a bilateral adrenalectomy was performed in one animal with NaCN tests carried out before and after the procedure.

All data were digitised using an AD converter (1401 micro, Cambridge Electronic Design), and captured and analysed using Spike2 v6.02 (Cambridge Electronic Design). For all cardiovascular parameters, peak change from baseline was measured. All statistical analyses were carried out using Prism 4 v4.03 (GraphPad Software, Inc.). To measure the effect of treatments on baseline values Student's two-tailed paired *t-test* was applied. To measure the effects of propranolol and time (using time matched controls, $n = 5$, body mass 260 ± 10 g), two way repeated measures ANOVA with Bonferroni multiple comparison test was applied. The criterion for statistical significance was $p < 0.05$. All results are expressed as mean ± SEM.

37.3 Results

Propranolol caused significant decreases in heart rate (−20.4 ± 1.2%) and HLC (−30.4 ± 2.3%), and a significant, paradoxical, increase in MAP (35.5 ± 4.4%) (Student's paired *t-test*, all $p < 0.0001$).

As previously reported (O'Connor et al. 2010), SLN stimulation with NaCN resulted in a hindlimb vasodilatation: there was a significant decrease in MAP (−15.7 ± 2.9%), and a significant increase in hindlimb conductance (6.6 ± 1.2%), both of which were decreased after propranolol administration (MAP response down to −11.1 ± 2.7%, HLC response down to 3.1 ± 1%). However, when compared with time matched controls, the MAP and HLC responses were unaffected by propranolol administration. There was a significant time dependent run down in the responses ($p = 0.0007$ and $p = 0.005$ respectively, 2 way repeated measures ANOVA. Results from the bilateral adrenalectomy experiment were similar to those obtained in propranolol treated animals (Fig. 37.1 and Table 37.1).

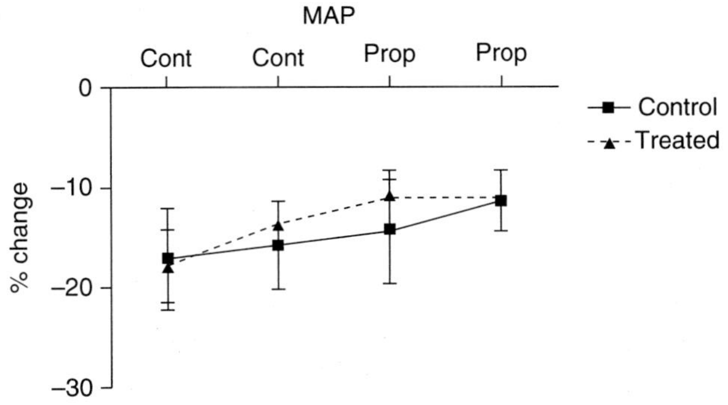

Fig. 37.1 Effect of SLN stimulation with NaCN on MAP in propranolol treated (n = 11) and control animals (n = 5). Peak changes from baseline were measured (pre to post) for the four NaCN challenges (two before and two after propranolol/saline administration)

Table 37.1 Cardiovascular effects of SLN stimulation with NaCN in propranolol treated and control animals

	Control				Propranolol			
Treated (n = 11)	Pre	Post	Pre	Post	Pre	Post	Pre	Post
MAP (mmHg)	91 ± 4	72 ± 6	86 ± 5	73 ± 5	117 ± 6	104 ± 9	115 ± 7	101 ± 10
HLC (μl/min/mmHg)	11.72 ± 1.1	12.54 ± 1.0	12.18 ± 1.2	12.96 ± 1.2	8.48 ± 0.5	8.61 ± 0.5	8.1 ± 0.6	8.42 ± 0.5
	Controls				Saline			
Control (n = 5)	Pre	Post	Pre	Post	Pre	Post	Pre	Post
MAP (mmHg)	99 ± 13	82 ± 9	95 ± 8	80 ± 7	98 ± 12	87 ± 11	98 ± 16	91 ± 12
HLC (μl/min/mmHg)	9.34 ± 1.2	10.27 ± 1.2	9.71 ± 1.0	10.46 ± 1.0	9.63 ± 1.0	10.52 ± 1.0	9.92 ± 1.0	10.52 ± 1.0

Peak changes from baseline were measured (pre to post) for the four NaCN challenges (two before and two after propranolol/saline administration)

37.4 Discussion

In the present study we confirmed the findings of our earlier work (O'Connor et al. 2010), which demonstrated that selective stimulation of the SLN with NaCN results in hindlimb vasodilatation. We aimed to investigate the role of adrenaline in this response, as this reflex somewhat resembles a defence like response, and have now shown that it does not appear to mediate the reflex hindlimb vasodilatation observed.

One might assume that a defence like response like that seen here, or in the Saffan anaesthetised rat (Marshall 1987), would be at least partly mediated by adrenaline, one of the primary fight-or-flight neurotransmitters. However, Michael de Burgh Daly's work in the dog (1968) showed that the reflex systemic vasodilator response elicited by lung inflation in the dog is unaffected by propranolol. This reflex, which also has a mainly vagal pathway to the medullary cardiorespiratory centre, was abolished by hexamethonium, dibenyline and bretylium tosylate, indicating that it was due to removal of sympathetic constrictor tone.

It is not clear what other mechanisms are involved in this reflex vasodilation; another possible humoral mediator is nitric oxide (NO), which has been shown to play an important role in neurotransmission at various sites in the body, and is a mediator of smooth muscle relaxation. An experiment investigating a role for NO should focus on specific synthase isoforms, with local application of antagonist if possible, as a systemic reduction of NO could prevent transmission of signals at every level, from the blood vessels to brain stem or the glomus cells. This reflex might also be neurally mediated through a reduction in the activity of sympathetic adrenergic vasoconstrictor fibres. Sympathetic denervation of the hindlimb vasculature could provide insight into the mechanisms controlling this response.

When studying responses of this nature, it is always important to consider the anaesthetic being used. Urethane was chosen for various reasons: it has been shown to have minimal effects on circulation and respiration whilst preserving reflex responses (Maggi and Meli 1986), making it ideal for investigating chemoreflexes. It does, however, have some undesirable effects. Propranolol, a β-adrenoceptor antagonist, is used clinically as a therapy for hypertension. However, as seen in the data above, i.v. propranolol (1 $mg.kg^{-1}$) causes a paradoxical rise in MAP in the urethane anaesthetised rat. First noted by Dasgupta (1968), this is due to elevated levels of circulating catecholamines causing strong β-vasodilator tone, which is then removed by the β-blockade. This phenomenon has been shown to vary across species (Himori and Ishimori 1988), so observations made on the role of adrenaline in urethane anaesthetised animals must be treated with caution.

Much work remains to be done to elucidate not only the mechanism of this reflex, but also its purpose. In the present study, we have shown that despite a decrease in the responses seen after β-adrenoceptor blockade, they are not significantly different to time matched controls, which all exhibit a time dependent run down in magnitude. Thus, we have excluded just one of many possible mechanisms of the depressor reflex associated with vagal paraganglia.

Acknowledgements We wish to acknowledge support of The Wellcome Trust (UK) and the School of Medicine and Medical Science UCD, Dublin.

References

Daly MD, Jones JF (1998) Respiratory modulation of carotid and aortic body reflex left ventricular inotropic responses in the cat. J Physiol 509(Pt 3):895–907

Daly MD, Robinson BH (1968) An analysis of the reflex systemic vasodilator response elicited by lung inflation in the dog. J Physiol 195(2):387–406

Daly M, Ungar A (1966) Comparison of the reflex responses elicited by stimulation of the separately perfused carotid and aortic body chemoreceptors in the dog. J Physiol 182(2):379–403
Dasgupta NK (1968) On the mechanism of the pressor response due to propranolol. Br J Pharmacol 34(1):200P–201P
Himori N, Ishimori T (1988) Different responses to beta-adrenoceptor blocking drugs of the blood pressure and heart rate in the urethane-anesthetized dog and rat. Jpn J Pharmacol 47(1):71–80
Hughes K, Pickering M, O'Leary DM, Bradford A, O'Regan RG, Jones JF (2003) The paraganglia of the rat superior laryngeal nerve. Adv Exp Med Biol 536:239–246
Maggi CA, Meli A (1986) Suitability of urethane anesthesia for physiopharmacological investigations in various systems. Part 2: cardiovascular system. Experientia 42(3):292–297
Marshall JM (1987) Analysis of cardiovascular responses evoked following changes in peripheral chemoreceptor activity in the rat. J Physiol 394:393–414
McDonald DM, Blewett RW (1981) Location and size of carotid body-like organs (paraganglia) revealed in rats by the permeability of blood vessels to Evans blue dye. J Neurocytol 10(4):607–643
O'Connor ET, O'Halloran KD, Jones JF (2010) The cardiorespiratory reflex effects of stimulation of superior laryngeal nerve paraganglia in the urethane anaesthetised rat. J Physiol Proc Physiol Soc 20:C19 and PC19
O'Leary DM, Murphy A, Pickering M, Jones JF (2004) Arterial chemoreceptors in the superior laryngeal nerve of the rat. Respir Physiol Neurobiol 141(2):137–144

Chapter 38
ATP Release from the Carotid Bodies of DBA/2J and A/J Inbred Mouse Strains

Pejmon Pashai, Eric W. Kostuk, Luis E. Pilchard, and Machiko Shirahata

Abstract The purposes of this study were to: (1) establish an effective method to measure the release of ATP from the mouse carotid body (CB) and (2) determine the release of ATP from the CB of the DBA/2 J (high hypoxic responder) and A/J (low hypoxic responder) mouse in response to hypoxia and hypercapnia. An incubation chamber was constructed utilizing a Costar® Spin-X Centrifuge Tube Filter. The filter was coated with low melting point agarose to hold 4 CBs or 4 superior cervical ganglia (SCG). Hypoxia did not increase ATP release from the CB of either strain. ATP increased in response to a normoxic/hypercapnic challenge in the DBA/2 J's CB but not in the A/J's CB. ATP release from the SCG was affected by neither hypoxia nor hypercapnia in both strains. Thus, we have concluded: (1) we successfully established a chamber system to measure ATP released from the mouse CB; (2) ATP may not be an excitatory neurotransmitter in the CB of these mice under hypoxia; (3) ATP may be a neurotransmitter in the CB of the DBA/2 J mouse strain during hypercapnia.

Keywords Carotid body • Hypoxia • Hyperoxia • Neurotransmitter

38.1 Introduction

The carotid body (CB) is a sensory organ that responds to hypoxia and hypercapnia. Previous studies performed in the cat and rat demonstrated that the release of ATP increased during hypoxia and that exposure to hyperoxia suppressed the release (Buttigieg and Nurse 2004; Fitzgerald et al. 2009). CB neural output in a co-culture model of glomus cells and petrosal ganglion neurons (Zhang et al. 2000) or *in vitro* preparation (He et al. 2006) were significantly inhibited by ATP channel blockers, Suramin and pyridoxalphosphate-6-azophenyl-2′,4′-disulfonic acid (PPADS). These results suggest that ATP is an excitatory neurotransmitter in the CB. However, this has not been shown in other species. Previous studies have shown that the DBA/2 J strain has markedly higher responses to hypoxia than the A/J strain (Campen et al. 2004, 2005; Rubin et al. 2003; Tankersley et al. 1994). The mechanisms involved with the differences in hypoxic responses between the strains have not been well understood.

P. Pashai • E.W. Kostuk • L.E. Pilchard • M. Shirahata (✉)
Division of Physiology, Department of Environmental Health Sciences, Bloomberg School of Public Health, Johns Hopkins University, 615 N. Wolfe Street, Baltimore, MD, 21205, USA
e-mail: ppashai@gmail.com; ekostuk@jhsph.edu; mshiraha@jhsph.edu

C.A. Nurse et al. (eds.), *Arterial Chemoreception: From Molecules to Systems*,
Advances in Experimental Medicine and Biology 758, DOI 10.1007/978-94-007-4584-1_38,

If ATP is an excitatory neurotransmitter in the CB of the mouse, we expect the mouse CB to release ATP and its release during hypoxia to increase in the DBA/2 J strain, but not in the A/J strain. In preliminary experiments, we have found that the methods previously employed to measure the ATP release from the CB of the cat and the rat cannot be directly applied to the mouse CB due to its small size. Thus, this study aimed to: (1) establish an effective chamber system to stimulate the mouse CB; (2) to determine the release of ATP from the CB of the DBA/2 J and A/J mice in response to hypoxia. Further, we examined if hypercapnia influenced the release of ATP from the CB of these mice. Hypercapnic cardiopulmonary responses in the DBA/2 J and A/J strain appear to differ (Campen et al. 2004, 2005; Tankersley et al. 1994). However, the contribution of the CB was not clear in these studies.

38.2 Methods

38.2.1 Chamber System

The major considerations in constructing the incubation system were: (1) the method to hold the CB in the chamber; (2) the PO_2 level in the incubation solution; (3) temperature control; (4) degradation of ATP; (5) bacterial growth during the experiments. After testing several different chambers, we established the following system (Fig. 38.1) that provided a stable measurement of ATP from the incubation medium. The CB was embedded in an agarose gel at the bottom of the chamber. Modified Krebs solution covered the agarose and was bubbled with different mixed gases. The Krebs solution used for the incubation was (in mM): 120 NaCl, 4.7 KCl, 1.8 $CaCl_2$, 1.2 Mg $SO_4 \cdot 7H_2O$, 1.2 Na_2HPO_4, 22 $NaHCO_3$, and 11.1 Glucose. An inhibitor of ecto-5′-nucleotidase, α, β-methyleneadenosine 5′-diphosphate (AOPCP), was added to experimental Krebs solutions to a final concentration of 200 μM. The addition of AOPCP within our incubation solution played an essential role in recovering ATP (Santos et al. 1999). Our preliminary data showed that without AOPCP, the measured ATP was about 1/3 compared to the sample with AOPCP. To avoid bacterial contamination, all solutions were sterilized using a membrane filter (0.2 μm). The entire apparatus was washed with 70% ethanol and dried prior to use.

38.2.2 Tissue Preparation

Male DBA/2J (n=54) and A/J (n=34) inbred mice, ages 4–6 weeks were used. They were purchased from Jackson Labs (Bar Harbor) or bred in the animal facilities at Johns Hopkins School of Public Health (Baltimore). Animals were kept at ~22°C with a regulated light cycle (12:12-h light–dark cycle). Food (Harlan TEKLAD) and water were provided ad libitum. All experiments were conducted in accordance with the NIH guidelines (Guide for the Care and Use of Laboratory Animals, National Academy Press Washington, D.C., 1996). Protocols were approved by Animal Care and Use Committee at Johns Hopkins University. Mice were anesthetized with 1.5 g/kg urethane (i.p.). Carotid bifurcations were excised and harvested as described before (Yamaguchi et al. 2004). Following the excision of the CB or SCG, four tissues (from two mice) were immersed within 80 μL of warmed agarose that had been pippetted onto the membrane within the incubation chamber. The agarose was allowed to solidify within the chamber by placing it in an ice bucket for 5 min.

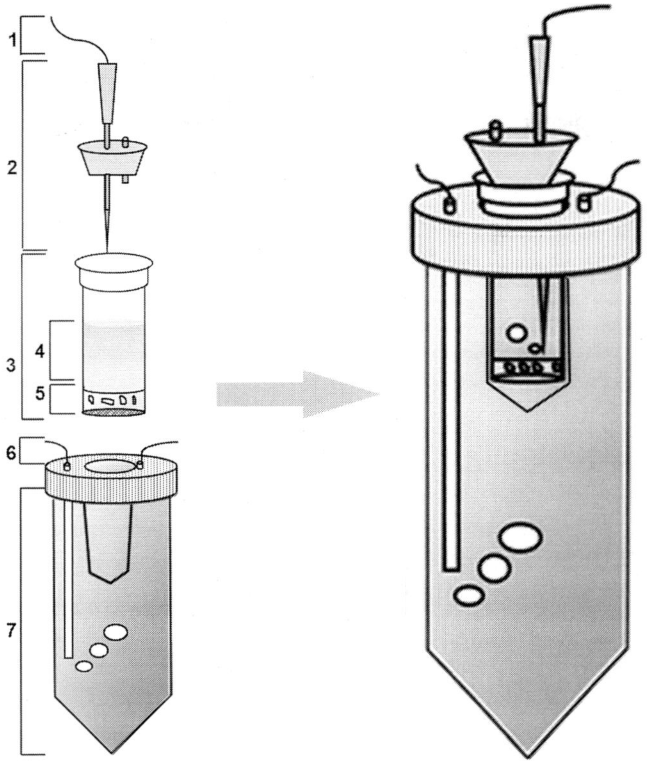

Fig. 38.1 Chamber system. Numbers correspond to the following: (1) gas line, (2) stopper/bubbling apparatus, (3) incubation chamber, (4) incubation solution, (5) agarose+4 tissues, (6) gas line, (7) fluid jacket containing H_2O+22 mM $NaHCO_3$. The incubation chamber was comprised of a sterile Costar® Spin-X centrifuge tube filter (Costar 8160) with a bubbling apparatus that ensured a quick (~25 s) and stable change in PO_2 within the chamber. The filter was covered with 80 µL of 2% low melting point agarose that contained four tissues. The incubation chamber was fitted in a fluid jacket made from a 50 mL conical tube. The fluid jacket was filled with filtered water containing 22 mM $NaHCO_3$ and continually bubbled with the same gas as the incubation chamber. The fluid jacket was immersed in a water bath that was set at 40°C to provide a stable temperature (~37°C) in the incubation solution

38.2.3 Protocol

Following the surgery and embedding, tissues were allowed to recover for 20 min in a hyperoxic Krebs solution. This was followed by two 7 min washes in Krebs containing AOPCP. Our protocol consisted of a series of 7 min exposures: a hyperoxic control (70% O_2/5% CO_2/25% N_2), followed by either a hypoxic (bubbled with 5% O_2/5% CO_2/85% N_2) or hypercapnic challenge (bubbled with 21% O_2/10% CO_2/69% N_2), and three successive hyperoxic recoveries. Before each exposure, 120uL Krebs solution containing AOPCP was added; the solution was collected after the exposure. Samples from the Krebs reservoir were collected before and after the experiment to track possible bacterial contamination. All samples were stored (−20 °C) until measured.

38.2.4 Measurement

Standards (200, 100, 50, 25, 10 and 0 pM ATP in Krebs) were measured to generate a standard curve. The lowest measurable amount of ATP was 10 pM. 50µL of solution (standard or experimental) was vortexed briefly with an equal volume of ENLITEN ATP Luciferase reagent (Promega Corp, FF2000) in a glass cuvette. The sample was measured in a luminometer (Berthold LB Lumat 9507). Measured RLU values are converted to pM concentration utilizing the standard curve.

38.2.5 Statistics

Statistical analysis was performed using *t* test (between the control and hypoxic or hypercapnic challenges) with the aid of Prism 4.0 (GraphPad Software). Differences were considered significant if $P<0.05$.

38.3 Results

38.3.1 Hypoxic Challenges

A measurable amount of ATP was recovered from the incubation solution in most experiments. No significant increases in ATP were observed in the CB and SCG of both strains (Fig. 38.2). Negligible traces of ATP were measured from the reservoir. Experiments conducted without CBs also showed negligible amount of ATP.

38.3.2 Hypercapnic Challenges

Hypercapnia significantly increased the ATP release from the DBA/2 J's CB compared to the control (Fig. 38.3**a**). No significant increases in ATP were observed in the CB and SCG of the A/J mice and in the SCG of the DBA/2 J mice.

38.4 Discussion

We have demonstrated, using the CB of the DBA/2 J and A/J strains of mice, that: (1) with our chamber system the release of ATP can be effectively measured; (2) hypoxia and hypercapnia differently influence the release of ATP; (3) strain variations may exist in the release of ATP in response to different stimuli. The acquired data further suggest that ATP may not be an excitatory neurotransmitter in the CB of the two strains in response to hypoxia. However, ATP may act as a neurotransmitter in the CB of the DBA/2 J strain during hypercapnia.

Previously, Buttigieg and Nurse (2004) employed three preparation methods (culture, slice, whole tissue) to measure ATP released from the CB of rats. However, their methods are impractical for our study. Dissociation of the mouse CB is possible (Yamaguchi et al. 2003), but the loss of viable cells during the dissociation and culture procedures makes this approach ineffective for measuring ATP with bioluminescence. Slicing the mouse CB is not technically feasible. We initially used an anchor to hold a whole bifurcation as shown by Yamaguchi et al. (2004). This approach held the tissue well, but we found that the release of ATP from the arterial tissues was much higher compared to that from the CB (data not shown). Suspending the entire CB directly within a Krebs incubation solution as

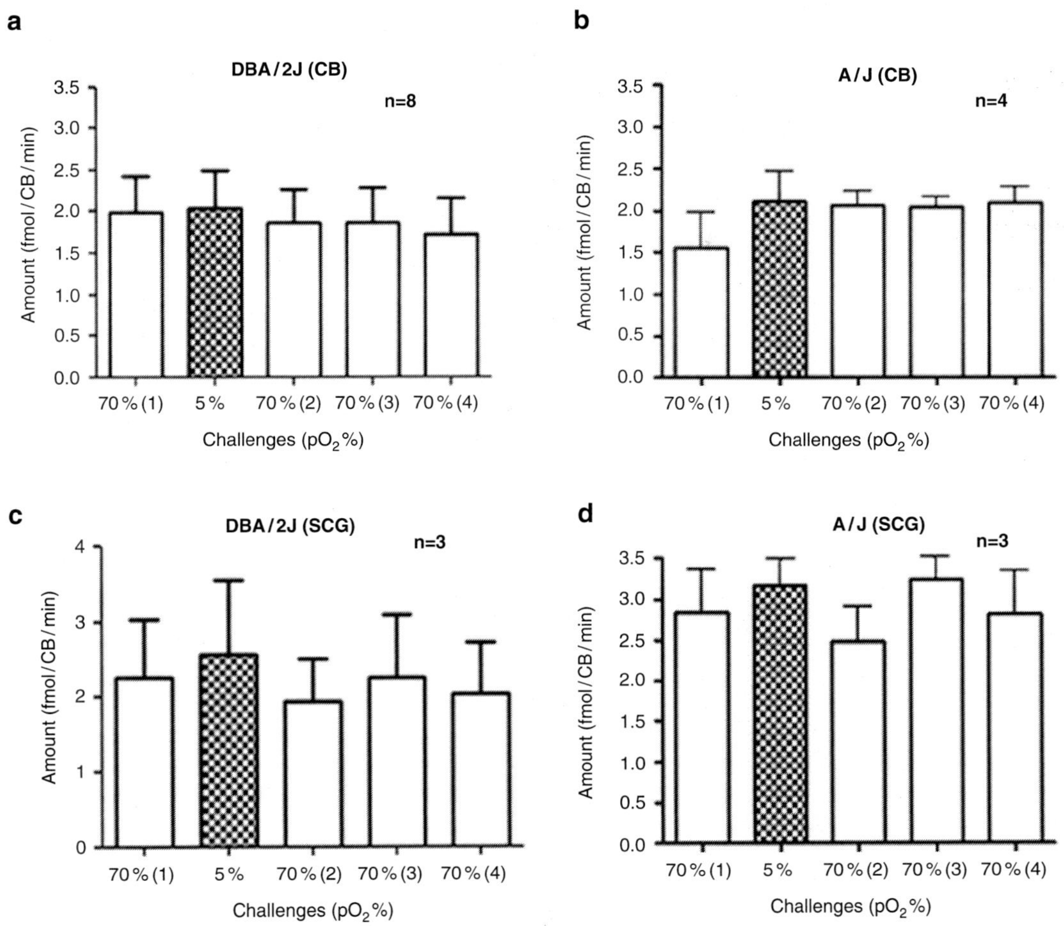

Fig. 38.2 Amount of ATP released from the CB and SCG. The response to hypoxia (5% O_2) was tested. The DBA/2 J's CB did not show a significant increase in the release of ATP during hypoxia (**a**). SCGs from the same mice showed no increase in the release of ATP during hypoxia (**c**). Similarly, in the A/J's CB (**b**) and SCG (**d**), the release of ATP during hypoxia was not significantly different when compared to baseline values. Additionally, no significant differences were seen between the strains in baseline conditions or in hypoxic challenges

used by Fitzgerald et al. (2009) is not feasible for the mouse CB. Because the CB of the mouse is barely visible, it is likely to be lost during the experiment. We found that embedding the whole CB within agarose is an effective method for holding the tissues during various exposures. A previous report suggested that the O_2 diffusivity of 2% agar gel is approximately 70% of that of water (0.107 $cm^2 \cdot h^{-1}$ at 30 °C) (Omar 1993). Since we used the challenge duration of 7 min, we do not think a reduced diffusion of O_2 in the agar gel compromised the ATP recovery.

To measure ATP with bioluminescence, a special care must be taken to prevent bacterial growth in the incubation medium. ATP is a byproduct of bacterial metabolism that can impede upon accurate assessment of ATP release from the CB. This is the rationale behind the asceptic techniques utilized in our chamber setup. During preliminary experiments, we sometimes experienced unreasonably high ATP levels in the reservoir of Krebs. After establishing sterilization techniques, we found negligible ATP levels in the reservoir. In addition, mock experiments without CBs also showed negligible amount of ATP.

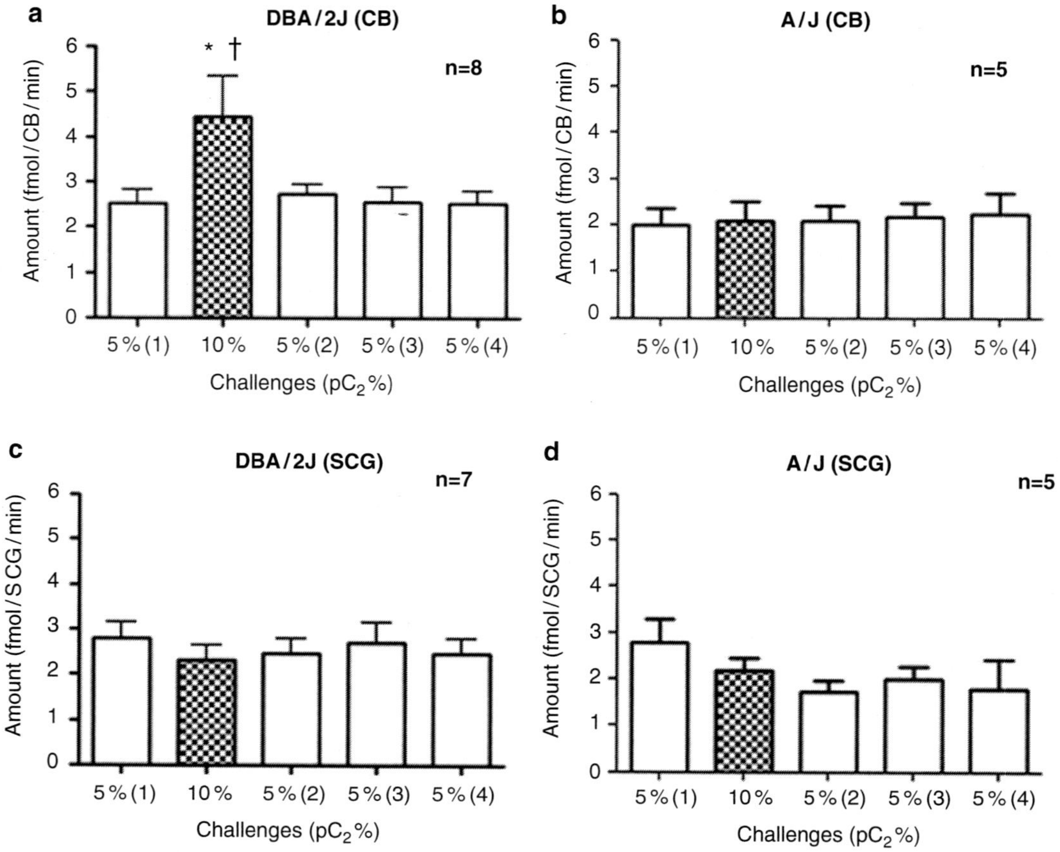

Fig. 38.3 Amount of ATP released from the CB and SCG. The response to hypercapnia (10%) was tested. Hypercapnia significantly increased the ATP release from the DBA/2 J's CB compared to the control (**a**,*) while no increase was seen from the A/J's CB (**b**). Additionally, the SCGs of both the DBA/2 J (**c**) and the A/J (**d**) mice showed no significant increase in the ATP release during hypercapnia compared to the baseline. The amount of ATP released from the DBA/2 J's CB during hypercapnia was significantly higher compared to the A/J strain (**a**, †)

We found no significant increase in the release of ATP from the CB in both the DBA/2 J and the A/J strains upon acute hypoxia. We do not believe this unresponsiveness was due to technical problems. As mentioned in the Fig. 38.1 legend, the change in PO_2 was quick and stable. Further, ATP from the CB of DBA/2 J mice significantly increased during hypercapnia (Fig. 38.3a). Because of the small size of the mouse CB, a possible contamination with other tissues, particularly SCG, was a concern. Thus we added experiments using SCG. Neither hypoxia nor hypercapnia changed the ATP release from the SCG. Therefore, the increased release of ATP during hypercapnia appears specific to the CB. Several studies (Campen et al. 2004, 2005; Rubin et al. 2003; Tankersley et al. 1994; Pichard et al. 2011) indicate that the DBA/2 J mice respond vigorously to hypoxia, but the A/J mice have muted response to hypoxia. Pichard et al. (2011) reported, that neural output from the DBA/2 J's CB was equivalent to other animals, but the neural output from the CB in the A/J mice did not increase until hypoxia became extremely severe. Thus, if ATP is an excitatory neurotransmitter, we can expect an increased release from the CB of DBA/2 J mice. Taken together, the results suggest that ATP may not be an excitatory neurotransmitter in the CB in these strains, although a possibility remains that our experimental procedures selectively impaired hypoxic sensing mechanisms of the CB. In addition, ATP may have a role as a neurotransmitter in the CB of DBA/2 J mice in response to hypercapnia.

Using similar bioluminescence methods, hypoxia increased ATP release from CB of the rat (Buttigieg and Nurse 2004). Whole tissue preparations of cats showed a significant increase in the release ATP under hypoxia from control levels. However, in the cat, the increase was not as robust as that in the rat. Our results showed that hypoxia did not increase ATP from the CB in the two strains, and that hypercapnia increased ATP only in the DBA/2 J strain. On the other hand, carotid nerve chemosensory excitation to hypoxia was greatly reduced by ATP receptor antagonists in rats (Zhang et al. 2000; He et al. 2006), cats (Reyes et al. 2007), and mice (C57BL/6 J-based; Rong et al. 2003). Moreover, mice deficient in P2X2 receptor subunit had significantly muted hypoxic response (Rong et al. 2003). Thus, it is possible that a role of ATP as a neurotransmitter in the CB differs among species and strains.

Funded by AHA 09GRNT2080158 and NHLBI HL81345.

References

Buttigieg J, Nurse CA (2004) Detection of hypoxia-evoked ATP release from chemoreceptor cells of the rat carotid body. Biochem Biophys Res Commun 322:82–87

Campen MJ, Tagaito Y, Li JG, Balbir A, Tankersley CG, Smith P, Schwartz A, O'Donnell CP (2004) Phenotypic variation in cardiovascular responses to acute hypoxic and hypercapnic exposure in mice. Physiol Genomics 20:15–20

Campen MJ, Tagaito Y, Jenkins TP, Balbir A, O'Donnell CP (2005) Heart rate variability responses to hypoxic and hypercapnic exposures in different mouse strains. J Appl Physiol 99:807–813

Fitzgerald RS, Shirahata M, Chang I, Kostuk E (2009) The impact of hypoxia and low glucose on the release of acetylcholine and ATP from the incubated cat carotid body. Brain Res 1270:39–44

He L, Chen J, Dinger B, Stensaas L, Fidone S (2006) Effect of chronic hypoxia on purinergic synaptic transmission in rat carotid body. J Appl Physiol 100:157–162

Omar SH (1993) Oxygen diffusion through gels employed for immobilization 1 in the absence of microorganisms. Appl Microbiol Biotechnol 40:1–6

Pichard L, Sgmabti F, Kostuk E, Pashai P, Fitzgerald RS, Shirahata M (2011) Murine carotid body responses to hypoxia: in vivo carotid sinus nerve recordings in the DBA/2J and A/J strains. FASEB 24:1026.19

Reyes EP, Fernández R, Larraín C, Zapata P (2007) Effects of combined cholinergic-purinergic block upon cat carotid body chemoreceptors in vitro. Respir Physiol Neurobiol 156:17–22

Rong WF, Gourine AV, Cockayne DA, Xiang ZH, Ford APDW, Spyer KM, Burnstock G (2003) Pivotal role of nucleotide P2X(2) receptor subunit of the ATP-gated ion channel mediating ventilatory responses to hypoxia. J Neurosci 23:11315–11321

Rubin AE, Polotsky VY, Balbir A, Krishnan JA, Schwartz AR, Smith PL, Fitzgerald RS, Tankersley CG, Shirahata M, O'Donnell CP (2003) Differences in sleep-induced hypoxia between a/J and DBA/2 J mouse strains. Am J Respir Crit Care Med 168:1520–1527

Santos PF, Caramelo OL, Carvalho AP, Duarte CB (1999) Characterization of ATP release from cultures enriched in cholinergic amacrine-like neurons. J Neurobiol 41:340–348

Tankersley CG, Fitzgerald RS, Kleeberger SR (1994) Differential control of ventilation among inbred strains of mice. Am J Physiol Regul Integr Comp Physiol 267:R1371–R1377

Yamaguchi S, Balbir A, Schofield B, Coram J, Tankersley CG, Fitzgerald RS, O'Donnell CP, Shirahata M (2003) Structural and functional differences of the carotid body between DBA/2 J and a/J strains of mice. J Appl Physiol 94:1536–1542

Yamaguchi S, Lande B, Kitajima T, Hori Y, Shirahata M (2004) Path clamp study of mouse glomus cells using whole carotid body. Neurosci Lett 357:155–157

Zhang M, Zhong H, Vollmer C, Nurse CA (2000) Co-release of ATP and ACh mediates hypoxic signalling at rat carotid body chemoreceptors. J Physiol 525:143–158

Chapter 39
Effect of Oxygen on Phosphodiesterases (PDE) 3 and 4 Isoforms and PKA Activity in the Superior Cervical Ganglia

Ana Rita Nunes, Vedangi Sample, Yang K. Xiang, Emília C. Monteiro, Estelle Gauda, and Jin Zhang

Abstract The cAMP-protein kinase A (PKA) signaling pathway is involved in regulating the release of transmitters from neurons and other cells. Multiple phosphodiesterase (PDE) isoforms regulate this pathway, however, the pattern of isoform expression and stimulus response across tissues has not been fully characterized.

Using fluorescent resonance energy transfer (FRET)-based imaging in primary superior cervical ganglia (SCG) neurons and real-time qPCR, we explored the role of PDE3 and PDE4 isoforms and oxygen tension in the activation of PKA and changes in gene expression. These primary neurons were infected with an adenovirus containing A-Kinase activity reporter (AKAR3) and assayed for responses to PDE inhibitors: rolipram (ROL, 1 μM), milrinone (MIL, 10 μM) and IBMX (100 μM), and adenylyl

A.R. Nunes
Department of Pediatrics, Johns Hopkins University, School of Medicine, CMSC 6-104, 600 N. Wolfe Street, Baltimore, MD 21287-3200, USA
e-mail: aritanunes@gmail.com

CEDOC, Departamento de Farmacologia, Faculdade de Ciências Médicas, Universidade NOVA de Lisboa, Lisbon, Portugal

V. Sample
Department of Pharmacology and Molecular Sciences, Johns Hopkins University, School of Medicine, 725 N. Wolfe Street, Hunterian 307, Baltimore, MD 21205, USA

Y.K. Xiang
Department of Molecular and Integrative Physiology, University of Illinois at Urbana, Urbana, IL, USA

E.C. Monteiro
CEDOC, Departamento de Farmacologia, Faculdade de Ciências Médicas, Universidade NOVA de Lisboa, Campo Mártires da Pátria 130, 1169-056, Lisbon, Portugal

E. Gauda (✉)
Department of Pediatrics, Johns Hopkins University, School of Medicine, CMSC 6-104, 600 N. Wolfe Street, Baltimore, MD 21287-3200, USA
e-mail: egauda@mail.jhmi.edu

J. Zhang (✉)
Department of Pharmacology and Molecular Sciences, Johns Hopkins University, School of Medicine, 725 N. Wolfe Street, Hunterian 307, Baltimore, MD 21205, USA

Department of Neuroscience and Oncology, Johns Hopkins University, School of Medicine, Baltimore, MD, USA
e-mail: jzhang32@jhmi.edu

C.A. Nurse et al. (eds.), *Arterial Chemoreception: From Molecules to Systems*, Advances in Experimental Medicine and Biology 758, DOI 10.1007/978-94-007-4584-1_39,

cyclase activator forskolin (FSK, 50 μM). Different PDE activity patterns were observed in different cells: high PDE4 activity (n = 3), high PDE3 activity (n = 3) and presence of activity of other PDEs (n = 3). Addition of PKA inhibitor H89 (10 μM) completely reversed the response. We further studied the effect of oxygen in the PKA activity induced by PDE inhibition. Both normoxia (20%O_2/5%CO_2) and hypoxia (0%O_2/5%CO_2) induced a similar increase in the FRET emission ratio (14.5 ± 0.8 and 14.7 ± 0.8, respectively).

PDE3a, PDE4b and PDE4d isoforms mRNAs were highly expressed in the whole SCG with no modulation by hypoxia.

CONCLUSION: Using a FRET-based PKA activity sensor, we show that primary SCG neurons can be used as a model system to dissect the contribution of different PDE isoforms in regulating cAMP/PKA signaling. The differential patterns of PDE regulation potentially represent subpopulations of ganglion cells with different physiological functions.

Keywords Superior cervical ganglia (SCG) • Phosphodiesterases (PDE) • Cyclic AMP (cAMP) • Hypoxia • Protein kinase A (PKA)

39.1 Introduction

Cyclic AMP is a ubiquitous second messenger involved in synaptic transmission and signal transduction. The levels of cAMP are regulated by the synthetic activity of adenylate cyclases (AC) and by the hydrolytic activity of phosphodiesterases (PDE); the latter determines the intracellular concentration of cAMP (Bender and Beavo 2006). Thus, PDEs have become important targets of study in the past years and have been pursued as therapeutic targets (Bender and Beavo 2006). There are 11 PDE isoforms that differ in their affinity for cAMP and/or cGMP and inhibitors, and in their responses to modulators (Bender and Beavo 2006).

Superior cervical ganglion (SCG) has a key role in the sympathetic nervous system and sends neuronal projections to the carotid body (CB) controlling the vascular tone. We have recently observed that acute hypoxia decreased cAMP levels induced by PDE inhibitors in these ganglia, an opposite effect to that observed in the CB (Nunes et al. 2010).

The aim of this work is to characterize the role of PDE3 and PDE4 regulation on cAMP/PKA activity in the SCG under different oxygen concentrations. We also wanted to explore whether the effect of hypoxia observed in cAMP production induced by IBMX is translated into apparent changes in PKA activity in primary SCG neurons.

To answer these questions we investigated the effect of PDE inhibitors on the activation of PKA in dissociated SCG neurons using fluorescence resonance energy transfer (FRET)-based imaging and studied the expression of PDE3 and PDE4 and their regulation by oxygen in the whole SCG by quantitative RT-PCR.

39.2 Methods

39.2.1 Surgical Procedures

The experiments were performed using SCGs isolated from Sprague–Dawley rats (Charles River Laboratories, Wilmington, MA) of both sexes at postnatal days 5–9. The Animal Care and Use Committee at the Johns Hopkins University School of Medicine approved all experimental protocols.

The animals were anesthetized briefly with isoflurane and immediately decapitated. The carotid bifurcation was isolated and the SCG dissected and cleaned of surrounding connective tissue under a dissecting microscope. The tissues removed were further prepared as follows.

39.2.2 *Measurement of Protein Kinase A (PKA) Activity in Primary Superior Cervical Ganglia Neurons by FRET*

SCGs were mechanically and enzymatically dissociated with an enzyme mixture consisting of trypsin and collagenase (~0.5 mg/ml). SCG cells were plated on poly-D-Lysine and laminin coated imaging dishes and cultured for 2–4 days in neurobasal-A supplemented with NGF and Ara-C (10 μM). To monitor PKA activity, we infected the SCG neurons with an adenovirus containing a 3rd-generation of a genetically encoded A-Kinase activity reporter (AKAR3) for 24 h.

In one set of experiments we imaged the SCG neurons at 37 °C and sequentially treated them with rolipram (ROL, specific PDE4 inhibitor, 1 μM), milrinone (MIL, specific PDE3 inhibitor, 10 μM), IBMX (non-specific inhibitor, 100 μM), Forskolin (FSK, transmembrane adenylyl cyclase activator, 50 μM) and H89 (PKA inhibitor, 10 μM).

In another set of experiments, we placed the imaging dish into a chamber (Warner Instruments) to allow superfusion of the cells with a constant flow rate of 2 ml/min (Ismatec, Cole Parmer instrument) with Kreb's modified solution containing (in mM): NaCl 116, $NaHCO_3$ 24, KCl 5, $CaCl_2$ 2, $MgCl_2$ 1.1, Hepes 10, Glucose 5.5 and adjusted to pH 7.40 (Perez-Garcia et al. 1990), equilibrated either with normoxia (20%O_2/5%CO_2) or hypoxia (0–5%O_2/5%CO_2) in the presence or absence of IBMX (100 μM).

The cells were imaged on a Zeiss Axiovert 200 M microscope (Carl Zeiss) with a 40X oil immersion objective and a cooled charge-coupled device camera (Roper Scientific) controlled by Metafluor 6.2 software (Molecular Devices). Dual emission ratio imaging used a 420DF20 excitation filter, a 450DRLP dichroic mirror and two emission filters (475DF40 for cyan fluorescent protein (CFP) and 535DF25 for yellow fluorescent protein (YFP)). Exposure time was 50–500 ms and images were acquired every 20 s. Background correction of the fluorescence images was performed by subtracting autofluorescent intensities of regions of the imaging dish with no cells. Graph curves were normalized by setting the emission ratio at baseline (before the addition of any drugs) as equal to one.

39.2.3 *PDE mRNA Gene Expression and Regulation*

In one set of experiments we compared PDE3 and PDE4 gene expression levels in the whole SCG. The tissues were isolated (3 rats per condition; n = 3 independent experiments), quick-frozen on dry ice and stored at −80 °C.

In another set of experiments, we studied the effect of changes in oxygen tension on the expression of PDE3 and PDE4 in the whole SCG. Tissues were pre-incubated in a 37 °C shaker bath for 30 min in Kreb's modified medium equilibrated with 20%O_2/5%CO_2 (normoxia). After the pre-incubation period, tissues were incubated in a fresh incubation medium, equilibrated in normoxia, hypoxia (5%O_2/5%CO_2) or hyperoxia (60%O_2/5%CO_2) for 60 min. Tissues were quick-frozen on dry ice and stored at −80 °C.

Frozen tissues were processed to obtain total RNA (about 1 μg, micro-to-Midi Total RNA Purification, Invitrogen) that was used for first-strand cDNA synthesis using an iSCRIPt cDNA synthesis kit (Bio-Rad Laboratories). The primer sequences used were: forward 5′-GTGGCCGTATTCTGAGCCAGGTG-3′ and reverse 5′-GCTGTGTTGTGAGATACCACACGGC-3′

for PDE3a, forward 5′-CGACCAATTCCTGGCTTACAGCAGC-3′ and reverse 5′- GCAGGAATG-TTTGAAGACAGGCAGCC-3′for PDE3b, forward 5′-CCTGCACGCAGCGGATGTGC-3′ and reverse 5′-GGAACTGGTTGGAG-ACGCCAGG-3′ for PDE4a, forward 5′-CCTGGCTGCCAT-TTTTGCAGC-TGC-3′ and reverse 5′-GAGCTTGAATCCCACAGCGAGGTG-3′ for PDE4b, forward 5′-CCAGGAGGAGCAGCTGGCTAAG-3′ and reverse 5′- CACGTTGGAGTGGTAGTGCCCTTC-3′ forPDE4c,forward5′-GCG-GAGCTGTCTGGCAACCG-3′andreverse5′-GACTGGACGACATCTG-CAGCATGG-3′ for PDE4d, and forward 5′- GATGGCCTTCTACCCG-AAGACACC-3′ and reverse 5′- GCCAGCTACATAGGAGTTACGGGC-3′ for G6PDH.

qRT-PCR was performed with a MyiQ iCycler qRT-PCR system (Bio-Rad Laboratories) with Syber Green detection. qRT-PCR conditions were: 5 min at 95 °C, followed by 40 cycles at 95 °C for 20 s, 60 °C for 20 s, 72 °C for 20 s, a terminal extension period (72 °C, 10 min) and a melting curve with 0.5 °C increments in temperature. Product formation during the exponential phase of the reaction was analyzed for relative quantification to the reference gene based on the threshold cycle (C_T) for amplification as $2^{(\Delta CT)}$, where $\Delta C_T = C_{T,reference} - C_{T,target}$.

39.3 Results

We investigated the role of PDE3 and PDE4 on the activation of PKA in primary SCG neurons using FRET-based imaging. Figure 39.1 shows a representative response in a SCG neuron induced by sequential addition of specific and non-specific PDE inhibitors. While ROL did not induce a change in emission ratio (ΔER), Mil induced an increase in ΔER with no further changes with IBMX and FSK. Addition of H89 (10 μM) completely reversed the response. This pharmacological profile suggests that there is high PDE3 activity in this neuron. However, this pattern was not consistent in all SCG neurons. PDE inhibitors differentially modulated PKA activity, suggesting that there are subpopulations of SCG neurons with different PDE activity patterns.

Figure 39.2 shows the profiles of PDE3 and PDE4 activity in regulating PKA in different SCG neurons: 1) high PDE4 activity with 20.0 ± 0.4% ΔER in response to ROL with no further increase with MIL, IBMX and FSK (n = 3, Fig. 39.2a), 2) high PDE3 activity with 1.9 ± 0.6% ΔER in response to ROL and 19.0 ± 7.2% ΔER to MIL with no further increase with IBMX and FSK (n = 3, Fig. 39.2b), and 3) activity of other PDEs than PDE3/PDE4, with 1.6 ± 0.8% ΔER in response to ROL, 5.7 ± 1.1% in response to MIL and 18.7 ± 2.7% in response to IBMX and no further increase in the presence of FSK (n = 3, Fig. 39.2c). Addition of H89 reversed the response in all neurons (Fig. 39.2).

To determine whether oxygen tension modifies PKA activity by affecting PDE activity, we perfused SCG neurons with Kreb's modified solution in the presence and absence of IBMX (100 μM) equilibrated either with normoxia or hypoxia.

ΔER increased as a consequence of perfusion with IBMX in normoxia and this effect was completely reversible after washout. No differences were observed in the ΔER between normoxia and hypoxia in the presence of IBMX (14.5 ± 0.8 ΔER and 14.7 ± 0.8 ΔER, respectively, n = 10).

We further characterized the gene expression of the PDE3a-b and PDE4a-d isoforms in the whole SCG. Even though all isoforms were expressed in the SCG: PDE3a, PDE4b and PDE4d were the most abundant (Fig. 39.3). This pattern of expression was similar in SCG incubated with control oxygen concentrations (hypoxia, normoxia and hyperoxia) (Fig. 39.3b) and those that were quick frozen without incubations (Fig. 39.3a). Exposing SCGs with different percentages of oxygen did not significantly modulate PDE expression (Fig. 39.3b).

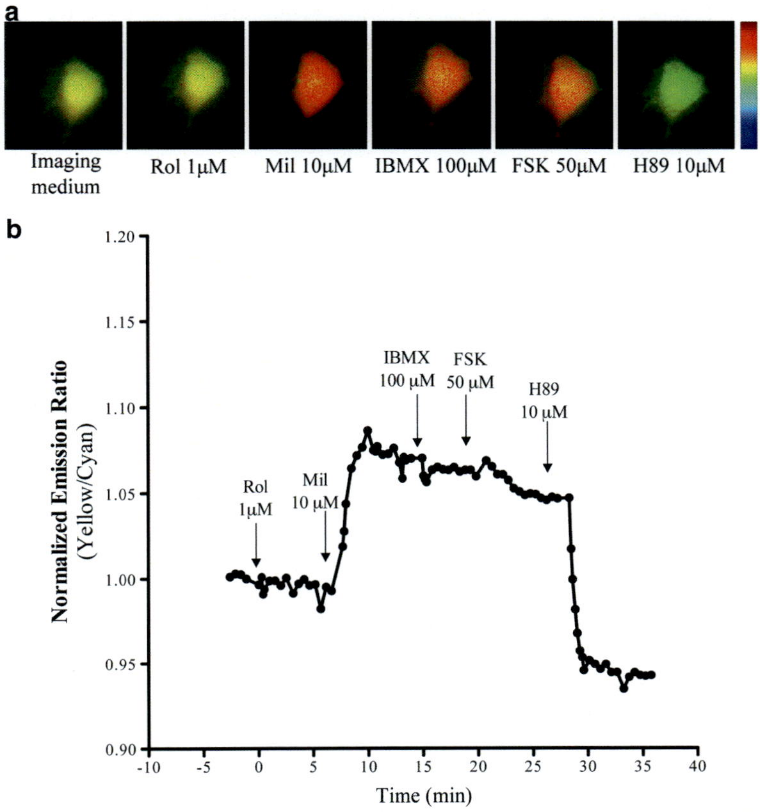

Fig. 39.1 (**a**) Pseudocolor images (400X) and (**b**) representative response curve of AKAR3 to rolipram (Rol, 1 μM), milrinone (Mil, 10 μM), IBMX (100 μM), FSK (50 μM) and H89 (10 μM) in a SCG neuron

39.4 Discussion

In this work, we characterized the PDE3 and PDE4 expression in the SCG and investigated the role of these PDE isoforms in the activation of PKA in ganglion neurons. We demonstrated that FRET-based imaging could be used to study the contribution of PDE isoforms in the regulation of cAMP/PKA signaling in primary SCG neurons as a model system. We observed differential patterns of PDE regulation in subpopulations of SCG neurons, which can potentially represent those subpopulations with different physiological functions. The level of oxygen tension did not modulate PDE expression and did not change the effect of PDE inhibition in the PKA activity.

SCG contains subpopulations of sympathetic neurons (secreto-, pilo-, vasomotor neurons), which receive input from the cervical sympathetic pre-ganglionic nerve fibers and provides innervations of the neck and head structures (Asamoto 2005). Each subpopulation of neurons projects exclusively to specific targets (Li and Horn 2006), which could explain the differences in PDE regulating PKA activity between SCG neurons described in our work.

We identified for the first time PDE3 and PDE4 mRNA expression patterns in the SCG. This finding adds to the literature by identifying PDE3 isoforms in this ganglion and is consistent with previous reports that characterize PDE4 by the effects of their specific inhibitors (Giorgi et al. 1994; Lakics et al. 2010; Li and Horn 2006; Nunes et al. 2010). We found that PDE3a, PDE4b and PDE4d

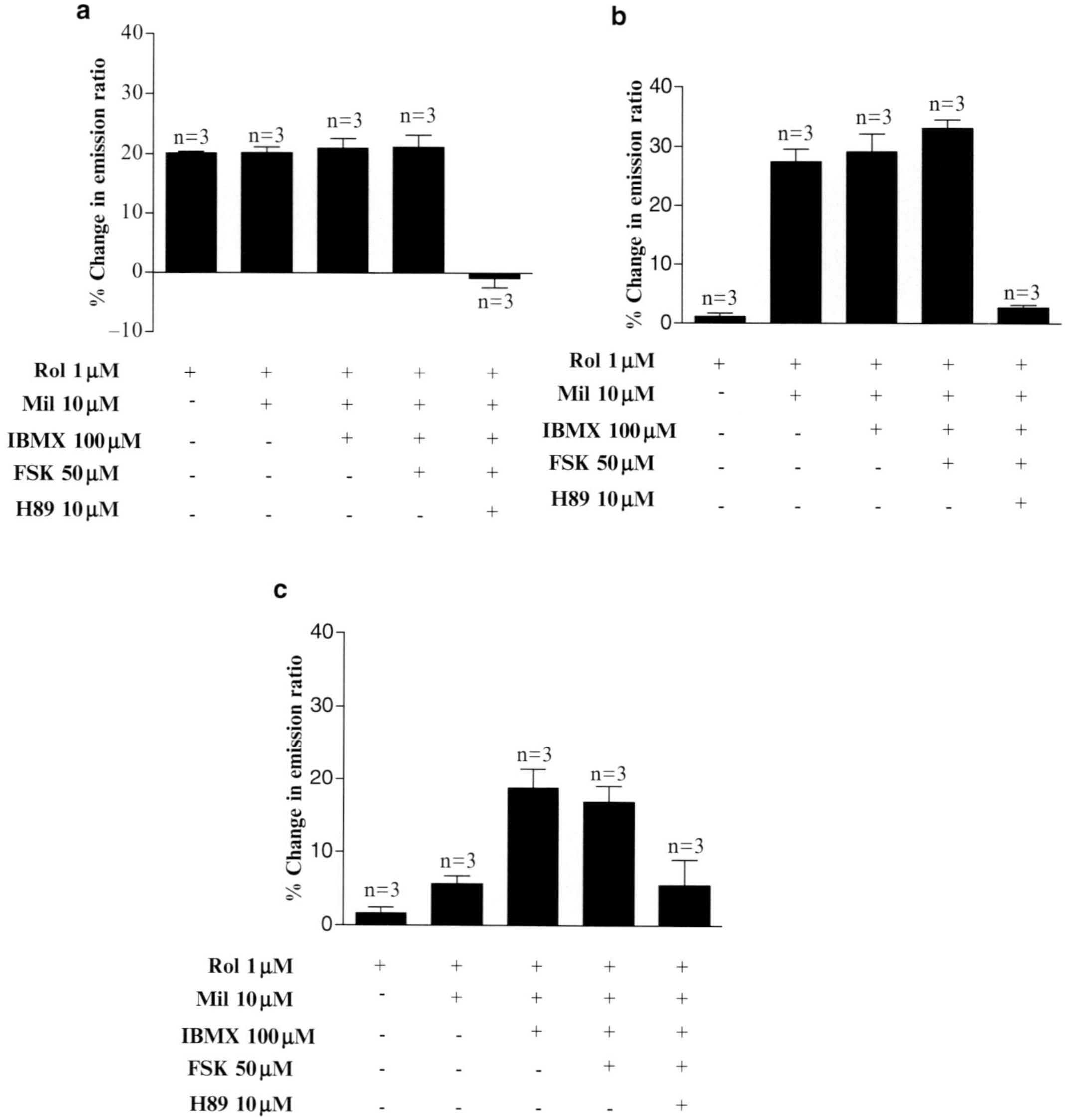

Fig. 39.2 Heterogenous responses in PKA activity induced by PDE inhibitors in SCG neurons. Composite data showing different response patterns of AKAR3 induced by PDE inhibitors, within a population of SCG neurons. SCG neurons containing (**a**) high PDE4 activity, (**b**) high PDE3 activity, and **c**) activity from other PDEs than PDE3 and PDE4. Values represent means ± SEM

were the most abundant isoforms in the SCG. It has been shown that these isoforms are highly expressed in specific tissues: PDE3a in the heart, PDE4b in the CNS and PDE4d in the muscle (Lakics et al. 2010).

We had previously observed that acute hypoxia induced a decrease in cAMP levels in the presence of the non-specific PDE inhibitor, IBMX and specific PDE2 and PDE4 inhibitors, in whole SCG of adult Wistar rats (Nunes et al. 2010). In the present work, hypoxia did not modulate either the levels of gene expression or PDE inhibition induced PKA activity. Taken together, these findings suggest that changes in IBMX-induced cAMP accumulation caused by changing oxygen concentrations are

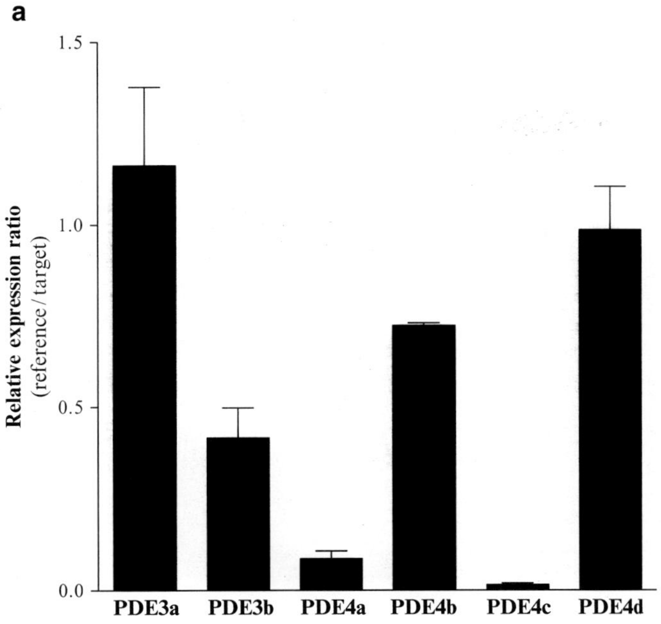

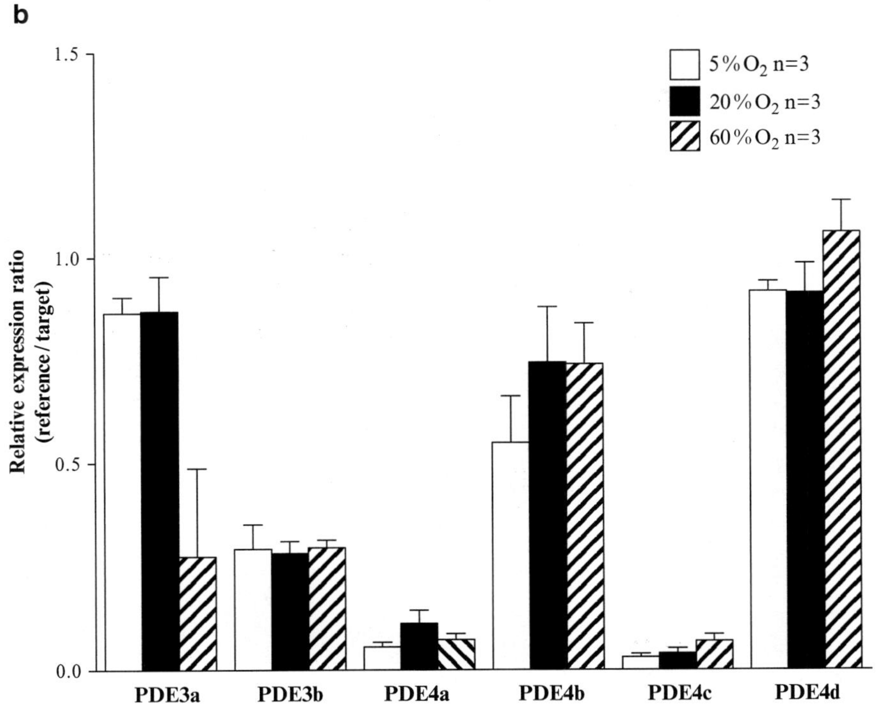

Fig. 39.3 Effect of different oxygen concentrations on PDE3 and PDE4 isoform gene expression in the whole superior cervical ganglia (SCG). (**a**) PDE3a-b and PDE4a-d isoform expression levels and their (**b**) regulation by hypoxia (5%O_2), normoxia (20%O_2) and hyperoxia (60%O_2)

not translated into changes in PKA activity. No effect of oxygen tension on changes in PKA activity could be due to saturation of PKA activity under high levels of cAMP caused by the PDE inhibition. Further experiments will test whether changes in oxygen concentrations directly modify PDE activity.

The results suggest selective distribution of PDE3a, PDE4b and PDE4d isoforms in subpopulations of ganglion cells that potentially represent subpopulations with different physiological functions.

Acknowledgment A.R. Nunes was supported by the Portuguese Fundacao de Ciẽncia e Tecnologia (FCT, SFRH/BD/39473/2007). This work is also supported by NIH R01 DK073368 (to JZ) and HL082846 (to YKX).

References

Asamoto K (2005) Network of the sympathetic nervous system: focus on the input and output of the cervical sympathetic ganglion. Anat Sci Int 80:132–140

Bender AT, Beavo JA (2006) Cyclic nucleotide phosphodiesterases: molecular regulation to clinical use. Pharmacol Rev 58:488–520

Giorgi M, Squitti R, Bonsi P, Paggi P, Toschi G (1994) Activities of 3'5' cyclic nucleotide phophodiesterases in the superior cervical ganglion of rat: characterization, compartmentalization and observation in young and old animals. Neurochem Int 25(5):493–500

Lakics V, Karran EH, Boess FG (2010) Quantitative comparison of phosphodiesterase mRNA distribution in human brain an peripheral tissues. Neuropharmacol 59:367–374

Li C, Horn JP (2006) Physiological classification of sympathetic neurons in the rat SCG. J Neurophysiol 95:187–195

Nunes AR, Batuca JR, Monteiro EC (2010) Acute hypoxia modifies cAMP levels induced by inhibitors of phosphodiesterase-4 in rat carotid bodies, carotid arteries and superior cervical ganglia. Br J Pharmacol 159:353–361

Perez-Garcia MT, Almaraz L, Gonzalez C (1990) Effects of different types of stimulation on cyclic AMP content in the rabbit carotid body: functional significance. J Neurochem 55:1287–1293

Chapter 40
Chronic Intermittent Hypoxia Alters Genioglossus Motor Unit Discharge Patterns in the Anaesthetized Rat

Deirdre Edge, Aidan Bradford, James F.X. Jones, and Ken D. O'Halloran

Abstract The respiratory control system is subject to diverse and considerable plasticity in health and disease. Intermittent hypoxia elicits expression of intrinsic plasticity within sensory and motor pathways involved in the control of breathing with potentially adaptive and maladaptive consequences for respiratory homeostasis. We and others have shown that chronic intermittent hypoxia (CIH) – a major feature of sleep-disordered breathing – has deleterious effects on rat upper airway dilator muscle contractile function and motor control. In the present study, we sought to test the hypothesis that CIH alters genioglossus (pharyngeal dilator) motor unit properties during basal breathing and obstructive airway events. Adult male Wistar rats were exposed to 20 cycles of normoxia and hypoxia (5% O_2 at nadir; $SaO_2 \sim 80\%$) per hour, 8 h a day for 7 days (CIH, N = 5). The sham group (N = 5) were subject to alternating cycles of air under identical experimental conditions in parallel. Following gas treatments, rats were anaesthetized with an i.p injection of urethane (1.5 g/kg; 20% w/v). Fine concentric needle electrodes were inserted into the genioglossus and the costal diaphragm. Genioglossus motor unit potentials, together with arterial blood pressure, tracheal pressure and arterial O_2 saturation were recorded during quiet basal breathing and nasal airway occlusion. During basal breathing, the amplitude of genioglossus motor units was significantly different in sham vs. CIH-treated rats (313 ± 32 μV vs. 430 ± 46 μV; mean ± SEM, Student's *t* test, p = 0.0415). The most common instantaneous firing frequency of individual units determined from auto correlograms was also significantly different in the two groups (53 ± 6 Hz vs. 37 ± 3 Hz; sham vs. CIH p = 0.0318). In addition, the amplitude of motor units recruited during airway obstruction was significantly decreased in CIH-treated rats (939 ± 102 μV vs. 619 ± 75 μV; sham vs. CIH p = 0.0267). Our results indicate that CIH causes remodelling in the central respiratory motor network with potentially maladaptive consequences for the physiological control of upper airway patency. We conclude that CIH could serve to exacerbate and perpetuate obstructive events in patients with sleep-disordered breathing.

Keywords Anaesthetized rat • Plasticity • Chronic intermittent hypoxia • Upper airway • Genioglossus • Single motor unit • Recruitment • Motor control • Oxidative injury.

D. Edge (✉) • J.F.X. Jones • K.D. O'Halloran
UCD School of Medicine and Medical Science, University College Dublin, Dublin 4, Ireland
e-mail: deirdre.edge@ucdconnect.ie

A. Bradford
Department of Physiology and Medical Physics, Royal College of Surgeons in Ireland, Dublin 2, Ireland

C.A. Nurse et al. (eds.), *Arterial Chemoreception: From Molecules to Systems*, Advances in Experimental Medicine and Biology 758, DOI 10.1007/978-94-007-4584-1_40,

40.1 Introduction

The pharyngeal dilator muscles, which are reflexely engaged by negative transmural airway pressure generated by the diaphragm upon inspiration, protect the patency of the oropharyngeal inlet. However, during sleep, the force produced by the pharyngeal dilator muscles of patients suffering with obstructive sleep apnoea (OSA) is insufficient to counterbalance the collapsing force generated by the diaphragm and accessory inspiratory muscles (Brouillette and Thach 1979), resulting in recurrent airway collapse. Intermittent hypoxia – a major feature of sleep-disordered breathing due to recurrent apnoea – elicits alterations at multiple levels of the respiratory control system (Peng et al. 2003; Fuller et al. 2000). We and others have shown that chronic intermittent hypoxia (CIH) has deleterious effects on rat upper airway dilator muscle contractile function (McGuire et al. 2002b; Bradford et al. 2005; Skelly et al. 2011), as well as motor control of the upper airway (O'Halloran et al. 2002; Veasey et al. 2004; Ray et al. 2007). Upper airway muscle and/or motor dysfunction may increase the susceptibility to further upper airway collapse and concomitant hypoxic insult – thus triggering a vicious cycle perpetuating obstructive events in OSA patients. We wished to characterize single motor unit activity of the genioglossus (GG) – a key pharyngeal dilator muscle. We hypothesized that exposure to CIH would impair control of the upper airway by altering GG motor unit discharge patterns, particularly during obstructive airway events.

40.2 Methods

40.2.1 Animal Model

Adult male Wistar rats (287 ± 3 g; mean ± SEM) were exposed to one of two chronic gas treatments. The gas supply to half of the environmental chambers alternated between air and nitrogen every 90s, reducing the ambient oxygen concentration to 5% at the nadir, resulting in arterial O_2 saturation values of ~80% (chronic intermittent hypoxia, CIH; N = 5). The sham group (N = 5) were subject to alternating cycles of normoxia under identical experimental conditions in parallel. Gas treatments were carried out for 20 cycles per hour, 8 h a day for seven consecutive days. Rats had free access to food and water.

40.2.2 Motor Unit Recordings

Following gas treatments, rats were anaesthetized with an i.p injection of urethane (1.5 g/kg; 20% w/v). Fine concentric needle electrodes were inserted into the genioglossus (pharyngeal dilator) and the costal diaphragm. Single genioglossus motor unit potentials, together with arterial blood pressure, tracheal pressure and arterial O_2 saturation were recorded during quiet basal breathing and nasal airway occlusion (lasting ~ 10 breaths) (Fig. 40.1).

40.2.3 Data Analysis

Using specialized software (Spike 2, Cambridge Electronic Design), single motor units were discriminated based on their shape and amplitude. Motor units were recorded during basal breathing, and periods of nasal airway occlusion. Units that were quiescent under basal conditions but became active during the

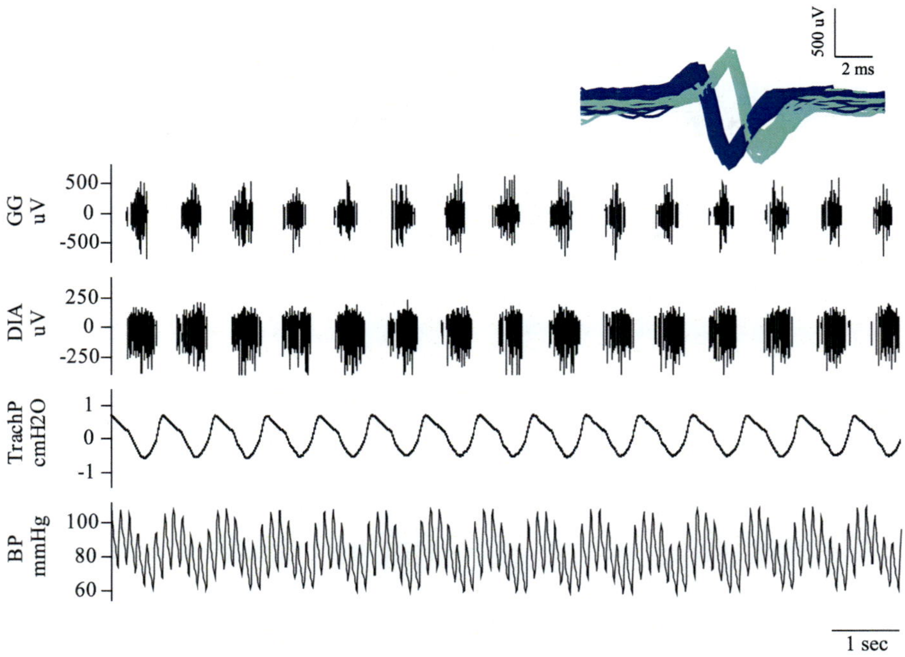

Fig. 40.1 Original record of genioglossus (GG) and diaphragm (DIA) motor multi-unit activity together with tracheal pressure (Trach P) and arterial blood pressure (BP), during quiet basal breathing in a spontaneously breathing urethane anaesthetized rat. Two discriminated GG single motor units are shown in the upper trace. Single motor units were discriminated on the basis of shape and amplitude

occlusion were termed 'recruited' units. Auto correlograms were constructed for GG units under basal conditions from sham and CIH-treated rats. The most common instantaneous frequencies of basal units were determined. Data were normally distributed therefore statistical comparisons were performed using a two tailed Student's *t* test; $p<0.05$ was considered statistically significant in all cases.

40.3 Results

During basal breathing, the amplitude of GG motor units was significantly different comparing sham (N=5) vs. CIH-treated (N=5) rats (313±32 μV vs. 430±46 μV; mean±SEM, Student's *t* test, $p=0.0415$). The "most common" discharge frequency of individual units determined from auto correlograms was also significantly different in the two groups (53±6 Hz vs. 37±3 Hz; sham vs. CIH $p=0.0318$). Additionally, the amplitude of motor units recruited during airway obstruction was significantly decreased in CIH-treated rats compared to sham controls (939±102 μV vs. 619±75 μV; sham vs. CIH $p=0.0267$) (Fig. 40.2).

40.4 Discussion

The upper airway dilator muscles maintain and defend upper airway patency. Obstructive sleep apnoea (OSA), a common sleep-related breathing disorder is characterized by recurrent collapse of the upper airway. The genioglossus (GG) is well recognized as a key pharyngeal dilator muscle and is strongly

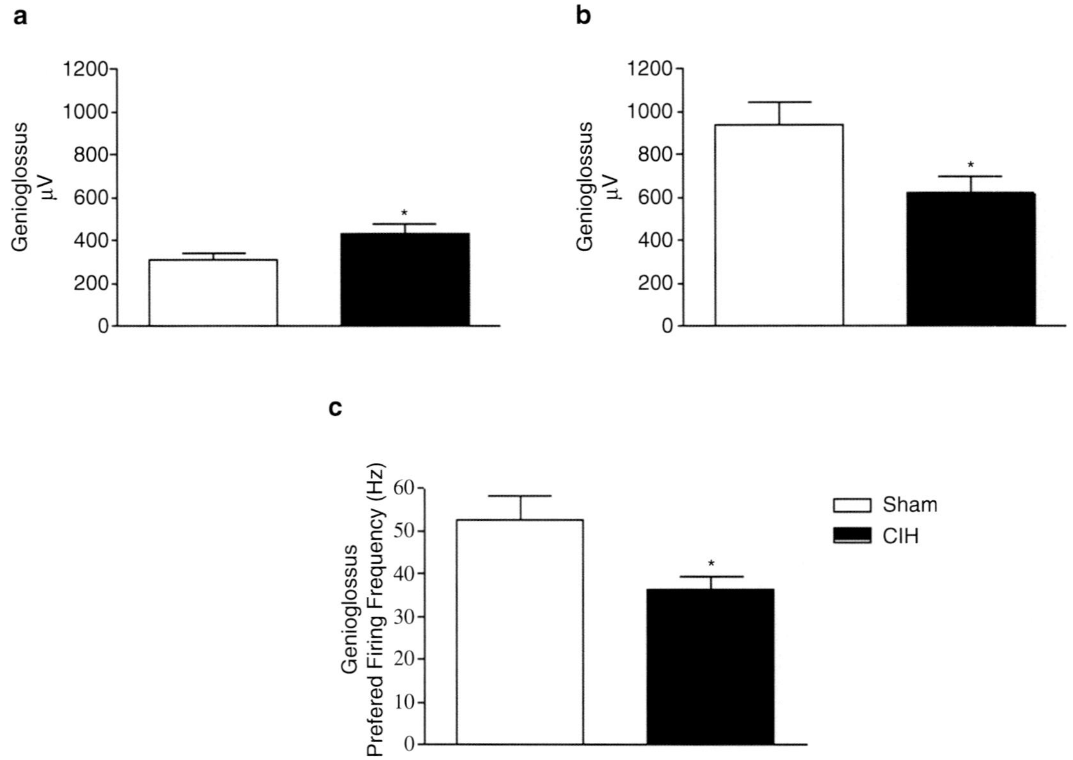

Fig. 40.2 Values (mean ± SEM) for genioglossus motor unit amplitude in sham and CIH-treated rats during (**a**) basal breathing and (**b**) nasal airway obstruction. Figure (**c**) shows the "most common" instantaneous firing frequency of genioglossus motor units in sham and CIH-treated rats. N = 5 for each group

implicated in reflex control of the upper airway (Remmers et al. 1978). Impaired function and control of the upper airway muscles is of serious consequence to the OSA patient, increasing the susceptibility to collapse of the airway perhaps exacerbating the condition. Dysfunction of the upper airway muscles has been identified as a possible mechanism facilitating the progression of the disorder (Boyd et al. 2004). Moreover, CIH has been shown to impair upper airway muscle contractile function (McGuire et al. 2002a; Skelly et al. 2011). As well as varying the intrinsic contractile properties of the muscles, CIH is also capable of altering motor control of the upper airway. Reduced responsiveness of central respiratory centres, including the hypoglossal motor nucleus has been described following exposure to CIH (Veasey et al. 2004; Ray et al. 2007). Our lab has previously demonstrated that chronic intermittent hypercapnic hypoxia impairs reflex activation of an upper airway dilator muscle to acute physiological stimuli (O'Halloran et al. 2002). In this study we examined the effects of CIH on GG single motor unit characteristics. During basal breathing, the amplitude of GG motor units of CIH-treated rats was significantly greater than that of sham controls. Although the amplitude of these units was greater, their "most common" firing frequency was significantly lower. Neurogenic changes in GG single motor units of OSA patients have also been reported, with an increased area of GG motor units and altered firing frequencies compared to control subjects during normal breathing (Saboisky et al. 2007). Viewed together these alterations are suggestive of remodelling in the central respiratory network – namely at the level of the hypoglossal motor nucleus, with potentially maladaptive consequences for the physiological control of upper airway patency. We were particularly interested in GG motor unit characterization during nasal obstruction, modelling an obstructive event in the OSA patient. The characteristics of recruited units (quiescent under basal conditions) during obstruction were different to those present during basal breathing; in accordance with the size principle (Henneman 1957), recruited units were typically of much greater amplitude. However, the amplitude of GG motor units active during nasal obstruction was significantly lower in CIH-treated rats compared to sham controls. The regulation of skeletal muscle force can be altered in two ways: altering the number of units that are active (by means of recruitment/derecruitment) and/or by varying the discharge frequencies of active motor units (rate coding) (Bailey 2011). Recently it has been demonstrated that recruitment is an important mechanism by which the GG muscle is controlled (Saboisky et al. 2010); single motor unit recordings in human subjects have indicated that increases in GG muscle activity during hypercapnia is due to motor unit recruitment rather than changes in firing rates (Nicholas et al. 2010). Recruitment of large motor units to the airway muscles during obstruction is essential to generate the necessary force to re-establish a patent airway. We argue that the decreased amplitude of motor units recruited to the airway following CIH exposure may affect the ability to recover an airway from an obstructive event, thus rendering the airway more susceptible to collapse and perpetuating obstructive events in mammals with collapsible airways such as humans. Many forms of IH-induced respiratory plasticity are dependent upon reactive oxygen species (ROS). Moreover IH has been shown to induce oxidative injury in many tissues (Row et al. 2003; Jun et al. 2008). Accordingly, we speculate that the CIH-induced altered control of the upper airway is a consequence of oxidative injury to hypoglossal motor neurons (Veasey et al. 2004). We conclude that CIH causes remodelling in the central respiratory network governing respiration that could serve to perpetuate obstructive events in patients with disordered breathing during sleep.

Acknowledgements Supported by the Health Research Board, Ireland (RP/2007/29). DE is enrolled in the School of Medicine and Medical Science Translational Medicine PhD training programme.

References

Bailey EF (2011) Activities of human genioglossus motor units. Respir Physiol Neurobiol 179:14–22

Boyd JH, Petrof BJ, Hamid Q, Fraser R, Kimoff RJ (2004) Upper airway muscle inflammation and denervation changes in obstructive sleep apnea. Am J Respir Crit Care Med 170:541–546

Bradford A, McGuire M, O'Halloran KD (2005) Does episodic hypoxia affect upper airway dilator muscle function? implications for the pathophysiology of obstructive sleep apnoea. Respir Physiol Neurobiol 147:223–234

Brouillette RT, Thach BT (1979) A neuromuscular mechanism maintaining extrathoracic airway patency. J Appl Physiol 46:772–779

Fuller DD, Bach KB, Baker TL, Kinkead R, Mitchell GS (2000) Long term facilitation of phrenic motor output. Respir Physiol 121:135–146

Henneman E (1957) Relation between size of neurons and their susceptibility to discharge. Science 126:1345–1347

Jun J, Savransky V, Nanayakkara A, Bevans S, Li J, Smith PL, Polotsky VY (2008) Intermittent hypoxia has organ-specific effects on oxidative stress. Am J Physiol Regul Integr Comp Physiol 295:R1274–R1281

McGuire M, Macdermott M, Bradford A (2002a) The effects of chronic episodic hypercapnic hypoxia on rat upper airway muscle contractile properties and fiber-type distribution. Chest 122:1400–1406

McGuire M, Macdermott M, Bradford A (2002b) Effects of chronic episodic hypoxia on rat upper airway muscle contractile properties and fiber-type distribution. Chest 122:1012–1017

Nicholas CL, Bei B, Worsnop C, Malhotra A, Jordan AS, Saboisky JP, Chan JK, Duckworth E, White DP, Trinder J (2010) Motor unit recruitment in human genioglossus muscle in response to hypercapnia. Sleep 33:1529–1538

O'Halloran KD, McGuire M, O'Hare T, Bradford A (2002) Chronic intermittent asphyxia impairs rat upper airway muscle responses to acute hypoxia and asphyxia. Chest 122:269–275

Peng YJ, Overholt JL, Kline D, Kumar GK, Prabhakar NR (2003) Induction of sensory long-term facilitation in the carotid body by intermittent hypoxia: implications for recurrent apneas. Proc Natl Acad Sci U S A 100:10073–10078

Ray AD, Magalang UJ, Michlin CP, Ogasa T, Krasney JA, Gosselin LE, Farkas GA (2007) Intermittent hypoxia reduces upper airway stability in lean but not obese Zucker rats. Am J Physiol Regul Integr Comp Physiol 293:R372–R378

Remmers JE, Degroot WJ, Sauerland EK, Anch AM (1978) Pathogenesis of upper airway occlusion during sleep. J Appl Physiol 44:931–938

Row BW, Liu R, Xu W, Kheirandish L, Gozal D (2003) Intermittent hypoxia is associated with oxidative stress and spatial learning deficits in the rat. Am J Respir Crit Care Med 167:1548–1553

Saboisky JP, Butler JE, McKenzie DK, Gorman RB, Trinder JA, White DP, Gandevia SC (2007) Neural drive to human genioglossus in obstructive sleep apnoea. J Physiol 585:135–146

Saboisky JP, Jordan AS, Eckert DJ, White DP, Trinder JA, Nicholas CL, Gautam S, Malhotra A (2010) Recruitment and rate-coding strategies of the human genioglossus muscle. J Appl Physiol 109:1939–1949

Skelly JR, Edge D, Shortt CM, Jones JFX, Bradford A, O'Hallorn KD (2011) Tempol ameliorates pharyngeal dilator muscle dysfunction in a rodent model of chronic intermittent hypoxia. Am J Respir Cell Mol Biol 46:139–148

Veasey SC, Zhan G, Fenik P, Pratico D (2004) Long-term intermittent hypoxia: reduced excitatory hypoglossal nerve output. Am J Respir Crit Care Med 170:665–672

Chapter 41
Upregulation of Pituitary Adenylate Cyclase Activating Polypeptide and Its Receptor Expression in the Rat Carotid Body in Chronic and Intermittent Hypoxia

S.Y. Lam, Y. Liu, E.C. Liong, G.L. Tipoe, and Man Lung Fung

Abstract The carotid body (CB) plays important roles in cardiorespiratory changes in chronic and intermittent hypoxia. Pituitary adenylate cyclase activating polypeptide (PACAP) is involved in the regulation of respiratory chemoresponse. We hypothesized an upregulation of the expressions of PACAP and its receptor (PAC1) in the rat CB in chronic and intermittent hypoxia. The CB expressions of PACAP and PAC1 were examined in rats breathing 10% O_2 (in isobaric chamber for chronic hypoxia, 24 h/day) or in intermittent hypoxia (cyclic between air and 5% O_2 per minute, 8 h/day) for 7 days. Immunohistochemical studies showed that the PACAP and PAC1 proteins were localized in CB glomic clusters containing tyrosine hydroxylase. The proportional amount of cells with positive staining of PACAP and PAC1 was significantly increased in both hypoxic groups when compared with the normoxic control. In addition, the mRNA level of PAC1 expression was markedly elevated in the hypoxic groups, despite no changes in the PACAP expression. These results suggest an upregulation of PACAP and its receptor expression in the rat CB under chronic and intermittent hypoxic conditions. The PACAP binding to its receptor could activate the PKA signaling pathway leading to an increased CB excitability under hypoxic conditions.

Keywords Carotid • Chemoreceptor • PACAP • PAC1 • Chronic • Intermittent • Hypoxia

41.1 Introduction

Carotid body (CB) is the major peripheral sensory organ for monitoring the arterial levels of oxygen and carbon dioxide. In acute hypoxia, activation of the chemoreceptor in the carotid body stimulates the brainstem respiratory activities leading to a reflexive increase in respiratory rate. In addition, the CB plays important roles in cardiorespiratory changes in chronic and intermittent hypoxia (Gonzalez et al. 1994; Prabhakar 2001).

Pituitary adenylate cyclase-activating polypeptide (PACAP) belongs to the vasoactive intestinal peptide (VIP)-glucagon-growth hormone releasing factor-secretin family. PACAP is widely expressed

S.Y. Lam • Y. Liu • M.L. Fung (✉)
Departments of Physiology, University of Hong Kong, Pokfulam, Hong Kong SAR, China
e-mail: fungml@hku.hk

E.C. Liong • G.L. Tipoe
Department of Anatomy, University of Hong Kong, Pokfulam, Hong Kong SAR, China

C.A. Nurse et al. (eds.), *Arterial Chemoreception: From Molecules to Systems*,
Advances in Experimental Medicine and Biology 758, DOI 10.1007/978-94-007-4584-1_41,

in the brain and peripheral organs. It is involved in the regulation of cardiorespiratory activities, including dilatation of the smooth muscle and mucus secretion in the airway (Sherwood et al. 2000; Vaudry et al. 2000). Also, PACAP receptors are presented in the brain regions which are involved in the regulation of respiratory rhythm and respiratory chemosensitivity (Hannibal 2002; Vaudry et al. 2000). It has been shown that PACAP stimulates respiration mediated by the carotid body (Runcie et al. 1995). Interestingly, the respiratory chemoresponse to hypoxia is reduced in PACAP-deficient mice with increased susceptibility to apnea, suggesting a role of the PACAP signaling pathway in the breathing disorders (Cummings et al. 2004). Yet, there is a paucity of information on the PACAP expression in the CB under hypoxic conditions. The aim of the present study was to examine the CB expressions and function of PACAP and PAC1 in rats breathing 10% O_2 for chronic hypoxia (CH) or in cyclic O_2 levels between 5–21% for intermittent hypoxia (IH) for 7 days. Our results suggest an upregulation of PACAP and its receptor expression, which may activate the intracellular signaling pathway leading to an increased carotid chemoresponse to acute hypoxia under CH and IH conditions.

41.2 Methods

The experimental protocol for this study was approved by the Committee on the Use of Live Animals in Teaching and Research of the University of Hong Kong. Twenty-eight days male Sprague–Dawley rats were randomly divided into groups for the treatment of CH or IH and for normoxic (Nx). The Nx controls were kept in room air and were freely accessible to water and regular chow, the CH and IH rats were kept in acrylic chambers for normabaric hypoxia (Tipoe and Fung 2003). The oxygen fraction inside the chamber was kept at 10 ±0.5%, 24 h per day for the CH group. For IH group, the oxygen was cyclic from 21 to 5±0.5% per min, 60 cycles/h, 8 h per day diurnally in which the inspired oxygen level fell to 4–5% (nadir arterial oxygen saturation ca. 70%) for about 15 s per min, which mimicks the recurrent episodic hypoxemia in patients with obstructive sleep apnoea (Prabhakar et al. 2001). The desired oxygen level was established by a mixture of room air and nitrogen hat was regulated and monitored by an oxygen analyzer (Vacumetrics Inc., CA, USA). Carbon dioxide was absorbed by soda lime granules and excess humidity was removed by a desiccator. The chamber was opened twice a week for an hour to clean the cages and replenish food and water. The rats were exposed to hypoxia for 7 days and were immediately used in experiments.

Following deep anesthesia, rats were decapitated and the carotid body bifurcation was excised rapidly. The carotid body was carefully dissected free from the bifurcation and was fixed in neutral buffered formalin for 72 h. Tissues were processed routinely for histology and embedded in paraffin blocks. Serial sections of 5 μm thickness were cut and mounted on silanized slides. Sections were kept in the oven overnight at 56 °C and were dewaxed with xylene and rehydrated with a series of decreasing grade of ethanol solution. Sections were immunostained with antiserum to the following proteins: PACAP (goat polyclonal antibody, 1:500 dilution, Cat # sc-7840, Santa Cruz, USA); PAC1 (goat polyclonal antibody, 1:250 dilution, Cat # sc-15965, Santa Cruz, USA). Sections were immersed in antigen retrieval solution (0.1 M citric acid buffer, pH 6.0) and then in 3% hydrogen peroxide for blocking endogenous peroxidase activity. Sections were pre-incubated with 20% normal serum to reduce non-specific binding for the anti-serum. Then sections were incubated with the corresponding primary antibodies in 0.05 M Tris–HCl buffer, respectively, containing 2% bovine serum albumin overnight. Sections were rinsed and then incubated with biotinylated link agent and streptavidin peroxidase. The peroxidase was visualized by immersing in 0.05% diaminobenidine containing 0.03% hydrogen peroxide in Tris–HCl buffer (pH7.5) for 3–5 min. Sections were rinsed in distilled water and counterstained with hematoxylin. Positive staining was indicated by a brown color. Control sections were incubated with either normal mouse or rabbit IgG and stained uniformly negative (data not shown). For double-labelling studies, DAKO Envision® Doublestain System (K-1395) and two sets of primary antibody were used. The first set was directed against PACAP (goat polyclonal antibody, 1:500 dilution, Cat # sc-7840, Santa Cruz, USA, conjugated to Rhodamine) and tyrosine hydroxylase (TH) (rabbit IgG antibody, 1:100 dilution, Cat # AB151, Chemicon International Inc., USA, conjugated to FITC). The second set of primary antibody was directed against PAC1 (goat polyclonal antibody, 1:250 dilution, Cat # sc-15965, Santa Cruz, USA, conjugated to Rhodamine) and tyrosine hydroxylase (TH) (rabbit IgG antibody, 1:100 dilution, Cat # AB151, Chemicon International Inc., USA, conjugated to FITC). After incubating with the corresponding first antibody (PACAP and PAC1 respectively), sections were incubated with peroxidase labeled polymer for 30 min. The peroxidase was visualized with substrate DAB chromogen for 5–10 min. Sections were then incubated in a Doublestain Block solution (DAKO Doublestain kit) which served to remove any potential cross-reactivity between reactions along with blocking endogenous alkaline phosphatase that may be present. The second antibody (TH) was then incubated for 1 h. The AP was visualized by Far Red solution for 5 min. Control sections were incubated with either normal goat or rabbit IgG and stained uniformly negative (data not shown).

For semi-quantitative reverse transcription-polymerase chain reaction (RT-PCR) studies, tissues were homogenized in 4-M guanidinium thiocyanate solution. The extracted total RNA was studied by gel electrophoresis and quantified by spectrophotometer. Total RNA (5 μg) was subjected to first strand cDNA synthesis using random hexamer primers and superscript II transcriptase (GIBCO-BRL, USA) in a final volume of 20 μl. After incubation at 42 °C for 1 h, the reaction mixture was treated with RNase H before proceeding to PCR analysis. The final mixture (2 μl) was directly used for PCR amplification. The procedures for RT-PCR were performed as reported previously (Fung et al. 2002). The primer sequences employed for the present study are: PACAP (sense: CATCTTCACAGACAGCTATAG; antisense: GTTTGGAAAGAACACATGAGT), PAC1 (sense: CTTGTACAGAAGCTGCAGTC; antisense: CGTGCTTGAAGTCCATAG) and β-actin (sense: AGT GTG ACG TTC ACA TCC GT; antisense: GAC TGA TCG TAC TCC TGC TT). RNA was tested to be free of DNA contamination by RT-PCR without addition of reverse transcriptase. The PCR conditions for: (1) PACAP was 28 cycles of denaturing, 94 °C, 1 min; annealing, 60 °C, 1 min; elongating, 72 °C, 2 min; (2) PAC1 was 35 cycles of denaturing, 94 °C, 1 min; annealing, 61 °C, 1 min; annealing, 72 °C, 2 min. (3) β-actin was 28 cycles of denaturing, 94 °C, 1 min; annealing, 61 °C, 1 min; elongating, 72 °C, 2 min; The amplified mixture was finally separated on agarose gel electrophoresis and the amplified DNA bands were detected using ethidium bromide staining. The fluorescence intensity of the amplified cDNA was quantified with an IMAGEQUANT software (Molecular Dynamics, USA).

41.3 Results

Immunohistochemical studies showed that the PACAP and PAC1 proteins were positively stained in the CB of rats with or without exposed to CH and IH treatment (n=5) (Fig. 41.1a). The proportional amount of cells with positive staining of PACAP and PAC1 was significantly increased in the CH and IH groups when compared with the normoxic control (Fig. 41.1b, c). Double-immunohistochemical studies show that both PACAP (Fig. 41.2a) and PAC1 (Fig. 41.2b) were co-localized with TH in the rat CB type-I glomus cells. The mRNA level of PACAP expression was not different among the normoxic and hypoxic groups; whereas the PAC1 expression was significantly increased in the hypoxic groups when compared with the control group (Fig. 41.3).

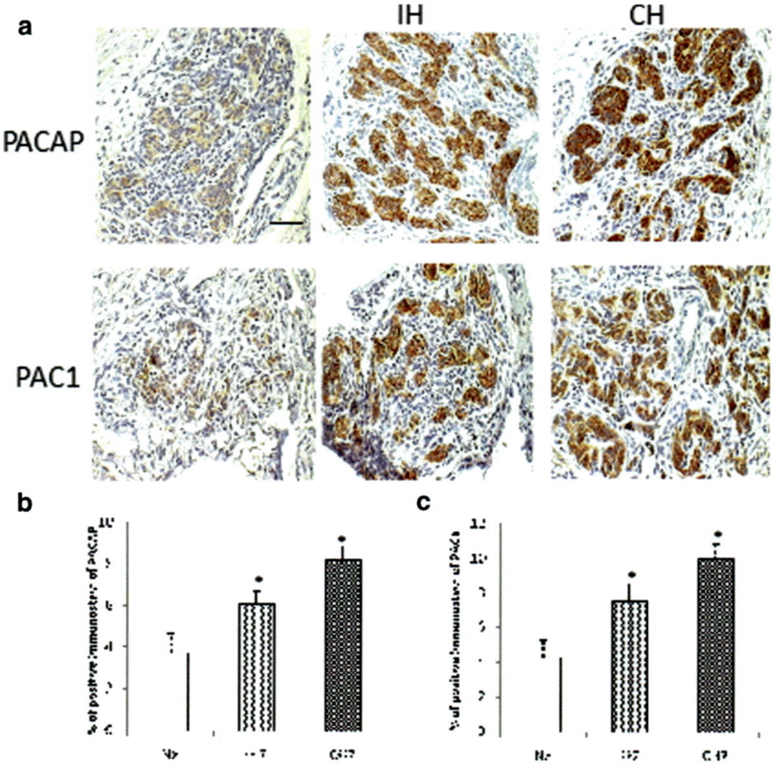

Fig. 41.1 Immunohistochemical localization of PACAP and PAC1 in the carotid body of rats exposed to normoxia (Nx) or intermittent hypoxia (IH) and chronic hypoxia (CH). Bar = 40 μm. (**a**) Protein expressions of PACAP and PAC1 in Nx, IH and CH rat CB. (**b, c**) Data are presented in % area with positive staining of PACAP and PAC1 immunoreactivity and are expressed as mean ± SEM (n = 5 for each group). * $p < 0.05$

41.4 Discussion

PACAP plays a role in the regulation of cardiorespiratory function. It has been reported that intravenous injection of PACAP induced an increase in ventilation in dogs, which was abolished by carotid sinus nerve sections (Runcie et al. 1995). More recently, Xu et al. (2007) found that PACAP stimulates carotid chemosensitive glomus cells via the PAC1 receptors coupled adenylate cyclase/PKA pathway, which decreases the background TASK-like K^+ current leading to membrane depolarization and an elevation of intracellular calcium levels. Consistent with these observations, our present study demonstrated that PACAP and its receptor PAC1 were expressed and localized in CB type-I glomus cells. The co-localization of the PACAP and it receptor PAC1 with TH-containing type-I glomus cells suggests that PACAP plays a role in the modulation of the chemotransduction in the carotid body.

Intriguingly, the expressions of PACAP and PAC1 immunoreactivities were significantly increased in the hypoxic groups. These findings support our hypothesis that the expressions of PACAP and PAC1 receptor were upregulated in the rat CB under chronically hypoxic conditions. The exact mechanism underlying the upregulation is not clear but it may be related to the increased activity of the glomus cells under hypoxic conditions. It has been shown that PACAP stimulates tyrosine hydroxylase activity and the effect is mediated by the adenylyl cyclase/cAMP/PKA transduction pathway (Marley et al. 1996). In this context, the upregulation of PACAP and its receptor in the type-I glomus cells could be an autocrine/paracrine pathway by which increases the chemoresponse to hypoxia. The activation of PKA pathway could elevate the intracellular calcium levels of the chemosensitive cells and the release

Fig. 41.2 The immunohistochemical co-localization of PACAP and TH in the carotid body of rats exposed to normoxia (Nx) or intermittent hypoxia (IH) and chronic hypoxia (CH); B: The co-localization of PAC1 and TH in Nx, IH and CH groups

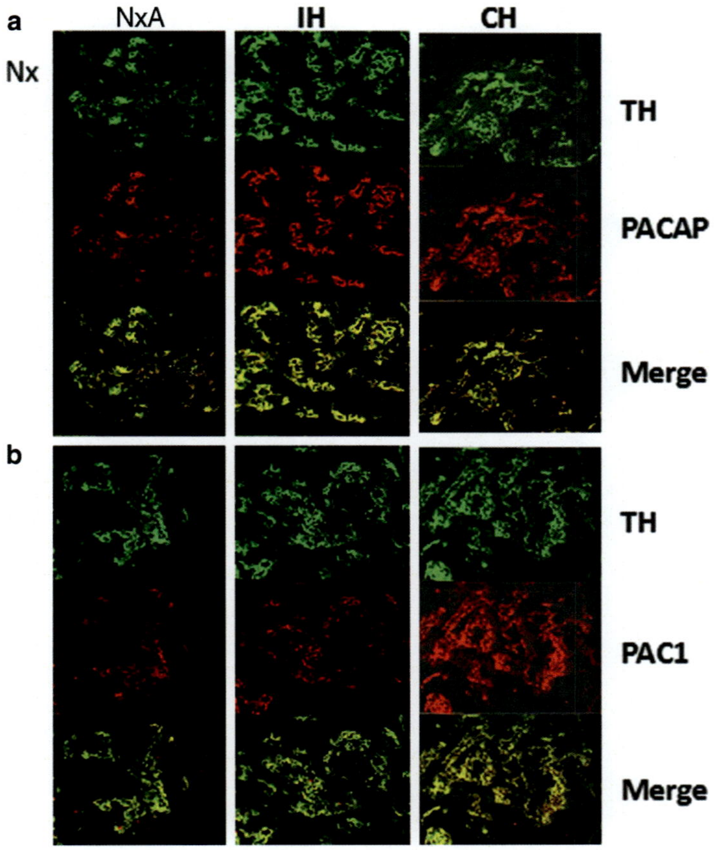

Fig. 41.3 RT-PCR analysis of the relative mRNA expression of PACAP and PAC1 in the carotid bodies of rats exposed to normoxia (Nx) or intermittent hypoxia (IH) and chronic hypoxia (CH)**.** The data are expressed as means plus/minus S.E.M. (n = 6 for the PAC1 group, n = 8 for the PACAP group). * p < 0.01

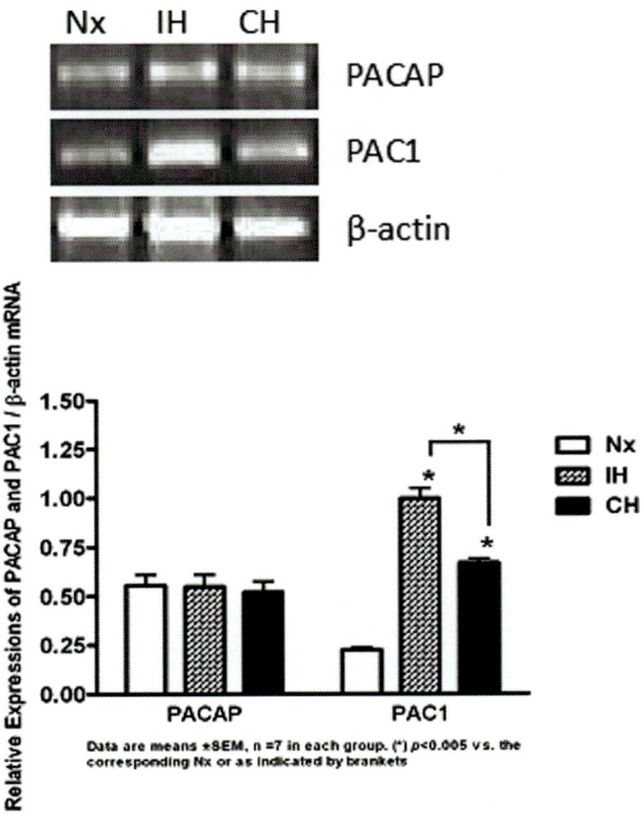

of neurotransmitters in response to hypoxia. This may be functionally significant to the increased excitability of the carotid chemoreceptor under chronically hypoxic conditions, which is clinically associated with patients with chronic cardiopulmonary diseases.

Acknowledgments We thank Mr. Y.M. Lo and Ms. K.M. Leung for their technical assistance. This work was supported by research grants (HKU766110M, HKU7510/06 M) from the Research Grants Council, HKSAR and CRCG Conference Grants (201007170441) from the University of Hong Kong.

References

Cummings KJ, Pendlebury JD, Sherwood NM, Wilson RJ (2004) Sudden neonatal death in PACAP-deficient mice is associated with reduced respiratory chemoresponse and susceptibility to apnoea. J Physiol 555:15–26

Fung ML, Lam SY, Dong X, Chen Y, Leung PS (2002) Postnatal hypoxemia increases angiotensin II sensitivity and up-regulates AT1a angiotensin receptors in rat carotid body chemoreceptors. J Endocrinol 173:305–313

Gonzalez C, Almaraz L, Obeso A, Rigual R (1994) Carotid body chemoreceptors: from natural stimuli to sensory discharges. Physiol Rev 74:829–898

Hannibal J (2002) Pituitary adenylate cyclase-activating peptide in the rat central nervous system: an immunohistochemical and in situ hybridization study. J Comp Neurol 453:389–417

Marley PD, Cheung CY, Thomson KA, Murphy R (1996) Activation of tyrosine hydroxylase by pituitary adenylate cyclase-activating polypeptide (PACAP-27) in bovine adrenal chromaffin cells. J Auton Nerv Syst 60:141–146

Prabhakar NR (2001) Oxygen sensing during intermittent hypoxia: cellular and molecular mechanisms. J Appl Physiol 90:1986–1994

Prabhakar NR, Fields RD, Baker T, Fletcher EC (2001) Intermittent hypoxia: cell to system. Am J Physiol Lung Cell Mol Physiol 281:L524–L528

Runcie MJ, Ulman LG, Potter EK (1995) Effects of pituitary adenylate cyclase-activating polypeptide on cardiovascular and respiratory responses in anaesthetised dogs. Regul Pept 60:193–200

Sherwood NM, Krueckl SL, McRory JE (2000) The origin and function of the pituitary adenylate cyclase-activating polypeptide (PACAP)/glucagon superfamily. Endocr Rev 21:619–670

Tipoe GL, Fung ML (2003) Expression of HIF-1alpha, VEGF and VEGF receptors in the carotid body of chronically hypoxic rat. Respir Physiol Neurobiol 138:143–154

Vaudry D, Gonzalez BJ, Basille M, Yon L, Fournier A, Vaudry H (2000) Pituitary adenylate cyclase-activating polypeptide and its receptors: from structure to functions. Pharmacol Rev 52:269–324

Xu F, Tse FW, Tse A (2007) Pituitary adenylate cyclase-activating polypeptide (PACAP) stimulates the oxygen sensing type I (glomus) cells of rat carotid bodies via reduction of a background TASK-like K^+ current. J Neurochem 101:1284–1293

Chapter 42
Rabbit Ventilatory Responses to Peripheral Chemoexcitators: Effects of Chronic Hypoxia

Julio Alcayaga, Rodrigo Del Rio, Esteban A. Moya, Matías Freire, and Rodrigo Iturriaga

Abstract Exposure to prolonged hypobaric hypoxia increases baseline ventilation and the ventilatory response to acute hypoxia, a phenomenon known as hypoxic ventilatory acclimatization (HVA). It is currently accepted that the carotid bodies and the reduced PO_2 levels are key elements in the generation of the HVA. However, because most of these experiments have been performed in hypobaric conditions, we studied the effects of 15 days of chronic normobaric hypoxia (CNH) on the rabbit ventilatory responses to hypoxia and chemoexcitatory molecules. New Zealand White rabbits were placed in a 0.3 m^3 chamber with controlled temperature and a mean O_2 content of 9.17 ± 0.09 %. Animals with or without CNH exposition (naïve) were anesthetized (ketamine/xylazine 75/7.5 mg/kg, i.m.), cannulated and air flow was measured. In naïve animals hypoxic challenges and NaCN increase ventilation, effect completely abolished after bilateral chemodenervation. However, ventilatory responses to nicotine, ATP and dopamine remained largely unchanged after bilateral chemodenervation, suggesting a centrally mediated effect for these drugs. Basal ventilation was reduced in CNH animals, but the dose dependent ventilatory increases induced by NaCN presented an increased sensibility. Further experiments are needed to elucidate the mechanisms responsible for these acclimatized responses.

Keywords Chronic hypoxia • Ventilatory acclimatization • Ventilatory reflexes • Acetylcholine • ATP • Dopamine • Cyanide •

42.1 Introduction

Exposure to high altitude augments the ventilatory response to acute hypoxia in humans and animals, a phenomenon known as ventilatory acclimatization (Bisgard 2000). The hypoxic ventilatory acclimatization (HVA) is a key physiological response in animals subjected to prolonged hypobaric

J. Alcayaga (✉) • M. Freire
Laboratorio de Fisiología Celular, Facultad de Ciencias,
Universidad de Chile, Casilla 653, Santiago, Chile
e-mail: jalcayag@uchile.cl

R. Del Rio • E.A. Moya • R. Iturriaga
Laboratorio de Neurobiología, Facultad de Ciencias Biológicas,
P Universidad Católica de Chile, Santiago, Chile

C.A. Nurse et al. (eds.), *Arterial Chemoreception: From Molecules to Systems*,
Advances in Experimental Medicine and Biology 758, DOI 10.1007/978-94-007-4584-1_42,

hypoxia. Ventilation increases progressively and continues increasing despite the maintenance of the same hypoxic level and finally the baseline ventilation is reset to a higher level than the pre hypobaric level. This HVA is accompanied by an increased ventilatory response to acute hypoxia, indicating an increased ventilatory sensibility and/or reactivity. It is currently accepted that the carotid body (CB) (Bisgard 2000; Powell et al. 2000a) and the reduced PO_2 levels (Bisgard et al. 1986a, b) are key components in the generation of the HVA. However, many of these experiments have been performed under hypobaric conditions (Aaron and Powell 1993; Dwinell and Powell 1999) assuming that O_2 *per se* is the only important variable in producing the HVA, despite of the fact that barometric pressure is substantially reduced. Thus, to separate the effects of hypoxia from the reduced barometric pressure, we studied the effects of chronic normobaric hypoxia (CNH) on basal ventilation and the ventilatory responses elicited by acute hypoxia in the rabbit, and the effects of several chemoexcitatory molecules.

42.2 Methods

42.2.1 Normobaric Hypoxic Exposure

Animals subjected to chronic normobaric hypoxia (CNH) were weighed and their hematocrit measured from blood withdrawn from an ear vein before they were placed in a 0.3 m³ (0.6 × 0.5 × 1.0 m) acrylic chamber. The chamber was initially purged with pure N_2 until the F_IO_2 reached ~ 8.5%. The F_IO_2 was continuously monitored in the chamber with an O_2 sensor (AX300, Teledyne Analytical Instruments, USA) connected to an electronic controller (Zelio SR2 B121BD, Schneider Electric, France), which admitted air or N_2 to maintain F_IO_2 between 8.42% and 10.24%, with a mean level of 9.17 ± 0.09% (mean ± SD; 5 h periods in 10 different experiments). The animals were maintained in this hypoxic environment for 15 days. Every 48 h the chamber was opened for a 5–10 min period necessary for cleaning and restitution of water and food. The ventilatory water and CO_2, and urine volatiles were trapped with a desiccant, $CaCO_3$ and boric acid, respectively. At the end of the CNH period the animals were weighed again and their hematocrit measured.

42.2.2 Physiological Recordings

Ventilatory and cardiovascular variables were recorded from male New Zealand White rabbits (1.77 ± 0.07 kg) without (naïve) or with exposure to CNH. The animals were initially anesthetized (ketamine/xylazine 75/7.5 mg/kg, i.m.) and intubated to measure air flow with a pneumotacograph. The right saphenous vein, the lingual artery and the left femoral artery were catheterized to maintain a surgical level of anesthesia (sodium pentobarbitone, 24 mg/h), apply drugs to the CB and measure arterial blood pressure (Pa), respectively. Tidal volume (V_T), minute volume (V_I) and ventilatory frequency (F_V) were derived from the air flow signal. Cardiac frequency (F_C) was assessed from the electrocardiographic signal recorded in the first derivative. All physiological signals were digitally recorded (PowerLab, ADInstruments) for later analysis. Intracarotid injections of nicotine (0.1–200 μg/kg), ATP (0.1–500 μg/kg), dopamine (0.1–200 μg/kg) and NaCN (0.1–100 μg/kg), and different F_IO_2 levels (0–100%) were used to evaluate the peripheral chemoreflexes. The carotid nerves were severed bilaterally and the vagus nerves cut central to the nodose ganglion.

At the end of the recording session the animals were sacrificed by an anesthetic overdose. Bio-Ethics Committees from the Facultad de Ciencias, Universidad de Chile, and Facultad de Ciencias Biológicas,

P. Universidad Católica de Chile approved the experimental protocols. All data expressed as mean ± SEM, excepted when stated. Statistical analyses performed according to data structure, and decision criteria set at $P<0.05$.

42.3 Results

42.3.1 Responses of Naïve Rabbits to Chemoreceptor Stimulation

Acute hypoxic challenges produced a dose dependent increase in V_I, while hyperoxia reduced it. The changes in V_I were produced by changes in V_T and F_V, although percent changes in volume were larger than dose of frequency in all experimental F_IO_2 values tested. Arterial pressure and heart rate were slightly reduced during hypoxic challenges. All these responses were completely abolished after bilateral section of the carotid and aortic nerves.

Nicotine produced a slight increase in V_I, mostly due to increases in V_T and large but variable changes in F_V. These changes were accompanied by a dose-dependent decrease in both Pa and F_C. Total peripheral chemodenervation had no significant effect on the ventilatory response induced by the maximal nicotine dose (Fig. 42.1a). Intracarotid applications of ATP reduced Pa and F_C in

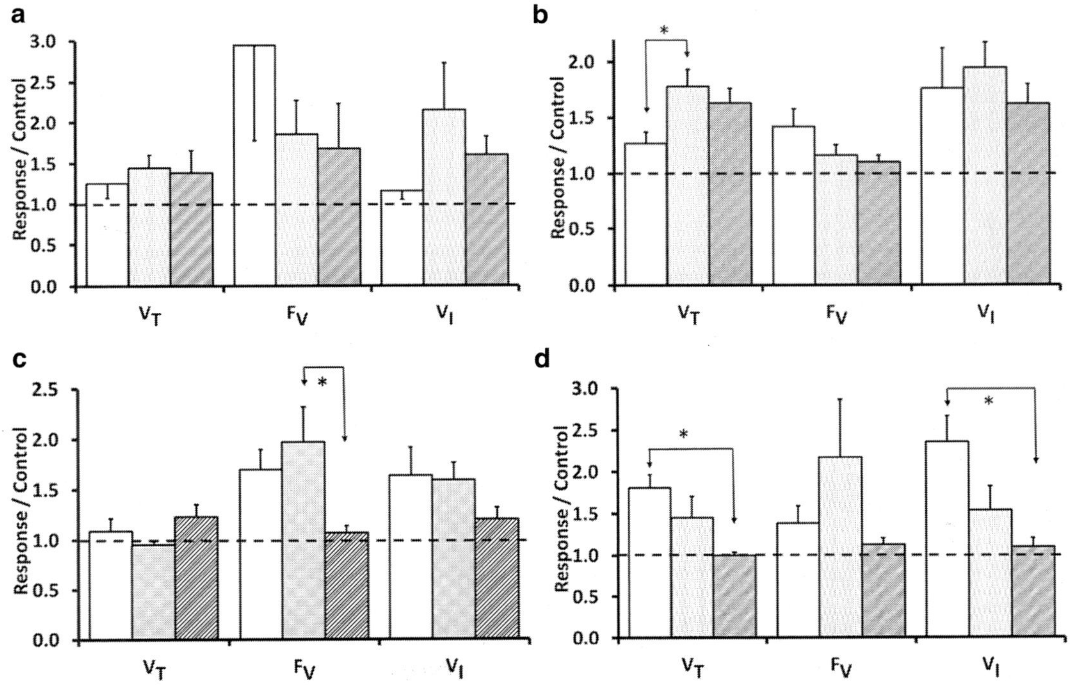

Fig. 42.1 Ventilatory effects of peripheral chemoreceptors chemoexcitants in control (empty bars) and after successive bilateral carotid (*light gray bars*) and aortic (*dark gray bars*) nerve section. (**a**) Responses elicited by nicotine maximal doses (100 μg/kg) were not significantly modified by denervation. (**b**) Responses induced by maximal ATP doses (100 μg/kg) were not significantly modified by denervation, except for a significant increase in V_T after carotid denervation that was reverted by total chemodenervation. (**c**) Responses induced by maximal DA doses (100 μg/kg) were not significantly modified by denervation, except for a significant decrease in F_V after total chemodenervation. (d) Responses induced by maximal NaCN doses (200 μg/kg) were abolished by total chemodenervation. *P<0.05 repeated measures ANOVA. N=5

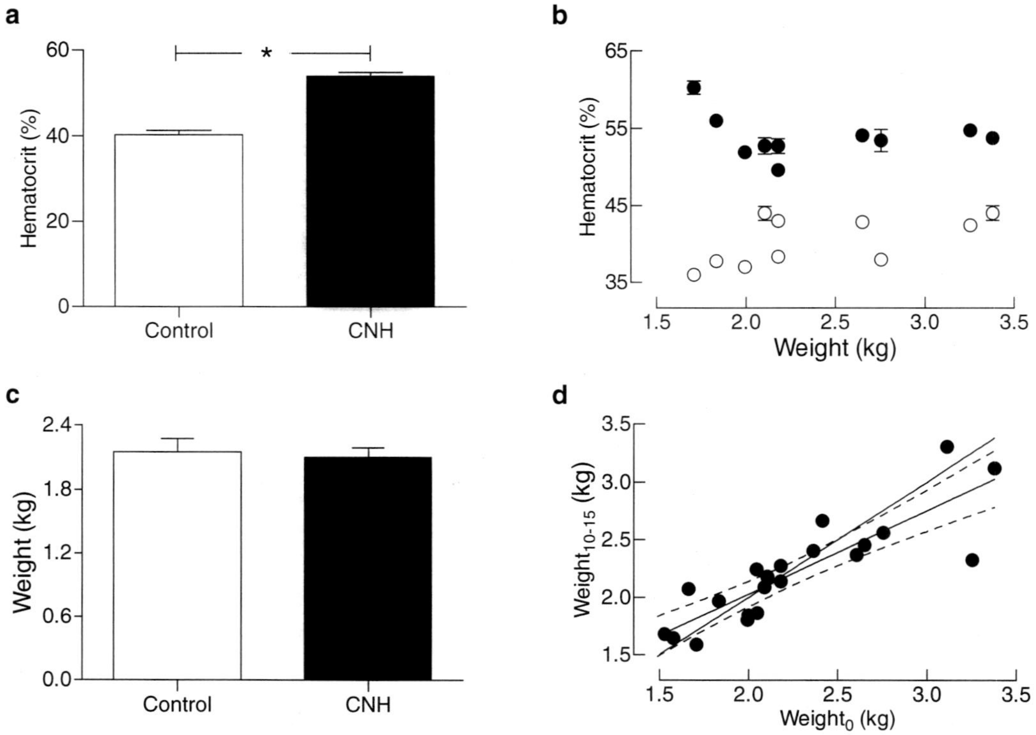

Fig. 42.2 Changes in hematocrit and weight induced by 15 days of chronic normobaric hypoxia (CNH). (**a**) Rabbits subjected to NCH (*filled bar*) significantly (*, $P<0.0001$; paired *t*-test; $n=10$) increased their hematocrit with respect to their initial values (*empty bar*). (**b**) The initial (empty circles) and final (*filled*) hematocrit were uncorrelated ($P>0.05$; F-test; $n=10$) with weight. (**c**) The animals gain no weight ($P>0.05$; Wilcoxon paired ranks test) during the CNH period. (**d**) The weight at the end of the CNH period was related to the initial weight ($P<0.001$; F-test; $n=19$) with a slope significantly lower than 1 ($P<0.001$; F-test)

a dose-dependent manner and slightly increased V_I by increasing both V_T and F_V. Denervation of carotid bodies slightly and significantly increased V_T responses induced by maximal ATP doses (Fig. 42.1b, $P<0.05$, repeated measures ANOVA), but ventilatory responses to maximal ATP doses remained largely unmodified after total peripheral chemodenervation (Fig. 42.1b). Dopamine increased V_I in a dose dependent manner, largely due to changes in F_V without major modifications in V_T. These changes were accompanied by reductions in Pa that were not accompanied by modifications of F_C. Bilateral total chemodenervation had no major impact on evoked ventilatory responses, except for a non-significant reduction of F_V with respect to controls but significant ($P<0.05$, repeated measures ANOVA) with respect to carotid denervation (Fig. 42.1c). NaCN produced a slight reduction of Pa and F_C and an increase in V_I, due to increases of V_T and F_V. Bilateral carotid chemodenervation reduced slightly but not significantly both V_T and V_I, while total peripheral chemodenervation significantly reduced V_T and V_I ($P<0.05$, repeated measures ANOVA), without significantly modifying F_V (Fig. 42.1d).

42.3.2 Physiological Modifications in Rabbits Subjected to CNH

After 15 days of CNH the hematocrit (Fig. 42.2a) of the animals increased significantly from 40.34±1.01% to 53.94±0.89% ($P<0.05$; Student paired *t*-test; $n=10$). The animals did not gain

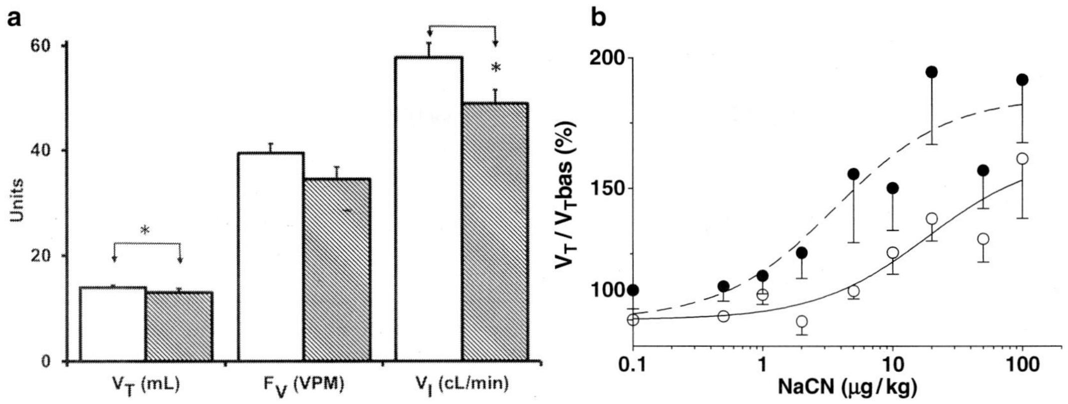

Fig. 42.3 Effects of CNH on rabbit basal ventilation in normoxia and responses to NaCN. (**a**) Basal (normoxic) V_T and V_I were significantly reduced (* $P<0.05$; Student *t*-test) in CNH (*lined bars*; n = 5) with respect to naïve (*empty bars*; n = 11) rabbits, without modification of F_V. (**b**) Dose response changes in V_T induced by NaCN in naïve (*empty dots, continuous line*; n = 5) and in CNH (*filled dots, segmented line*; n = 7) animals. ED_{50} was significantly reduced from 18.6 ± 10.4 to 3.6 ± 2.3 ($P<0.01$; 2 way ANOVA) in CNH rabbits

weight during the CNH exposure (Fig. 42.2c). We found a significant linear correlation between the initial and final weight, although the slope was significantly lower than unity (Fig. 42.2d). There was no significant correlation between the weight and the hematocrit at the beginning or the end of the CNH period (Fig. 42.2b), or between the initial and final hematocrit.

Basal V_I and V_T were significantly reduced ($P<0.05$; Student *t*-test) in CNH animals with respect to naïve ones, while F_V was reduced without reaching significance level (Fig. 42.3a).

42.3.3 Responses of CNH Rabbits to Chemoreceptor Stimulation

Intracarotid injections of NaCN produced a dose dependent increase in V_T and V_I that was dependent on both carotid and aortic chemosensory innervation as in naïve rabbits (Fig. 42.1d). However, the dose–response curve of CNH rabbits was significantly shifted to the left with respect to naïve animals. Thus, ED_{50} was reduced from 18.6 ± 10.4 μg/kg in naïve animals to 3.6 ± 2.3 μg/kg in CNH animals ($P<0.01$; 2 way ANOVA), without significant modification of the slope or maximal response ($P>0.01$; 2 way ANOVA).

42.4 Discussion

The ventilatory and cardiovascular responses to acute hypoxia (Chalmers et al. 1967; Korner and Edwards 1960) and NaCN (Docherty and McQueen 1979; Matsumoto 1986) were similar to those previously reported. However, CB denervation reduced the responses that were abolished only after total peripheral chemodenervation. In non-anesthetized rabbits, the hypoxic ventilatory responses are completely dependent on carotid afferents with negligible participation of the aortic chemoreceptors (Bouverot et al. 1973; Chalmers et al. 1967; Korner and Edwards 1960). The observed differences could be partly explained by the use of anesthetic.

In naïve rabbits, nicotine and dopamine increased V_I, increasing V_T and F_V the former and mostly F_V the latter. Carotid denervation had no significant effect on the ventilatory responses induced by maximal doses of nicotine or dopamine, while total peripheral chemodenervation only significantly reduced F_V responses induced by maximal dopamine doses. Nicotine increases (Docherty and McQueen 1979) while dopamine reduces chemosensory discharges (Docherty and McQueen 1979; Ponte and Sadler 1989). ATP increased V_T, F_V and V_I, and reduced Pa and F_R, effects that were not abolished by total peripheral chemodenervation. It noteworthy that carotid denervation significantly augmented F_V increases induced by ATP. Thus, the increases in ventilation elicited by maximal nicotine, dopamine and ATP doses which were resistant to peripheral chemodenervation may result from centrally mediated effects of these drugs.

The 15 day CNH period produced a significant increase of the rabbit hematocrits. The initial hematocrit was similar to that previously reported in neonatal (Baker et al. 1997) and adult rabbits (Harcourt-Brown and Baker 2001), and the final levels were similar to those obtained after 30 days of CNH in rabbits subjected to similar F_IO_2 (Baker et al. 1997). On the other hand, the rabbits show no increase in weight during the CNH exposure, similar to neonatal rabbits subjected to a similar F_IO_2 level for 30 days (Baker et al. 1997). Similar changes have been reported in cats subjected to CNH for 28 days (Barnard et al. 1987). Thus, the observed changes in these variables indicate that the hypoxic challenge was sufficient to evoke a physiologically adaptive response similar to that previously reported in rabbits and cats.

In CNH rabbits V_T and V_I were significantly reduced with respect to naïve animals, without a significant reduction in F_V. Thus, opposed to other animal or chronic hypobaric hypoxia models (Aaron and Powell 1993; Chen et al. 2002) basal ventilation appeared depressed in CNH rabbits, suggesting that other components of the ventilatory regulatory pathway are modulated in a different way in this species. Basal ventilation increases in cats subjected to CNH for 28 days, with a concomitant increase in CB chemosensory afferent responses to acute hypoxia (Barnard et al. 1987). However, CB chemosensory afferent activity at normoxic and hyperoxic levels was not significantly modified by CNH (Barnard et al. 1987). Thus, normoxic afferent chemosensory activity could remain unchanged in CNH rabbits, and the changes in basal ventilation may result from modifications at different levels of the ventilatory control (Powell et al. 2000b). Conversely, discharges recorded from *in vitro* isolated CB preparations from chronic hypobaric hypoxic animals show a time dependent increase of discharge at normoxia and hyperoxia that reaches a plateau after about 9 days of treatment (Chen et al. 2002). Altogether, these data suggest that chemosensory hypoxic sensitivity could be differentially modulated by PO_2 and the atmospheric pressure. Recording the afferent chemosensory activity in CNH rabbits could provide further insight on the mechanisms involved in the CNH reduced normoxic ventilation.

Because most of the ventilatory effects induced by the chemoexcitants tried in this study in naïve animals appear to exert their effects through central mechanisms, we evaluated the ventilatory reflexes in CNH rabbits only with intracarotid NaCN injections. The relationship between the elicited ventilatory responses and the NaCN dose in CNH rabbits presented similar maximal increases and a similar slope than in naïve animals, suggesting that similar mechanisms operate in both CNH and naïve animals. However, the ED_{50} of the curve was reduced in CNH with respect to naïve animals, suggesting an increased sensitivity in the ventilatory response. Chemosensory denervation completely abolished this NaCN-induced response indicating that chemosensory reflex sensitivity is increased in CNH, and suggests that this increase is at the peripheral chemoreceptor level. Changes in transmitter and receptor expression (Chen et al. 2002) or transmitter relevance (He et al. 2005) have been reported to occur in the rat CB as a result of chronic hypoxia. Moreover, electrical properties of CB chemoreceptor cells are also modified in chronic hypoxic cell cultures (Kääb et al. 2005; Stea et al. 1995). Thus, several changes at the transduction and/or transmission processes in the CB can be modified by CNH leading to an increase in the peripheral chemoreceptor sensitivity to NaCN stimulation.

In summary, in naïve rabbits ventilatory responses to hypoxia and NaCN are dependent on peripheral chemosensory inputs, while nicotine, dopamine and ATP exert their effects through other peripheral afferents or central mechanisms. Rabbits subjected to CNH present reduced ventilation, but their responses to NaCN are increased in sensitivity. Further experiments are needed to elucidate the mechanisms responsible for these acclimatized responses.

Acknowledgements Study supported by Grant 1090157 from the Fondo Nacional de Desarrollo Científico y Tecnológico (FONDECYT), Chile.

References

Aaron EA, Powell FL (1993) Effect of chronic hypoxia on hypoxic ventilatory response in awake rats. J Appl Physiol 74:1635–1640

Baker JE, Curry BD, Olinger GN, Gross GJ (1997) Increased tolerance of the chronically hypoxic immature heart to ischemia Contribution of the KATP channel. Circulation 95:1278–1285

Barnard P, Andronikou S, Pokorski M, Smatresk N, Mokashi A, Lahiri S (1987) Time-dependent effect of hypoxia on carotid body chemosensory function. J Appl Physiol 63:685–691

Bisgard GE (2000) Carotid body mechanisms in acclimatization to hypoxia. Respir Physiol 121:237–246

Bisgard GE, Busch MA, Forster HV (1986a) Ventilatory acclimatization to hypoxia is not dependent on cerebral hypocapnic alkalosis. J Appl Physiol 60:1011–1015

Bisgard GE, Busch MA, Daristotle L, Berssenbrugge AD, Forster HV (1986b) Carotid body hypercapnia does not elicit ventilatory acclimatization in goats. Respir Physiol 65:113–125

Bouverot P, Candas V, Libert JP (1973) Role of the arterial chemoreceptors in ventilatory adaptation to hypoxia of awake dogs and rabbits. Resp Physiol 17:209–219

Chalmers JP, Korner PI, White SW (1967) The relative roles of the aortic and carotid sinus nerves in the rabbit in the control of respiration and circulation during arterial hypoxia and hypercapnia. J Physiol 188:435–450

Chen J, He L, Dinger B, Stensaas L, Fidone S (2002) Role of endothelin and endothelin a-type receptor in adaptation of the carotid body to chronic hypoxia. Am J Physiol Lung Cell Mol Physiol 282:L1314–L1324

Docherty RJ, McQueen DS (1979) The effects of acetylcholine and dopamine on carotid chemosensory activity in the rabbit. J Physiol 288:411–423

Dwinell MR, Powell FL (1999) Chronic hypoxia enhances the phrenic nerve response to arterial chemoreceptor stimulation in anesthetized rats. J Appl Physiol 87:817–823

Harcourt-Brown FM, Baker SJ (2001) Parathyroid hormone, haematological and biochemical parameters in relation to dental disease and husbandry in rabbits. J Small Anim Pract 42:130–136

He L, Dinger B, Fidone S (2005) Effect of chronic hypoxia on cholinergic chemotransmission in rat carotid body. J Appl Physiol 98:614–619

Kääb S, Migel-Velado E, López-López JR, Pérez-García MT (2005) Down regulation of Kv3.4 Channels by chronic hypoxia increases acute oxygen sensitivity in rabbit carotid body. J Physiol Lond 566:395–408

Korner PI, Edwards AWT (1960) The immediate effects of acute hypoxia on the heart rate, arterial pressure, cardiac output and ventilation of the unanaesthetized rabbit. Exp Physiol 45:113–122

Matsumoto S (1986) Effects of carotid body chemoreceptor stimulating and depressing agents on internal intercostal muscle activity in the rabbit. Jap J Physiol 36:1001–1013

Ponte J, Sadler CL (1989) Interactions between hypoxia, acetylcholine and dopamine in the carotid body of rabbit and cat. J Physiol 410:395–410

Powell FL, Dwinell MR, Aaron EA (2000a) Measuring ventilatory acclimatization to hypoxia: comparative aspects. Respir Physiol 122:271–284

Powell FL, Huey KA, Dwinell MR (2000b) Central nervous system mechanisms of ventilatory acclimatization to hypoxia. Respir Physiol 121:223–236

Stea A, Jackson A, Macintyre L, Nurse CA (1995) Long-term modulation of inward currents in O_2 chemoreceptors by chronic hypoxia and cyclic AMP *in vitro*. J Neurosci 15:2192–2202

Chapter 43
Effect of Chronic Caffeine Intake on Carotid Body Catecholamine Dynamics in Control and Chronically Hypoxic Rats

Silvia V. Conde, Ana Obeso, Emília C. Monteiro, and Constancio Gonzalez

Abstract Caffeine is the most commonly psychoactive drug, an habitual drink in high altitude sporting, and when acutely taken, it causes profound alterations in carotid body (CB) function and ventilation via adenosine receptors antagonism. In the present work we have investigated the effects of chronic caffeine ingestion in catecholamine (CA) dynamics in the carotid body of control and chronic hypoxic rats. Four groups of animals were used: normoxic (N), caffeine-treated normoxic (1 mg/mL in drinking water 15 days; CafN), chronic hypoxic (CH, 12%O_2, 15 days) and chronically hypoxic-caffeine-treated (CafH).. Caffeine intake in controls rats did not modify CA content, synthesizing, and releasing responses, and the expression of tyrosine hydroxylase. CH increased dopamine content, synthesis, and basal and acute hypoxia-induced release; chronic caffeine ingestion augmented CH effects. Findings indicate that chronic caffeine ingestion in normoxic rats did not modify dopamine dynamics at the CB, but increases dopaminergic system during chronic hypoxia.

Keywords Catecholamines • Carotid body • Caffeine • Chronic hypoxia

43.1 Introduction

Caffeine is the most widely consumed behaviourally active substance in the world. Acute and, especially, chronic caffeine intake appears to have minor negative consequences on health. Nevertheless, caffeine is described as a "model drug of abuse" (Fredholm et al. 1999).

Caffeine stimulates ventilation and decrease apneic episodes in premature infants (Aranda and Turmen 1979; Bairam et al. 1987), an effect attributed to a direct action in the central respiratory neurons (Eldridge et al. 1983) by inhibiting A_2 adenosine receptors (Sawynok 1995). However, consistent

S.V. Conde (✉) • E.C. Monteiro
CEDOC, Departamento de Farmacologia, Faculdade de Ciências Médicas
Universidade Nova de Lisboa, Campo Mártires da Pátria 130, 1169-056 Lisbon, Portugal
e-mail: silvia.conde@fcm.unl.pt; emilia.monteiro@fcm.unl.pt

A. Obeso • C. Gonzalez
Departamento de Bioquímica y Biología Molecular y Fisiología, Universidad de Valladolid, Facultad de Medicina, Instituto de Biología y Genética Molecular, CSIC.
Ciber de Enfermedades Respiratorias, CIBERES, Instituto de Salud Carlos III, Valladolid 47005, Spain
e-mail: aobeso@ibgm.uva.es; constanc@ibgm.uva.es

C.A. Nurse et al. (eds.), *Arterial Chemoreception: From Molecules to Systems*, Advances in Experimental Medicine and Biology 758, DOI 10.1007/978-94-007-4584-1_43,

with an excitatory effect of adenosine observed in ventilation (Monteiro and Ribeiro 1987) and in carotid sinus nerve chemosensory discharges (McQueen and Ribeiro 1983), it has been shown in unanaesthetised monkeys that caffeine, antagonist nonselective to adenosine receptor isoforms, significantly attenuated the CB-mediated hyperventilation occurring while the animals were breathing 10% O_2 (Howell and Landrum 1995). In fact, caffeine when applied acutely inhibits rat CB through an action on post-synaptic A_{2A} and pre-synaptic A_{2B} receptors (Conde et al. 2006b). Recently, it has been shown that chronic exposure to caffeine alters both CB dopaminergic and adenosinergic systems and central regulation of breathing under baseline conditions and in response to hypoxia in newborn pups (Julien et al. 2011). Nevertheless, little is known about the long-term effects of caffeine in adult CB chemoreceptors. In the present work we have investigated the effect of chronic caffeine intake on CA dynamics in the rat CB in control and CH rats. We found that chronic caffeine treatment in control rats did not modify CB catecholamine synthesis, content and release. Also, we have found that chronic caffeine treatment augments dopamine synthesis, content and acute hypoxia-induced release in the CB of CH rats.

43.2 Methods

43.2.1 Protocols and Surgical Procedures

Experiments were performed in Wistar adult rats (250–350 g) obtained from *vivarium* of the Faculty of Medicine of the University of Valladolid. The Institutional Committee of the University of Valladolid for Animal Care and Use approved the protocols. Control animals were maintained in room air atmosphere and CH rats were kept for 15 days in a chamber equilibrated with a gas mixture of 11–12%O_2 in 88–89%N_2; $PO_2 \approx 80$–84 mmHg and a constant a flow of 3 L/min. To study the chronic caffeine intake we subdivided the groups into two (with and without caffeine), and kept the rats drinking 1 mg/mL of caffeine for 15 days (Conde et al. 2012). For the experiments aimed to measure endogenous CA content and synthesis of ^{3}H-CA we used 4–8 CBs/experiment and 4–8 superior cervical ganglion (SCGs)/experiment. Due to the small size of the CB (≈50 μg) (Conde et al. 2006b) in the experiments where the release of endogenous CA were analysed it was necessary to make pools of 4 CBs in each experiment to reach limits of detection of analytical techniques used. In all instances animals were killed by an intracardiac overdose of sodium pentobarbital. Plasma caffeine concentrations were measured as described by Conde et al. (2012).

43.2.2 Measurement of Catecholamines Content, Synthesis and Release

For the measurement of CA content, CBs and SCG were cleaned and transferred to cold Eppendorfs containing 0.4 N perchloric acid. Tissues were weighed and homogenized. The homogenates of CBs and SCGs were centrifuged and supernatants were collected to quantify CA content. For study the release of CA from CB, the CBs were incubated in Tyrode solution. The incubation media were kept at 37 °C and continuously bubbled with 20%O_2/5%CO_2/75%N_2 saturated with water vapour, except when hypoxic stimuli were applied. Specific protocols for stimulus (7 and 2%O_2-equilibrated solutions) are provided in the Results. The collected fractions were acidified with PCA. At the end of the experiment the CBs were immersed in PCA and weighed. The collected fractions were centrifuged and supernatants were recovered and reserved to quantify CA. Endogenous CA release and tissue content was quantified by HPLC (Conde et al. 2006b). For measurement of rate of synthesis of ^{3}H-CA, protocols have been described in detail elsewhere (Vicario et al. 2000; Conde et al. 2006b).

43.2.3 Western Blot Analysis of Tyrosine Hydroxylase Expression in the Carotid Body

CBs and SCGs were homogenized in Zurich medium (Tris–HCl 10 mM; EDTA 1 mM; NaCl 150 mM; Triton X-100 1%; Sodium cholate 1%; SDS 0.1%) and a cocktail of protease inhibitors. Samples and molecular weight markers were separated by SDS-Page (10% with a 5% concentrating gel) and electro-transferred to PVDF membranes. After blocking, the membranes were incubated overnight with a mouse anti-TH (1:20,000, Abcam, Cambridge, UK). After 3 washing periods of 10 min with TTBS, the membranes were incubated with anti-mouse HRP (1:10,000, Sta Cruz Biotechnology, USA) in TTBS for 1 h at room temperature. After 3 washes of 10 min with TTBS the membranes were incubated with enhanced chemiluminescence (ECL, Sta Cruz Biotechnology, USA). The membranes were then re-probed and tested for βactin immunoreactivity (bands in the 42 kDa region) in order to compare and normalise the expression of proteins with the amount of protein loaded. The density of the bands on Western blots was quantified by a PDI Scanner and by Origin 7.0.

43.2.4 Data Analysis

Data were evaluated using Graph Pad Prism Software, version 4 and was presented as mean ± SEM. The significance of the differences between the means was calculated by One and Two-Way Analysis of Variance (ANOVA) with Dunnett's and Bonferroni multiple comparison tests, respectively. *P* values of 0.05 or less were considered to represent significant differences.

43.3 Results

No differences were observed in weight gain and fluid intake between the chronic caffeine groups in normoxic and CH rats (data not shown). In both N and CH rats we did not detect the presence of caffeine or its metabolites. Plasma caffeine concentration in CafN group was 3.69 ± 1.02 μg/ml being the value in CafH not significantly different. These values are within the range of 0.37–5.95 μg/ml previously described by Gasior et al. (2002) for doses of 0.25–1 mg/ml of caffeine in drinking water in rats.

43.3.1 Carotid Body CA Metabolism in Control and Chronic Hypoxic Rats. Effect of Chronic Caffeine Intake

Figure 43.1a shows mean DA content of CB in the four groups of animals. Mean DA content in control CB (N group) was 311.1 ± 57.60 pmol/mg tissue (n = 10). Chronic caffeine intake in normoxic rats (CafN) did not modify the DA content. Chronic hypoxia for 15 days (CH) resulted in an increase in DA content in the CB to reach 705.7 ± 54.42 pmol/mg tissue (n = 8). When rats drank caffeine while exposed to hypoxia (CafH) the DA content reached 1,021.00 ± 79.06 pmol/mg ($p<0.001$ and $p<0.01$ vs. CH group; Fig. 43.1a). The rate of CB ^{3}H-DA synthesis in the four experimental groups calculated as described in Conde et al. (2006b) is shown in Fig. 43.1b. In N rats it amounted to 31.65 ± 4.02 pmol/mg tissue/hour (n = 8) and CH exposure nearly doubled it ($p<0.001$). The effects of caffeine ingestion on ^{3}H-DA synthesis were comparable to those observed in DA content, i.e., caffeine in normoxia did

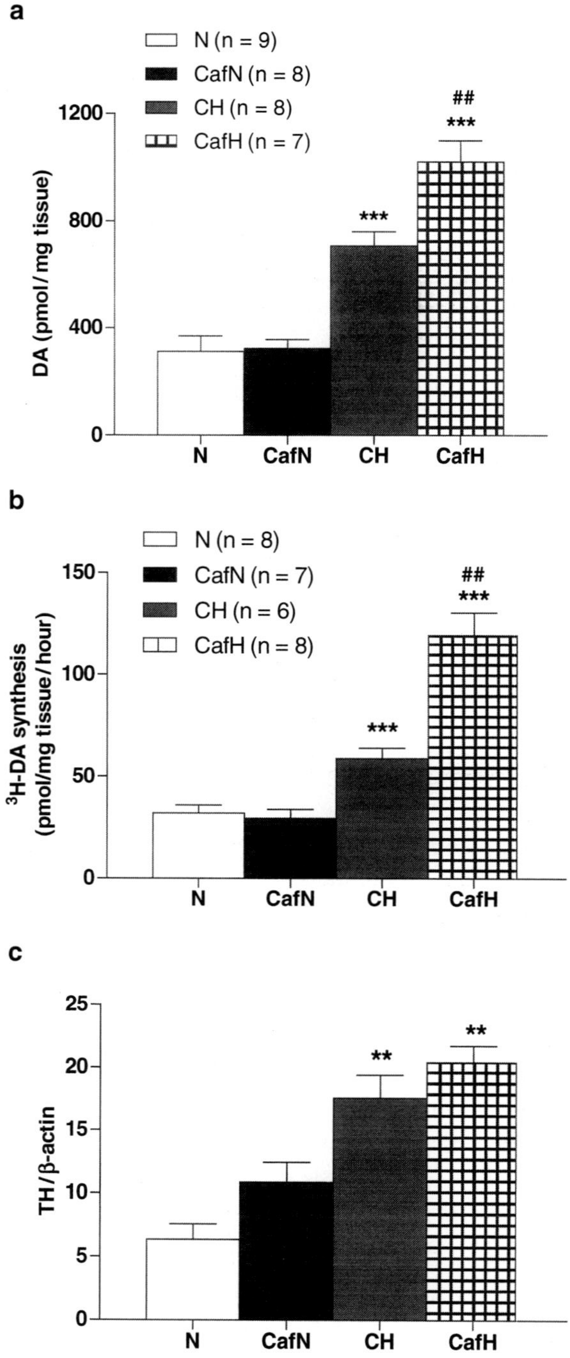

Fig. 43.1 Effect of 15 days of chronic hypoxia and/or caffeine intake (1 g/l) on dopamine content (**a**), DA synthesising capacity (**b**) and tyrosine hydroxylase expression (**c**) in the carotid body of control and chronically hypoxic animals. N- normoxic animals; CafN – normoxic rats with 15 days of caffeine ingestion; CH – rats submitted to 15 days of chronic hypoxia; CafH – rats that drink caffeine during the exposure to chronic hypoxia. Data represent means ± SEM. *P < 0.05;***P < 0.001 vs normoxic rats, ##P < 0.01 vs chronic hypoxic rats; One-Way ANOVA with Bonferroni multi-comparison test

not alter the rate of synthesis, and caffeine intake plus chronic hypoxia augmented further the effects of CH applied alone ($p < 0.001$ CH and CafH vs. N group and **$p < 0.01$ CafH vs CH; Fig. 43.1b). The DA turnover time was estimated by dividing endogenous content by the rate of DA synthesis (i.e., {pmol/mg tissue/[(pmol/mg tissue)/hour]}). The turnover time for DA did not vary significantly among experimental conditions as in all groups oscillated between 9–12h indicating that DA storage and synthesis capacity of chemoreceptor cells varied proportionally.

Table 43.1 Effect of chronic caffeine intake in norepinephrine metabolism and DA/NE ratio in control and chronic hypoxic rats

	Normoxia	Normoxia + Caffeine	Chronic hypoxia	Chronic hypoxia + Caffeine
NE content	88.99 ± 13.03	92.84 ± 5.90	207.8 ± 14.58 ***	175.4 ± 20.99
	(n = 10)	(n = 8)	(n = 8)	(n = 7) ***
^{3}H-NE synthesis	1.82 ± 0.32	1.27 ± 0.17	1.22 ± 0.18	2.64 ± 0.36
	(n = 8)	(n = 7)	(n = 6)	(n = 8) ###
Turnover time	51.61 ± 3.69	73.24 ± 4.65	170.96 ± 11.99	66.20 ± 7.92
	(n = 9)	(n = 8)	(n = 8)***	(n = 7) ###
DA/NE ratio	3.97 ± 0.32	3.52 ± 0.18	3.55 ± 0.34	6.53 ± 0.42
	(n = 9)	(n = 8)	(n = 8) ***	(n = 7) ***

NE content is expressed as pmol/mg tissue. ^{3}H-NE synthesis is expressed as (pmol/mg tissue)/h. Turnover times are expressed in hours and were calculated by dividing each individual datum of NE by the mean rate of synthesis. DA/NE ratio was calculated by dividing DA content (Fig. 43.1) by the NE content for each CB. *** $p<0.001$ vs normoxic values; ### $p<0.001$ vs chronic hypoxic values.

Caffeine is a nonselective adenosine receptor antagonist and Kobayashi and Milhorn (1999) have suggested that adenosine could modulate CB gene expression. Therefore, we have performed Western-Blot for TH, the rate limiting enzyme in CA biosynthesis in CBs and SCG of the four experimental groups. Figure 43.1**c** shows amount of TH protein (normalized against densities of correspondent β-actin bands). Data indicate that in CafN group there was a non-significant increase in TH ($p=0.1$) while in CH and CafH groups TH increased, respectively, by factors of 2.8 and 3.3 ($p<0.01$).

Norepinephrine (NE) levels were similar in N and CafN groups (Table 43.1). Chronic hypoxia increased significantly the levels of NE however, contrary to the situation with DA, there were not differences in NE levels between CH and CafH groups (Table 43.1). Chronic caffeine intake did not modify the rate of ^{3}H-NE synthesis in control normoxic rats but significantly increased it in CH rats (Table 43.1). From the calculation of mean (DA + DOPAC)/NE ratios in all paradigms, it is evident that the rat CB remains a dopaminergic organ in all experimental situations studied, and that the dopaminergic trait increased when rats are exposed to chronic hypoxia conjunctly with chronic caffeine intake (Table 43.1). In the case of NE there was a statistically significant increase in the NE turnover time with chronic hypoxia exposure (Table 43.1). The intake of caffeine during CH returned the NE turnover time to a value similar to control value (Table 43.1).

Comparative studies in the SCG indicated that the effects of CH and caffeine + CH are rather specific for the CB as neither DA and NE content and ^{3}H-DA and ^{3}H-NE rate synthesis were different among the groups of animals (data not shown).

43.3.2 Carotid Body Dopamine Release in Normoxic and Chronic Hypoxic Animals. Effects of Caffeine Ingestion

General protocol for release experiments consisted in a sequential incubation of pools of 4 CBs in normoxic solutions (10 min), moderate hypoxia (7%O_2-saturated), 2 additional normoxic incubations, a strong hypoxic incubation (2% O_2-saturated) and a final normoxic incubation. Normoxic means 20%O_2-equilibrated solutions for N and CafN groups and 12%O_2-equilibrated for CH and CafH groups. Figure 43.2a show, mean normoxic (basal) release of DA in the four experimental groups.

Figure 43.2a shows normoxic DA release in N group that amounted to 10.44 ± 1.54 pmol/mg tissue/10 min ($n=6$), was nearly identical in CafN group, and more than doubled in CH and CafH groups, tending to be larger in CafH ($p<0.001$). Figures 43.2b and c show the release of DA induced

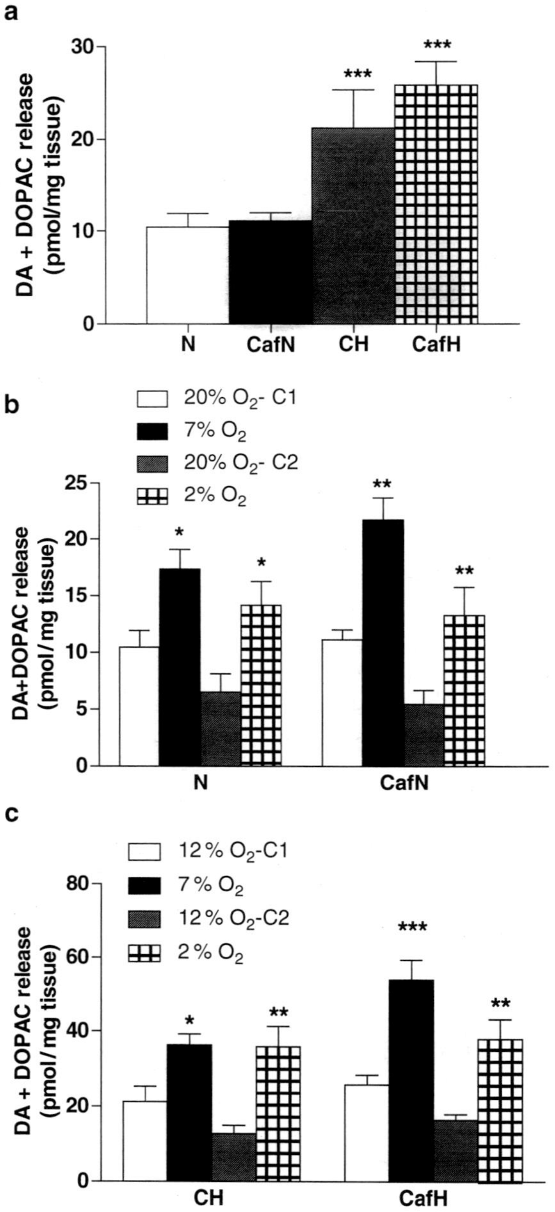

Fig. 43.2 Effect of chronic caffeine intake (1 g/l) on basal and evoked release of catecholamines from rat carotid bodies in control and chronic hypoxic animals. (**a**) Release of dopamine+DOPAC in normoxia. Note that basal release corresponds in normoxic rats to the release while incubating the organs in 20%O_2-equilibrated solutions and in chronic hypoxic animals to the release while incubating the organs in response 12% O_2-equilibrated solutions. (**b, c**) Release of dopamine and DOPAC from CB in response to acute hypoxia (7% and 2% O_2) in normoxic and chronic hypoxia rats, respectively. C1 and C2 represent basal values before the application of the acute hypoxic stimulus, corresponding to the release of adenosine from CB in response to 20% O_2 in normoxic rats and 12% O_2 in chronic hypoxic rats. N – normoxic rats (n=6), CafN - normoxic submitted to chronic caffeine intake (n=7), CH – chronic hypoxic rats (n=5), CafH – chronic hypoxic rats submitted to chronic caffeine intake (n=6). Data represent means±SEM. *P<0.05, ***P<0.01 Normoxic rats vs. all other paradigms. One-Way ANOVA and Two-Way ANOVA with Bonferroni multi-comparison test

by acute hypoxic challenges in CBs of the four experimental groups. Figures also show that the normoxic (basal) release immediately prior to hypoxic challenges decreases as the experiment proceeds so that the second period of basal incubation yields a smaller release than the first one (empty vs. grey columns). Similarly, in N group, the release induced by hypoxic stimuli also tends to decrease as the experiment proceeds, even if the second hypoxic stimulus is more intense; but the ratio of evoked/basal release is greater for the more intense hypoxia. Acute moderate hypoxia (7%O_2) produced an increase in the release of DA that percentagewise was comparable in N and in CH (≈80% above basal), but in absolute amounts was much greater in CH (36.36±2.87 pmol/mg tissue) than in N animals (17.36±1.73 pmol/mg tissue). Acute intense hypoxic stimulation, 2%O_2/10 min, elicited a release response that represented (vs. the correspondent basal) a 123.36% increase in CB of N rats and a 119.17% in CB of CH (Fig. 43.2**b**, **c**); i.e., in absolute amounts the DA released in the CB of CH animals was about double than in those of N animals. Chronic caffeine did not produce significant effects of basal and acute hypoxia induced release in normoxic rats, although a tendency to increase the hypoxia-induced release was noticed. In CH rats chronic caffeine increased release induced by acute moderate hypoxia (7%O_2) but not elicited by intense hypoxia.

43.4 Discussion

The present study demonstrates that chronic caffeine ingestion does not modify catecholaminergic system in the CB of normoxic rats but increases dopaminergic dynamics in the CB of chronic hypoxic rats. Findings show that chronic caffeine intake: (1) in normoxic rats did not modify CB CA content, synthesis and DA basal release, showing a tendency to increase low PO_2 induced release of DA and TH expression; and that (2) in chronic hypoxic rats significantly increased CB CA content, synthesis and low PO_2 induced release.

The concentrations of caffeine found in the plasma in the present study indicate that most (if not all) effects of caffeine herein observed should be mediated via inhibition of adenosine receptors, implying that observed effects result from antagonism of adenosine receptors (Fredholm et al. 1999).

The effect of chronic caffeine on DA metabolism observed both in the CB of N and CH animals is opposite to those previously found in response to acute caffeine. Acute caffeine produces a marked inhibition of basal and low intensity hypoxia induced release of DA (Conde et al. 2006a, b) while in present experiments we observe that chronic caffeine does not affect basal release and increases the release induced by moderate hypoxia in N rats and increase the release induced by moderate hypoxia in chronic hypoxic rats (Fig. 43.2). Additionally, although there are no data available on the acute effects of caffeine on DA synthesis in chemoreceptor cells, due to interaction between D2 dopamine receptors and A_{2B} receptors (Conde et al. 2008) it should be expected that acute caffeine produces a decrease in the rate of DA synthesis, and presumably content; as Fig. 43.1 show chronic caffeine ingestion in N animals caused no change in either rate of synthesis or content in normoxic rats and increase these parameters in chronic hypoxic rats. We do not have molecular explanations to these observations, however, we can postulate that an "opposite effect" in acute vs. chronic caffeine treatment could be happen as discussed by Jacobson et al. (1996) or that tolerance to chronic caffeine could be developed as described in the shell of rat nucleus accumbens (Quarta et al. 2004). Yet, we can speculate that chronic inhibition of adenosine receptors by caffeine should tend to produce an upregulation of adenosine receptors and it can be suggested that in normoxia the effects of the low adenosine concentrations would be minimized by the direct blockade by caffeine, while during the chronic hypoxia the augmented release of adenosine can partially override the inhibitory action of caffeine.

Also, the results described in the present paper are, in a sense, opposite to those recently described by Julien et al. (2011). These authors found that in newborn pups chronic exposure to caffeine in basal conditions alters dopaminergic system as the increase in ventilation produced by domperidone, a D2 antagonist, is abolished. These results underlie that the antagonistic interaction between A_2 adenosine receptors and dopamine D2 receptors found in adulthood in the CB (Conde et al. 2006a, 2008) also seems to be present in newborn pups. In fact, Julien et al. (2011) suggest that an interaction between A_{2A} and D2 receptors exist in newborn pups and that the alteration produced by chronic caffeine is due to desensitisation of dopamine D2 receptors and/or downregulation of their expression level, hence decreasing their tonic control of carotid body activity. We do not have an explanation for the discrepancy between our results and those of Julien, however we know that A_2 and D2 receptors are developmentally regulated in the CB (Gauda et al. 2000) and therefore we can postulate that differences in the expression of D2 receptors and A_2 receptors can account to the discrepancy observed.

To summarize, we can say that exposure to chronic caffeine treatment in normoxia in adulthood as no harmful effects but alters dopamine dynamics in the CB in chronic hypoxia leading probably to an altered respiratory control.

Acknowledgements We want to thank Mª de los Llanos Bravo for technical assistance. The work was supported by BFU2007-61848 (DGICYT), JCyL-GR242, CIBER (FISS-ICiii) and Spanish MEC/Portugal CRUP 2009–0172.

References

Aranda JV, Turmen T (1979) Metylxanthines in apnea of prematurity. Clin Prerinatol 6:87–108

Bairam A, Boutroy MJ, Badonnel Y, Vert P (1987) Theophylline versus caffeine: comparative effects in treatment of idiophatic apnea in the preterm infant. J Pediatr 110:636–639

Conde SV, Obeso A, Vicario I, Rigual R, Rocher A, Gonzalez C (2006a) Caffeine inhibition of rat carotid body chemoreceptors is mediated by A2A and A2B adenosine receptors. J Neurochem 98:616–628

Conde SV, Obeso A, Rigual R, Monteiro EC, Gonzalez C (2006b) Function of rat carotid body chemoreceptors in ageing. J Neurochem 99:711–723

Conde SV, Gonzalez C, Batuca JR, Monteiro EC, Obeso A (2008) An antagonistic interaction between A2B adenosine and D2 dopamine receptors modulates the function of rat carotid body chemoreceptor cells. J Neurochem 107:1369–1381

Conde SV, Nunes da Silva T, Gonzalez C, Mota CM, Monteiro EC, Guarino MP (2012) Chronic caffeine intake decreases circulating catecholamines and prevents diet induced insulin resistance and hypertension in rats. Br J Nutr 107:86–95

Eldridge FL, Milhorn DE, Waldrop TG, Kiley JP (1983) Mechanism of respiratory effects of methylxanthines. Respir Physiol 53:239–261

Fredholm BB, Battig K, Holemn J, Nehlig A, Zvartau EE (1999) Actions of caffeine in the brain with special reference to factors that contribute to its widespread use. Pharmacol Rev 51:83–133

Gasior M, Jaszyma M, Munzar P, Witkin J, Goldberg SR (2002) Caffeine potentiates the discriminative–stimulus effects of nicotine in rats. Psycopharmacol 162:385–395

Gauda EB, Northington FJ, Linden J, Rosin DL (2000) Differential expression of a(2a), A(1)-adenosine and D(2)-dopamine receptor genes in rat peripheral arterial chemoreceptors during postnatal development. Brain Res 872:1–10

Howell LL, Landrum AM (1995) Attenuation of hypoxia-induced increases in ventilation by adenosine antagonists in rhesus monkeys. Life Sci 57:773–783

Jacobson KA, Von Lubitz DKJE, Daly JW, Fredholm BB (1996) Adenosine receptor ligands: differences with acute versus chronic treatment. TIPS 17:108–113

Julien CA, Joseph V, Bairam A (2011) Alteration of carotid body chemoreflexes after neonatal intermittent hypoxia and caffeine treatment in rat pups. Respir Physiol Neurobiol 177:301–312

Kobayashi S, Millhorn DE (1999) Stimulation of expression for the adenosine A2A receptor gene by hypoxia in PC12 cells. J Biol Chem 274:20358–20365

McQueen DS, Ribeiro JA (1983) On the specificity and type of receptor involved in carotid body chemoreceptor activation by adenosine in the cat. Br J Pharmacol 80:347–354

Monteiro EC, Ribeiro JA (1987) Ventilatory effects of adenosine mediated by carotid body chemoreceptors in the rat. Naunyn-Schmiedeberg's Arch Pharmacol 335:143–148

Quarta D, Ferré S, Solinas M, You ZB, Hockemeyer J, Popoli P, Goldberg SR (2004) Opposite modulatory roles for adenosine A1 and A2A receptors on glutamate and dopamine release in the shell of the nucleus accumbens effects of chronic caffeine exposure. J Neurochem 88:1151–1158
Sawynok J (1995) Pharmacological rationale for the clinical use of caffeine. Drugs 49:37–50
Vicario I, Rigual R, Obeso A, Gonzalez C (2000) Characterization of the synthesis and release of catecholamine in the rat carotid body in vitro. Am J Physiol Cell Physiol 278:C490–C499

Chapter 44
Effects of Cigarette Smoke and Chronic Hypoxia on Ventilation in Guinea Pigs. Clinical Significance

Elena Olea, Elisabet Ferrer, Jesus Prieto-Lloret, Carmen Gonzalez-Martin, Victoria Vega-Agapito, Elvira Gonzalez-Obeso, Teresa Agapito, Victor Peinado, Ana Obeso, Joan Albert Barbera, and Constancio Gonzalez

Abstract Ventilatory effects of chronic cigarette smoke (CS) alone or associated to chronic hypoxia (CH), as frequently occurs in chronic obstructive pulmonary disease (COPD), remain unknown. We have addressed this problem using whole-body plethysmography in guinea-pigs, common models to study harmful effects of CS on the respiratory system. Breathing frequencies (Bf) in control (2–5 months old) guinea pigs is 90–100 breaths/min, their tidal volume (TV) increased with age but lagged behind body weight gain and, as consequence, their minute volume (MV)/Kg decreased with age. MV did not change by acutely breathing 10% O_2 but doubled while breathing 5% CO_2 in air. Exposure to chronic sustained hypoxia (15 days, 12% O_2, CH) did not elicit ventilatory acclimatization nor adaptation. These findings confirm the unresponsiveness of the guinea pig CB to hypoxia. Exposure to CS (3 months) increased Bf and MV but association with CH blunted CS effects. We conclude that CS and CH association accelerates CS-induced respiratory system damage leading to a hypoventilation that can worsen the ongoing COPD process.

Keywords Guinea pig • Ventilation • Tobacco • Hypoxia • Carotid body.

E. Olea • J. Prieto-Lloret • C. Gonzalez-Martin • V. Vega-Agapito
E. Gonzalez-Obeso • T. Agapito • A. Obeso • C. Gonzalez (✉)
Department of Biochemistry and Molecular Biology and Physiology,
IBGM Universidad de Valladolid and CSIC, Valladolid, Spain

Ciber de Enfermedades Respiratorias, Valladolid, Spain
e-mail: constanc@ibgm.uva.es

E. Ferrer
Department of Pulmonary Medicine Hospital Clínic-Institut d'Investigacions Biomèdiques
August Pi i Sunyer (IDIBAPS), Universitat de Barcelona, Valladolid, Spain

V. Peinado • J.A. Barbera
Department of Pulmonary Medicine Hospital Clínic-Institut d'Investigacions Biomèdiques
August Pi i Sunyer (IDIBAPS), Universitat de Barcelona, Valladolid, Spain

Ciber de Enfermedades Respiratorias,Valladolid, Spain

C.A. Nurse et al. (eds.), *Arterial Chemoreception: From Molecules to Systems*,
Advances in Experimental Medicine and Biology 758, DOI 10.1007/978-94-007-4584-1_44,

44.1 Introduction

Guinea pigs exposed to cigarette smoke (CS) are valuable models for chronic obstructive pulmonary disease (COPD) to study lung and respiratory pathology (Wright and Churg 2002). To underscore the mechanisms involved in the pathology found in COPD (Churg et al. 2011) most studies focus their interest on the effects of CS on lung parenchyma and more recently on pulmonary circulation (Ferrer et al. 2011). However, a systematic study of the effects of CS on ventilation, including responses to hypoxia and hypercapnia, has not been performed in any laboratory model. In humans, Tobin et al. (1983) found that healthy chronic smokers had resting MV augmented, being the increase due to Bf and TV.

In middle to advanced stages of COPD, CS and chronic hypoxia (CH) coexist in many patients. The study of the interaction between CS and CH results of great interest in order to find if association of CS and CH accelerates the progression of alterations produced by CS alone. The present work has specifically addressed the study in the guinea pig of the ventilatory effects of CH and CS exposure individually and in combination. Our experimental design consisted in exposing the guinea pigs to a smoking protocol of 4 cigarettes/day for 3 months; this smoking protocol does not produce emphysema (Ferrer et al. 2009) nor produces alterations in blood gases (Wright and Churg 1991). Therefore, simultaneous exposure of smoking animals to CH was an absolute requirement to define the combined effects of both stressors, CS and CH, at these early stages of CS induced lung damage.

Prior to the effects of CS and CH we have aimed to settle a basic aspect in the respiratory physiology of guinea pigs, namely, if hypoxia increases their MV as it does in most mammals. Most authors have found that carotid body (CB) mediated ventilation or electrical activity in the carotid sinus nerve in response to hypoxia is very small or totally absent when compared to rat and other laboratory rodents (Blake and Banchero 1985; Curran et al. 1995; Yilmaz et al. 2005; Schwenke et al. 2007). Fernandez et al. (2003) have shown that guinea pigs exhibit a chemosensory drive both in normoxia and hypoxia comparable to that seen in the rat. Blake and Banchero (1985) found that acute hypoxia did not augment ventilation. On the other hand, if the animals were maintained (>3 months) at 4,600 m of altitude they observed a small increase in ventilation (36% above sea level MV; see also Yilmaz et al. 2005). We have performed a basic characterization of the ventilatory pattern of guinea pigs, including the responses to acute hypoxic and hypercapnic tests and the effects of exposure to CH on ventilation.

44.2 Material and Methods

44.2.1 Animals and Experimental Groups

Forty eight male Hartley guinea pigs (7 weeks of age) purchased to Harlam Iberica were housed (3/cage) in the vivarium in a 12 h light–dark cycle at room temperature (20–23 °C). Animals were provided standard chow and water supplemented with vitamin C (1 g/l) *ad libitum*. After 1 week of adaption (i.e., at 2 months of age), animals were distributed in 2 groups of 24 animals and divided in subgroups. Twenty four animals were exposed to cigarette smoke for 3 months: 8 of them were maintained in normal atmosphere (smoking animals, CS) and 16 were exposed to a hypoxic atmosphere the last two smoking weeks (smoking hypoxic animals, CSCH). The other group was similarly divided: 8 animals remained in normal atmosphere for the entire 3 months (control animals, Control) and 16 were exposed to hypoxia for the last 2 weeks (chronic hypoxic animals, CH). Body weight was measured weekly. Animal protocols were approved by the University of Valladolid Institutional Committee for Animal Care and Use following international laws and policies (Guide for the Care and Use of Laboratory Animals, National Institutes of Health, 85–23, 1985).

44.2.2 Smoking Protocol

Animals were daily exposed to the smoke of 4 cigarettes (2R4F; Kentucky University Research; Lexington, KY, USA, 11 mg tar, 0.8 mg nicotine per cigarette), 5 days/week, using a nose-only inhalation system (Protowerx Design Inc; Langley, British Columbia, Canada). Control animals were equally sham-exposed.

44.2.3 Exposure to Chronic Hypoxia

The chamber and experimental maneuvers for CH exposure (12% O_2 in N_2; $PO_2 \approx 85$ mmHg; equivalent to ≈ 4,300 m; 15 days) have been described in prior studies (Caceres et al. 2007).

44.2.4 Plethysmography

The plethysmographic system allows recording pressure fluctuations within the chamber with a high gain differential transducer. Ideally, the frequency of pressure fluctuations would correspond to breathing frequency (Bf, breaths/min). The spurious fluctuations due to animal movements were electronically rejected. The TV is provided by the software from the integration of the inspiratory curve. The system was calibrated automatically by software after steady injection into the chamber of 5 ml air. Ventilatory parameters were measured in conscious, freely moving guinea pigs by whole body plethysmography while breathing room air, 10% O_2, and 5% CO_2 in air. The system (Emka Technologies, Paris, France with software IOX version 1.8.9.4) and recording conditions have been described elsewhere (Agapito et al. 2009). Parameters measured or computed include Bf, TV, and MV.

44.2.5 Data Presentation and Statistics

Data were evaluated using a Graph Pad Prism Software, version 4 (GraphPad Software Inc., San Diego, CA, USA) and were presented as mean ± SEM. The significance of the differences between the means was calculated by One and Two-Way Analysis of Variance (ANOVA) with Newman-Keuls and Bonferroni multiple comparison tests for repeated measurements, respectively. P values of 0.05 or less were considered to represent significant differences.

44.3 Results

44.3.1 Animal's General Status, Body Weight and Hematocrit

The general status of the smoking animals was apparently not different from controls. Weight gain in all groups ran parallel up to the 3 months of age or 1 month of exposure to CS. As exposure increased, weight gain in CS and CSCH was less than in control so that at completion of exposure they weighed 19.5% less than control animals.

Hematocrit was measured at the end of the study was significantly increased (by 5–10%) in all experimental groups, being CS and CH nearly additive in eliciting polycythemia.

Table 44.1 Ventilatory parameters in growing guinea pigs maintained in control atmosphere and in animals exposed during 15 days to (12 O_2) (statistical differences in the main text)

Breathing atmosphere	Age				
	Months, 2	Months, 3	Months, 4	Months, 5	Months, 4. 5;0.5, 12% O_2
Room Air					
Bf (breaths/min)	95.6±2.6	101.8±2.6	94.1±2.2	96.0±6.4	83.2±4.4
TV (ml)	2.1±0.1	3.0±0.1	3.4±0. 1	4.1±0.1	4.0±0.1
MV (ml/Kg/min)	600.6±15.9	528.2±17.4	401.7±9.7	446.4±22.8	388.9±2.4
Hypoxia (10% O_2)					
Bf (breaths/min)	95.5±2.8	100.6±2.7	95.5±2.6	89.9±5.0	93.8±4.5
TV (ml)	2.0±0.1	2.9±0.1	3.3±0.1	4.2±0.2	4.5±0.2
MV (ml/Kg/min)	580.2±19.9	511.6±17.4	403.4±12.4	455.7±28.7	506.0±19.2
Hypercapnia (5% CO_2 in air)					
Bf (breaths/min)	116.1±3.4	111.8±2.4	105.1±2.8	110.2±7.1	86.6±3.7
TV (ml)	4.2±0.1	5.4±0.1	5.8±0.1	6.9±0.4	7.7±0.3
MV (ml/Kg/min)	1,542.0±48.4	1,065.8±33.3	772.3±24.0	852.8±58.1	760.0±30.1

44.3.2 Effects of Acute Hypoxia and Hypercapnia on Ventilation. Effects of CH Exposure

Table 44.1 shows ventilatory parameters in Control and CH while animals breathe room air, hypoxia (10% O_2, 10 min), and hypercapnia (10 min, 5% CO_2 in air).

Normoxia. Bf in normoxia was nearly identical at all ages: around 100 breaths/min. TV doubled from 2 to 5 months of age, from 2.06±0.06 ml to 4.15±0.14 ml ($p<0.001$ vs. 2 months). This increase runs behind body weight gain so that MV/Kg of body weight decreased with age from 601±16 at 2 months down to 446±23 ml/min/Kg at 5 months ($p<0.001$ vs. 2 months).

Hypoxia. Acute hypoxia did not alter the Bf, TV nor MV/Kg encountered in normoxia indicating that guinea pigs do not hyperventilate in response to acute hypoxic hypoxia.

Hypercapnia. Bf in hypercapnia was statistically identical at all ages. Hypercapnia caused a moderate increase (vs. normoxia) in the Bf at all ages but 5 months ($p<0.001$ and $p<0.05$). TV in hypercapnic atmosphere increased with age being at 3, 4, and 5 months statistically higher than at 2 months ($p<0.001$). At every age TV in hypercapnic atmosphere was nearly double than in air ($p<0.001$). Consequently MV/Kg in hypercapnia was nearly double normoxic value ($p<0.001$; i.e., hypercapnia nearly doubled ventilation), but also declined with age being maximum at 2 months and minimum at 5 months ($p<0.001$ at all ages vs. 2 months).

Last column of Table 44.1 shows the effects of CH: neither Bf, TV nor MV/Kg were significantly modified in any conditions. Alternatively stated, CH did not elicit any adaptive response.

44.3.3 Ventilatory Effects of CS Exposure and its Association with CH

Exposure to CS changed the respiratory pattern observed in Control animals (Table 44.2, compare with Table 44.1). In CS animals Bf increased with age and duration of the smoke exposure in normoxia, hypoxia, and hypercapnia ($p<0.01$ and $p<0.001$ at all times vs. 0 exposure or 2 months old animals). Bf in all atmospheres was statistically higher than in their age-matched controls ($p<0.001$). Close inspection of the data suggests some positive interaction between CS and acute hypoxia as Bf tended

Table 44.2 Ventilatory parameters in guinea pigs of different ages and different times of cigarette smoking (CS) exposure maintained in normal atmosphere or exposed to chronic hypoxia (12% O_2) for 15 days (statistical differences are given in the main text)

	Age, CS				
	Months	Months	Months	Months	Months
Breathing Atmosphere	Age, 2; CS, 0	Age, 3; CS, 1	Age, 4; CS, 2	Age, 5; CS, 3	4.5; CS 2,5; 0.5, CS + 12% O_2
Room air					
Bf (breaths/min)	81.8 ± 2.3	116.6 ± 8.0	117.7 ± 7.9	155.6 ± 22.7	102.37 ± 5.24
TV (ml)	2.1 ± 0.1	3.2 ± 0.1	3.2 ± 0.1	4.1 ± 0.3	3.9 ± 0.3
MV (ml/Kg/min)	600.6 ± 15.9	676.4 ± 50.9	548.9 ± 64.7	901.4 ± 185.6	571.36 ± 50.47
Hypoxia (10% O_2)					
Bf (breaths/min)	86.0 ± 2.6	125.6 ± 9.7	141.3 ± 11.9	168.9 ± 19.3	141.5 ± 9.3
TV (ml)	2.0 ± 0.1	3.2 ± 0.1	3.6 ± 0.2	4.8 ± 0.4	4.2 ± 0.2
MV (ml/Kg/min)	580.2 ± 19.9	695.8 ± 42.1	749.5 ± 82.5	1086.6 ± 152	813.9 ± 49.7
Hypercapnia (5% CO_2 in air)					
Bf (breaths/min)	101.1 ± 2.4	137.0 ± 9.0	131.6 ± 25.0	188.4 ± 25.0	120.3 ± 7.0
TV (ml)	4.5 ± 0.1	5.0 ± 0.2	5.3 ± 0.3	6.61 ± 0.6	5.5 ± 0.3
MV (ml/Kg/min)	1,542.0 ± 48.4	1,180.5 ± 47.0	995.3 ± 73.1	1,649.0 ± 207	931.5 ± 44.5

to be higher in hypoxia than in normoxia, while in controls Bf in hypoxia was nearly identical to that seen in normoxia. In fact, at 4 months of age (2 of CS exposure) Bf was higher in hypoxia than in normoxia ($p<0.05$).

In CS animals, TV increased in normoxia, hypoxia, and hypercapnia with the duration of tobacco exposure (and age) following a trend comparable to control animals (i.e., there were significant increases in TV at 3, 4, and 5 months of age vs. 2 months; $p<0.05$; $p<0.01$, and $p<0.001$). However, there was a trend for CS animals to have slightly higher hypoxic TV than age-matched Controls. Additionally, CS animals have lower hypercapnic TV than their age-matched Controls, at 1 and 2 months of CS exposure ($p<0.05$; $p<0.001$).

While in Control group MV/Kg decreases with age in all atmospheres (Table 44.1), in CS animals the age-dependent decrease in MV is nearly abolished in all atmospheres, implying that compared to age-matched controls, CS hyperventilate in all atmospheres ($p<0.05$; $p<0.01$, and $p<0.001$). Even further, in normoxia and in hypoxia animals have MV that augment with age and the duration of the exposure to CS ($p<0.05$; $p<0.001$ vs. 2 month old, 0 time exposure). CS animals tend to breathe more in hypoxic than in normoxic atmosphere, appearing that CS time-dependently sensitized CB-mediated hypoxic ventilation. Alternatively, CS might have triggered a centrally mediated hypoxic ventilatory response. As stated above CS animals exposed to hypercapnia had higher MV/Kg than their age-matched controls: the marked age-dependent decrease in MV/Kg seen in control animals was lessened and after 3 months of exposure it recovered levels encountered at 2 months of age.

Assuming that dead space in guinea pigs is 1/3 of TV (Crosfill and Widdicombe 1961) it is possible to estimate alveolar ventilation (ml/min/Kg). It was found that alveolar ventilation in Control animals breathing air ranged from 260 to 350 ml/min/Kg at different ages. Following this simple reasoning CS animals should have alveolar ventilations 125% (1 month of exposure), 134% (2 months of exposure), and 192% (3 months of exposure) higher than age matched Control animals. However these alveolar ventilations in CS group are in all likelihood underestimations as in these CS animals TV would have increased without a parallel increase in dead space. Although we have not measured functional residual capacity (FRC) nor residual volume (RV) we can exclude significant alterations of these parameters, as Ferrer et al. (2009) have shown that the smoking protocol here employed does not cause emphysema after 3 months.

In CSCH group (last column in Table 44.2) Bf were intermediate between CS and CH groups in the three conditions studied ($p<0.05$ and $p<0.001$ vs. CH group and $p<0.01$ vs. CS). TV in CSCH

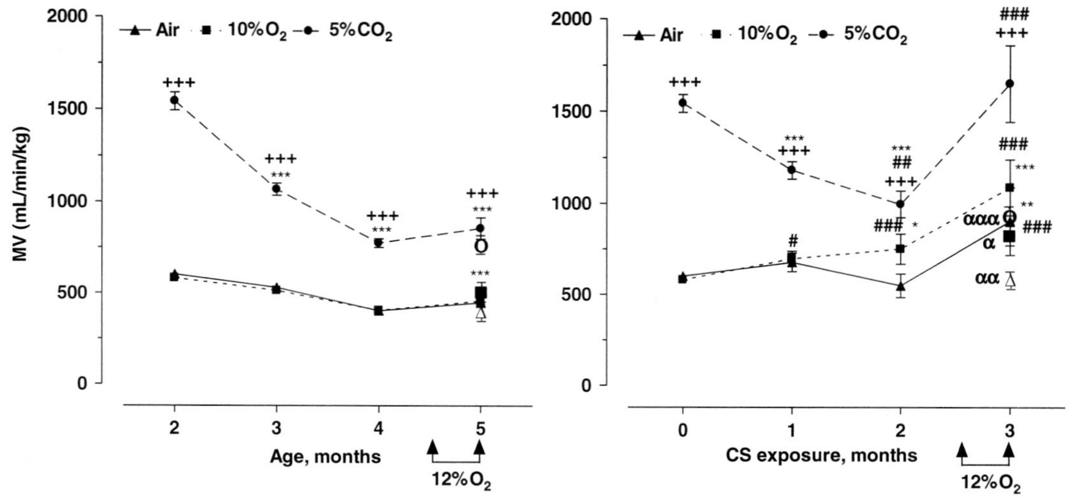

Fig. 44.1 Age-dependent variation in MV/Kg body weight in control (C, left) and cigarette smoke exposed guinea pigs (CS, right) while breathing in the atmospheres shown in the Figure labels. Empty symbols correspond to animals that have been exposed to chronic hypoxia during 2 weeks to conform groups CH and CSCH. Statistics: *,**,*** $p<0.05$, 0.01, and 0.001, respectively vs. 2 months of age. +,++,+++$p<0.05$, 0.01, and 0.001, respectively vs. air breathing. # and ### $p<0.05$ and 0.001, respectively, CS vs. C; α, αα, and ααα $p<0.05$, 0.01, and 0.001, respectively, CSCH vs.CS

group was slightly smaller than Control, CS, and CH groups (at 5 months of age) in the three atmospheres studied. Again, the interaction between cigarette smoke and chronic hypoxia in CSCH group resulted in an intermediate MV/Kg between Control and CS groups in the three conditions ($p<0.05$, $p<0.01$, and $p<0.001$ vs. CS; $p<0.001$ vs. CH in FiO_2 0.10), implying that CH partially blunts the stimulatory action of CS on ventilation.

For the sake of clarity Fig. 44.1 depicts the age dependent variation in MV/Kg in Control and CS animals in the three atmospheres as well as the effects of CH exposure on those relationships.

44.4 Discussion

Our findings include: (1) CS exposure causes a moderate time-dependent decrease in the body weight and an increase in haematocrit. (2) CH increases haematocrit acting in a synergistic manner with CS to increase the haematocrit. (3) Ventilatory pattern of control guinea pigs in the 2–5 months window is: a constant Bf, an increase in TV with age, and an age-dependent decrease in MV/Kg. (4) Acute hypoxia does not alter the ventilatory pattern at any age, while acute hypercapnia causes a marked TV-dependent increase in MV at all ages. (5) CS exposure increases Bf proportionally to exposure duration in air, 10% O_2, and 5% CO_2 atmospheres, preventing the age-dependent decrease in MV/Kg seen in Control. (6) Exposure to CH does not increase Bf, TV, and MV/Kg when data are compared with age matched controls. (7) Exposure of CS animals to CH, CSCH group, eliminates the Bf increase produced by CS causing a very significant decrease in MV/Kg when compared with CS group.

Consistent with findings in smokers humans (Yanbaeva et al. 2007), CS group showed an increased hematocrit, probably resulting from a high carboxyhemoglobinemia. The increased hematocrit in CH group should be considered a HIF-1α mediated adaptive response to hypoxia. In CSCH group both, hypoxic hypoxia and anaemic hypoxia (carboxyhemoglo- binemia), would generate a higher adaptive hematopoietic response.

Absolute values of Bf, TV, and MV are comparable to those found by other authors (Yilmaz et al. 2005; Wiester et al. 2005). The age-dependent variation observed in ventilatory pattern is common in all mammals (Mortola 2001) and parallels the mass specific metabolic rate. Yet, maintenance of Bf and increase in TV would increase alveolar ventilation, i.e., O_2 uptake/Kg decreases less than the observed MV/Kg.

Our data indicate a lack of hypoxia driven CB chemoreflex in adult guinea pigs and a minimal or absent acclimatization to CH which is known to be CB mediated (Gonzalez et al. 1994). In other words, guinea pigs, contrary most mammals, lack hypoxic hypoxia driven CB chemoreflex. However, since acute hypoxia in guinea pigs at the ambient temperature used in present experiments nearly halves O_2 consumption (Hill 1959), our data would imply that in hypoxia guinea pigs hyperventilate for their metabolic needs (i.e., MV/min O_2 consumption would nearly double). Ventilatory response to hypercapnia is comparable to that observed in the rat (Agapito et al. 2009) suggesting it is mediated by CB ($\approx$30%) and central ($\approx$70%) chemoreceptors (Gonzalez et al. 1994). Consistent with that CSN denervation in guinea pigs caused a *ca.* 28% decrease in the hyperventilation produced by 8% CO_2 breathing (Schwenke et al. 2007). As a whole data suggest that brainstem integration of CB signals is appropriate and that guinea pigs lack the oxygen-sensing machinery (Gonzalez et al. 2007, 2010) in their CB.

Our findings constitute the first set of data on the effects of CS on ventilation in animal models. CS MV/Kg mostly via an increase in Bf while in humans, both Bf and TV contributed to it (Tobin et al. 1983). CS in guinea pigs generated some signs of the activation of the CB chemoreflex evidenced by the discrete increase in MV/Kg produced by acute hypoxia in CS. This last finding would suggest that the origin of changes produced by chronic cigarette smoke probably is multifactorial with central and peripheral structures and mechanisms involved.

As stated before CH exposure alter minimally control breathing parameters in non-smoking animals, but associated to CS very significantly reversed the facilitating effects of tobacco smoke on ventilation. Thus, it would appear that hypoxia associated with nicotine and/or with lung damage, as it occurs in many COPD patients, would precipitate a decrease in ventilation with the subsequent worsening of oxygenation and CO_2 wash out. Yet, we must recall the singularity guinea pigs as they do not hyperventilate in response to hypoxia. This would limit, *a priori*, the extrapolation of our findings to other mammals including humans. Although we shall also note that severe COPD patients exhibit depressed ventilatory responses to hypoxia and hypercapnia (Gorini et al. 1996). Within limitations of extrapolation, our findings would indicate is that even limited lung and lung airways damage (Ferrer et al. 2009; Olea et al. 2011), as found in our animals, association of hypoxia (i.e., ascension to high altitude or mild pneumonia) can precipitate hypoventilation and a worsening of blood oxygenation. These changes would accelerate the appearance or deterioration of the cardiovascular components of COPD (e.g., increased pulmonary hypertension and risk of right and left heart dysfunction or even failure). This pathophysiological picture constitutes a feed-forward process that would negatively affect survival rate in COPD patients and would advise prompt oxygen therapy as soon as hypoxemia appears aiming to break the vicious circle (Coleta et al. 2008). The mechanisms of the negative nicotine-CH interation on ventilation are unknown. Yet, it is well founded that in the sudden infant death syndrome such association (smoking mothers or smoke ambient frequent respiratory infections or sleep related apneas) constitutes a precipitating factor for the respiratory misregulation leading to loss of the arousal reaction to hypoxia and sudden dead (Kinney and Thach 2009).

In summary, guinea pigs lack functional O_2-driven ventilatory chemoreflex. Exposure of guinea pigs to CS during 3 months causes an increase in ventilation. Co-exposure to CS and CH blunts the ventilation increase induced by tobacco. From the clinical standpoint we consider that the appearance of hypoxemia in COPD patients would represent a sign that should be promptly corrected to avoid the triggering of feed-forward mechanisms that would endanger the survival of patients.

Acknowledgements We want to thank Mª de los Llanos Bravo and Elena Gonzalez for technical assistance. The work was supported by the "Ministerio de Ciencia e Innovaci6n of Spain"(grant number BFU2007-61848) and by the "Instituto Carlos III"(grant number CIBER CB06/06/0050).

References

Agapito MT, Sanz-Alfayate G, Gomez-Niño A, Gonzalez C, Obeso A (2009) General redox environment and carotid body chemoreceptor function. Am J Physiol Cell Physiol 296:C620–C631

Blake and Banchero (1985) Effects of cold and hypoxia on ventilation and oxygen consumption in awake guinea pigs. Respir Physiol 61:357–368

Caceres AI, Obeso A, Gonzalez C, Rocher A (2007) Molecular identification and functional role of voltage-gated sodium channels in rat carotid body chemoreceptor cells regulation of expression by chronic hypoxia in vivo. J Neurochem 102:231–245

Churg A, Sin DD, Wright JL (2011) Everything prevents emphysema: are animal models of cigarette smoke-induced chronic obstructive pulmonary disease any use? Am J Respir Cell Mol Biol 45(6):1111–1115

Coleta KD, Silveira LV, Lima DF, Rampinelli EA, Godoy I (2008) Predictors of first-year survival in patients with advanced COPD treated using long-term oxygen therapy. Respir Med 102:512–518

Crosfill ML, Widdicombe JG (1961) Physical characteristics of the chest and lungs and the work of breathing in different mammalian species. J Physiol 158:1–14

Curran AK, O'Halloran KD, Bradford A (1995) Effects of superior laryngeal nerve section on ventilation in neonatal guinea-pigs. Respir Physiol 101:23–29

Fernández R, Arriagada I, Garrido AM, Larraín C, Zapata P (2003) Ventilatory chemosensory drive in cats, rats and guinea-pigs. Adv Exp Med Biol 536:489–495

Ferrer E, Peinado VI, Díez M, Carrasco JL, Musri MM, Martínez A, Rodríguez-Roisin R, Barberà JA (2009) Effects of cigarette smoke on endothelial function of pulmonary arteries in the guinea pig. Respir Res 10(1):76

Ferrer E, Peinado VI, Castañeda J, Prieto-Lloret J, Olea E, González-Martín MC, Vega-Agapito MV, Díez M, Domínguez-Fandos D, Obeso A, González C, Barberà JA (2011) Effects of cigarette smoke and hypoxia on the pulmonary circulation in the guinea pig. Eur Respir J 38:617–627

Gonzalez C, Almaraz L, Obeso A, Rigual R (1994) Carotid body chemoreceptors: from natural stimuli to sensory discharges. Physiol Rev 74:829–898

Gonzalez C, Agapito MT, Rocher A, Gonzalez-Martin MC, Vega-Agapito V, Gomez-Niño A, Rigual R, Castañeda J, Obeso A (2007) Chemoreception in the context of the general biology of ROS. Respir Physiol Neurobiol 157:30–44

Gonzalez C, Agapito MT, Rocher A, Gomez-Niño A, Rigual R, Castañeda J, Conde SV, Obeso A (2010) A revisit to O2 sensing and transduction in the carotid body chemoreceptors in the context of reactive oxygen species biology. Respir Physiol Neurobiol 174(3):317–330

Gorini M, Misuri G, Corrado A, Duranti R, Iandelli I, De Paola E, Scano G (1996) Breathing pattern and carbon dioxide retention in severe chronic obstructive pulmonary disease. Thorax 51(7):677–683

Hill JR (1959) The oxygen consumption of new-born and adult mammals its dependence on the oxygen tension in the inspired air and on the environmental temperature. J Physiol 149:346–373

Kinney HC, Thach BT (2009) The sudden infant death syndrome. N Engl J Med 361(8):795–805

Mortola JP (2001) Respiratory physiology of newborn mammals: a comparative perspective. The Johns Hopkins University Press, Baltimore

Olea E, Ferrer E, Prieto-Lloret J, Gonzalez-Martin C, Vega-Agapito V, Gonzalez-Obeso E, Agapito T, Peinado V, Obeso A, Barbera JA, Gonzalez C (2011) Effects of cigarette smoke and chronic hypoxia on airways remodeling and resistance. Clinical significance. Respir Physiol Neurobiol 179(2–3):305–313

Schwenke DO, Bolter CP, Cragg PA (2007) Are the carotid bodies of the guinea-pig functional? Comp Biochem Physiol A Mol Integr Physiol 146:180–188

Tobin MJ, Chadha TS, Jenouri G, Birch SJ, Gazeroglu HB, Sackner MA (1983) Breathing patterns. 2. Diseased subjects. Chest 84:286–294

Wiester MJ, Costa DL, Tepper JS, Winsett DW, Slade R (2005) Agonist-mediated airway challenge: cardiopulmonary interactions modulate gas exchange and recovery. Respir Physiol Neurobiol 145:183–199

Wright JL, Churg A (1991) Effect of long-term cigarette smoke exposure on pulmonary vascular structure and function in the guinea pig. Exp Lung Res 17(6):997–1009

Wright JL, Churg A (2002) Animal models of cigarette smoke-induced COPD. Chest 122:301S–306S

Yanbaeva DG, Dentener MA, Creutzberg EC, Wesseling G, Wouters EF (2007) Systemic effects of smoking. Chest 131:1557–1566

Yilmaz C, Hogg DC, Ravikumar P, Hsia CC (2005) Ventilatory acclimatization in awake guinea pigs raised at high altitude. Respir Physiol Neurobiol 145:235–242

Chapter 45
Some Reflections on Intermittent Hypoxia. Does it Constitute the Translational Niche for Carotid Body Chemoreceptor Researchers?

Constancio Gonzalez, Sara Yubero, M. Angela Gomez-Niño, Teresa Agapito, Asuncion Rocher, Ricardo Rigual, Ana Obeso, and Jose M. Montserrat

Abstract The views presented in this article are the fruit of reflections and discussion with my colleagues at Valladolid and with the members of the Sleep Apnea Hypopnea Syndrome Group of the CIBERES (Spain). We have assembled the article in three sections. In the first one we provide a mechanistic description of obstructive sleep apnea (OSA) and all of its components, including the repetitive episodes of upper airways (UA) obstruction and accompanying hypoxic hypoxia, the respiratory efforts to fight and overcome the obstruction, and the sleep fragmentation due to the hypoxia-triggered arousal reactions, all events occurring during sleep hours with frequencies that might reach up >40–50 episodes/sleep hour. When OSA is accompanied by some of the elements of a big cohort of associated pathologies (vascular, metabolic, and neuropsychiatric) it conforms the obstructive sleep apnea syndrome (OSAS). The high frequency of OSAS in adults (>35 years old) and the costs in every regard of the treatment makes the syndrome a primary importance socio-sanitary problem. In the second section, we describe the experimental models of OSAS, basically the episodic repetitive hypoxic model described by Fletcher and coworkers in 1992, today named in short intermittent hypoxia (IH). From these lines, we want to call for some kind of consensus among researchers to lessen the dispersion of IH protocols. Finally, in the last section we intend to share our optimism with all ISAC members. The optimism is based on the recognition that carotid body (CB) chemoreceptors are critical elements of one of the main pathophysiologic loops in the genesis of OSAS. Therefore, we believe that all of us, as ISAC members, are well qualified to contribute in multidisciplinary research teams with well defined translational interests.

Keywords Carotid body • Intermittent hypoxia • Obstructive sleep apnea • Obstructive sleep apnea syndrome • Translational research

---

C. Gonzalez (✉) • S. Yubero • M.A. Gomez-Niño • T. Agapito • A. Rocher • R. Rigual • A. Obeso
CIBER Enfermedades Respiratorias, Valladolid, Spain

Department of Biochemistry and Molecular Biology and Physiology, and IBGM,
Universidad de Valladolid and CSIC, Valladolid, Spain
e-mail: constanc@ibgm.uva.e

J.M. Montserrat
CIBER Enfermedades Respiratorias, Valladolid, Spain

Laboratori de la Son, Pneumologia Hospital Clínic-IDIBAPS, Barcelona, Spain

C.A. Nurse et al. (eds.), *Arterial Chemoreception: From Molecules to Systems*,
Advances in Experimental Medicine and Biology 758, DOI 10.1007/978-94-007-4584-1_45,

45.1 Obstructive Sleep Apnea (OSA) and Obstructive Sleep Apnea Syndrome (OSAS)

Obstructive sleep apnea (OSA) refers to a repetitive obstruction of UA occurring during sleep. It can be caused by either complete (obstructive apnea) or partial obstruction of UA (obstructive hypopnea). The duration of each obstruction is variable, rarely reaching 1 min of duration, being the frequencies quite uneven from patient to patient. The frequency of obstructive episodes (number/h), the apnea-hypopnea index (AHI), is in fact the most commonly used criterion to define the severity of OSA. Using a simplified classification we can say that patients having an AHI between 5 and 15 have a mild OSA, if the AIH ranges between 15 and 30, OSA is moderate, and it is severe if AHI is >30.

Each obstruction produces a fall in arterial PO_2 (PaO_2) and arterial hemoglobin saturation (SaO_2), and generates an inspiratory effort, commanded by the carotid body (CB) chemoreceptors in its greatest part, aimed to resolve the obstruction. Another criterion to classify the severity of OSA is the night degree of SaO_2 measured either as average oxygen saturation ($ASaO_2$), lowest oxygen saturation ($LSaO_2$), and the percentage of total time with oxygen saturation level lower than 90% (T<90%). Chaudhary et al. (1998) classified his patients as belonging to Groups 1–3, with correspondent AHI of <20, >20 to <50, and >50, and found a good correlation among $ASaO_2$ and $LSaO_2$ and AHI index, and a poor or no correlation between AHI and T<90%. Excellent correlations have been reported between AHI and Oxygen Desaturation Index (ODI, or number of desaturations greater than 3%/hour, taking as basal-normal saturation the mean saturation before the apneic episodes) (Torre-Bouscoulet et al. 2007) (Fig. 45.1).

Airways obstructions are caused by a misregulation of UA muscle tone, so that an inadequate functioning of the openers or dilators of the UA is at the heart of the obstruction. Muscles of the larynx, pharynx, tongue and soft palate are multifunctional, participating in breathing, vocalization, swallowing and reflexes, such as cough and sneezing, and their activity varies during the respiratory cycle (van Lunteren 1997). For example, the posterior cricoarytenoids (openers of the vocal cords and therefore maintaining the glottis open), the genioglossi (which enlarge the pharyngeal lumen and the communication between the oropharynx and laryngopharinx by depressing and protruding the tongue), and the geniohioids (which open the larynx) are activated during inspiration increasing the patency of the airways and facilitating the airflow at any given level of diaphragm contraction. The activity of these muscles precedes that of the diaphragm and dilates and gives strength to the UA before the airflow starts; their contraction is maintained during the entire inspiration and is proportional to the inspiratory drive and airflow. Hypoxia, hypercapnia, and all pharmacological agents that activate the CB chemoreceptors (Gonzalez et al. 1994) activate these muscles in parallel or even at higher strength than diaphragm itself; the counterpart is also true, a decreased drive from the CB chemoreceptors decreases the activity of the openers of the UA. On the contrary, the thyroarythenoid (constrictors of the glottis) and the pharyngeal constrictors are activated during expiration and narrow the UA slowing the expiratory airflow due to the lung recoil, and enlarge the duration of the expiration to facilitate the gas exchange. During sleep, particularly in phases 3 and 4 of non-REM and mainly during the REM periods, there is a decrease drive to air pumping muscles (diaphragm and external intercostals muscles), as well as to the airway opener muscles (Kubin et al. 1998) with the result of a diminished air-pumping power and increased airways resistance. This behavior is physiological, but an abnormally great decrease in the drive to airways dilator muscles, or an abnormal response of the muscles, produces obstruction of the respiratory ways (Morrison et al. 1993). In this regard, it is well documented a decrease in the electromyographic activity of the genioglossi in patients with OSA: this reduced activity causes the tongue to drop backwards and to close the airways (Eisele et al. 2003; Ryan and Bradley 2005).

Each episode of obstruction causes a decrease in arterial PO_2, an increase in arterial PCO_2, and consequently an activation of the CB chemoreceptors. The increased drive of chemoreceptors impinges

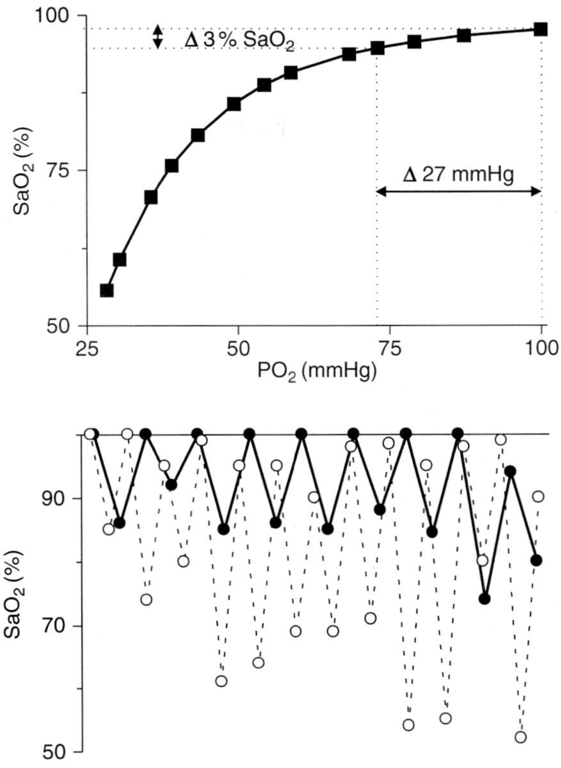

Fig. 45.1 *Upper part*: Zoomed drawing of the security lung segment of the haemoglobin dissociation curve to evidence that a decrease in a 3% in SaO_2 causes a 27 mmHg drop in arterial blood PO_2 and a proportional change in tissue PO_2. *Lower part* shows recordings obtained from two patients with a low (*continuous line*) and a high (*dotted line*) desaturations levels (this lower panel has been redrawn from a figure of Marrone O and Bonsignore MR. *Sleep Med Rev.* 6(3):175–93, 2003)

on the brainstem centers, both, regulators of the respiratory rhythm and controllers of the sleep-wake cycle. This drive generated in the CB causes an increased output to the air pumping and airways dilator muscles, as well as to the centers controlling the sleep-wake cycle and causing a transient wakening reaction. Increased orders to dilator muscles originated as a result of CB activation plus increased orders originated on arousal finally open the airways and the obstruction is overcome (Dempsey et al. 1997). Once blood gases are restored, another cycle of UA obstruction is generated and it repeats to frequencies that might reach >40–60 apneas/h.

The respiratory (inspiratory) effort with obstructed airways, causes swings in the thoracic pressures and hemodynamic changes: the strong negative pressure generated in the thoracic cavity on inspiration with the UA closed favors the venous return and, at the same time, distends the intrapulmonary vessels and cardiac cavities, producing an overfilling of all intrathoracic high capacitance vessels. The net result is a decrease in left heart output during the apneic episode and an increase immediately after the obstruction resolution. These hemodynamic changes would surely contribute to the genesis of arterial blood gases oscillations seen in OSA, as well as to the correspondent changes in tissue PO_2.

The final aspect of the OSA is the obstruction related arousal reaction. Although the wakefulness reaction is probably multifactorial, it has been demonstrated, using hypoxic protocols mimicking OSA, that hypoxia *per se* is probably the main trigger of the awakening reaction (Hamrahi et al. 2001; but see Montserrat et al. 1996).

In sum, OSA is an entity of unknown origin, which via an abnormally decreased drive towards the UA dilator muscles or their abnormal response, produces frequent UA obstructions during the sleep hours. Each obstruction causes a hypoxic hypoxia with desaturation of hemoglobin (and CO_2 retention), a hypoxia (and CO_2)-triggered CB-mediated increased inspiratory drive, and an inspiratory effort that solves the obstruction. Important hemodynamic changes associated to the respiratory effort, and waking reactions associated to each obstructive-hypoxic episode are also significant events of OSA.

The repetitive awakenings produce fragmentation of sleep, lessen the normal restoring character of sleep, and produces next day diurnal sleepiness, discomfort, and anxiety. Additionally, the respiratory efforts with narrowed airways cause snoring, with the obvious discomfort for the night partner of the snorer.

In the long run, OSA quite frequently generates OSAS. This syndrome is the combination of OSA and a cohort of associated pathologies: cardiovascular (hypertension, augmented acute vascular accidents), metabolic-hepatic (insulin resistance, glucose intolerance fatty liver disease), and neuro-psychiatric (anxiety, depression and cognitive impairment) (Sateia 2003; Almendros et al. 2010). The number and severity of associated pathologies varies among patients.

Some epidemiological data of interest include: 1) at least 2% of women and 4% of men >35 years old suffer from OSAS; 2) at least 60–70% of OSAS patients have body mass indexes $\geq$25 kg/m^2 of body surface; if body mass index is greater than 35, the frequency of OSAS is 12–30 times higher than in lean age-matched population; 3) 60% of OSAS patients <60 years old are hypertensive; 4) whether cause or consequence, it is estimated that 50% of patients with acute cerebrovascular accidents suffer from OSAS; 5) there is a well recognized, but not precisely quantified, relationship between OSAS and diabetes, due to the fact that there are not longitudinal studies in this regard (however, see above point 2, obesity, and consider the tight association obesity-diabetes); 6) the prevalence of neuropsychiatric disorders among OSAS patients varies in different series studied: 7–63% of OSAS patients suffer from depression and 11 to 70% from anxiety.

Age is another factor to consider in the study of OSAS. In aged subjects (>60 years) the clinical picture of OSAS is less sharp that in younger patients, and according to Martinez-Garcia et al. (2010) it has the following characteristics: a) it offers higher diagnostic difficulties, as diurnal somnolence that represents one of the main reasons to visit the physician in younger patients. In older patients it is taken as a normal age tribute; yet, even using very restrictive criteria (AHI>30), 26% of men and 21% of women >70 years old have AHI>10; b) they present the same cardiovascular associated pathologies than younger subjects but at much higher frequencies; c) metabolic disorders follow the thumb rule "*older heavier may experience the highest rate of AHI increase over time*" and viceversa. d) neuro-psychiatric disorders: it is estimated that 50% of aged subjects with OSAS suffer anxiety; in addition the sleep fragmentation, which causes diurnal somnolence, aggravates physiologically age related diminution of psychomotor and cognitive functions.

These epidemiological data have the merit of showing the unequivocal interest from a sanitary point of view. In addition, taking into account the economical and social costs of the treatment of these patients (e.g. Banno et al. 2009), working days losses, reduction of efficiency at work, and the high rate of car accidents among OSAS, (Teran-Santos et al. 1999) make OSAS a sanitary and socio-economic problem of prime interest.

45.2 Intermittent Hypoxia, Paradigm of OSAS

The introduction of intermittent hypoxia (IH) as a model of OSAS followed reasonably soon after the recognition of OSAS as a clinical entity. Charles Dickens in 1836 described Joe's disorder, the Fat Boy who eats and then falls asleep. Joe's disorder is the origin of the medical term Pickwickian Syndrome, introduced by Sir William Osler who also noted the association between obesity and hypersomnolence when he described the obesity-hypoventilation syndrome or Pickwickian syndrome in 1918. However, it was not until 1956 that Charles S. Burwell (Bickelmann et al. 1956) brought recognition to obstructive sleep apnea syndrome, which he termed "Pickwickian syndrome". Ten years later 1965, Gastaut and coworkers documented polygraphic features of apnea in a group of Pickwickian patients. Guilleminault in 1973 described most of the polysomnographic features of OSA, although he referred in the paper as all-night polygraphic recordings. The term polysomnography was introduced by Jerome Holland a year later, after the seminal paper by Guilleminault upon his arrival to Stanford. Holland added to the routine night exploration of sleep (EEG, EOG, and EMG) sensors to monitor cardiac and respiratory parameters and coined the term polysomnography. Since these early times, the cohort of alterations conforming the syndrome (congestive heart failure as well as other cardiovascular pathologies "that could not be assigned causally to apnea", extreme sleepiness or fatigue, as well as improper airflow to the lungs, or respiratory failure) were recognized and slowly added to the syndrome to conform current notions on OSAS as described in the previous section.

Although literature is full of articles referring to IH before 1992, none of them have the aiming of mimicking, and they did not, the hypoxia of OSAS. Most protocols used several hours of continuous hypoxia alternating with similar periods of normoxia and different periods of time, from a few days to weeks. Others applied short lasting (minutes) periods of hypoxia alternating with normoxia, acutely, for a few hours. We must reach 1992 when Fletcher et al. published three articles setting IH in the rat as a potential model of OSAS. In the abstract of one of the papers (Fletcher et al. 1992a) it reads: *"An association between chronic high blood pressure and obstructive sleep apnea has been described. We hypothesized that repetitive episodic hypoxia patterned after the hypoxia seen in sleep apnea could contribute to diurnal elevation of blood pressure. Using 12-s infusions of nitrogen into daytime sleeping chambers, four groups of male rats were subjected to IH (3–5% nadir ambient oxygen) every 30 s, 7 h per day for up to 35 days... This duration-of-exposure-related blood pressure [increase] response to hypoxia along with increased left ventricular size after 35 days indicates that chronic IH could be a mechanism directly contributing to diurnal arterial blood pressure elevation".* In another article in 1995 (Fletcher et al. 1995) it was shown that the relevant, hypertension generating factor, was the episodic hypoxia which in itself cause hypocapnia, and that the addition of CO_2 to make the episodic hypoxia eucapnic or moderately hypercapnic (as it occurs in OSAS patients) does not augment the hypertension. They concluded: neither episodic eucapnic nor hypercapnic hypoxia had any additional effect on the changes in chronic diurnal blood pressure compared with hypocapnic hypoxia. Interestingly enough, in a later paper (Bao et al. 1997) they found that when applied acutely (10 min), eucapnic intermittent hypoxia is a more potent stimulus to produce an acute blood pressure elevation and to cause bradycardia than is hypocapnic hypoxia.

The point is: does IH mimic OSAS in full? We believe that the relevant primary factors of OSAS, i.e. IH, arousal reactions and next day somnolence and anxiety are well mimicked by the IH models. The respiratory effort to overcome the obstruction is not present but the hypoxia-induced hyperventilation causes itself a respiratory effort that we cannot decide if it matches in properties that encountered in OSA. In addition most IH models cause hypocapnia. To this phenomenological description of the differences it appears that there are more profound differences, at least in acute models. For example, quite recently we have developed a model using anaesthetized animals that allows comparing genuine OSA vs. IH (Almendros et al. 2010, 2011). Analogies and differences between models were noticed: a) the magnitude of oscillations in blood PO_2 and the mean level of SaO_2 attained during the hour of application of both paradigms were stable and nearly identical in both models; b) the transmission of the oscillation of PO_2 from blood to muscle and adipose tissue was stable so that oscillating tissue PO_2 during the hour of the experiment was constant and also indistinguishable between IH and OSA models; c) by contrast, in brain tissue in the OSA model PO_2 exhibited a different pattern from that of arterial PO_2, muscle and adipose tissue. The oscillating, minimum and maximum values of brain tissue PO_2 gradually increased over the course of the 60 min studied; or in other words, OSA model elicits a clearly compensating mechanisms to defeat brain hypoxia while IH does not; d) paradoxically, OSA was more pro-oxidant in brain tissue that IH. These findings, we believe call for some caution in directly extrapolating findings from IH following Fletcher model to OSA.

It should be added that the IH model is capable of generating behavioural and learning alterations in the rat as Gozal group has shown (e.g., Row et al. 2002, 2003). These alterations might well represent the counterparts of the neuropsychiatric pathologies of OSAS.

Finally, we want to call the attention on the great variety IH protocols used by different laboratories (see Gonzalez-Martín et al. 2009), which are at least as wide as the clinical OSAS when defined by AHI or ODI. This dispersion of protocols is rendering difficult the comparison of results from laboratory to laboratory. Would it not be desirable to reach a consensus among researchers in the field to define a few protocols mimicking for example mild, moderate, and severe OSA?

45.3 Involvement of the CB in the Pathogenesis of OSAS: An Opportunity to CB Physiologists to Perform Translational Research

Up to year 2000 very few laboratories used the IH as a model for OSAS, and therefore it is advisable to keep with Fletcher notions. Along these lines, it was in the 1997 paper (Lesske et al. 1997; but see already Fletcher et al. 1992b) that they concluded: "*Our data imply that repetitive hypoxemia in OSA is probably the cause of the high prevalence of systemic hypertension in this population, and that peripheral chemoreceptors and the sympathetic nervous system play important roles in this pathophysiologic process*". And in 2003 Fletcher wrote: "*There are vascular reactivity changes in humans with sleep apnea ... and complimentary to the above evidence in humans, there is indirect evidence of sympathetic over activity as well as differences in vascular reactivity in IH challenged rats. ...Future research will need to look at exact mechanism of sympathetic nervous system over activity, particularly how central nervous system pathways may undergo facilitation, leading to daytime over activity. Furthermore, the mechanisms of sustained hypertension in sleep apnea patients is almost certainly of multiple aetiologies... Perhaps endothelial cell molecular markers could help to identify patients at risk for cardiovascular change associated with snoring and apnea, as well to guide treatment... finally, studies demonstrating microvascular changes ... promise to yield important knowledge about cellular mechanisms and results of long-term treatment of sleep apnea on cardiovascular disease*"

As it happens with OSA itself, the mechanisms of genesis of the pathologies associated to OSA to conform OSAS are not fully understood. However, we can assert that a well recognized and admitted pathogenic mechanism in the cardiovascular pathology of the OSAS patients is as follows: daily activation of CB chemoreceptors → sensitization of CB chemoreceptors → permanent sympathetic activation → hypertension and augmented acute vascular accidents (see preceding paragraph). However, when CB chemoreceptor function is evaluated as the level of the chemoreflex output (i.e. hypoxic ventilation) such sensitization is not evident as in many studies with patients and experimental animals models, it has been found that hypoxic ventilation is normal or diminished (Table 45.1a). On the other hand, when CB function is evaluated as activity in the carotid sinus nerve (CSN) activity, the consensus is unanimous: all three laboratories working with animal models found increased CSN responses to hypoxia (Table 45.1b). In our laboratory (Gonzalez-Martín et al. 2011) and using a model of IH of low intensity (AHI equivalent =30; nadir O_2 in the rats chamber =10%), we observed increased CB responses to hypoxia, measured at the chemoreceptor cell (release of dopamine) and at the CSN level (action potential frequency) and, at the same time, a diminished ventilation to hypoxia. Since increased sympathetic output to several vascular beds has been recorded both, from OSAS patients and animal models by many laboratories, we concluded that IH produces a bias in the integration of the input arising from the CB. This biased integration would generate a diminished drive of ventilation and an exaggerated activation of brainstem sympathetic neurons that we monitored as an increased rate of CA utilization by renal arteries. In fact many authors have suggested a biased integration of the sensory input to brainstem in IH (e.g. Kline 2010). Additionally, the state of anxiety generated by poor sleep can contribute to the increased sympathetic tone, which in addition to cardiovascular

Table 45.1 (**a**) Chemoreflex responses in OSAS patients and IH animals according to different authors. (**b**) Hypoxic responses obtained in the carotid sinus nerve of IH animals

Author	Model/patient	Hypoxic response monitored
(a) Parameter measured: Carotid body reflex output (Ventilation/Phrenic activity)		
Kimoff et al. (1997)	Dog: 57 occlusions/h, 4.5 h, 15.5 weeks	Attenuated
Osanai et al. (1999)	Patients: (AHI=35)	Attenuated
García-Río et al. (2000)	Patients (AHI=47.3)	Attenuated
O'Halloran et al. (2007)	Rat (90s, air/90s, 5% O_2; 8 h/day; 10 days)	Attenuated/no change
Gonzalez-Martin et al. (2011)	Rat (80s, air/40s, 10%O_2 or 5%O_2; 8 h/day; 15 days)	Attenuated
Costes et al. (1995) Marcus et al. (1994) Greenberg et al. (1999) Wilcox et al. (1998) Zoccal et al. (2008)	OSA Patients and IH animals (wide AHI rage and different IH protocols)	No change
Ling et al. (2001)	Rat: 5 min, 11% O_2/5 min air; 12 h/day, 7 days	Increased Integrated Phrenic Amplitude
Peng et al. (2006)	Mice: 15 s, 5% O_2/5 min air; 8 h/day, 10 days	Increased: Integrated Phrenic Integrated Amplitude/MV
Del Rio et al. (2010)	Rat (5%O_2, 20 s/air, 280 s; 12/h; 8 h/day; 21 days.)	Increased MV
(b) Parameter measured: CSN electrical activity		
Peng et al. (2003)	Rat: 15 s, 5%O_2, 5 min air, 8 h/day; 10 days	Ex vivo: Increased hypoxic response. Long term sensory facilitation
Rey et al. (2004)	Cat: N_2, 90 s/air 270 s. Nadir $PO_2 \approx 75$ mmHg. 10/h, 8 h/day, 4 days	In vivo: Increased hypoxic hypoxia, not NaCN, response
Del Rio et al. (2010)	Rat: 5% O_2, 20 s/air, 280 s; 8 h/day, 21 days	In vivo CSN recordings
Gonzalez-Martin et al. (2011)	Rat (80s, air/40s, 10%O_2; 8 h/day; 15 days)	Ex vivo: increased hypoxic, not hypercapnic response

diseases seems to be linked to the high prevalence of type II diabetes and other metabolic pathologies (Chasens et al. 2003).

By space constrains we are not going to consider other potential pathogenic or physiopathologic mechanisms such as oxidative stress resulting from the ischemia-reperfusion reaction of hypoxic episodes, or metabolic disregulations, obesity and associated alterations, that are seen as causes and consequences of OSAS, etc. (Fig. 45.2). Similarly we do not want to enter in cellular and/or molecular mechanisms that can lead to CB sensitization or into the CB-activated central mechanisms that cause the exaggerated sympathetic activation. These are the topics we all, as ISAC members, can successfully tackle in our animal models or in collaboration with clinicians. Then, the important point, in the frame of present reflections, is that ISAC members have in nowadays a clinical entity, OSAS, with a very important sanitary, social and economic impact, where we can find our niche in multidisciplinary research teams to perform translational research. It only rests in our reflection to provide an operative definition of translational research. In the context of biomedical sciences, translational research could be defined as a full spectrum research carried on by multidisciplinary teams that originating in the patients (usually in a specific clinical entity), deals with or runs trough multiple steps including basic, clinical, epidemiological, and pathological research, and returns to the patients with some type of improvement whether diagnostic, therapeutic or preventive health care. ISAC members can fit in many places in these multidisciplinary teams dedicated to perform translational research on OSAS.

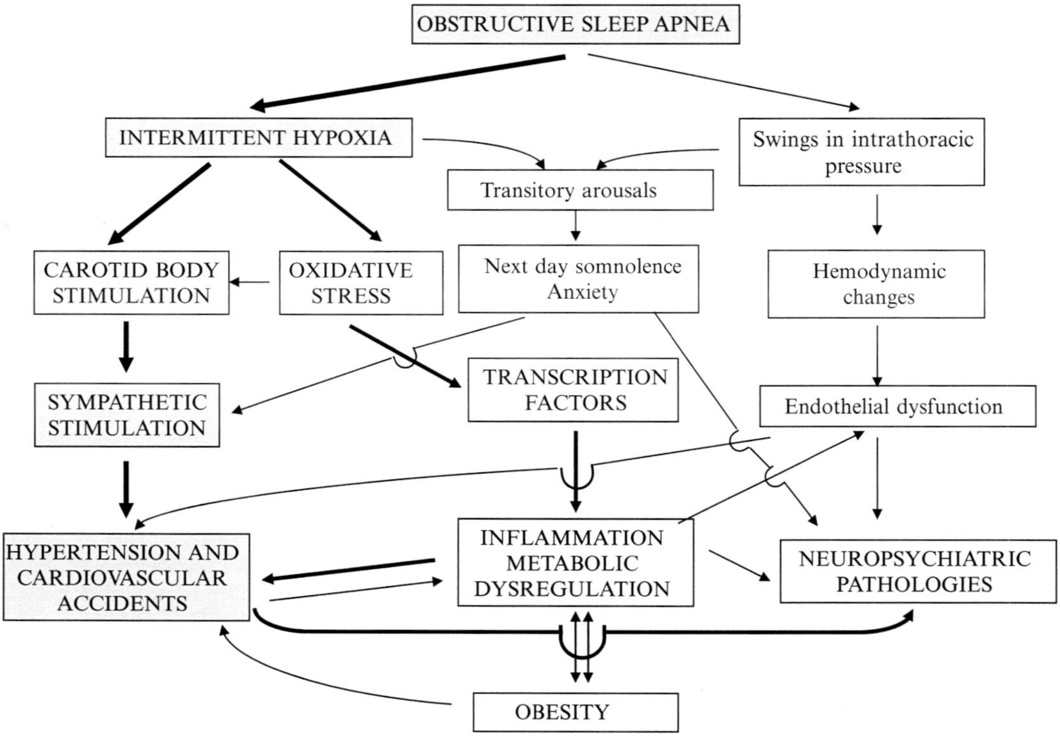

Fig. 45.2 Main elements in the pathophysiology of OSAS. The pathway involving CB chemoreceptors is highlighted (grey rectangles). Thickness of the connecting arrows would correspond to the importance of the pathogenic pathway

Acknowledgements Supported by the MICINN grant (BFU2007-61848) and Accion Integrada PT2009-0172 and ICiii-CIBERES CB06/06/0050.

References

Almendros I, Montserrat JM, Torres M, Gonzalez C, Navajas D, Farré R (2010) Changes in oxygen partial pressure of brain tissue in an animal model of obstructive apnea. Respir Res 11:3

Almendros I, Farré R, Planas AM, Torres M, Bonsignore MR, Navajas D, Montserrat JM (2011) Tissue oxygenation in brain, muscle, and fat in a rat model of sleep apnea: differential effect of obstructive apneas and intermittent hypoxia. Sleep 34:1127–1133

Banno K, Ramsey C, Walld R, Kryger MH (2009) Expenditure on health care in obese women with and without sleep apnea. Sleep 32(2):247–252

Bao G, Randhawa PM, Fletcher EC (1997) Acute blood pressure elevation during repetitive hypocapnic and eucapnic hypoxia in rats. J Appl Physiol 82(4):1071–1078

Bickelmann AG, Burwell CS, Robin ED, Whaley RD (1956) Extreme obesity associated with alveolar hypoventilation; a pickwickian syndrome. Am J Med 21:811–818

Chasens ER, Weaver TE, Umlauf MG (2003) Insulin resistance and obstructive sleep apnea: is increased sympathetic stimulation the link? Biol Res Nurs 5(2):87–96

Chaudhary B, Dasti S, Park Y, Brown T, Davis H, Akhtar B (1998) Hour-to-hour variability of oxygen saturation in sleep apnea. Chest 113(3):719–722

Costes F, Court-Fortune I, Fournel P, Vergnon JM, Emonot A, Geyssant A (1995) Study of chemosensitivity in patients believed to have sleep apnea syndrome. Rev Mal Respir 12:359–364

Del Rio R, Moya EA, Iturriaga R (2010) Carotid body and cardiorespiratory alterations in intermittent hypoxia: the oxidative link. Eur Respir J 361:143–150

Dempsey JA, Harms CA, Morgan BJ, Badr MS, Skatrud JB (1997) Sleep effects on breathing and breathing stability. In: Crystal RG, West JB, Weibel ER, Barnes PJ (eds) The lung: scientific foundations. Philadelphia, Lippincott-Raven, pp 2063–2072
Eisele DW, Schwartz AR, Smith PL (2003) Tongue neuromuscular and direct hypoglossal nerve stimulation for obstructive sleep apnea. Otolaryngol Clin North Am 36(3):501–510
Fletcher EC (2003) Sympathetic over activity in the etiology of hypertension of obstructive sleep apnea. Sleep 26(1):15–19
Fletcher EC, Lesske J, Behm R, Miller CC 3rd, Stauss H, Unger T (1992a) Carotid chemoreceptors, systemic blood pressure, and chronic episodic hypoxia mimicking sleep apnea. J Appl Physiol 72:1978–1984
Fletcher EC, Lesske J, Qian W, Miller CC 3rd, Unger T (1992b) Repetitive, episodic hypoxia causes diurnal elevation of blood pressure in rats. Hypertension 19:555–561
Fletcher EC, Bao G, Miller CC 3rd (1995) Effect of recurrent episodic hypocapnic, eucapnic, and hypercapnic hypoxia on systemic blood pressure. J Appl Physiol 78(4):1516–1521
García-Río F, Racionero MA, Pino JM, Martínez I, Ortuño F, Villasante C, Villamor J (2000) Sleep apnea and hypertension. Chest 117:1417–1425
Gastaut H, Tassinari CA, Duron B (1965) Polygraphic study of diurnal and nocturnal (hypnic and respiratory) episodal manifestations of Pickwick syndrome. Rev Neurol (Paris) 112(6):568–579
Gonzalez C, Almaraz L, Obeso A, Rigual R (1994) Carotid body chemoreceptors: from natural stimuli to sensory discharges. Physiol Rev 74:829–898
Gonzalez-Martín MC, Vega-Agapito V, Prieto-Lloret J, Agapito MT, Castañeda J, Gonzalez C (2009) Effects of intermittent hypoxia on blood gases plasma catecholamine and blood pressure. Adv Exp Med Biol 648:319–328
Gonzalez-Martín MC, Vega-Agapito MV, Conde SV, Castañeda J, Bustamante R, Olea E, Perez-Vizcaino F, Gonzalez C, Obeso A (2011) Carotid body function and ventilatory responses in intermittent hypoxia evidence for anomalous brainstem integration of arterial chemoreceptor input. J Cell Physiol 226(8):1961–1969
Greenberg HE, Sica A, Batson D, Scharf SM (1999) Chronic intermittent hypoxia increases sympathetic responsiveness to hypoxia and hypercapnia. J Appl Physiol 86:298–305
Guilleminault C, Eldridge FL, Dement WC (1973) Insomnia with sleep apnea: a new syndrome. Science 181(102): 856–858
Hamrahi H, Stephenson R, Mahamed S, Liao KS, Horner RL (2001) Selected contribution: regulation of sleep-wake states in response to intermittent hypoxic stimuli applied only in sleep. J Appl Physiol 90(6):2490–2501
Kimoff RJ, Brooks D, Horner RL, Kozar LF, Render-Teixeira CL, Champagne V, Mayer P, Phillipson EA (1997) Ventilatory and arousal responses to hypoxia and hypercapnia in a canine model of obstructive sleep apnea. Am J Respir Crit Care Med 156:886–894
Kline DD (2010) Chronic intermittent hypoxia affects integration of sensory input by neurons in the nucleus tractus solitarii. Respir Physiol Neurobiol 174(1–2):29–36
Kubin L, Davies RO, Pack AI (1998) Control of upper airway motoneurons during REM sleep. News Physiol Sci 13:91–97
Lesske J, Fletcher EC, Bao G, Unger T (1997) Hypertension caused by chronic intermittent hypoxia–influence of chemoreceptors and sympathetic nervous system. J Hypertens 15:1593–1603
Ling L, Fuller DD, Bach KB, Kinkead R, Olson EB Jr, Mitchell GS (2001) Chronic intermittent hypoxia elicits serotonin-dependent plasticity in the central neural control of breathing. J Neurosci 21:5381–5388
Marcus CL, Gozal D, Arens R, Basinski DJ, Omlin KJ, Keens TG, Ward SL (1994) Ventilatory responses during wakefulness in children with obstructive sleep apnea. Am J Respir Crit Care Med 149:715–721
Martínez-García MA, Durán-Cantolla J, Montserrat JM (2010) Sleep apnea-hypopnea syndrome in the elderly. Arch Bronconeumol 46(9):479–488
Montserrat JM, Kosmas EN, Cosio MG, Kimoff RJ (1996) Mechanism of apnea lengthening across the night in obstructive sleep apnea. Am J Respir Crit Care Med 154:988–993
Morrison DL, Launois SH, Isono S, Feroah TR, Whitelaw WA, Remmers JE (1993) Pharyngeal narrowing and closing pressures in patients with obstructive sleep apnea. Am Rev Respir Dis 148(3):606–611
O'Halloran KD, McGuire M, Bradford A (2007) Respiratory plasticity following chronic intermittent hypercapnic hypoxia in conscious rats. In: Strbak V (ed) Medimond SRl Proceedings of the Joint Meeting of the Slovak Physiological Society the Physiological Society and the Federation of European Physiological Societies, Italy, pp 99–103
Osanai S, Akiba Y, Fujiuchi S, Nakano H, Matsumoto H, Ohsaki Y, Kikuchi K (1999) Depression of peripheral chemosensitivity by a dopaminergic mechanism in patients with obstructive sleep apnoea syndrome. Eur Respir J 13:418–423
Peng YJ, Overholt JL, Kline D, Kumar GK, Prabhakar NR (2003) Induction of sensory long-term facilitation in the carotid body by intermittent hypoxia: implications for recurrent apneas. Proc Natl Acad Sci U S A 100:10073–10078
Peng YJ, Yuan G, Ramakrishnan D, Sharma SD, Bosch-Marce M, Kumar GK, Semenza GL, Prabhakar NR (2006) Heterozygous HIF-1alpha deficiency impairs carotid body-mediated systemic responses and reactive oxygen species generation in mice exposed to intermittent hypoxia. J Physiol 577:705–716

Rey S, Del Rio R, Alcayaga J, Iturriaga R (2004) Chronic intermittent hypoxia enhances cat chemosensory and ventilatory responses to hypoxia. J Physiol 560 (Pt 2):577–586
Row BW, Kheirandish L, Neville JJ, Gozal D (2002) Impaired spatial learning and hyperactivity in developing rats exposed to intermittent hypoxia. Pediatr Res 52(3):449–453
Row BW, Goldbart A, Gozal E, Gozal D (2003) Spatial pre-training attenuates hippocampal impairments in rats exposed to intermittent hypoxia. Neurosci Lett 339(1):67–71
Ryan CM, Bradley TD (2005) Pathogenesis of obstructive sleep apnea. J Appl Physiol 99(6):2440–2450
Sateia MJ (2003) Neuropsychological impairment and quality of life in obstructive sleep apnea. Clin Chest Med 24(2):249–259
Teran-Santos J, Jimenez-Gomez A, Cordero-Guevara J (1999) The association between sleep apnea and the risk of traffic accidents cooperative group Burgos-Santander. N Engl J Med 340(11):847–851
Torre-Bouscoulet L, Castorena-Maldonado A, Baños-Flores R, Vázquez-García JC, Meza-Vargas MS, Pérez-Padilla R (2007) Agreement between oxygen desaturation index and apnea-hypopnea index in adults with suspected obstructive sleep apnea at an altitude of 2240 m. Arch Bronconeumol 43(12):649–654
Van Lunteren E (1997) Upper airway effects on breathing. In: Crystal RG, West JB, Weibel ER, Barnes PJ (eds) The lung: scientific foundations. Philadelphia, Lippincott-Raven, pp 2073–2084
Wilcox I, McNamara SG, Dodd MJ, Sullivan CE (1998) Ventilatory control in patients with sleep apnoea and left ventricular dysfunction: comparison of obstructive and central sleep apnoea. Eur Respir J 11:7–13
Zoccal DB, Simms AE, Bonagamba LG, Braga VA, Pickering AE, Paton JF, Machado BH (2008) Increased sympathetic outflow in juvenile rats submitted to chronic intermittent hypoxia correlates with enhanced expiratory activity. J Physiol 586:3253–3265

Chapter 46
Role of Central/Peripheral Chemoreceptors and Their Interdependence in the Pathophysiology of Sleep Apnea

Jerome A. Dempsey, Curtis A. Smith, Gregory M. Blain, Ailiang Xie, Yuansheng Gong, and Mihaela Teodorescu

Abstract Unstable periodic breathing with intermittent ventilatory overshoots and undershoots commonly occurs in chronic heart failure, in hypoxia, with chronic opioid use and in certain types of obstructive sleep apnea. Sleep promotes breathing instability because it unmasks a highly sensitive dependence of the respiratory control system on chemoreceptor input, because transient cortical arousals promote ventilatory overshoots and also because upper airway dilator muscle tonicity is reduced and airway collapsibility enhanced. We will present data in support of the premise that carotid chemoreceptors are essential in the pathogenesis of apnea and periodicity; however it is the hyper-additive influence of peripheral chemoreceptor sensory input on central chemosensitivity that accounts for apnea and periodic breathing. This chemoreceptor interdependence also provides a significant portion of the normal drive to breathe in normoxia (i.e. eupnea) and in acute hypoxia. Finally, we discuss the effects of preventing transient hypocapnia (via selective increases in $FICO_2$) on centrally mediated types of periodic breathing and even some varieties of cyclical obstructive sleep apnea.

Keywords Chemoreceptors • Controller/plant gain • Apnea treatments

46.1 Introduction

Sleep-induced apneas and periodic breathing, such as the types depicted in Fig. 46.1, occur with significant prevalence in humans even in the general, non-clinical population. If the resultant intermittent hypoxemia is sufficiently severe (e.g. > 20–30 apnea events per hour with the percent HbO_2 saturation < 90%) serious consequences often ensue in the form of excessive sympathetic vasoconstrictor outflow, fractionated sleep stages impacting daytime functioning, insulin insensitivity and mortality (Dempsey et al. 2010). Further, these negative consequences of periodic breathing when superimposed on chronic heart failure will exacerbate the progress of the disease (Javaheri et al. 1998).

J.A. Dempsey (✉) • C.A. Smith • G.M. Blain • A. Xie • Y. Gong
Departments of Population Health Sciences, University of Wisconsin – Madison, 1300 University Ave., Room 4245 MSC, Madison 53706, WI, USA
e-mail: jdempsey@wisc.edu

M. Teodorescu
Departments of Medicine, University of Wisconsin – Madison, 1300 University Ave., Room 4245 MSC, Madison 53706, WI, USA

C.A. Nurse et al. (eds.), *Arterial Chemoreception: From Molecules to Systems*, Advances in Experimental Medicine and Biology 758, DOI 10.1007/978-94-007-4584-1_46,

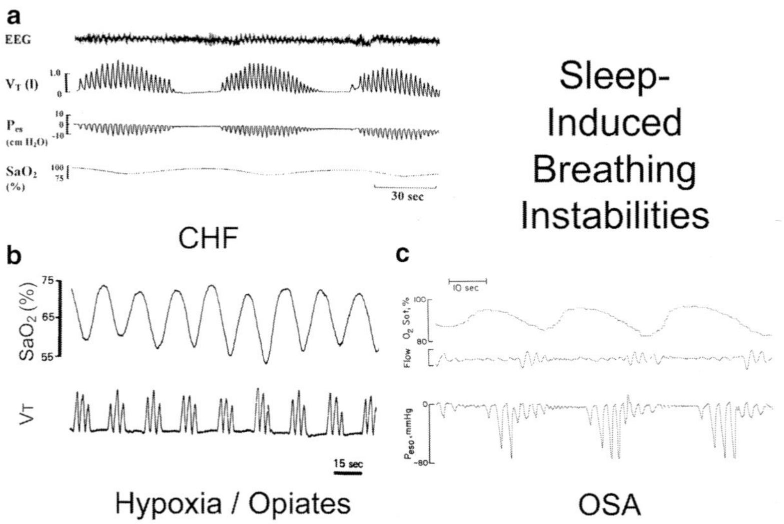

Fig. 46.1 Common types of periodic breathing during NREM sleep. (**a**) *Upper left*: Apnea and periodicity in CHF patients consists of waxing and waning tidal volume with purely central (as in the example shown) or often airway obstructive apnea events. Note the cortical EEG speeding at the termination of each apnea indicating a transient arousal. (**b**) *Lower left*: Periodic breathing in the hypoxia of high altitudes or with chronic opiate use consists of repeated "clusters" of very large tidal volumes interspaced with apneas – usually of a central variety – at regular intervals. (**c**) *Lower right*: The obstructive sleep apnea (OSA) example shows each apnea beginning with central apnea (note the absence of inspiratory efforts by the unchanging esophageal pressure) followed by an obstructive apnea (note the large progressive P_{es} swings). Speeding of cortical EEG frequency (not shown here) showed that a transient arousal occurred at the termination of each obstructive apnea, leading to ventilatory overshoot, hypocapnia and the next apnea and airway closure

We briefly discuss the hypothesis that chemoreceptors play a significant role in the pathogenesis of many types of sleep apnea – even some forms of obstructive sleep apnea.

46.2 Sleep and Chemoreceptor Gain Effects on Breathing Stability

The three major effects of sleep – or a removal of the "wakefulness" stimulus – on ventilatory control include: a) an inhibition of motor output to dilator musculature of the pharyngeal airway, resulting in significant increases in upper airway resistance; b) critical dependence of ventilatory control on $PaCO_2$ and brain stem mechanisms – as manifested in the unmasking of a hypocapnic-induced apneic threshold residing within a few mmHg below waking levels of eupneic $PaCO_2$. The propensity for unstable breathing in the form of ventilatory overshoots and undershoots during sleep varies with changes in the two principle gains regulating ventilatory control, namely "controller" (or chemoreceptor) gain above and below eupnea ($\Delta\dot{V}A/\Delta PaCO_2$) and plant gain (slope of $\Delta PaCO_2/\Delta\dot{V}A$) as determined principally by the position of eupneic $PaCO_2$ on the isometabolic line relating $PaCO_2$ to $\dot{V}A$ (Cherniack and Longobardo 2006; Khoo et al. 1982). When these gains are manipulated experimentally in sleeping humans or dogs as shown in Fig. 46.2, the apneic threshold is displaced closer to or further away from eupneic PCO_2 and breathing is destabilized or stabilized, respectively (Chenuel et al. 2006; Nakayama et al. 2002; Xie et al. 2001, 2009). These overshoots and undershoots are "centrally" driven by variations in medullary respiratory output – but if they occur in sleeping subjects whose upper airways are already anatomically compromised and susceptible to closure, then cyclical obstructive sleep apnea will also occur (see Fig, 46.1c). Indeed, bronchoscopic imaging of the pharyngeal airway during sleep in humans has shown substantial narrowing and even closure within a few seconds after the onset of a central sleep apnea and in the absence of inspiratory effort or negative esophageal pressure (Badr et al. 1995). Once airway obstruction occurs, whether it will reoccur in a cyclical fashion with repeated

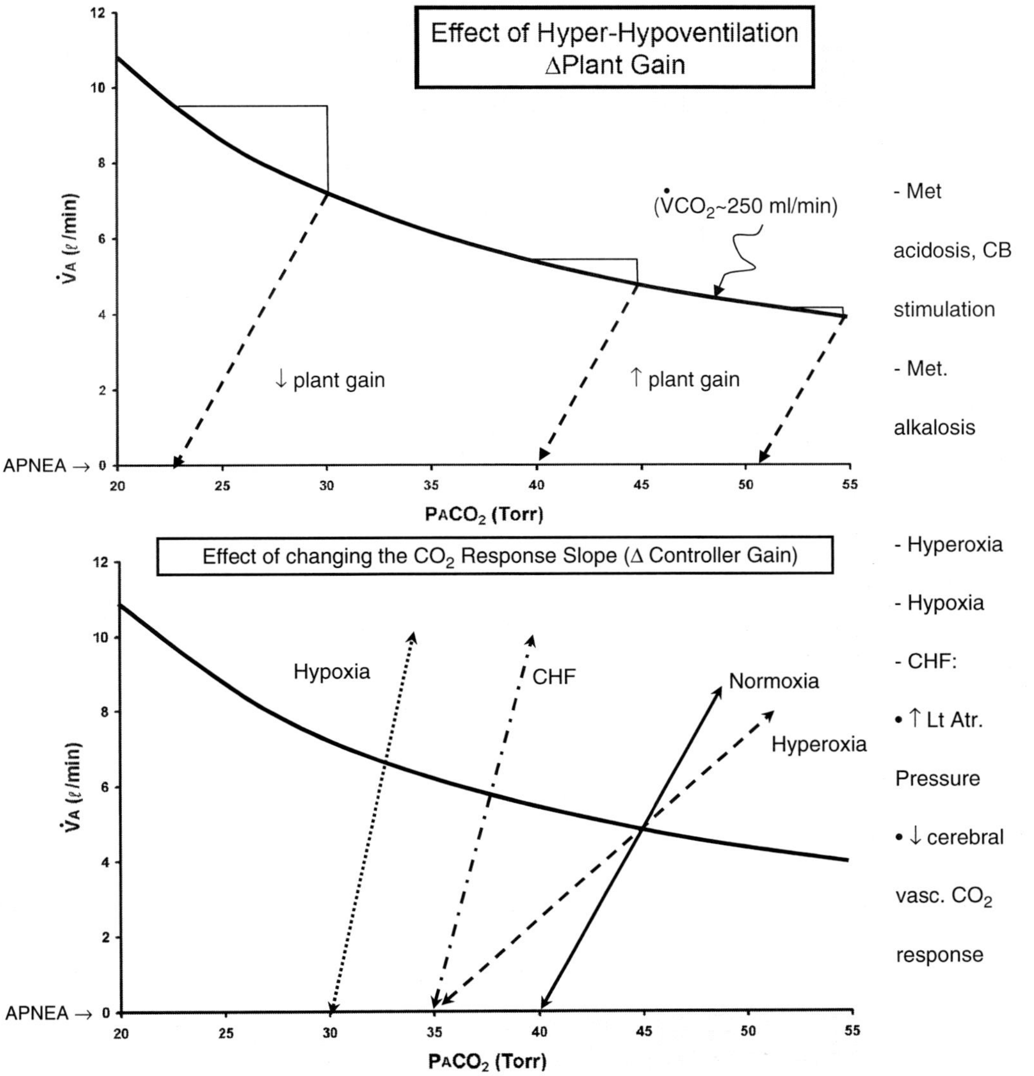

Fig. 46.2 The graphical presentation of the relationship of $\dot{V}_A$ to $PaCO_2$ at a fixed resting $\dot{V}CO_2$, to illustrate how alterations in plant gain or controller gain effect the CO_2 reserve (eupneic $PaCO_2$ – apneic threshold $PaCO_2$) and the propensity to apnea and instability. *Top panel*: changing plant gain by stimulating or reducing eupneic ventilation displaces $PaCO_2$ along the iso-metabolic line relating $\dot{V}_A$ to $PaCO_2$, thereby altering the $\Delta PaCO_2/\Delta\dot{V}_A$ ratio and changing the CO_2 reserve and the susceptibility to apnea/periodicity. *Bottom panel*: altering controller gain ($\Delta\dot{V}_A/\Delta PaCO_2$ slopes) via acute hyperoxia or acute hypoxia in normals or in many CHF patients (via a combination of increased chemosensitivity, raised lt. arterial pressure and reduced cerebrovascular response) changes CO_2 reserve and susceptibility to apnea/periodicity. Note that often controller and plant gains will both change together – for example in hypoxia. In this case the stabilizing effect of reduced plant gain (decreased eupneic $PaCO_2$) is outweighed by the increased controller gain and the CO_2 reserve is markedly reduced, leading to instability. Listed in the right hand panel are the experimental conditions used to change plant gain (*top*) or controller gain (*bottom*)

arousals, also depends on chemoreceptor function and sensitivity – only in this case the important motor pathways responding to chemoreceptor stimuli are those innervating dilator muscles of the upper airway, with the aim of opening the airway sufficiently in a short time period before cortical arousal and sleep disruption occurs (Younes 2008). If cortical arousal occurs before airway patency is restored then this will cause a substantial ventilatory overshoot and transient hypocapnia thereby promoting cyclical periods of airway obstruction (see Fig. 46.1c).

These principles governing instability may be applied to the treatment of sleep apnea. For example, in CHF patients, acetazolamide treatment causes a chemoreceptor-induced hyperventilation(via metabolic acidosis), which reduces plant gain, widens the CO_2 reserve below eupnea (see Top, Fig. 46.2) and removes most central apneas and breathing periodicity (Javaheri 2006). Furthermore, rebreathing CO_2 prevents $PaCO_2$ from reaching the apneic threshold and eliminates apnea and periodic breathing in CHF and in high altitude-induced periodic breathing (Berssenbrugge et al. 1983; Khayat et al. 2003; Xie et al. 1997). Theoretically, we would also predict that many obstructive sleep apneas could be treated effectively via preventing CO_2-induced central instabilities – especially in those patients with airways that are anatomically only moderately susceptible to closure. Our preliminary results over the past year using a "selective" rebreathe system which prevents transient hypocapnia (without causing significant hypercapnia) suggest that stabilizing CO_2 and central motor outputs may effectively diminish obstructive sleep apnea in some patients with high controller gain (and narrow CO_2 reserves) and airways that are only moderately collapsible. Administering supplemental O_2 during sleep has also been shown to reduce the severity of obstructive sleep apnea especially in some OSA patients with high loop gain (Wellman et al. 2008). Given the enhanced carotid chemoreceptor sensitivity in CHF patients (Ponikowski et al. 2001; Sun et al. 1999) it is important that methods for diminishing their high controller gain be pursued to treat their apnea/periodicity. The use of acute hyperoxia and/or pharmacologic agents to reduce high carotid chemoreceptor gain has not yet been studied systematically (Javaheri 2003).

46.3 Sites of Chemoreception Mediating Sleep Apnea/Instability

At which chemoreceptor sites are transient changes in hypocapnia and hypercapnia predominantly sensed leading to apnea/periodicity? Given that only very transient periods of hypocapnia (i.e. only 2 or three breaths of hyperventilation) are required to cause apnea it is likely that carotid chemoreceptors are involved.[1] Indeed, when carotid bodies are denervated in dogs, mechanical ventilation-induced reductions in $PaCO_2$ do *not* cause apnea/periodicity (Nakayama et al. 2003). So intact carotid chemoreceptors are clearly *required* to elicit apnea and periodic breathing, but can carotid chemoreceptors – by themselves – be responsible for apnea? We used a sleeping canine model with an isolated extracorporeal perfused carotid body to show that even substantial carotid chemoreceptor hypocapnia (up to −15 mmHg $P_{cb}CO_2$) – by itself – does not cause apnea/periodic breathing (Smith et al. 1995); furthermore if the isolated perfused carotid body was maintained with normal PO_2 and PCO_2, reducing arterial (and therefore CNS) PCO_2 via mechanical ventilation also did *not* cause apneas/periodic breathing (Smith et al. 2007). So, marked inhibition of either set of chemoreceptors produced limited reductions in tidal volume, but they were unable to cause apnea or significantly alter breath timing.

A solution to these apparently paradoxical findings may be found in the recent suggestion by Guyenet et al. that peripheral and central chemoreceptors are anatomically linked. Neuroanatomical evidence in the rodent model shows that Phox 2b gene expression delineates "an uninterrupted chain of sensors and neurons involved in the integration of peripheral and central chemoreception. This circuit includes the carotid bodies, chemoreceptor afferents, chemoresponsive NTS projections to the ventrolateral medulla (VLM), and pFRG/RTN central chemoreceptors" (Stornetta et al. 2006; Dauger et al. 2003; Takakura et al. 2006).

[1]Another explanation for this lack of effectiveness of carotid body hypocapnia, per se, in eliciting apnea may be that afferent feedback from lung stretch interacts with carotid body inhibition (perhaps at the level of the NTS (Bajic et al. 1994) to produce sufficient inhibition at the medullary level to cause apnea. This possibility has not yet been tested under physiological conditions.

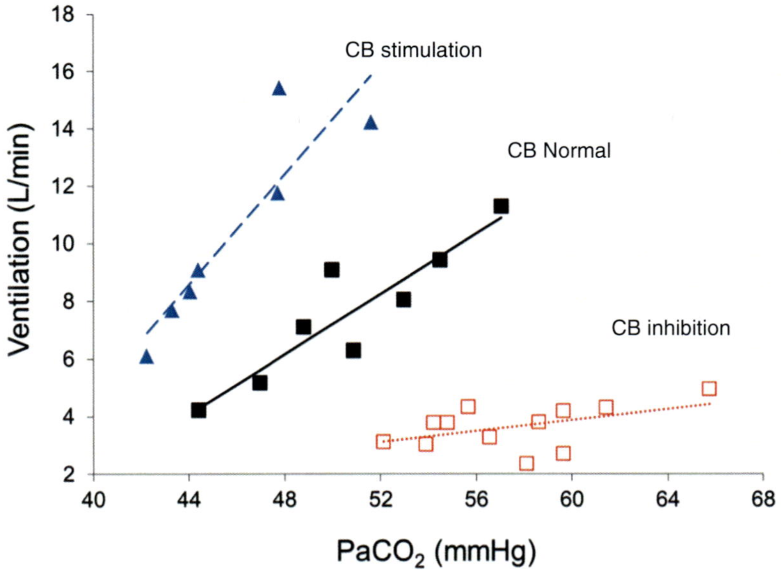

Fig. 46.3 Effects of steady state stimulation of isolated peripheral carotid body ($P_{cb}O_2$ ~ 40 mmHg, $P_{cb}CO_2$ ~ 40 mmHg) or inhibiting isolated perfused carotid body ($P_{cb}O_2 > 500$, $P_{cb}CO_2$ ~20 mmHg) and the response to superimposed systemic hypercapnia (achieved via increased $FICO_2$). Carotid body inhibition caused an average reduction in the systemic CO_2 response slope to one-fifth normal and carotid chemoreceptor stimulation caused an average two fold increase in the central CO_2 response slope (Blain et al. 2010)

To address the functional meaning of these neuroanatomical connections we again used the isolated perfused carotid body canine preparation, only now we held the isolated CB at normal, stimulated or inhibited levels of PO_2 and PCO_2 while we superimposed a progressive systemic hypercapnia via increased $FICO_2$ (Blain et al. 2010). As summarized in Fig. 46.3, we found a hyperadditive effect on ventilation resulting from peripheral and central chemoreceptor interdependence, whereby stimulating CBs with hypoxia increased and inhibiting CB input with hyperoxia plus hypocapnia greatly reduced the gain of the central CO_2 response.

These hyperadditive, interdependent effects are consistent with the suppressive effect of carotid body denervation on the ventilatory response to focal medullary respiratory acidosis in awake goats (Hodges et al. 2005). They also agree with the longitudinal reports in awake dogs (Rodman et al. 2003) and in human patients (Dahan et al. 2007) before and after CBX which showed marked reductions in the ventilatory response to hyperoxic hypercapnia and/or in the "late phase, central" ventilatory response to inhaled CO_2. On the other hand our findings differ with the hypoadditive effects of changes in central pH on carotid body responsiveness, as shown in decerebrate, vagotomized rats (Day and Wilson 2009). Several potential causes of these divergent results were recently discussed (Blain et al. 2010).

So, in addition to the conventional concept that *each* of the chemoreceptors effect breathing by responding reflexively to ionic changes in their own specific environments, the response of central chemoreceptors also appears to be markedly influenced by the magnitude of sensory input from the carotid chemoreceptors. We suggest that this interdependence is an important – indeed obligatory – contribution to apnea and periodic breathing experienced during sleep (Blain et al. 2010). This interdependence may also likely explain why apnea cannot be produced via reduced PCO_2 in *either* the isolated CB *or* CNS alone (see above). Furthermore, a step inhibition of the isolated CB via hyperoxia plus hypocapnia caused a nearly 60% immediate reduction in eupneic VE – a sizeable effect on respiratory motor output which we attributed to the loss of sensory input directly on the medullary controller output as well as indirectly on medullary chemoreceptor gain (Blain et al. 2009).

46.4 Summary – Future Considerations

Evidence accumulated over the past few years has provide an exciting new paradigm-shifting view of chemoreceptor interdependence and its role in the control of breathing. More limited evidence suggests further that chemoreceptors and especially their interdependence likely play a significant although underappreciated role in the pathogenesis of many types of sleep apnea. Several major fundamental problems remain. For example:

- What are the important phenotypical characteristics of OSA patients in whom stabilization of central respiratory motor output might provide an effective treatment for cyclical airway obstructions? … and how might this stabilization be most effectively achieved in patients with obstructive and/or central types of apnea (e.g. manipulating $PaCO_2$, reducing loop gain, specifically targeting carotid chemoreceptor hypersensitivity, etc.).
- What is the relative importance of peripheral – central chemoreceptor interdependence in the ventilatory and sympathetic afferent responses to chemoreceptor stimuli. …and will the magnitude and nature of this carotid chemoreceptor input on central chemoreceptor gain be dependent on the types of peripheral chemoreceptor stimuli applied?

References

Badr MS, Toiber F, Skatrud JB, Dempsey J (1995) Pharyngeal narrowing/occlusion during central sleep apnea. J Appl Physiol 78:1806–1815

Bajic J, Zuperku EJ, Tonkovic-Capin M, Hopp FA (1994) Interaction between chemoreceptor and stretch receptor inputs at medullary respiratory neurons. Am J Physiol 266:R1951–R1961

Berssenbrugge A, Dempsey J, Iber C, Skatrud J, Wilson P (1983) Mechanisms of hypoxia-induced periodic breathing during sleep in humans. J Physiol 343:507–526

Blain GM, Smith CA, Henderson KS, Dempsey JA (2009) Contribution of the carotid body chemoreceptors to eupneic ventilation in the intact, unanesthetized dog. J Appl Physiol 106:1564–1573

Blain GM, Smith CA, Henderson KS, Dempsey JA (2010) Peripheral chemoreceptors determine the respiratory sensitivity of central chemoreceptors to CO(2). J Physiol 588:2455–2471

Chenuel BJ, Smith CA, Skatrud JB, Henderson KS, Dempsey JA (2006) Increased propensity for apnea in response to acute elevations in left atrial pressure during sleep in the dog. J Appl Physiol 101:76–83

Cherniack NS, Longobardo GS (2006) Mathematical models of periodic breathing and their usefulness in understanding cardiovascular and respiratory disorders. Exp Physiol 91:295–305

Dahan A, Nieuwenhuijs D, Teppema L (2007) Plasticity of central chemoreceptors: effect of bilateral carotid body resection on central CO2 sensitivity. PLoS Med 4:e239

Dauger S, Pattyn A, Lofaso F, Gaultier C, Goridis C, Gallego J, Brunet JF (2003) Phox2b Controls the development of peripheral chemoreceptors and afferent visceral pathways. Development 130:6635–6642

Day TA, Wilson RJ (2009) A negative interaction between brainstem and peripheral respiratory chemoreceptors modulates peripheral chemoreflex magnitude. J Physiol 587:883–896

Dempsey JA, Veasey SC, Morgan BJ, O'Donnell CP (2010) Pathophysiology of sleep apnea. Physiol Rev 90:47–112

Hodges MR, Opansky C, Qian B, Davis S, Bonis JM, Krause K, Pan LG, Forster HV (2005) Carotid body denervation alters ventilatory responses to ibotenic acid injections or focal acidosis in the medullary raphe. J Appl Physiol 98:1234–1242

Javaheri S (2003) Pembrey's Dream: the time has come for a long-term trial of nocturnal supplemental nasal oxygen to treat central sleep apnea in congestive heart failure. Chest 123:322–325

Javaheri S (2006) Acetazolamide improves central sleep apnea in heart failure: a double-blind, prospective study. Am J Respir Crit Care Med 173:234–237

Javaheri S, Parker TJ, Liming JD, Corbett WS, Nishiyama H, Wexler L, Roselle GA (1998) Sleep apnea in 81 ambulatory male patients with stable heart failure types and their prevalences, consequences, and presentations. Circulation 97(21):2154–2159

Khayat RN, Xie A, Patel AK, Kaminski A, Skatrud JB (2003) Cardiorespiratory effects of added dead space in patients with heart failure and central sleep apnea. Chest 123:1551–1560

Khoo MC, Kronauer RE, Strohl KP, Slutsky AS (1982) Factors inducing periodic breathing in humans: a general model. J Appl Physiol 53:644–659

Nakayama H, Smith CA, Rodman JR, Skatrud JB, Dempsey JA (2002) Effect of ventilatory drive on carbon dioxide sensitivity below eupnea during sleep. Am J Respir Crit Care Med 165:1251–1260

Nakayama H, Smith CA, Rodman JR, Skatrud JB, Dempsey JA (2003) Carotid body denervation eliminates apnea in response to transient hypocapnia. J Appl Physiol 94:155–164

Ponikowski P, Chua TP, Anker SD, Francis DP, Doehner W, Banasiak W, Poole-Wilson PA, Piepoli MF, Coats AJ (2001) Peripheral chemoreceptor hypersensitivity: an ominous sign in patients with chronic heart failure. Circulation 104:544–549

Rodman JR, Henderson KS, Smith CA, Dempsey JA (2003) Cardiovascular effects of the respiratory muscle metaboreflexes in dogs: rest and exercise. J Appl Physiol 95:1159–1169

Smith CA, Saupe KW, Henderson KS, Dempsey JA (1995) Ventilatory effects of specific carotid body hypocapnia in dogs during wakefulness and sleep. J Appl Physiol 79:689–699

Smith CA, Chenuel BJ, Henderson KS, Dempsey JA (2007) The apneic threshold during non-REM sleep in dogs: sensitivity of carotid body vs. Central chemoreceptors. J Appl Physiol 103:578–586

Stornetta RL, Moreira TS, Takakura AC, Kang BJ, Chang DA, West GH, Brunet JF, Mulkey DK, Bayliss DA, Guyenet PG (2006) Expression of Phox2b by brainstem neurons involved in chemosensory integration in the adult rat. J Neurosci 26:10305–10314

Sun SY, Wang W, Zucker IH, Schultz HD (1999) Enhanced activity of carotid body chemoreceptors in rabbits with heart failure: role of nitric oxide. J Appl Physiol 86:1273–1282

Takakura AC, Moreira TS, Colombari E, West GH, Stornetta RL, Guyenet PG (2006) Peripheral chemoreceptor inputs to retrotrapezoid nucleus (RTN) CO_2-sensitive neurons in rats. J Physiol 572:503–523

Wellman A, Malhotra A, Jordan AS, Stevenson KE, Gautam S, White DP (2008) Effect of oxygen in obstructive sleep apnea: role of loop gain. Respir Physiol Neurobiol 162:144–151

Xie A, Rankin F, Rutherford R, Bradley TD (1997) Effects of inhaled CO_2 and added dead space on idiopathic central sleep apnea. J Appl Physiol 82:918–926

Xie A, Skatrud JB, Dempsey JA (2001) Effect of hypoxia on the hypopnoeic and apnoeic threshold for CO(2) in sleeping humans. J Physiol 535:269–278

Xie A, Skatrud JB, Barczi SR, Reichmuth K, Morgan BJ, Mont S, Dempsey JA (2009) Influence of cerebral blood flow on breathing stability. J Appl Physiol 106:850–856

Younes M (2008) Role of respiratory control mechanisms in the pathogenesis of obstructive sleep disorders. J Appl Physiol 105:1389–1405

Chapter 47
Physiologic Basis for Intermittent Hypoxic Episodes in Preterm Infants

R.J. Martin, J.M. Di Fiore, P.M. MacFarlane, and C.G. Wilson

Abstract Intermittent hypoxic episodes are typically a consequence of immature respiratory control and remain a troublesome challenge for the neonatologist. Furthermore, their frequency and magnitude are commonly underestimated by clinically employed pulse oximeter settings. In extremely low birth weight infants the incidence of intermittent hypoxia [IH] progressively increases over the first 4 weeks of postnatal life, with a subsequent plateau followed by a slow decline beginning at weeks six to eight. Over this period of unstable respiratory control, increased oxygen-sensitive peripheral chemoreceptor activity has been associated with a higher incidence of apnea of prematurity. In contrast, infants with bronchopulmonary dysplasia [chronic neonatal lung disease] exhibit decreased peripheral chemosensitivity, although the effect on respiratory stability in this population is unclear. Such episodic hypoxia/reoxygenation in early life has the potential to sustain a proinflammatory cascade with resultant multisystem, including respiratory, morbidity. Therapeutic approaches for intermittent hypoxic episodes comprise careful titration of baseline or supplemental inspired oxygen as well as xanthine therapy to prevent apnea of prematurity. Characterization of the pathophysiologic basis for such intermittent hypoxic episodes and their consequences during early life is necessary to provide an evidence-based approach to their management.

Keywords Intermittent hypoxia • Apnea • Prematurity

47.1 Characterization of IH Events

Immature respiratory control with resultant apnea of prematurity and periodic breathing is a well recognized complication of preterm birth (Abu-Shaweesh and Martin 2008). The resultant respiratory pauses may occur in isolation or assume a repetitive or periodic pattern (Fig. 47.1). When these pauses are longer (e.g., >15 s duration) they are frequently prolonged by obstructed inspiratory efforts, presumably secondary to loss of upper airway tonic or phasic activity (Fig. 47.1). However, in contradistinction to adult sleep apnea, obstructed inspiratory efforts appear not to be the primary

R.J. Martin (✉) • J.M. Di Fiore • P.M. MacFarlane • C.G. Wilson
Department of Pediatrics Division of Neonatology, Rainbow Babies & Children's Hospital, Case Western Reserve University, 11100 Euclid Avenue, Cleveland, OH 44106-6010, USA
e-mail: rxm6@case.edu

C.A. Nurse et al. (eds.), *Arterial Chemoreception: From Molecules to Systems*, Advances in Experimental Medicine and Biology 758, DOI 10.1007/978-94-007-4584-1_47,

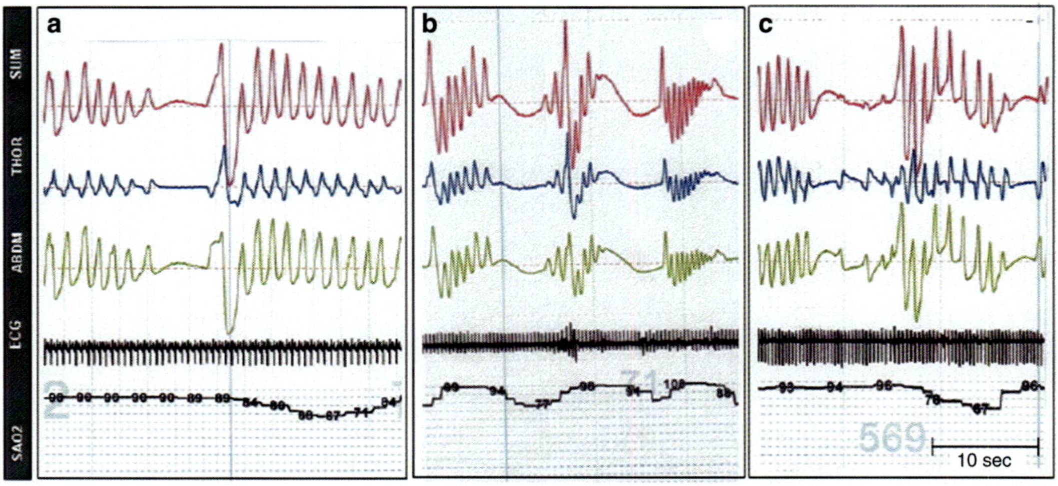

Fig. 47.1 Three representative examples of desaturation, secondary to apnea or respiratory pauses, in preterm infants monitored via respiratory inductance plethysmography. The left panel demonstrates an isolated central apnea, the middle panel demonstrates periodic breathing with repetitive IH, and the right panel demonstrates a mixed apnea with obstructed inspiratory efforts preceding resumption of breathing

precipitating event. The result is intermittent desaturation or hypoxia, often accompanied by episodes of bradycardia. Current neonatal practice attempts to minimize the use of mechanical ventilation via endotracheal intubation and favors the widespread use of continuous positive airway pressure [CPAP]. Nonetheless, mechanical ventilation may be needed and IH episodes are also common in intubated, ventilated infants as a consequence of ineffective ventilatory support and loss of the infant's lung volume (Dimaguila et al. 1997; Esquer et al. 2007).

While hypoventilation is the trigger for IH, several physiologic parameters likely contribute to the resultant desaturation. The most important of these is probably pulmonary oxygen stores which reflect lung volume. Preterm infants are at risk for a low basal lung volume as a consequence of both atelectasis and high chest wall compliance. Therapy with CPAP clearly benefits these infants by both splinting the upper airway and stabilizing lung volume, thereby minimizing the risk of desaturation. Other physiologic parameters may also be implicated in IH, including blood oxygen capacity comprising blood volume and hemoglobin content as well as peripheral tissue oxygen consumption (Sands et al. 2009). Regardless of etiology, the magnitude of IH episodes in preterm infants has generated interest in their role in neonatal pathophysiology (Martin et al. 2011).

As the incidence of intermittent hypoxic events is a function of maturation, we sought to identify the natural history of intermittent hypoxic episodes. We, therefore, followed a cohort of preterm infants over the first weeks of postnatal life. This required modification of the algorithm used for clinical pulse oximetry, as the latter typically employs a prolonged averaging time and low sample rate to filter out resultant monitor alarms. Intermittent hypoxia [and hyperoxia] were documented using high-resolution pulse oximetry data [2-s sample rate and 2-s averaging time] recorded continuously from day 1 to 8 weeks' postnatal age. In this cohort of 79 infants [24–28 weeks' gestation] desaturation events, defined as a drop in arterial oxygen saturation [SaO_2] to <=80% for between 10 s and 3 min duration, were identified using custom software [RTI, Research Triangle Park, NC, USA]. For each infant, the number of hypoxemic events was calculated for each week of postnatal age (Di Fiore et al. 2010).

As seen in Fig. 47.2, there was a marked change in intermittent hypoxemic events over time with relatively few hypoxemic episodes occurring during the first week of life, then a progressive increase in weeks 2–3, followed by a plateau around weeks 4–6, and finally a decrease in weeks 6–8. These

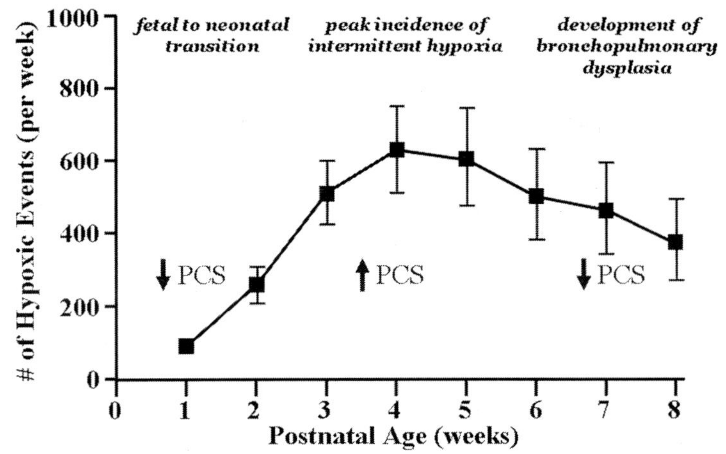

Fig.47.2 A proposed relationship between O_2 sensitive peripheral chemosensitivity [PCS] and the natural history of intermittent hypoxia [IH] in a cohort of preterm infants studied over the first 8 weeks of life (Di Fiore et al. 2010). The fetal to neonatal transition is associated with decreased PCS and a low incidence of IH. Subsequently there is an increased incidence of IH associated with increased PCS, while later maturation is accompanied by a declining incidence of IH and decreased PCS in the presence of bronchopulmonary dysplasia [BPD]

trends were apparent in the presence and absence of assisted ventilation, and consistent with historical beliefs that apnea of prematurity is less common in the first postnatal days. The remarkably high incidence of intermittent hypoxic episodes in this population [e.g., 50–100/day] would likely be underestimated at conventional pulse oximeter settings employing prolonged averaging times. These data provide a basis for evaluating potential morbidity associated with such episodes.

47.2 Proposed Respiratory Morbidity Associated with Neonatal IH

The transition from fetal to neonatal life is accompanied by a dramatic increase in PaO_2 from around 25 to at least 80 Torr within 5–10 min of birth. This postnatal onset of continuous breathing is probably primarily the result of arousal and thermal, rather than chemical, stimuli. The abrupt increase in PaO_2 is widely believed to inhibit O_2 sensitive peripheral chemoreceptors in the early postnatal period. However, in human infants onset of spontaneous breathing is significantly delayed in the presence of supplemental oxygen, suggesting that some peripheral chemosensitivity is present at this time (Fig. 47.3).

While CO_2 is the major chemical driver of breathing, ventilatory responses to hypoxia have been extensively studied and well characterized in human infants and newborn animal models (Rigatto and Brady 1972). A transient increase in ventilation in response to hypoxia of 1–2 min duration, and mediated via oxygen-sensitive peripheral chemoreceptors, is followed by a decline in ventilation that may even fall below baseline ventilation, and is presumably centrally mediated. This pattern persists in preterm infants into the second postnatal month (Martin et al. 1998).

While absent peripheral chemosensitivity may inhibit breathing, upregulated peripheral chemoreceptors may also destabilize breathing (Fig. 47.3). Neonatal rat pups exposed to repetitive IH exhibit a marked and sustained increase in sensory activity of the carotid body, and this response is of greater magnitude than that exhibited by IH-exposed mature rats (Pawar et al. 2008). Reactive oxygen species have been implicated in these findings. Such an increase in carotid body excitation in response to hypoxic exposure may destabilize breathing, as has been shown in human preterm infant studies.

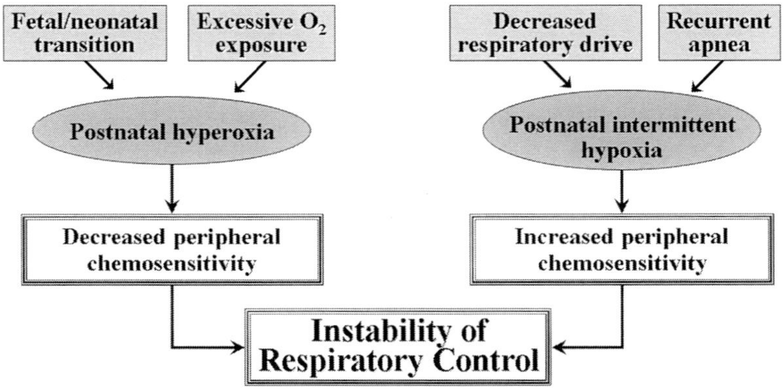

Fig. 47.3 Both decreased peripheral chemosensitivity, as occurs during the fetal to neonatal transition, and increased peripheral chemosensitivity, as may occur secondary to neonatal IH, have the potential to destabilize cardiorespiratory control

Two studies of preterm infants have shown a direct relationship between apnea frequency and increased peripheral chemosensitivity to hypoxia (Nock et al. 2004; Cardot et al. 2007). The mechanisms underlying this relationship are unclear, although if baseline $PaCO_2$ and the CO_2 threshold for apnea are close, fluctuations in ventilation associated with increased oxygen sensitive peripheral chemosensitivity may readily lower $PaCO_2$ to below the apneic threshold (Al-Matary et al. 2004). Therefore, increased peripheral oxygen chemosensitivity may be a marker rather than cause of respiratory instability.

Two groups of investigators have studied peripheral chemosensitivity in preterm infants of advanced postnatal age who have developed bronchopulmonary dysplasia [BPD] or chronic neonatal lung disease (Katz-Salamon et al. 1995; Calder et al. 1994). Such infants with BPD may be exposed to more chronic hypoxia rather than acute IH, although these may coexist. In both studies BPD was associated with decreased peripheral chemosensitivity possibly associated with a declining rate of IH episodes (Fig. 47.3). While these data suggest an association between increased peripheral chemosensitivity, respiratory instability and a higher incidence of IH, other factors may come into play. For example, the interaction between peripheral and central chemosensitivity is attracting considerable current interest. Data from mature animal models suggest a synergistic interdependence between medullary chemoreceptors and peripheral chemoreceptor input (Smith et al. 2010), although it is unknown how maturation might affect such an interrelationship. In neonatal rodent models, it appears that the CO_2 response is strong immediately after birth, but decreases during subsequent postnatal days (Darnall 2010). If birth is associated with increased central CO_2 chemosensitivity and decreased peripheral O_2 chemosensitivity, this would be inconsistent with data from adult animal models and worthy of future study.

Sleep-disordered breathing is increasingly recognized as a potential source of developmental and behavioral problems in the pediatric population, although the role of intermittent hypoxia in these problems is controversial. Two studies have demonstrated that preterm infants are at increased risk for sleep-disordered breathing in childhood and young adulthood (Rosen et al. 2003; Paavonen et al. 2007). The mechanisms underlying this relationship between preterm birth and later sleep disordered breathing are unknown. Prior exposure to xanthines in the neonatal period appeared to increase the later risk, suggesting that apnea of prematurity and resultant intermittent hypoxia in early life may serve as a risk factor for later perturbations of sleep and breathing (Hibbs et al. 2008). Future study should focus on the possible relationship between neonatal intermittent hypoxia, subsequent oxidant stress, and the genesis of unstable breathing patterns in later life.

47.3 Physiologic Basis for Therapeutic Interventions to Prevent IH

47.3.1 Xanthine Therapy

Given that apnea is the major cause of intermittent hypoxic episodes in preterm infants, CPAP and xanthine therapy should be the major therapeutic approaches (Martin et al. 2011 (in press)). CPAP is a safe and effective proven therapy that serves to maintain functional residual capacity [FRC] and resultant oxygenation as well as splinting the upper airway and preventing its closure in the face of decreased respiratory drive. Xanthine therapy has been used to prevent and treat apnea of prematurity since the 1970s. Its primary mechanism of action in the perinatal period is thought to be blockade of inhibitory adenosine A_1 receptors with resultant excitation of respiratory neural output (Herlenius et al. 2002). An alternative mechanism of caffeine action is blockade of excitatory adenosine A_{2A} receptors at GABAergic neurons and resultant decrease in GABA output, also resulting in excitation of respiratory neural output (Mayer et al. 2006) (Fig. 47.4). It is widely believed that xanthine-induced phosphodiesterase inhibition does not occur at the concentration of xanthines used clinically, however, such a contribution cannot be excluded. It is also unclear why xanthine therapy for apnea is so effective in the neonatal period, while it is largely ineffective and impractical in adults with sleep apnea. It is unclear whether peripheral oxygen-sensitive chemosensor effects contribute to the efficacy of xanthine therapy. Of interest is the observation in neonatal piglets that xanthine-induced reversal of hypoxic ventilatory depression is dependent on the presence of peripheral chemoreceptors (Cattarossi et al. 1995). More recent data in rat pups exposed to chronic IH show that caffeine has no effect on hypoxic ventilatory reflexes, but does enhance baseline ventilation via increased long term facilitation (Julien et al. 2010).

The complex neurotransmitter interactions elicited by caffeine led to concerns regarding its safety and a large multicenter trial was undertaken in the 1990s. The results of this study have demonstrated that caffeine treatment is effective in decreasing the rate of BPD and improving neurodevelopmental outcome at 18–21 months, especially in those receiving respiratory support (Davis et al. 2010; Schmidt et al. 2007). It is possible that this benefit is secondary to decrease in apnea and resultant IH episodes; however, this is speculative. A small earlier study actually did not demonstrate a decrease in hypoxic episodes, as measured by transcutaneous PO_2, in response to xanthine therapy (Bucher and Duc 1988).

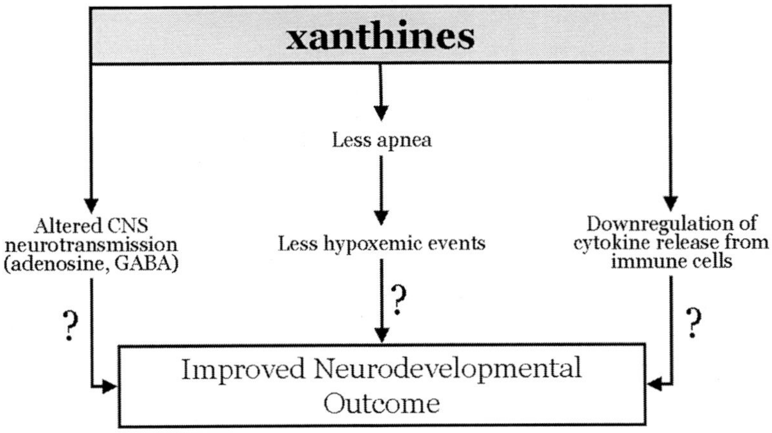

Fig. 47.4 The proposed mechanisms whereby xanthine therapy improves neurodevelopmental outcome in preterm infants include resolution of apnea and IH episodes, downregulation of cytokine mediated immune mechanisms and alteration in CNS neurotransmitter function, in particular adenosine and GABA-mediated pathways

Xanthine therapy may also play a role in modulating inflammatory mechanisms in preterm infants as demonstrated in cord blood monocytes (Chavez-Valdez et al. 2009). Relief of both IH and proinflammatory mechanisms may be interrelated mechanisms, given that hypoxia/reoxygenation is a likely oxidant and proinflammatory stress. The clinical benefits of xanthine therapy in preterm infants should trigger interest in a "bedside to bench" approach to enhance our understanding of underlying mechanisms.

47.3.2 Optimizing Baseline Oxygen Saturation

It has long been suspected that targeting a lower baseline oxygen saturation in infants with BPD results in more desaturation. Meanwhile, multiple large trials, some of which are ongoing, have randomized infants to two different levels of baseline oxygen saturation in order to identify resultant morbidity. Based on a need for prolonged oxygen supplementation when levels of 95% were targeted, the current focus is on 85–89% vs 91–95% oxygen in preterm infants <28 weeks' gestation. While the lower targeted range is associated with less retinopathy of prematurity [ROP], there appears to be a significantly higher mortality in this group (SUPPORT Study Group 2010; Stenson et al. 2011). In a subgroup of infants from the SUPPORT Trial we have now preliminarily identified a higher incidence of intermittent hypoxic episodes in the low oxygen targeted group (Di Fiore et al. 2011). It is conceivable that the resultant IH and/or oxidative stress from these desaturation episodes may predispose to this adverse mortality outcome. An additional speculation is that high risk preterm infants may need to be subjected to a stepwise increase in baseline oxygen during their postnatal course initially to minimize the risk of retinopathy and subsequently to minimize desaturation episodes or IH. Clearly this approach needs careful further study before being considered in clinical practice.

47.3.3 Automated Control of Inspired Oxygen

Preterm infants receiving mechanical ventilation frequently have substantial fluctuations in oxygen saturation. These episodes of desaturation are a real challenge to personnel trying to maintain baseline oxygen saturation in a predetermined range. Automated adjustments of inspired oxygen would have the potential to reduce the time-consuming task of repetitively changing supplemental oxygen. This automated technique has now been compared to routine adjustments of inspired oxygen as performed by clinical personnel in infants of 24–27 weeks' gestation (Claure et al. 2011). During the automated period, time with oxygen saturation within the intended range of 87–93% increased significantly and times in the hyperoxic range were significantly reduced. This was not associated with a clear benefit for hypoxic episodes. While times with an oxygen saturation <80% did not differ, the automated system actually increased time with oxygen saturation between 80% and 86%. Therefore, this novel technique may not be an obvious solution to minimize episodic intermittent hypoxia.

47.4 Conclusion

Despite the sophistication of modern neonatal intensive care, many basic pathophysiologic questions remain to be addressed. While the frequency of intermittent episodes of hypoxia or desaturation has been identified, it is unclear whether IH represents a source of acute or chronic morbidity. It is even possible that intermittent hypoxic events in early life may provide later benefit by way of oxygen tolerance in humans, as has been proposed in rodent studies (Adle-Biassette et al. 2011). There is widespread agreement that apnea and ineffective ventilation are the main precipitants of intermittent hypoxic events. It is possible that episodic hypoxia/reoxygenation establishes or promotes a proinflammatory/prooxidant cascade with adverse effects on multiple organ systems, including respiratory control. Greater understanding of the underlying pathophysiology is needed in order to provide a rationale for preventing or treating these episodes with either targeted supplemental oxygen or aggressive pharmacotherapy in the form of xanthine treatment.

Acknowledgements Supported by: NIH Grants R21 HL098628 [RJM] and RO3 HD064830 [CGW] grant

References

Abu-Shaweesh JM, Martin RJ (2008) Neonatal apnea: what's new? Pediatr Pulmonol 43:937–944

Adle-Biassette HBM, Matrot B, Favrais G, Gallego J, Gressens P (2011) Moderate intermittent hypoxia protects neurodevelopment in newborn mice. *E-PAS* 4534–4510

Al-Matary A, Kutbi I, Qurashi M, Khalil M, Alvaro R, Kwiatkowski K, Cates D, Rigatto H (2004) Increased peripheral chemoreceptor activity may be critical in destabilizing breathing in neonates. Semin Perinatol 28:264–272

Bucher HU, Duc G (1988) Does caffeine prevent hypoxaemic episodes in premature infants? a randomized controlled trial. Eur J Pediatr 147:288–291

Calder NA, Williams BA, Smyth J, Boon AW, Kumar P, Hanson MA (1994) Absence of ventilatory responses to alternating breaths of mild hypoxia and air in infants who have had bronchopulmonary dysplasia: implications for the risk of sudden infant death. Pediatr Res 35:677–681

Cardot V, Chardon K, Tourneux P, Micallef S, Stephan E, Leke A, Bach V, Libert JP, Telliez F (2007) Ventilatory response to a hyperoxic test is related to the frequency of short apneic episodes in late preterm neonates. Pediatr Res 62:591–596

Cattarossi L, Haxhiu-Poskurica B, Haxhiu MA, Litmanovitz I, Martin RJ, Carlo WA (1995) Carotid bodies and ventilatory response to hypoxia in aminophylline-treated piglets. Pediatr Pulmonol 20:94–100

Chavez-Valdez R, Wills-Karp M, Ahlawat R, Cristofalo EA, Nathan A, Gauda EB (2009) Caffeine modulates TNF-alpha production by cord blood monocytes: the role of adenosine receptors. Pediatr Res 65:203–208

Claure N, Bancalari E, D'Ugard C, Nelin L, Stein M, Ramanathan R, Hernandez R, Donn SM, Becker M, Bachman T (2011) Multicenter crossover study of automated control of inspired oxygen in ventilated preterm infants. Pediatrics 127:e76–e83

Darnall RA (2010) The role of CO(2) and central chemoreception in the control of breathing in the fetus and the neonate. Respir Physiol Neurobiol 173:201–212

Davis PG, Schmidt B, Roberts RS, Doyle LW, Asztalos E, Haslam R, Sinha S, Tin W (2010) Caffeine for apnea of prematurity trial: benefits may vary in subgroups. J Pediatr 156:382–387

Di Fiore JM, Bloom J, Orge F, Schutt A, Schluchter M, Cheruvu VK, Walsh M, Finer N, Martin RJ (2010) A higher incidence of intermittent hypoxemic episodes is associated with severe retinopathy of prematurity. J Pediatr 157:69–73

Di Fiore J, Walsh M, Finer N, Carlo W, Martin RJ, SUPPORT Study Group of the NICHD Neonatal Network (2011) Low oxygen saturation target range is associated with increased incidence of intermittent hypoxemia. J Pediatr 2012 in press

Dimaguila MA, Di Fiore JM, Martin RJ, Miller MJ (1997) Characteristics of hypoxemic episodes in very low birth weight infants on ventilatory support. J Pediatr 130:577–583

Esquer C, Claure N, D'Ugard C, Wada Y, Bancalari E (2007) Role of abdominal muscles activity on duration and severity of hypoxemia episodes in mechanically ventilated preterm infants. Neonatology 92:182–186

Herlenius E, Aden U, Tang LQ, Lagercrantz H (2002) Perinatal respiratory control and its modulation by adenosine and caffeine in the rat. Pediatr Res 51:4–12

Hibbs AM, Johnson NL, Rosen CL, Kirchner HL, Martin R, Storfer-Isser A, Redline S (2008) Prenatal and neonatal risk factors for sleep disordered breathing in school-aged children born preterm. J Pediatr 153:176–182

Julien CA, Joseph V, Bairam A (2010) Caffeine reduces apnea frequency and enhances ventilatory long-term facilitation in rat pups raised in chronic intermittent hypoxia. Pediatr Res 68:105–111

Katz-Salamon M, Jonsson B, Lagercrantz H (1995) Blunted peripheral chemoreceptor response to hyperoxia in a group of infants with bronchopulmonary dysplasia. Pediatr Pulmonol 20:101–106

Martin RJ, Di Fiore JM, Jana L, Davis RL, Miller MJ, Coles SK, Dick TE (1998) Persistence of the biphasic ventilatory response to hypoxia in preterm infants. J Pediatr 132:960–964

Martin RJ, Wang K, Köroglu Ö, Di Fiore J, Prabha KC (2011) Intermittent hypoxic episodes in preterm infants: Do they matter? review. Neonatology 100:303–310
Mayer CA, Haxhiu MA, Martin RJ, Wilson CG (2006) Adenosine A_{2A} receptors mediate GABAergic inhibition of respiration in immature rats. J Appl Physiol 100:91–97
Nock ML, Di Fiore JM, Arko MK, Martin RJ (2004) Relationship of the ventilatory response to hypoxia with neonatal apnea in preterm infants. J Pediatr 144:291–295
Paavonen EJ, Strang-Karlsson S, Raikkonen K, Heinonen K, Pesonen AK, Hovi P, Andersson S, Jarvenpaa AL, Eriksson JG, Kajantie E (2007) Very low birth weight increases risk for sleep-disordered breathing in young adulthood: the Helsinki study of very Low birth weight adults. Pediatrics 120:778–784
Pawar A, Peng YJ, Jacono FJ, Prabhakar NR (2008) Comparative analysis of neonatal and adult rat carotid body responses to chronic intermittent hypoxia. J Appl Physiol 104:1287–1294
Rigatto H, Brady JP (1972) Periodic breathing and apnea in preterm infants I. Evidence for hypoventilation possibly due to central respiratory depression. Pediatrics 50:202–218
Rosen CL, Larkin EK, Kirchner HL, Emancipator JL, Bivins SF, Surovec SA, Martin RJ, Redline S (2003) Prevalence and risk factors for sleep-disordered breathing in 8- to 11-year-old children: association with race and prematurity. J Pediatr 142:383–389
Sands SA, Edwards BA, Kelly VJ, Skuza EM, Davidson MR, Wilkinson MH, Berger PJ (2009) A model analysis of arterial oxygen desaturation during apnea in preterm infants. PLoS Comp Biol 5:1–14
Schmidt B, Roberts RS, Davis P, Doyle LW, Barrington KJ, Ohlsson A, Solimano A, Tin W (2007) Long-term effects of caffeine therapy for apnea of prematurity. N Engl J Med 357:1893–1902
Smith CA, Forster HV, Blain GM, Dempsey JA (2010) An interdependent model of central/peripheral chemoreception: evidence and implications for ventilatory control. Respir Physiol Neurobiol 173:288–297
Stenson B, Brocklehurst P, Tarnow-Mordi W (2011) Increased 36-week survival with high oxygen saturation target in extremely preterm infants. N Engl J Med 364:1680–1682
SUPPORT Study Group of the Eunice Kennedy Shriver NICHD Neonatal Research Network, Carlo WA, Finer NN, Walsh MC, Rich W, Gantz MG, Laptook AR, Yoder BA, Faix RG, Das A, Poole WK, Schibler K, Newman NS, Ambalavanan N, Frantz ID, Piazza AJ, Sánchez PJ, Morris BH, Laroia N, Phelps DL, Poindexter BB, Cotten CM, Van Meurs KP, Duara S, Narendran V, Sood BG, O'Shea TM, Bell EF, Ehrenkranz RA, Watterberg KL, Higgins RD (2010) Target ranges of oxygen saturation in extremely preterm infants. N Engl J Med 362:1959–1969

Chapter 48
Chronic Intermittent Hypoxia Increases Apnoea Index in Sleeping Rats

Deirdre Edge, Aidan Bradford, and Ken D. O'Halloran

Abstract Intermittent hypoxia (IH) is the dominant feature of sleep-disordered breathing which is very common. It is recognized that IH elicits plasticity in the respiratory control system. Recently it was reported in humans that IH destabilizes breathing during sleep increasing the susceptibility to apnoea. Many forms of respiratory plasticity are dependent upon reactive oxygen species (ROS), and NADPH oxidase has been identified as an important source of ROS necessary for IH-induced plasticity. In the present study, we sought to examine the effects of chronic IH (CIH) on the propensity for spontaneous apnoea during sleep. Adult male Wistar rats were exposed to 20 cycles of normoxia and hypoxia (5% O_2 at nadir; $SaO_2 \sim 80\%$) per hour, 8 h a day for 7 consecutive days (CIH group, N=6). The sham group (N=6) were subject to alternating cycles of air under identical experimental conditions in parallel. Two additional groups of CIH-treated rats were given either the superoxide dismutase mimetic – tempol (1 mM, N=8), or the NAPDH oxidase inhibitor – apocynin (2 mM, N=8) in their drinking water throughout the study. Following gas exposures, breathing during sleep was assessed in unrestrained animals using the technique of whole-body plethysmography. CIH significantly increased apnoea index during sleep (4.7 ± 0.8 vs. 11.3 ± 1.6 events/h; mean ± SEM, sham vs. CIH, Student's *t* test, $p=0.0035$). Apnoea duration was unaffected by CIH treatment. The CIH-induced increase in the occurrence of apnoea was completely reversed by antioxidant supplementation (4.9 ± 0.9 events/h for CIH+tempol and 5.6 ± 0.9 events/h for CIH+apocynin). CIH-induced increase in the propensity for apnoea may have clinical relevance and may explain the phenomenon of 'complex' apnoea in sleep apnoea patients. Our results suggest that oxidative stress is implicated in CIH-induced respiratory disturbance during sleep. We conclude that antioxidants may be a realistic adjunct therapy in the treatment of sleep-disordered breathing.

Keywords Chronic intermittent hypoxia • Sleep disordered breathing • Animal model • Plasticity • Apnoea index • Reactive oxygen species • Oxidative stress • Respiratory neurons • NADPH-oxidase • Antioxidants

---

D. Edge (✉) • K.D. O'Halloran
UCD School of Medicine and Medical Science, University College Dublin, Dublin 4, Ireland
e-mail: deirdre.edge@ucdconnect.ie

A. Bradford
Department of Physiology and Medical Physics, Royal College of Surgeons in Ireland, Dublin 2, Ireland

C.A. Nurse et al. (eds.), *Arterial Chemoreception: From Molecules to Systems*, Advances in Experimental Medicine and Biology 758, DOI 10.1007/978-94-007-4584-1_48,

48.1 Introduction

Intermittent hypoxia (IH), the dominant feature of sleep-disordered breathing, has been shown to induce plasticity in the respiratory control system at multiple levels. At a sensory level, enhanced output of the carotid body has been described (Peng et al. 2003; Iturriaga et al. 2006). In addition, acute IH can facilitate phrenic motor output (Fuller et al. 2000), leading to long-term facilitation (LTF) of breathing. In healthy humans IH has been shown to destabilize breathing during sleep, leading to dysrhythmic breathing increasing the susceptibility to apnoea (Chowdhuri et al. 2010). Increased propensity for apnoea has also been reported in patients with obstructive sleep apnoea (OSA) (Salloum et al. 2009), and in neonatal rat pups exposed to IH (Julien et al. 2008). Thus, IH can be maladaptive for respiratory homeostasis. Many forms of respiratory plasticity are dependent upon reactive oxygen species (ROS) (Prabhakar et al. 2007). Recently, NADPH oxidase, a superoxide generating enzyme has been identified as a major source of ROS necessary for IH-induced plasticity (Peng et al. 2009). Of interest, supplementation with antioxidants has proven beneficial in ameliorating IH-induced maladaptive responses (Skelly et al. 2011; Dunleavy et al. 2008; Peng et al. 2003). Thus we sought to examine the effects of chronic intermittent hypoxia (CIH) on the propensity for apnoea during sleep. We hypothesized that CIH would increase apnoea index in sleeping rats and that this would be a ROS-dependent mechanism.

48.2 Methods

48.2.1 Animal Model of Chronic Intermittent Hypoxia

Adult male Wistar rats (280 ± 3 g; mean ± SEM) were exposed to one of two chronic gas treatments. The gas supply to half of the environmental chambers alternated between air and nitrogen every 90s, reducing the ambient oxygen concentration to 5% at the nadir, resulting in arterial O_2 saturation values of ~80% (chronic intermittent hypoxia, CIH; N = 6). The sham group (N = 6) were subject to alternating cycles of normoxia under identical experimental conditions in parallel. Gas treatments were carried out for 20 cycles per hour, 8 h a day for 7 consecutive days. Rats had free access to food and water throughout the exposures. A subset of CIH-treated rats were given either tempol (1 mM) – a superoxide dismutase mimetic (N = 8), or apocynin (2 mM) – a NAPDH oxidase inhibitor (N = 8) in their drinking water throughout CIH exposure (Fig. 48.1).

48.2.2 Whole-Body Plethysmography

Following the chronic gas treatments, respiratory recordings were performed in unrestrained, sleeping rats using the technique of whole-body plethysmography. An acute intermittent hypoxic (AIH) challenge was performed within the plethysmograph, consisting of 10 cycles of hypoxia (10% O_2, 5 min) interspersed with normoxia (21% O_2, 7 min). Following the last episode of hypoxia rats were monitored for 1 h breathing room air. Within this normoxic hour, respiratory flow signals were analyzed for the occurrence of spontaneous apnoeas. Two types of apnoeas were distinguished: (1) post-sigh apnoeas preceded by a "sigh" or augmented breath (a breath twice the amplitude of the normal tidal volume), and (2) spontaneous apnoeas, which were not preceded by a sigh, but were prolonged expiratory pauses between breaths (see Fig. 48.2). The Mendelson criteria (Mendelson et al. 1988) were followed for scoring apnoeas and apnoeas consisting of 2 missed breaths were included in the analysis.

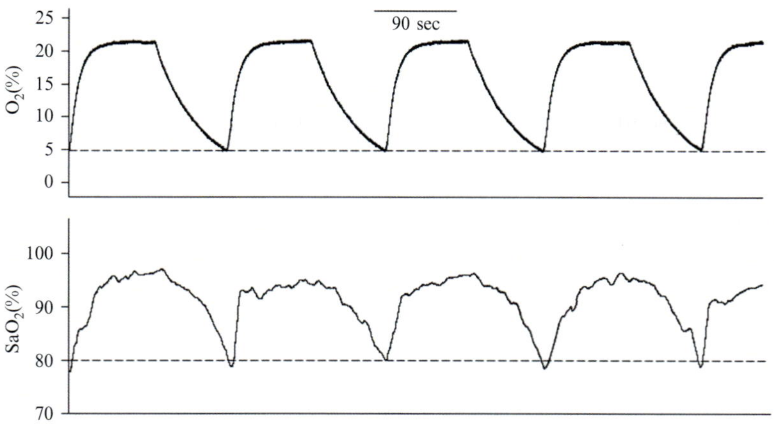

Fig. 48.1 *Top trace* is an original record of chamber O_2 concentration during CIH treatment. Gas was sampled from an exit port in the animal chamber. The *bottom trace* shows the corresponding arterial O_2 saturation in an unrestrained, unanaesthetized rat during quiet rest

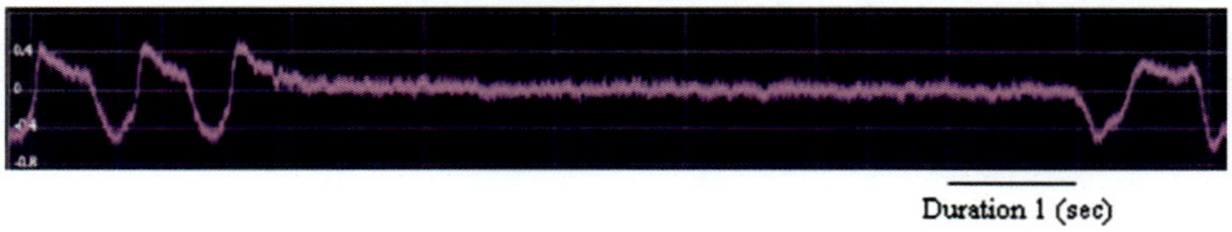

Fig. 48.2 An original trace of a prolonged spontaneous apnoea in an unrestrained, sleeping rat

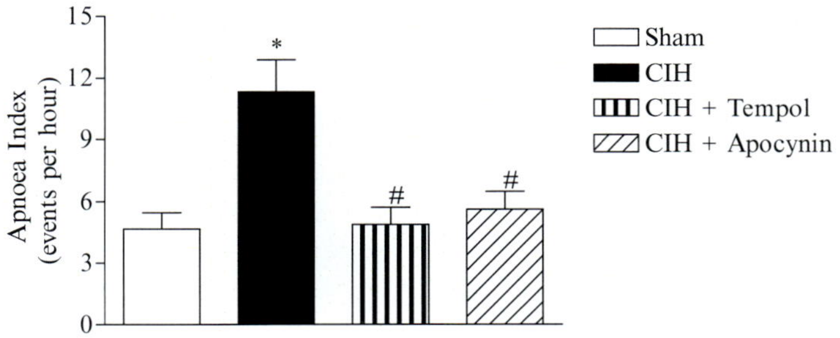

Fig. 48.3 The frequency of combined apnoeas expressed as events per hour in sham, CIH, CIH+tempol, and CIH+apocynin groups. * Significantly different from sham ($p<0.01$); # significantly different from CIH ($p<0.01$); one-way ANOVA, with Tukey post-hoc test

Apnoea index was expressed as the total number of apnoeas per hour. The events were apparently periods of complete cessation of respiratory effort i.e. central apnoeas. Indeed, obstructive events would generate an increased plethysmographic signal instead of a pause.

48.3 Results

CIH significantly increased the frequency of central apnoeas during sleep (4.7±0.8 vs. 11.3±1.6 events/h; mean±SEM, sham vs. CIH, Student's *t* test, $p=0.0035$). The duration of central apnoeas was unaffected by CIH treatment. The CIH-induced increase in apnoea index was completely reversed by antioxidant supplementation: 4.9±0.9 events/h for CIH+tempol and 5.6±0.9 events/h for CIH+apocynin, one-way ANOVA with Tukey post-hoc test (Fig. 48.3).

48.4 Discussion

Chronic intermittent hypoxia (CIH) – a central feature of sleep-disordered breathing, has myriad effects on the respiratory control system. We have shown that 1 week of CIH significantly increases the frequency of central apnoeas in sleeping rats. OSA is characterized by obstructive apnoeas due to collapse of the upper airway, whereas central apnoeas occur during sleep due to a lack of respiratory drive. These two disorders may overlap; indeed OSA patients receiving CPAP treatment often present with periodic breathing and central apnoeas – these patients are then said to be suffering from 'complex' apnoea (Morgenthaler et al. 2006). We speculate that 'complex' apnoea in OSA may be a consequence of CIH-induced remodelling in the central respiratory network governing respiration. To our knowledge this is the first report of an increased apnoea index following CIH in adult sleeping rats; it is yet another demonstration of the maladaptive nature of CIH for respiratory homeostasis. OSA is almost universally recognized as an oxidative stress disorder (Lavie 2009) and many forms of IH-induced plasticity are reactive oxygen species (ROS)-dependent (MacFarlane and Mitchell 2008). Pre-treatment with antioxidants has been shown to ameliorate CIH-induced alterations in neurocognition (Row et al. 2003; Xu et al. 2004), hypoglossal motor output (Veasey et al. 2004b) and chemoafferent discharge (Peng et al. 2003). With this in mind we examined whether the increased frequency of central apnoeas in our model, was also a ROS-dependent phenomenon. Administration of the superoxide dismutase mimetic-tempol significantly ameliorated the CIH-induced increase in apnoea index; from this we conclude that ROS are implicated in the underlying mechanism. NADPH oxidase (Nox), the ubiquitously expressed superoxide-generating enzyme has been identified as a major source of ROS involved in IH-induced plasticity (MacFarlane et al. 2009; Peng et al. 2009). To determine if Nox is a potential source of ROS in our model, we administered apocynin – a Nox 2 inhibitor. Apocynin has been shown to have anti-inflammatory and antioxidant effects in cell and animal models (Hougee et al. 2006; MacFarlane et al. 2009). Apocynin significantly blocked the CIH-induced increase in central apnoeas suggesting the involvement of-Nox-derived superoxide in this effect. In consideration of the locus of the CIH-induced respiratory phenotype, it is plausible to suggest that CIH is acting at the level of the carotid body. Indeed, IH has been shown to alter chemo-responses by sensitizing the hypoxic ventilatory response (HVR) (Del Rio et al. 2010; Julien et al. 2008; Peng et al. 2003). An enhancement of the HVR may lead to ventilatory overshoot and in this way may destabilize breathing – possibly by increasing the propensity for apnoea (Peng et al. 2003; Julien et al. 2008). However, in our model the HVR in unrestrained sleeping rats is unaltered (data not shown); therefore we argue this is not the case. Alternatively, the impairment may lie in defuse sites of the central respiratory network (Zhang et al. 2010) or in more discreet sites, for example in the putative rhythm-generating neurons of the medulla – the preBötzinger Complex (preBötC). Indeed, unilateral ablation of the preBötC induces apnoea in sleeping rats (McKay and Feldman 2008). Therefore we speculate that the preBötC is a key site implicated in the CIH-induced increase in apnoea index in our rats. Whilst a certain level of ROS are required for normal signalling cascades as well as activity-dependent plasticity in the hippocampus, in excess ROS can induce oxidative tissue injury and ultimately cell death. CIH exposure has been shown to induce oxidative neural injury to many key sites in the brain (Veasey et al. 2004a; Xu et al. 2004; Row et al. 2003). We plan to investigate preBötC neurons for evidence or not of oxidative injury following CIH treatment. We conclude that CIH, a dominant feature of OSA, is maladaptive for respiratory homeostasis and could exacerbate the condition in some patients. We argue that antioxidants may be a useful and realistic adjunct therapy in disorders characterized by CIH.

Acknowledgements Supported by the Health Research Board, Ireland (RP/2007/29). DE is enrolled in the School of Medicine and Medical Science Translational Medicine PhD training programme.

References

Chowdhuri S, Shanidze I, Pierchala L, Belen D, Mateika JH, Badr MS (2010) Effect of episodic hypoxia on the susceptibility to hypocapnic central apnea during NREM sleep. J Appl Physiol 108:369–377
del Rio R, Moya EA, Iturriaga R (2010) Carotid body and cardiorespiratory alterations in intermittent hypoxia: the oxidative link. Eur Respir J 36:143–150
Dunleavy M, Bradford A, O'Halloran KD (2008) Oxidative stress impairs upper airway muscle endurance in an animal model of sleep-disordered breathing. Adv Exp Med Biol 605:458–462
Fuller DD, Bach KB, Baker TL, Kinkead R, Mitchell GS (2000) Long term facilitation of phrenic motor output. Respir Physiol 121:135–146
Hougee S, Hartog A, Sanders A, Graus YM, Hoijer MA, Garssen J, van den Berg WB, van Beuningen HM, Smit HF (2006) Oral administration of the NADPH-oxidase inhibitor apocynin partially restores diminished cartilage proteoglycan synthesis and reduces inflammation in mice. Eur J Pharmacol 531:264–269
Iturriaga R, Rey S, Alcayaga J, del Rio R (2006) Chronic intermittent hypoxia enhances carotid body chemosensory responses to acute hypoxia. Adv Exp Med Biol 580:227–232, discussion 351–9
Julien C, Bairam A, Joseph V (2008) Chronic intermittent hypoxia reduces ventilatory long-term facilitation and enhances apnea frequency in newborn rats. Am J Physiol Regul Integr Comp Physiol 294:R1356–R1366
Lavie L (2009) Oxidative stress – a unifying paradigm in obstructive sleep apnea and comorbidities. Prog Cardiovasc Dis 51:303–312
Macfarlane PM, Mitchell GS (2008) Respiratory long-term facilitation following intermittent hypoxia requires reactive oxygen species formation. Neuroscience 152:189–197
Macfarlane PM, Satriotomo I, Windelborn JA, Mitchell GS (2009) NADPH oxidase activity is necessary for acute intermittent hypoxia-induced phrenic long-term facilitation. J Physiol 587:1931–1942
McKay LC, Feldman JL (2008) Unilateral ablation of pre-botzinger complex disrupts breathing during sleep but not wakefulness. Am J Respir Crit Care Med 178:89–95
Mendelson WB, Martin JV, Perlis M, Giesen H, Wagner R, Rapoport SI (1988) Periodic cessation of respiratory effort during sleep in adult rats. Physiol Behav 43:229–234
Morgenthaler TI, Kagramanov V, Hanak V, Decker PA (2006) Complex sleep apnea syndrome: is it a unique clinical syndrome? Sleep 29:1203–1209
Peng YJ, Overholt JL, Kline D, Kumar GK, Prabhakar NR (2003) Induction of sensory long-term facilitation in the carotid body by intermittent hypoxia: implications for recurrent apneas. Proc Natl Acad Sci U S A 100:10073–10078
Peng YJ, Nanduri J, Yuan G, Wang N, Deneris E, Pendyala S, Natarajan V, Kumar GK, Prabhakar NR (2009) NADPH oxidase is required for the sensory plasticity of the carotid body by chronic intermittent hypoxia. J Neurosci 29:4903–4910
Prabhakar NR, Kumar GK, Nanduri J, Semenza GL (2007) ROS signaling in systemic and cellular responses to chronic intermittent hypoxia. Antioxid Redox Signal 9:1397–1403
Row BW, Liu R, Xu W, Kheirandish L, Gozal D (2003) Intermittent hypoxia is associated with oxidative stress and spatial learning deficits in the rat. Am J Respir Crit Care Med 167:1548–1553
Salloum A, Rowley JA, Mateika JH, Chowdhuri S, Omran Q, Badr MS (2009) Increased propensity for central apnea in patients with obstructive sleep apnea: effect of nasal continuous positive airway pressure. Am J Respir Crit Care Med 181:189–193
Skelly JR, Edge D, Shortt CM, Jones JFX, Bradford A, O'Halloran KD (2012) Tempol ameliorates pharyngeal dilator muscle dysfunction in a rodent model of chronic intermittent hypoxia. Am J Respir Cell Mol Biol 46:139–148
Veasey SC, Davis CW, Fenik P, Zhan G, Hsu YJ, Pratico D, Gow A (2004a) Long-term intermittent hypoxia in mice: protracted hypersomnolence with oxidative injury to sleep-wake brain regions. Sleep 27:194–201
Veasey SC, Zhan G, Fenik P, Pratico D (2004b) Long-term intermittent hypoxia: reduced excitatory hypoglossal nerve output. Am J Respir Crit Care Med 170:665–672
Xu W, Chi L, Row BW, Xu R, Ke Y, Xu B, Luo C, Kheirandish L, Gozal D, Liu R (2004) Increased oxidative stress is associated with chronic intermittent hypoxia-mediated brain cortical neuronal cell apoptosis in a mouse model of sleep apnea. Neuroscience 126:313–323
Zhang JH, Fung SJ, Xi M, Sampogna S, Chase MH (2010) Apnea produces neuronal degeneration in the Pons and medulla of guinea pigs. Neurobiol Dis 40:251–264

Chapter 49
Contribution of TASK-Like Potassium Channels to the Enhanced Rat Carotid Body Responsiveness to Hypoxia

Fernando C. Ortiz, Rodrigo Del Rio, Rodrigo Varas, and Rodrigo Iturriaga

Abstract A major hallmark of obstructive sleep apnea is the potentiation of the carotid body (CB) chemosensory response to acute hypoxia, as result of the chronic intermittent hypoxia (CIH) exposition. Several mechanisms have been involved in this CB chemosensory potentiation, but the primary target of CIH remains elusive. In physiological conditions, hypoxia depolarized CB chemoreceptor cells, trigger an increase of intracellular Ca^{2+}, and the subsequent transmitter's release. Since the depolarization is initiated by the inhibition of a TASK-like K^+ channel, we studied if CIH may increase the amplitude of the hypoxic-induced depolarization in the chemoreceptor cells, due to an enhanced inhibition of the TASK-like current.

CBs obtained from adult rats exposed to CIH (5% O_2, 12 times/hr for 8 hr/day) for 7 days were acute dissociated, and the membrane potential and TASK-like current were recorded from isolated chemoreceptor cells. Resting membrane properties were not modified by CIH, but the amplitude of the hypoxic-evoked depolarization increases ~2-fold. The same result was obtained when all the voltage-dependent K^+ currents were pharmacologically blocked. Accordingly, the inhibition of the TASK-like current induced by acute hypoxia (PO_2 ~5 torr) increased from ~62% in control cells to ~96% in the CIH cells.

Present results show that acute hypoxic inhibition of TASK-like K^+ channel is potentiated by CIH exposure, suggesting that the enhancing effect of CIH on CB chemosensory responsiveness to hypoxia occurs at the initial step of the oxygen transduction in the CB chemoreceptor cells.

Keywords Carotid body • Chemosensory discharge • Glomus cells • TASK channels • Acute hypoxia • Single channel recording • Intermittent hypoxia • Obstructive sleep apnea

49.1 Introduction

Chronic intermittent hypoxia (CIH) is the main characteristic of the obstructive sleep apnea (OSA) syndrome, a growing health problem that affects 5% of the worldwide population (Somers et al. 2008). The OSA syndrome is recognized as an independent risk factor for sympathetic overactivation

F.C. Ortiz • R. Del Rio • R. Varas • R. Iturriaga (✉)
Laboratorio de Neurobiología, Facultad de Ciencias Biológicas, Pontificia Universidad Católica de Chile, Santiago, Chile
e-mail: radelrio@uc.cl; riturriaga@bio.puc.cl

C.A. Nurse et al. (eds.), *Arterial Chemoreception: From Molecules to Systems*, Advances in Experimental Medicine and Biology 758, DOI 10.1007/978-94-007-4584-1_49,

and hypertension (Somers et al. 2008; Garvey et al. 2009). A major contributing mechanism to the cardiovascular alterations induced by CIH is the potentiation of the carotid body chemosensory responses to acute hypoxia (Peng et al. 2003; Rey et al. 2004; Del Rio et al. 2010). Oxidative stress, endothelin-1 and pro-inflammatory molecules have been involved in the carotid chemosensory potentiation, but the primary target for CIH in the carotid body remains elusive (Iturriaga et al. 2009). The most accepted model for carotid body chemoreception states that in response to acute hypoxia the CB chemoreceptors (type I) cells depolarize, triggering an intracellular calcium increase and the subsequent release of one or more excitatory neurotransmitters (Iturriaga et al. 2007). In neonatal rats it has been shown that the depolarization is initiated by the hypoxic inhibition of a background TASK-like potassium channel, probably a TASK-1/TASK-3 heterodimer (Buckler 1999, 2010; Kim et al. 2009). We hypothesized that the enhanced carotid chemosensory responses to hypoxia is partially explained by the potentiation of the hypoxia-induced depolarization, due to a greater inhibition of the TASK-like current by acute hypoxia. Therefore, we evaluated if CIH may increase the amplitude of the hypoxic-induced depolarization by enhancing the inhibition of the TASK-like current in chemoreceptor cells from rats exposed to CIH for 7 days.

49.2 Methods

49.2.1 Animals and Exposure to Intermittent Hypoxia

Experiments were performed on adult male Sprague-Dawley rats (200–250 g), fed with standard chow diet *ad libitum*, and kept on a 12-h light/dark schedule (8:00 a.m.–8:00 p.m.). Animals were randomly assigned to CIH or to Control (sham) conditions. The experimental procedures were approved by the Bio-Ethical Committee of the Biological Sciences Faculty, P. Universidad Católica de Chile, and were performed according to the National Institutes of Health Guide for the Care and Use of Laboratory Animals. Rats were housed in individual chambers and exposed to either hypoxic cycles of 5% inspired O_2 for 20 s, followed by room air for 280 s, applied 12 times/h; 8 h/day or control (sham) group exposed to air/air cycles for 7 days (Del Rio et al. 2010). The O_2 level in the chambers was continuously monitored with an oxygen analyzer and the CO_2 was maintained low by continuous air extraction. The room temperature was kept at 23–25°C.

49.2.2 Recordings of Ventilatory Reflex and Carotid Chemosensory Responses to Acute Hypoxia

Rats were anesthetized with sodium pentobarbitone (40 mg/kg, i.p.). The trachea was cannulated for recording the airflow signal with appropriate pneumotachographs. Minute inspiratory volume (V_I), tidal volume and respiratory frequency were digitally obtained from the airflow signal. The physiological data was analyzed using the LabChart 7.2 Pro software (AD Instruments, Australia). To assess the effects of CIH on the sensitivity and reactivity of peripheral chemoreceptor reflexes, we studied the ventilatory responses elicited by several isocapnic levels of PO_2, maintained during 30 s. At the end of the ventilatory physiological recordings, one carotid sinus nerve was dissected and placed on a pair of platinum electrodes, and covered with warm mineral oil. The neural signal was pre-amplified (Grass P511, USA), filtered (30 Hz to 1 kHz) and fed to an electronic spike-amplitude discriminator, allowing the selection of action potentials of given amplitude above the noise to be counted with a frequency meter to measure the frequency of carotid chemosensory discharge (*f*x), expressed in Hz. Carotid sinus barosensory fibers were eliminated by crushing the common carotid artery wall between the carotid sinus and the carotid body. The contralateral carotid sinus nerve was cut to prevent vascular and ventilatory reflexes evoked by the activation of the CB. The chemosensory discharge was measured at different (isocapnic) levels of PO_2.

49.2.3 Electrophysiological Recordings

The carotid bodies extracted from anesthetized rats were enzymatic (trypsin/collagenase) and mechanically dissociated as previously described (Buckler 1999). Cellular suspension was plated into poly-D-lysine coated glass coverslip. The acute dissociated cells were studied upon either cell-attached or whole-cell configuration of the voltage clamp technique. Cells were bathed with standard HCO_3-buffered saline contained (in mM) 117 NaCl, 4.5 KCl, 23 $NaHCO_3$, 1.0 $MgCl_2$, 2.5 $CaCl_2$ and 11 glucose; bubbled with either 5% CO_2 and 95% air (normoxia, PO_2 ~140 mmHg) or 5% CO_2 and 95% N_2 (during hypoxic stimuli, PO_2 ~5 mmHg by 30 s). Pipette solution for cell attached recordings contained (in mM) 140 KCl, 4 $MgCl_2$, 1 EGTA, 10 HEPES, 10 tetraethylammonium (TEA)-Cl and 5 4-aminopyridine (4-AP), equilibrated to pH = 7.43, and for whole-cell recordings (in mM) 140 KCl, 1.5 $MgCl_2$, 10 EGTA, 2 $CaCl_2$, 10 HEPES (pH 7.21). Channel activity was reported as the open probability times the number of active channels in a given patch (NPo). Membrane potential recordings were performed under whole-cell current clamp conditions with no imposed current. Mean values were expressed as MEAN ± SEM.

49.3 Results

49.3.1 Effect of CIH on Ventilatory and Carotid Chemosensory Responses to Acute Hypoxia

Rats exposed to CIH for 7 days showed potentiated reflex ventilatory responses to hypoxia when compared with control rats (Fig. 49.1a, $p < 0.001$, two-way ANOVA). The carotid chemosensory responses to acute hypoxia were also enhanced by the exposure to CIH (Fig. 49.1b). The two-way ANOVA analysis showed that the overall carotid chemosensory curve for PO_2 was different in CIH rats ($p < 0.001$) compared with the control group. The posthoc test showed that carotid chemosensory discharges were higher ($p < 0.01$) not only in the hypoxic range, but also in normoxia (Fig. 49.1b).

49.3.2 Effects of CIH on the Resting Membrane Potential of Isolated Chemoreceptor Cells

Exposure of rats to CIH for 7 days did not modify the resting membrane potential of chemoreceptor cells. The resting membrane potential measured in chemoreceptor cells of the CIH group was −54.3 ± 2.4 mV (n = 30), while in the control cells was −55.1 ± 1.4 mV (n = 28) ($p > 0.05$, Student T test).

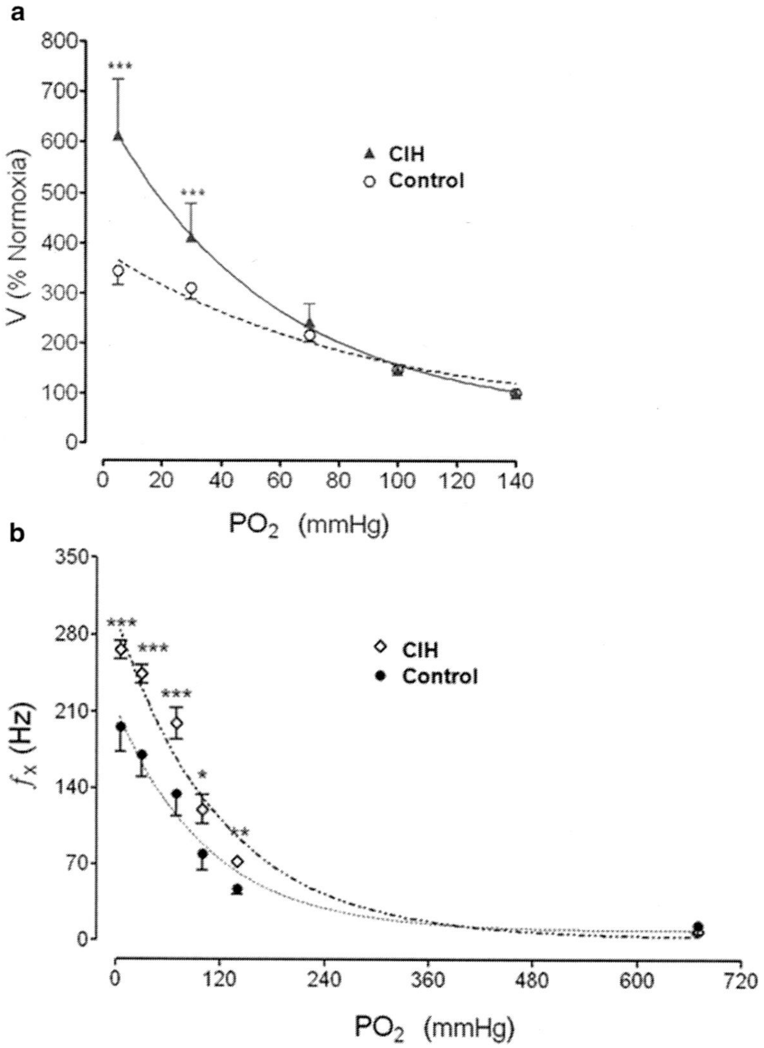

Fig. 49.1 Effects of intermittent hypoxia exposure for 7 days on ventilatory and carotid chemosensory responses to acute hypoxia. **a** Ventilatory response (V, minute volume) to acute hypoxia in rats exposed to CIH for 7 days (CIH, n = 10) and control rats (Control, n = 10). **b** Carotid chemosensory responses to hypoxia (two way ANOVA, followed by Bonferroni posthoc test. *** $p < 0.001$, ** $p < 0.01$ and * $p < 0.05$). fx, frequency of carotid chemosensory discharge expressed in Hz

49.3.3 Effects of CIH on the Depolarization Induced by Acute Hypoxia in Chemoreceptor Cells

The depolarization (ΔVm_{Hx}) induced by acute hypoxia (PO_2 ~5 Torr) was enhanced by the CIH exposure (Fig. 49.2). Indeed, we recorded a ΔVm_{Hx} of 27.5 ± 2.8 mV in 16 chemoreceptor cells from CIH-treated rats versus ΔVm_{Hx} of 15.7 ± 1.5 mV in 20 control cells ($p < 0.05$, unpaired T-test). The same result was obtained when the voltage-dependent K^+ currents were pharmacologically blocked with tetraethyl ammonium and 4-aminopiridine (not shown).

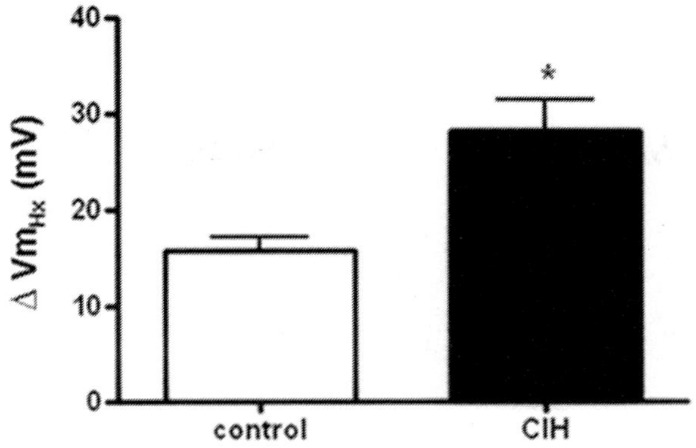

Fig. 49.2 Depolarization induced by acute hypoxia (ΔVm_{Hx}) in chemoreceptor cells was enhanced by CIH exposition. (CIH, n = 16) as compared to control cells (control, n = 20). * $p < 0.05$, Unpaired T-test

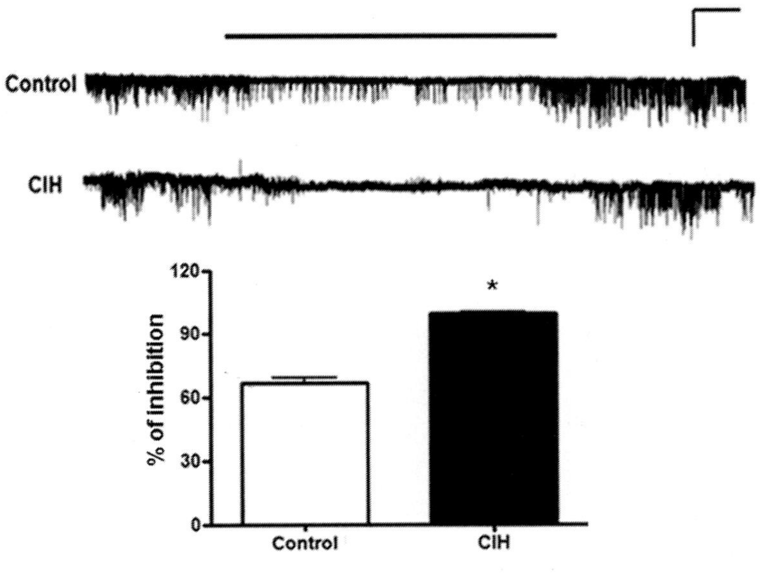

Fig. 49.3 TASK-like current inhibition by acute hypoxia is enhanced by CIH exposition. *Upper panel*, representative cell-attached record of TASK-like channel activity from cells of control and CIH groups during acute hypoxia stimulation (pipette potential 0 mV). Scale: 1 pA, 5 s Bar, hypoxic estimulation. *Bottom panel* summarized the effects of hypoxia inhibition in 16 control cells and 12 cells from rats exposed to CIH, respectively. * $p < 0.05$, Mann-Whitney Test

49.3.4 Effect of CIH on TASK-Channel Inhibition Induced by Acute Hypoxia

In cell-attached recordings under resting conditions (pipette potential of 0 mV) we observed robust TASK-like channel activity in patches from cells of control and CIH groups. Interestingly, we found that TASK-like current inhibition by acute hypoxia was enhanced by CIH exposition. Fig. 49.3 shows representative traces of TASK-like channel activity recorded from cell-attached patches of chemoreceptor cells in response to acute hypoxia (PO_2 ~ 5 Torr). The hypoxic inhibition, measured as percentage of NPo in normoxic conditions was 62.5 ± 2.4% (n = 16) in the control group while in the cells from rats exposed to CIH the inhibition was 96.4 ± 1.5% (n = 12). Thus, present results showed that the hypoxic inhibition of TASK-like K^+ channel is potentiated by the CIH exposure, suggesting that the enhancing effects of CIH on carotid chemosensory responsiveness to hypoxia occurs at the initial step of the oxygen transduction in the chemoreceptor cells.

49.4 Discussion

Exposure to CIH induces a potentiation of the rat carotid body chemosensory and ventilatory responses to hypoxia (Del Rio et al. 2010). In the present report the enhanced carotid chemosensory response to hypoxia was associated with a large inhibition of the TASK-like current in chemoreceptor cells in response to acute hypoxia.

It is worth noting that the present study was performed in adult rats, which functionally expresses a TASK-like current in the carotid body chemoreceptor cells. This current was reversibly inhibited by acute hypoxia, and therefore, could be involved in the adult carotid hypoxic response in this animal model. The resting TASK-like channel activity in normoxia was not modified by the CIH treatment. This result agrees with the data showing no effects of CIH in the resting membrane potential. However, the TASK-like current inhibition by acute hypoxia clearly increased, ~96% in CIH group versus ~62% of inhibition in control group. In order to establish a relationship between the enhanced TASK-like current inhibition and the augmented depolarization observed in chemoreceptor cells from CIH-treated rats, it was studied the hypoxia-induced depolarization when the main voltage-gated K^+ conductances were blocked with TEA and 4-AP. In this condition the depolarization induced by acute hypoxia in both control and CIH group were roughly the same, suggesting that the enhanced depolarization observed in CIH cells is explained at least in part by a potentiation in the inhibition of TASK-like current. It is interesting to note that while the normoxic TASK-like current remains unaffected by the CIH treatment, its sensitivity to hypoxia enhances. One explanation for these observations is that CIH treatment changes the CB type-I cell metabolic state, triggering modifications on TASK-like channel regulatory factors involved in the hypoxic response (Varas et al. 2007) rather than a direct effect on channel structure or expression.

In conclusion, present results showed that the chemoreceptor cells depolarization to acute hypoxia was enhanced by the CIH treatment, due to the increase of the acute hypoxia inhibitory effect on TASK-like potassium channel. This raises a new mechanism to explain the effects of chronic intermittent hypoxia on carotid body acclimatization.

Acknowledgements This work was supported by grant 1100405 from the National Fund for Scientific and Technological Development of Chile (FONDECYT).

References

Buckler KJ (1999) Background leak K^+-currents and oxygen sensing in carotid body type 1 cells. Respir Physiol 115:179–187

Buckler KJ (2010) Two-pore domain K^+ channels and their role in chemoreception. Adv Exp Med Biol 66:15–30

Del Rio R, Moya EA, Iturriaga R (2010) Carotid body and cardiorespiratory alterations in intermittent hypoxia: the oxidative link. Eur Resp J 36:143–150

Garvey JF, Taylor CT, McNicholas WT (2009) Cardiovascular disease in obstructive sleep apnoea syndrome: the role of intermittent hypoxia and inflammation. Eur Resp J 33:1195–1205

Iturriaga R, Varas R, Alcayaga J (2007) Electrical and pharmacological properties of petrosal ganglion neurons that innervate the carotid body. Resp Physiol Neurobiol 157:130–139

Iturriaga R, Moya EA, Del Rio R (2009) Carotid body potentiation induced by intermittent hypoxia: implications for cardiorespiratory changes induced by sleep apnoea. Clin Exp Pharmacol Physiol 36:1197–1204

Kim D, Cavanaugh EJ, Kim I, Carroll JL (2009) Heteromeric TASK-1/TASK-3 is the major oxygen-sensitive background K^+ channel in rat carotid body glomus cells. J Physiol 587:2963–2975

Peng YJ, Overholt JL, Kline D, Kumar GK, Prabhakar NR (2003) Induction of sensory long-term facilitation in the carotid body by intermittent hypoxia: implications for recurrent apneas. Proc Natl Acad Sci U S A 100:10073–10078

Rey S, Del Rio R, Alcayaga J, Iturriaga R (2004) Chronic intermittent hypoxia enhances cat chemosensory and ventilatory responses to hypoxia. J Physiol 560:577–586

Somers VK, White DP, Amin R, Abraham WT, Costa F, Culebras A, Daniels S, Floras JS, Hunt CE, Olson LJ, Pickering TG, Russell R, Woo M, Young T (2008) Sleep apnea and cardiovascular disease. J Am Coll Cardiol 52:686–717

Varas R, Wyatt CN, Buckler KJ (2007) Modulation of TASK-like background potassium channels in rat arterial chemoreceptor cells by intracellular ATP and other nucleotides. J Physiol 583:521–536

Chapter 50
Antioxidation and the Hypoxic Ventilatory Response

Mieczyslaw Pokorski, Agnieszka Rekawek, Izabela Zasada, Justyna Antosiewicz, and Rene Delgado

Abstract Reactive oxygen species favor the reductive state of iron. Antioxidation, by depleting biologically active ferrous iron, could then have a stabilizing effect, akin to hypoxia, on HIF-1α; the process which controls the genetic responses to hypoxia. However, the influence of antioxidation on the hypoxic ventilatory responses (HVR) is unclear. In this study we set out to determine the influence of mangiferin, a natural polyphenolic compound present in mango trees, with strong antioxidant and iron chelating properties, on the HVR. The study was performed in awake Wistar rats. Acute HVR to 12% and 8% FiO_2 before and 40 min after mangiferin (300 mg/kg, i.p.) pretreatment were recorded plethysmographically. We found that mangiferin significantly dampened the HVR over its course. To distinguish between the scavenging and chelating mechanisms of mangiferin we reinvestigated its effects on the HVR in a separate group of rats after chronic antecedent iron chelation with ciclopirox olamine (20 mg/kg daily for 1 week). The dampening effect on the HVR of mangiferin was preserved in the pre-chelated rats, which points to the preponderance of the antioxidant over chelating properties of mangiferin in its ventilatory effects. Although the exact determinants of mangiferin action remain unclear, the study suggests a role for oxidative signaling in the peripheral chemosensory processing of the HVR. The study also implies the possible clinical use of the antioxidant mangiferin in the regulation of lung ventilation.

Keywords Antioxidation • Iron chelation • Hypoxic ventilatory response • Mangiferin

M. Pokorski (✉)
Department of Respiratory Research, Medical Research Center, Polish Academy of Sciences, Pawinskiego 5 St, 02-106, Warsaw, Poland

Department of Psychology, Institute of Psychology, Opole University, Opole, Poland
e-mail: m_pokorski@hotmail.com

A. Rekawek • I. Zasada • J. Antosiewicz
Department of Respiratory Research, Medical Research Center, Polish Academy of Sciences, Pawinskiego 5 St, 02-106, Warsaw, Poland

R. Delgado
Laboratory of Molecular Pharmacology, Drug Research and Development Center (CIDEM), Havana, Cuba

C.A. Nurse et al. (eds.), *Arterial Chemoreception: From Molecules to Systems*,
Advances in Experimental Medicine and Biology 758, DOI 10.1007/978-94-007-4584-1_50,

50.1 Introduction

Reactive oxygen species (ROS) provide a facilitatory environment for iron to stay in the biologically active, reduced state. Ferrous iron (Fe^{2+}), in turn, along with the availability of oxygen, is essential for hypoxia inducible factor (HIF-1α) degradation by prolyl hydroxylases (Taylor 2008). The corollary is that antioxidation, by depleting ferrous iron, could have a stabilizing effect on HIF-1α, akin to hypoxia; the process which initiates the adaptive bodily responses to hypoxia at the transcriptional level. However, the functional role of antioxidation in shaping the hypoxic ventilatory response (HVR) is unsettled. This role is of considerable interest in view of the widespread use of supplemental antioxidants, even more notable in case of health problems.

Mangiferin, a polyphenolic compound extracted from mango trees, is used in the Caribbean region as a natural medicine panacea. The compound has antioxidative propeties (Pardo-Andreu et al. 2006a) and a spate of beneficial health effects, from antiinflammatory (Beltrán et al. 2004), antidiabetic (Muruganandan et al. 2005) to anticancer (Yoshimi et al. 2001). Respiration, in particular hypoxic reactivity, is essential for the maintenance of health. From the rational perspective, mangiferin seemed a good tool to study the effect on the HVR of antioxidation, since the molecule shows both free radical scavenging and iron chelating properties (Andreu et al. 2005) and does not penetrate into the brain after systemic injection (Pokorski et al. 2010). It is thus a reasonable assumption that the ventilatory alterations after mangiferin should mostly stem from its peripheral effects exerted at the carotid body level, an organ engaged in the generation of the HVR. In the present study, we seek to determine the effects of mangiferin on the HVR in unsedated rats. We hypothesized that mangiferin would be well suited to downregulate the HVR, which could have to do with HIF-1α stabilization through either depleting ROS or iron chelation. The basic results confirmed the hypothesis. To distinguish between these two modes of action we reinvestigated the HVR on the background of prior chronic iron chelation. The dampening effect on the HVR of mangiferin was preserved in prechelated rats, which points to the preponderance of the antioxidant over chelating pathway in the ventilatory effects of mangiferin. Part of this work has been reported elsewhere in the abstract form (Pokorski et al. 2010).

50.2 Methods

50.2.1 Animals and Study Protocol

All animal procedures were conducted in accordance with the ethical guidelines of the European Committee Council Directives 86/609/EEC. The study was approved by a local Ethics Committee for Animal Research.

A total of 17 male adult Wistar rats 12–16 weeks old, weighing 270–412 g were used for the study. The animals were fed commercial chow and had water *ad libitum*. They were kept at an artificial 12/12 dark/light cycle, with the lights on at 08:00, room temperature of 22 ±1°C and humidity of 50–60%.

There were four experimental groups of animals. Three groups were devoted to study the hypoxic ventilatory response in conscious animals in the non-invasive routine. The first group consisted of 5 rats of the mean weight 319.6 ± 14.5(SE) g in which the HVR to 12% and 8% FiO_2 were studied before and 40 min after injection of mangiferin in a dose of 300 mg/kg, i.p., dissolved in 0.3 ml DMSO, in the animals not subjected to prior iron chelation. The second group consisted of 6 age-matched rats of the mean weight of 343.6 ± 26.6 g in which the HVR to the same two levels of hypoxia were studied, before and after mangiferin as above outlined, on the background of chronic pretreatment with the intracellular iron chelator olamine ciclopirox (CPX; 20 mg, i.p., daily for 7 days). The third group, consisting of 3 rats, was devoted to study the possible ventilatory effects of the solvent alone.

DMSO, given in 0.3 ml volume, i.p., had no effects on the HVR (data not shown). In the fourth group, consisting of another 3 rats, the animals were lightly anesthetized (α-chloralose and urethane, 60 and 600 mg/kg, respectively; sedation for *ca* 1.5 h) to record the influence of mangiferin on arterial blood pressure and to perform gasometric measurements (IDEXX VetStat analyzer, Diamond Diagnostics, Holliston, MA) in the blood withdrawn from a femoral artery in normoxia and during the last 30 s of 3-min hypoxic exposures. All reagents used in the study were purchased from Sigma-Aldrich (Poznan, Poland).

50.2.2 Ventilatory Measurements

After the rat had acclimated to the chamber, basal ventilation was recorded. Next, the rat was randomly exposed to two levels of decreased inspired oxygen fraction: 12% and 8% FiO_2 in N_2. Hypoxic tests took 3 min and were separated by a 10 min recovery interval in room air.

Ventilation was investigated in a whole body rodent single-chamber plethysmograph (PLY3223, Buxco Electronics, Wilmington, NC). The system consisted of a chamber equipped with two pneumotachographs. Pressure difference between the experimental and reference chambers was measured with a differential pressure transducer. The pressure signal was amplified and integrated by data analysis software (Biosystem XA for Windows, Buxco Electronics, Wilmington, NC). Minute ventilation (V_E ml min, BTPS), tidal volume (V_T), and respiratory frequency (*f*) were computed breath-by-breath and stored for off-line analysis. Ten-second averages of each variable just preceding the sequential 30-s time marks along the hypoxic response course were taken for analysis.

50.2.3 Data Evaluation

Data were expressed as means ± SE. Ventilatory variables were normalized to weight in kg. The profiles of hypoxic response courses were analyzed with two-way repeated-measures ANOVA to test for between-factor (mangiferin) and within-factor (time) effects, and an interaction between the two. There were three main points of interest along the hypoxic response course which were subjected to further analysis: resting V_E, peak hypoxic V_E at 30 s from hypoxia onset, and end-test depressant nadir of hypoxic V_E. Here, differences between each two corresponding points before and after mangiferin within the two main groups, i.e., without and with chronic iron chelation were compared with a two-tailed paired *t*-test. A paired *t*-test also was used to evaluate the effect of mangiferin on the increment in peak hypoxic V_E in response to the stronger 8% FiO_2. Statistical significance was accepted at $p<0.05$. We used a commercial software package (SPSS version 18).

50.3 Results

Acute hypoxia decreased PaO_2 to 64.0 ± 3.1 and 48.0 ± 3.5 mmHg at 12% and 8% FiO_2, respectively, the decreases were significant compared with the baseline level and also differed from each other ($P<0.04$), with no appreciable changes in $PaCO_2$ and pHa (Table 50.1). Mangiferin alone caused insignificant changes in the basal values of PaO_2, 96.3 ± 1.2 *vs.* 94.3 ± 6.7 mmHg, and of the mean arterial blood pressure, 89.2 ± 19.1 *vs.* 81.0 ± 18.5 mmHg; before and 40 min after mangiferin, respectively.

Table 50.1 Arterial blood gasometry

	FiO_2 21%	FiO_2 12%	FiO_2 8%
PaO_2 (mmHg)	106.0 ± 13.6	64.0 ± 3.1*	48.0 ± 3.5*†
$PaCO_2$ (mmHg)	44.3 ± 1.1	42.7 ± 2.2	41.3 ± 3.3
pHa	7.36 ± 0.01	7.36 ± 0.04	7.37 ± 0.04

Values are means ± SE of three experiments
*Different from normoxia and †different from 12% FiO_2 ($p<0.05$)

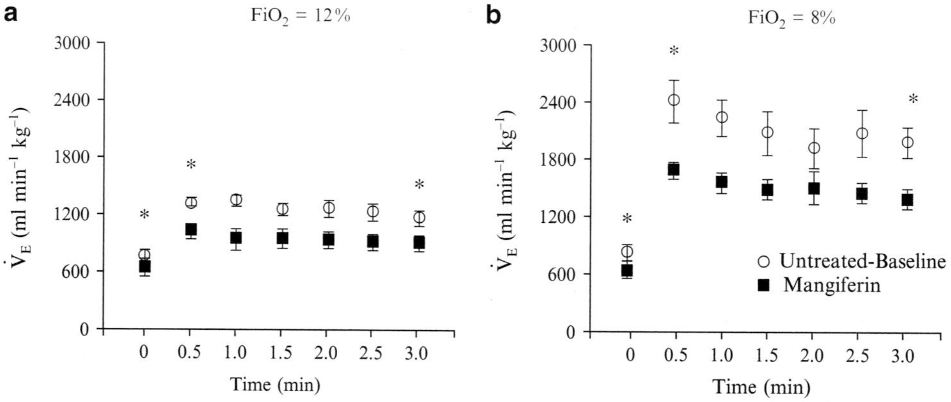

Fig. 50.1 Minute ventilation responses to successive levels of hypoxia before and 40 min after administration of mangiferin in conscious, non-chelated rats. *$p<0.05$ between the corresponding (*vertical*) time points of interest before *vs*. after mangiferin

Mangiferin depressed moderately, albeit significantly, resting ventilation, measured before the commencement of the responses hypoxia; the decreases amounted to 133 ± 43 and 192 ± 48 ml min^{-1} kg^{-1} before the introduction of 12% and 8% FiO_2, respectively ($p<0.01$ both). The profiles of the HVR were more distinctly depressed over the 3 min test at both oxygen fractions studied. These profiles were typically stimulatory/inhibitory, with the peak hypoxic V_E reached at 30 s from onset of hypoxia. There were significant mangiferin- and time-dependent effects for 12% FiO_2 ($p<0.001$ both) (Fig. 50.1a) and for 8% FiO_2 ($p<0.02$ and $p<0.001$, respectively) (Fig. 50.1b), but no interaction between the two effects. Peak hypoxic ventilation decreased by 303 ± 88 and 733 ± 145 ml min^{-1} kg^{-1} in response to 12% and 8% FiO_2, respectively ($P<0.01$ both); the decrease in peak hypoxic V_E was significantly greater at the stronger hypoxic stimulus ($p<0.02$). Likewise, at the depressant nadir of hypoxic responses, V_E was lower by 271 ± 77 and 619 ± 171 ml min^{-1} kg^{-1} at 12% and 8% FiO_2, respectively ($p<0.01$ both), after mangiferin.

To distinguish between the scavenging and chelating mechanisms of mangiferin we examined the effects on the HVR of mangiferin administered on the background of chronic antecedent iron chelation with CPX (see Sect. 50.2). The dampening effect on ventilation of mangiferin was preserved in pre-chelated rats all along the course of the ventilatory responses to the same two levels of hypoxia (Fig. 50.2a and b); the responses ran closely akin to those in the non-chelated animals above outlined. Again, there were significant mangiferin- and time-dependent effects for both 12% FiO_2 and 8% FiO_2 ($p<0.001$ all). In addition, there was a significant interaction between the two effects for the stronger hypoxia ($p<0.01$). Neither did mangiferin nor CPX chelation change the biphasic profile of the HVR.

Both frequency (*f*) and tidal (V_T) components of ventilation contributed to the hypoxic responses before and after mangiferin in both non-chelated and chelated rats. The proportion of the contribution was not appreciably affected by antecedent chelation, which may be best assessed when analyzing the

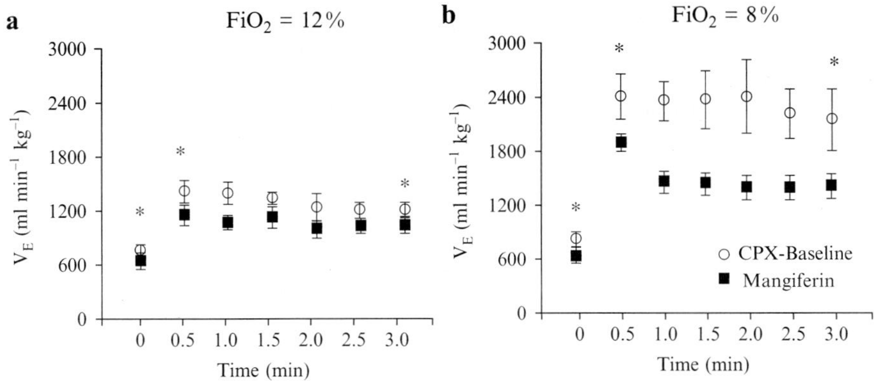

Fig. 50.2 Minute ventilation responses to successive levels of hypoxia before and 40 min after administration of mangiferin in conscious, CPX-chelator pretreated rats. *$p<0.05$ between the corresponding (*vertical*) time points of interest before *vs*. after mangiferin

stronger ventilatory effects of the stronger 8% FiO_2 stimulus. On average, in response to this level of hypoxia peak *f* increased by 80% and V_T by 86% before mangiferin in the non-chelated rats and by 91% and 102%, respectively, in the chronically chelated rats. However, the decline in peak hypoxic V_E after mangiferin tended to be due more to inhibition of frequency, with a smaller decline or no change in tidal volume. Here, in response to 8% FiO_2, peak *f* increase was stymied to an average of 59%, while the increase in V_T was sustained at 96% in the non-chelated rats, compared with by 67% and 82% increases in *f* and V_T, respectively, in the chelated rats. Interestingly, the depressant nadir of the hypoxic response was due uniformly to a decline in V_T, with no change in *f* in all conditions studied, which suggests different regulatory mechanisms of both ventilatory components.

We also evaluated the effect of mangiferin on the hypoxic ventilatory sensitivity, as the gain in peak hypoxic V_E achieved with increased stimulus strength. This evaluation is graphically presented in Fig. 50.3. Mangiferin administered in the non-chelated rats caused an appreciably stronger decrement in peak V_E at the stronger than milder stimulus. This loss in gain resulted in a significant downward shift of the slope of the response line between the two stimulus strengths, describing hypoxic sensitivity (Fig. 50.3a). In contrast, mangiferin decreased peak hypoxic V_E by about the same magnitude at both 12% and 8% FiO_2 on the background of prior iron chelation (Fig. 50.3b), whereby a parallel downward shift was noted in the response line, indicating no appreciable change in hypoxic sensitivity.

50.4 Discussion

The major finding of this study was that the antioxidant mangiferin decreased ventilation; the inhibitory effect was particularly expressed in relation to the stimulatory HVR. Mangiferin, apart from being an antioxidant, is an iron chelator. Therefore, to distinguish between the two properties, either of which might be at play in the mechanisms of the hypoxic chemoreflex, we reinvestigated the effects of mangiferin on the background of antecedent iron chelation. The inhibitory effect on the HVR of mangiferin was preserved, which implicates antioxidation rather than chelation carried by the compound as the underlying cause of chemoreflex alterations. Actually, a lesser availability of iron weakened the action of mangiferin on the hypoxic ventilation as the decrease in hypoxic sensitivity after mangiferin in the non-chelated rats was gone after prior iron chelation. A weaker ventilatory effect of mangiferin on the background of prior iron chelation is in accord with the studies showing potentiation of the antioxidant capacity of mangiferin in the presence of iron (Pardo-Andreu et al. 2006b, c).

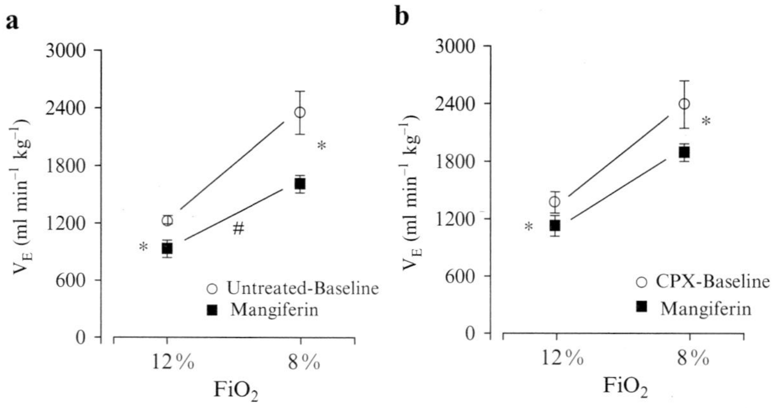

Fig. 50.3 Hypoxic ventilatory gain assessed from the increment in peak hypoxic V_E achieved between 12% and 8% hypoxia before and after mangiferin in the two experimental conditions: non-chelated (**a**) and CPX-chelated rats (**b**). * Significant decreases in peak hypoxic V_E after mangiferin (p-values as detailed in Sect. 50.3); # denotes a greater decrease in peak V_E at the stronger 8% FiO_2 than that at 12% FiO_2 after mangiferin in non-chelated rats ($p<0.02$). Thus, mangiferin decreased not only peak hypoxic ventilation, but also hypoxic ventilatory gain

Mangiferin does not cross the blood–brain barrier after systemic administration (Pokorski et al. 2010). Ventilatory changes due to mangiferin are presumably exerted at the carotid body level; an organ that generates the HVR in mammals (Lahiri et al. 2007). The influence of mangiferin on the carotid body mechanisms is further supported by its inhibiting the frequency rather than tidal component of ventilation; which is fairly characteristic for the carotid body-mediated effects (Izumizaki et al. 2004). The basis for inducing the HVR is stabilization of HIF-1α by hypoxia in carotid chemoreceptor cells; the master protein having control over the transcriptional expression of genes involved in the adaptive responses to insufficient arterial oxygenation (Fandrey and Gassmann 2009). Ferrous iron is a cofactor in HIF-1α degradation in normoxia. ROS could play an enhancing role in this process as they are liable to facilitate the reductive, easily engaging in reactions, and thus more bioactive, state of iron. A decline in ferrous iron due to either the CPX chelator used in the study or a shift in the Fenton reaction, in which ferrous iron is the substrate, to the right due to mass action law, when the hydroxyl radical is scavenged by mangiferin, ought to have a stabilizing effect on HIF-1α, emulating that of hypoxia. Such an effect on the innate carotid body mechanisms of iron chelation is well established (Daudu et al. 2002; Roy et al. 2004). At the functional level, the effects on the HVR of iron chelation are less clear. In a previous study we have found that CPX, an intracellular ferrous iron chelator, decreases the HVR in awake rats (Pokorski et al. 2009). Others, however, have reported that desferrioxamine, an extracellular Fe^{3+} iron chelator, increases the HVR in humans (Nguyen et al. 2007). Although the intracellular bioactive ferrous iron (Fe^{2+}) seems a better candidate for being operational in shaping the HVR by carotid chemoreceptor cells, which also is supported by the experimental evidence (Roy et al. 2004), the exact determinants of the role of either pool of iron remain obscure. At any rate, the lack of appreciable effects on the HVR of chelating properties of mangiferin in the present study is explicable by its being mostly the extracellular Fe^{3+} iron chelator (Andreu et al. 2005; Pardo-Andreu et al. 2006c).

Functional effects on the HVR of antioxidation are a contentious issue. The present study demonstrates that mangiferin causes a downturn in the HVR, which, we surmise, would stem from its facilitating the HIF-1α stabilization, resulting in a sort of 'pseudo-hypoxic environment' already in normoxia. On top of that, true hypoxia would have less of an effect, which finds support in other experimental studies (Daudu et al. 2002; Roy et al. 2004). Inhibition of the HVR by mangiferin stands, however, in contrast to previous findings which demonstrate an enhancement of the HVR by vitamin C in the aged humans (Pokorski and Marczak 2003). Apart from the difference in physico-chemical properties of both antioxidants, it appears that antioxidation may have opposing ventilatory effects depending on the basal oxidative status. An antioxidant stimulates the HVR when the antioxidative capacity is deficient in the body but hardly otherwise; this notion is supported by the literature reports (Teppema and Dahan 2010). In vitamin C study above outlined, the old subjects were liable to have the vitamin C deficiency. In contrast, in the present study, mangiferin was supplemented to the animals on top of their being in the normal oxidative state, which resulted in HVR inhibition.

A major limitation of functional studies, particularly like the present one designed to be minimally invasive, is the incapability to get insight into exact cellular mechanisms of an observation. Nevertheless, we believe we have shown that antioxidation interacts with peripheral chemosensory pathways which are central to the processing of hypoxic hyperventilation. The results of this study also show the hitherto unrecognized influence on lung ventilation of the nutrient antioxidant mangiferin, which may be of therapeutic relevance in respiratory pathologies and thus warrants further exploration to this end.

Acknowledgements: This study was born out of stimulating conceptual collaboration between Polish (MP) and Cuban (RD) neuroscientists. JA was supported by grant NN 401153938 from the Polish Ministry of Science and Higher Education.

Conflicts of Interest: The authors declare no conflicts of interest in relation to this article.

References

Andreu GP, Delgado R, Velho JA, Curti C, Vercesi AE (2005) Iron complexing activity of mangiferin, a naturally occurring glucosylxanthone, inhibits mitochondrial lipid peroxidation induced by Fe^{2+}-citrate. Eur J Pharmacol 513:47–55

Beltrán AE, Alvarez Y, Xavier FE, Hernanz R, Rodriguez J, Núñez AJ, Alonso MJ, Salaices M (2004) Vascular effects of the Mangifera indica L. extract (Vimang). Eur J Pharmacol 499:297–305

Daudu PA, Roy A, Rozanov C, Mokashi A, Lahiri S (2002) Extra- and intracellular free iron and the carotid body responses. Respir Physiol Neurobiol 130:21–31

Fandrey J, Gassmann M (2009) Oxygen sensing and the activation of the hypoxia inducible factor 1 (HIF-1). Adv Exp Med Biol 648:197–206

Izumizaki M, Pokorski M, Homma I (2004) Role of the carotid bodies in chemosensory ventilatory responses in the anesthetized mouse. J Appl Physiol 97:1401–1407

Lahiri S, Antosiewicz J, Pokorski M (2007) A common oxygen sensor regulates the sensory discharge and glomus cell HIF-1α in the rat carotid body. J Physio Pharmacol 58(Suppl 5):327–333

Muruganandan S, Srinivasan K, Gupta S, Gupta PK, Lal J (2005) Effect of mangiferin on hyperglycemia and atherogenicity in streptozotocin diabetic rats. J Ethnopharmacol 97:497–501

Nguyen MVC, Pouvreau S, El Hajjaji FZ, Denavit-Saubie M, Pequignot JM (2007) Desferrioxamine enhances hypoxic ventilatory response and induces tyrosine hydroxylase gene expression in the rat brainstem *in vivo*. J Neurosci Res 85:1119–1125

Pardo-Andreu GL, Philip SJ, Riaño A, Sánchez C, Viada C, Núñez-Sellés AJ, Delgado R (2006a) *Mangifera indica* L. (Vimang) protection against serum oxidative stress in elderly humans. Arch Med Res 37:158–164

Pardo-Andreu GL, Sánchez-Baldoquín C, Avila-González R, Delgado R, Naal Z, Curti C (2006b) Fe(III) improves antioxidant and cytoprotecting activities of mangiferin. Eur J Pharmacol 547:31–36

Pardo-Andreu GL, Sánchez-Baldoquín C, Avila-González R, Yamamoto ET, Revilla A, Uyemura SA, Naal Z, Delgado R, Curti C (2006c) Interaction of Vimang (Mangifera indica L. extract) with Fe(III) improves its antioxidant and cytoprotecting activity. Pharmacol Res 54:389–395

Pokorski M, Marczak M (2003) Ascorbic acid enhances hypoxic ventilatory reactivity in elderly subjects. J Int Med Res 31:448–457

Pokorski M, Antosiewicz J, Di Giulio C, Lahiri S (2009) Iron chelation and the ventilatory response to hypoxia. Adv Exp Med Biol 648:215–221

Pokorski M, Delgado Hernandez R, Rekawek A (2010) Mangiferin: a neuroactive polyphenol 4th International Congress of Pharmacology and Therapeutics; Havana, Cuba. *Vacci Monitor* 19(Suppl 2):161

Roy A, Li J, Baby SM, Mokashi A, Burek DG, Lahiri S (2004) Effects of iron-chelators on ion-channels and HIF-1α in the carotid body. Respir Physiol Neurobiol 141:115–123

Taylor CT (2008) Mitochondria and cellular oxygen sensing in the HIF pathway. Biochem J 409:19–26

Teppema LJ, Dahan A (2010) The ventilatory response to hypoxia in mammals: mechanisms, measurement, and analysis. Physiol Rev 90:675–754

Yoshimi N, Matsunaga K, Katayama M, Yamada Y, Kuno T, Qiao Z, Hara A, Yamahara J, Mori H (2001) The inhibitory effects of mangiferin, a naturally occurring glucosylxanthone, in bowel carcinogenesis of male F344 rats. Cancer Lett 163:163–170

Chapter 51
Differential Regulation of Tyrosine Hydroxylase by Continuous and Intermittent Hypoxia

Gayatri Raghuraman, Nanduri R. Prabhakar, and Ganesh K. Kumar

Abstract Although continuous hypoxia (CH) and intermittent hypoxia (IH) cause reduction in oxygen availability, organisms adapt to the effects of chronic CH whereas IH adversely impacts autonomic functions. Catecholamines are expressed both in the central and peripheral nervous systems and they play important roles in the regulation of cardio-respiratory functions during hypoxia. Tyrosine hydroxylase (TH) is the rate-limiting enzyme for catecholamine synthesis. Several studies have examined the effects of hypoxia on catecholamines by focusing on the regulation of TH. In this article, we present a brief overview of the impact of chronic CH and IH on TH expression, activity and the associated cellular mechanism(s).

Keywords Chronic intermittent hypoxia • Catecholamines • Tyrosine hydroxylase • Protein phosphorylation • Protein kinases • Protein phosphatases • Reactive oxygen species

51.1 Introduction

Humans encounter two forms of chronic hypoxia; continuous hypoxia (CH) which is experienced during high altitude sojourns and intermittent hypoxia (IH) that is encountered under a variety of conditions including sleep disordered breathing manifested as recurrent apneas. Although both CH and IH result in reduced oxygen supply, humans and experimental animals adapt to the effects of chronic CH by partially compensating for the lack of oxygen. On the other hand, chronic IH associated with recurrent apneas leads to autonomic abnormalities including cardio-respiratory morbidities. Catecholamines are expressed in many regions of the brain, adrenal medulla, as well as the carotid body, a peripheral neuronal sensory organ, specialized for detecting changes in arterial blood O_2. In addition to their roles in development and energy metabolism, catecholamines play important roles in regulation of cardio-respiratory functions during hypoxia (Soulier et al. 1997; Díaz-Cabiale et al. 2007). Since tyrosine hydroxylase (TH) is the rate-limiting enzyme in catecholamine synthesis,

G. Raghuraman • N.R. Prabhakar • G.K. Kumar (✉)
Department of Medicine, Institute for Integrative Physiology, University of Chicago, 5841 S Maryland Ave, Chicago, IL 60637, USA
e-mail: nanduri@uchicago.edu; gkumar@medicine.bsd.uchicago.edu

C.A. Nurse et al. (eds.), *Arterial Chemoreception: From Molecules to Systems*, Advances in Experimental Medicine and Biology 758, DOI 10.1007/978-94-007-4584-1_51,

several studies have examined the effects of hypoxia on catecholamines by focusing on the regulation of TH. In this article, we will briefly review the impact of chronic CH and IH on TH and the cellular mechanism(s) that underlie the responses of TH to these two forms of hypoxia.

51.2 Features of TH Regulation

TH, a non-heme iron protein, catalyzes the hydroxylation of L-tyrosine to form L-dihydroxy-phenylalanine. Several mechanisms contribute to the short- and long-term regulation of TH. Changes in the redox state of the iron and feed-back product inhibition contribute to the short-term regulation, whereas alterations in either transcriptional activation of *TH* resulting in *de novo* protein synthesis or cellular compartmentalization or cofactor synthesis mediate long-term regulation of TH. Post-translational modification of TH involving phosphorylation of serine residues at the N-terminal regulatory domain seems to involve both short- and long-term regulation of TH (Fitzpatrick 1999). Phosphorylation alters the K_m and V_{max} leading to a more active form of TH with increased activity (Dunkley et al. 2004). Available evidence suggests that cyclic AMP-dependent phosphorylation of serine-40 is critical for the activation of TH (Bobrovskaya et al. 2007). Phosphorylation of serine-19 is dependent on intracellular Ca^{2+} levels whereas serine-31 phosphorylation requires activation of phosphoinositol/protein kinase C pathway (Dunkley et al. 2004). Studies in cell cultures and with purified enzyme showed that site-specific TH phosphorylation is physiologically regulated. Since molecular oxygen is required for eliciting the catalytic activity of TH, several studies have examined the impact of hypoxia on TH activity and determined the underlying mechanisms.

51.3 CH and Regulation of TH

CH (lasting several days) enhanced TH activity in the rat carotid body (Hanbauer 1977) and brain cortex (Gozal et al. 2005). The CH-induced increase in TH activity in the carotid body was associated with an up-regulation of TH mRNA (Paulding et al. 2002) and protein expression (Hui et al. 2003). In pheochromocytoma (PC)-12 cell cultures, CH not only activated hypoxia-inducible transcription factors (HIFs) that interact with a specific hypoxia-responsive element in the TH promoter region (Schnell et al. 2003) contributing to increased TH transcription but also increased TH mRNA stability thereby facilitating TH protein expression. In addition, CH increased TH phosphorylation at serine-19, serine-31 and serine-40 residues in the rat carotid bodies (Hui et al. 2003). A similar increase in TH phosphorylation was also reported in the cortex and brainstem of rats exposed to CH (Gozal et al. 2005). Collectively, these studies suggest that the increase in TH activity by CH is due to coordinated increases in transcription, protein expression and serine phosphorylation of TH. The mechanisms by which CH alters TH phosphorylation, however, remain to be elucidated.

51.4 IH and Regulation of TH

The effects of IH on TH activity were determined in cell cultures and rodents exposed to different patterns of IH. Initial studies conducted in PC12 cell cultures have showed that IH (15 s of 1% O_2 and 3 min of 21% O_2 per cycle for 60 cycles) augmented TH activity in a cycle-dependent manner with a concomitant increase in dopamine levels. Although IH augmented HIF-1-dependent TH mRNA expression (Yuan et al. 2005), TH protein expression per se was not altered (Kumar et al. 2003; Yuan et al. 2005) suggesting a possible adverse effect of IH on protein stability. The IH-evoked increase in TH activity seems to be due in part to enzyme activation via increased phosphorylation of TH at serine-40 mediated by calcium/calmodulin-dependent protein kinase (CaMK) and cAMP-dependent protein kinase A (PKA) (Kumar et al. 2003). The above results from cell culture studies suggest that IH-induced increase in TH activity involves enzyme activation. Later studies in brainstem medullary regions of rats treated with IH (15 s 5% O_2 and 4 min 21% O_2; 8 h/day; 10 days) further confirmed IH-induced TH activation which occurs via increased serine-31 and serine-40 phosphorylation without altering TH protein expression (Raghuraman et al. 2009).

Exposure of rats to an IH paradigm (90 s 10% O_2 and 90 s 21% O_2) that differs in the severity of hypoxia and duration of hypoxia-reoxygenation from the IH pattern used in the study of Raghuraman et al. (2009) differentially affected serine-40 phosphorylation of TH in various brain regions. Thus, IH decreased TH phosphorylation in the brainstem whereas it induced a modest increase in the cerebral cortex with a concomitant rise in TH activity (Gozal et al. 2005). However, this "90 s IH" paradigm was shown to increase serine-19, serine-31 and serine-40 phosphorylation of TH selectively in the rat carotid body, without altering TH phosphorylation in other catecholamine expressing tissues such as superior cervical ganglion and adrenal medulla (Hui et al. 2003). The stimulatory effect of "90 s IH" on TH activity in the carotid body can be attributed to a greater hypoxic sensitivity of the carotid body relative to other tissues. Taken together, the above results suggest that IH, depending on the type of tissue, duration and severity of hypoxia and period of re-oxygenation, activates TH via increased serine phosphorylation. Thus, the effects of IH on TH activity are primarily mediated via activation of post-translational mechanisms and are not associated with changes in protein expression.

51.5 Cellular Mechanisms of TH Regulation

51.5.1 Involvement of Protein Kinases and Protein Phosphatases

A balance between phosphorylation (via the activities of protein kinases) and dephosphorylation (via protein phosphatases) reactions defines the steady state level of serine phosphorylation of TH in vivo (Dunkley et al. 2004). Studies in brainstem showed that IH not only increases the levels of active forms of multiple protein kinases including PKA, CaMKII, and ERK1/2 but also decreases the activity and protein expression of protein phosphatase 2A (Raghuraman et al. 2009). However, the activities of other protein phosphatases such as PP2C, PP4, PP5 and PP6 were not altered by IH. Since down regulation of PP2A has been shown to activate protein kinases including CaMK IV (Truttmann et al. 2004) and ERK (Nyunoya et al. 2005), it is likely that PP2A down regulation by IH is functionally coupled to activation of protein kinases that are implicated in TH phosphorylation. This may provide a positive feed-forward mechanism for sustained TH phosphorylation, a possibility which needs to be further investigated. Taken together, the above results suggest that IH via tilting the balance between protein kinases and protein phosphatases facilitates robust TH phosphorylation leading to TH activation.

51.5.2 Evidence for ROS Signaling

IH differs from CH in that periods of hypoxia is interspersed with re-oxygenation phase which is similar to that seen during ischemia-reperfusion. It is therefore, likely that reactive oxygen species (ROS) are generated during the re-oxygenation phase of IH triggering IH-induced cellular and systemic effects.

The activity of aconitase, an enzyme of the citric acid cycle is sensitive to intracellular levels of ROS especially the superoxide anion ($O_2^{\cdot-}$) and the activity is inversely related to ROS levels (Gardner 2002). Using aconitase activity and thio-barbituric acid reactive substance (T-BARS) levels as indices of ROS, IH has been shown to elevate ROS levels in PC12 cell cultures (Yuan et al. 2004; Khan et al. 2011), rat carotid body (Peng et al. 2003), adrenal medulla (Kumar et al. 2006; Souvannakitti et al. 2010) and brainstem (Raghuraman et al. 2009). The possible role of ROS in mediating IH-induced TH activation was investigated using antioxidants. Systemic administration of either MnTMPyP, a superoxide dismutase (SOD) mimetic which scavenges superoxide anion or N-acetylcysteine, an antioxidant that scavenges ROS prevented IH-induced increase in TH activity and the tilting of balance between protein kinases and phosphatases leading to serine phosphorylation of TH (Raghuraman et al. 2009). The above observations suggest that ROS-mediated signaling contributes to IH-induced serine phosphorylation and the ensuing activation of TH. Further studies are, however, needed to identify the nature ($O_2^{\cdot-}$or H_2O_2 or OH^-) and source of ROS generation (cytosolic oxidases or mitochondrial complexes of the electron transport chain) that mediate IH-induced alterations in TH activity.

51.6 Functional Implications

Exposure to CH elevated dopamine (DA) and norepinephrine (NE) levels in the rat carotid body (Hanbauer 1977; Pequignot et al. 1987). Likewise, IH treatment also increased DA levels in the rat brainstem (Raghuraman et al. 2009) and in PC12 cell cultures (Kumar et al. 2003) as well as NE levels in the adrenal medulla (Kumar et al. 2006; Souvannakitti et al. 2009). Since IH-mediated activation of TH is coupled to elevation of catecholamines in the brainstem (Raghuraman et al. 2009) and adrenal medulla (Kumar et al. 2006; Souvannakitti et al. 2009) it is conceivable that IH-evoked changes in catecholamines may, in part, contribute to cardio-vascular abnormalities associated with recurrent apneas.

Acknowledgements This work was supported by grants from the National Heart, Lung, and Blood Institute (PO1HL-90554 to NRP and RO1HL-89616 to GKK).

References

Bobrovskaya L, Gilligan C, Bolster EK, Flaherty JJ, Dickson PW, Dunkley PR (2007) Sustained phosphorylation of tyrosine hydroxylase at serine 40: a novel mechanism for maintenance of catecholamine synthesis. J Neurochem 100:479–489

Díaz-Cabiale Z, Parrado C, Fuxe K, Agnati L, Narváez JA (2007) Receptor-receptor interactions in central cardiovascular regulation. Focus on neuropeptide/alpha(2)-adrenoreceptor interactions in the nucleus tractus solitarius. J Neural Transm 114:115–125

Dunkley PR, Bobrovskaya L, Graham ME, von Nagy-Felsobuki EI, Dickson PW (2004) Tyrosine hydroxylase phosphorylation: regulation and consequences. J Neurochem 91:1025–1043

Fitzpatrick PF (1999) Tetrahydropterin-dependent amino acid hydroxylase. Ann Rev Biochem 68:355–381

Gardner PR (2002) Aconitase: sensitive target and measure of superoxide. Methods Enzymol 349:9–23

Gozal E, Shah ZA, Pequignot J-M, Pequignot J, Sachleben LR, Czyzyk-Krzeska MF, Li RC, Guo S-Z, Gozal D (2005) Tyrosine hydroxylase expression and activity in the rat brain: differential regulation after long-term intermittent or sustained hypoxia. J Appl Physiol 99:642–649

Hanbauer I (1977) Regulation of tyrosine hydroxylase in carotid body. Adv Biochem Psychopharmacol 16:275–280

Hui AS, Striet JB, Gudelsky G, Soukhova GK, Gozal E, Beitner-Johnson D, Guo SZ, Sachleben LR Jr, Haycock JW, Gozal D, Czyzyk-Krzeska MF (2003) Regulation of catecholamines by sustained and intermittent hypoxia in neuroendocrine cells and sympathetic neurons. Hypertension 42:1130–1136

Khan SA, Nanduri J, Yuan G, Kinsman B, Kumar GK, Joseph J, Kalyanaraman B, Prabhakar NR (2011) NADPH oxidase 2 mediates intermittent hypoxia-induced mitochondrial complex I inhibition: relevance to blood pressure changes in rats. Antioxid Redox Signal 14:533–542

Kumar GK, Kim DK, Lee MS, Ramachandran R, Prabhakar NR (2003) Activation of tyrosine hydroxylase by intermittent hypoxia: involvement of serine phosphorylation. J Appl Physiol 95:536–544

Kumar GK, Rai V, Sharma SD, Ramakrishnan DP, Peng YJ, Souvannakitti D, Prabhakar NR (2006) Chronic intermittent hypoxia induces hypoxia-evoked catecholamine efflux in adult rat adrenal medulla via oxidative stress. J Physiol 575:229–239

Nyunoya T, Monick MM, Powers LS, Yarovinsky TO, Hunninghake GW (2005) Macrophages survive hyperoxia via prolonged ERK activation due to phosphatase down-regulation. J Biol Chem 280:26295–26302

Paulding WR, Schnell PO, Bauer AL, Striet JB, Nash JA, Kuznetsova AV, Czyzyk-Krzeska MF (2002) Regulation of gene expression for neurotransmitters during adaptation to hypoxia in oxygen-sensitive neuroendocrine cells. Microsc Res Tech 59:178–187

Peng YJ, Overholt JL, Kline D, Kumar GK, Prabhakar NR (2003) Induction of sensory long-term facilitation in the carotid body by intermittent hypoxia: implications for recurrent apneas. Proc Natl Acad Sci USA 100:10073–10078

Pequignot JM, Cottet-Emard JM, Dalmaz Y, Peyrin L (1987) Dopamine and norepinephrine dynamics in rat carotid body during long-term hypoxia. J Auton Nerv Syst 21:9–14

Raghuraman G, Rai V, Peng YJ, Prabhakar NR, Kumar GK (2009) Pattern-specific sustained activation of tyrosine hydroxylase by intermittent hypoxia: role of reactive oxygen species-dependent downregulation of protein phosphatase 2A and upregulation of protein kinases. Antioxid Redox Signal 11:1777–1789

Schnell PO, Ignacak ML, Bauer AL, Striet JB, Paulding WR, Czyzyk-Krzeska MF (2003) Regulation of tyrosine hydroxylase promoter activity by the von Hippel-Hindau tumor suppressor protein and hypoxia-inducible transcription factors. J Neurochem 85:483–491

Soulier V, Gestreau C, Borghini N, Dalmaz Y, Cottet-Emard JM, Pequignot JM (1997) Peripheral chemosensitivity and central integration: neuroplasticity of catecholaminergic cells under hypoxia. Comp Biochem Physiol A Physiol 118:1–7

Souvannakitti D, Kumar GK, Fox A, Prabhakar NR (2009) Neonatal intermittent hypoxia leads to long-lasting facilitation of acute hypoxia-evoked catecholamine secretion from rat chromaffin cells. J Neurophysiol 101:2837–2846

Souvannakitti D, Nanduri J, Yuan G, Kumar GK, Fox AP, Prabhakar NR (2010) NADPH oxidase-dependent regulation of T-type Ca2+ channels and ryanodine receptors mediate the augmented exocytosis of catecholamines from intermittent hypoxia-treated neonatal rat chromaffin cells. J Neurosci 30:10763–10772

Truttmann AC, Ashraf Q, Mishra OP, Delivoria-Papadopoulos M (2004) Effect of hypoxia on protein phosphatase 2A activity, subcellular distribution and expression in cerebral cortex of newborn piglets. Neuroscience 127:355–363

Yuan G, Adhikary G, McCormick AA, Holcroft JJ, Kumar GK, Prabhakar NR (2004) Role of oxidative stress in intermittent hypoxia-induced immediate early gene activation in rat PC12 cells. J Physiol 557:773–783

Yuan G, Nanduri J, Bhasker CR, Semenza GL, Prabhakar NR (2005) Ca2+/Calmodulin kinase-dependent activation of hypoxia inducible factor 1 transcriptional activity in cells subjected to intermittent hypoxia. J Biol Chem 280:4321–4328

Chapter 52
Heart Failure and Carotid Body Chemoreception

Harold D. Schultz and Noah J. Marcus

Abstract There is substantial evidence to implicate a role of the carotid body (CB) chemoreflex in sympathetic and breathing dysregulation in several cardio-respiratory diseases, drawing renewed interest in its potential implications for clinical treatment and management. Evidence from both chronic heat failure (CHF) patients and animal models indicates that the CB chemoreflex is enhanced in CHF and contributes to the tonic elevation in sympathetic nerve activity (SNA) and periodic breathing associated with the disease. Although this maladaptive change likely derives from altered function at all levels of the reflex arc, a change in afferent function of the CB is likely to be a main driving force. This review will focus on recent advances in our understanding of the physiological mechanisms that alter CB function in CHF and their potential translational impact on treatment of CHF.

Keywords Chemoreflex • Sympathetic nervous system • Sleep apnea • Cardiovascular disease • Blood flow • Exercise • Kruppel like factor

52.1 Introduction

The role of the carotid body (CB) in cardio-respiratory homeostasis under normal healthy conditions is clear. The CB chemoreflex controls alveolar ventilation to ensure that gas exchange rate in the lungs continually matches metabolic demand for O_2 uptake. Another important component to CB chemoreflex activation is an increase in sympathetic outflow to the vascular beds. This response aids in the maintenance of arterial pressure in the face of the direct vasodilatory effects of hypoxemia, and thus assists in the maintenance of arterial pressure for adequate blood flow and gas exchange in tissues.

H.D. Schultz (✉)
Department of Cellular and Integrative Physiology, University of Nebraska Medical Center, 985850 Nebraska Medical Center, Omaha, NE 68198-5850, USA

College of Medicine, University of Nebraska, 985850 NE Med Center, Omaha, NE 68198-5850, USA
e-mail: hschultz@unmc.edu

N.J. Marcus
Department of Cellular and Integrative Physiology, University of Nebraska Medical Center, 985850 Nebraska Medical Center, Omaha, NE 68198-5850, USA
e-mail: nmarcus@unmc.edu

C.A. Nurse et al. (eds.), *Arterial Chemoreception: From Molecules to Systems*, Advances in Experimental Medicine and Biology 758, DOI 10.1007/978-94-007-4584-1_52,

A maladaptive or pathophysiological role for the CB in certain cardiovascular and respiratory diseases is supported by many studies, but it is less well understood. The CB chemoreflex is known to be exaggerated in sleep apnea, hypertension, and chronic heart failure (CHF) (Schultz et al. 2007) as well as other disease states (Schmidt et al. 2005), and has been implicated as a contributing factor to the autonomic dysregulation associated with these disorders.

In particular, CHF is characterized by tonic over-activation of sympathetic neural activity (SNA) that exacerbates the progression of the cardiac failure (Esler 2010). CHF is also characterized the development of breathing instability with Cheyne-Stokes breathing and central apneas that further negatively impact autonomic and metabolic homeostasis. A potential role of the CB in these pathophysiological changes is self-evident. Animal models (Schultz and Li 2007) and patients (Ponikowski and Banasiak 2001) with CHF exhibit increased CB chemoreflex drive to increase sympathetic outflow and ventilation under both normoxic and hypoxic conditions. Moreover, the exaggerated CB chemoreflex is correlated with poor prognosis as well as the pathophysiology of CHF in patients (Ponikowski et al. 2001; Schmidt et al. 2005).

The aim of this review is to summarize present knowledge of the causes and consequences of altered CB function in CHF and its potential impact on the treatment and management of CHF patients.

52.2 What Drives the Enhanced CB Function in CHF?

There is an enhanced afferent input from CB chemoreceptors in the CHF rabbits (Sun et al. 1999), which provides a primary contribution to the augmentation of reflex function. The baseline discharge in the normoxic state and the magnitude of the afferent response to corresponding levels of isocapnic hypoxia are greater in CHF rabbits than in sham rabbits. These alterations are observed both in the intact (blood perfused) and isolated CB preparations (Sun et al. 1999). This observation suggests that an intrinsic alteration within the CB, rather than a circulating factor, drives the altered afferent sensitivity in the CHF state.

A striking feature of this tonic enhanced afferent activity arising from the CB in CHF animals is its similarity to that which occurs after exposure of animals to chronic intermittent hypoxia (Peng et al. 2011), termed CB sensory long term facilitation (LTF). Recent studies in rats and mice have revealed that intermittent hypoxia-induced LTF in the CB is mediated, at least in part, by serotonin and angiotensin II (Ang II) that activate of NADPH oxidase (NOX), which in turn enhances the excitability of the CB afferents via O_2^- mediated effects (Peng et al. 2011, 2006).

It is not known whether serotonin plays a role in facilitation of CB activity in CHF, but Ang II is involved (Schultz 2011). A locally generated Ang II system exists in the CB, and systemic and tissue Ang II levels are increased in CHF patients and animal models.

Afferent recordings from the isolated CB confirm that a functional AT_1R in the CB contributes to the elevated chemoreceptor afferent activity in CHF rabbits (Li et al. 2006). Pharmacological blockade of AT_1R decreases CB chemoreceptor responses to hypoxia in the isolated CB in CHF rabbits but not in sham rabbits. In addition, Ang II concentration and mRNA expression and protein levels of AT_1R in the CB are increased in CHF animals. These results suggest that locally produced Ang II in the CB is capable of altering CB chemoreceptor function and that elevation of tissue Ang II with upregulation of AT_1R in the CB plays a major role in enhancing CB chemoreceptor sensitivity to hypoxia.

The mechanism by which Ang II enhances the hypoxic sensitivity of the CB chemoreceptors involves, at least in part, an interaction with oxygen sensitive potassium channels to suppress their voltage-gated currents (I_{Kv}) in CB glomus cell (Li and Schultz 2006). Hypoxia inhibits I_{Kv} and this effect is enhanced in isolated CB glomus cells from CHF rabbits (Li and Schultz 2006). Blockade of AT_1R alone is capable of reversing this enhanced hypoxic sensitivity of glomus cells I_{Kv}. In addition, exposing normal rabbit CB glomus cells to Ang II mimics this effect of CHF on I_{Kv}. The specific types of K^+ channels involved in glomus cell function varies by species, but in the rabbit, the suppression of I_{Kv} observed in CHF appears to involve the Ca^{++}-sensitive K^+ channel (BK) (Li et al. 2004), Kv3.4 and Kv4.3 (Li and Schultz 2006; Schultz and Li 2007). Other channels also may play a role but have not been examined.

In the CB, expression of NOX2 subunits and $O_2^{\bullet-}$ production are enhanced in CHF rabbits (Li et al. 2007). An AT_1R antagonist, NOX inhibitor, and the $O_2^{\bullet-}$ scavenger tempol normalized the elevated $O_2^{\bullet-}$ (Li et al. 2007). Similarly in CHF rabbits, NOX inhibitors and tempol normalize the enhanced CB chemoreceptor discharge to hypoxia, and normalize the enhanced sensitivity of I_{Kv} to hypoxia in CB glomus cells (Li et al. 2007). These results confirm that the NOX-$O_2^{\bullet-}$ pathway mediates the effects of Ang II in the CB to enhance chemoreceptor sensitivity in the CHF state.

Other angiotensin metabolites also may play a role in modulating CB function in CHF. Ang II production is controlled by the rate-limiting angiotensin converting enzyme (ACE). Ang-(1–7) is produced from Ang II by angiotensin converting enzyme 2 (ACE2) or from Ang I by either neutral endopeptidases or propylendopeptidases. Ang-(1–7) is of particular interest because its cardiovascular and central neural actions counteract those of Ang II. The vasodilatory and sympatho-inhibitory effects of Ang-(1–7) are mediated by the Mas receptor (MasR). Recent studies have demonstrated that Ang-(1–7) activation of MasR increases nitric oxide (NO) via neural NOS (nNOS) in cultured catecholaminergic neuronal cells (CATH.a) (Yang et al. 2010) and within autonomic regions of the brain stem (Feng et al. 2010). Ang-(1–7) activation of NO leads to increased I_{Kv} in CATH.a neurons (Yang et al. 2010). This action may explain its sympatho-inhibitory effect.

Ang-(1–7) similarly enhances I_{Kv} in CB glomus cells via MasR activation of nNOS and NO production: an effect that opposes the inhibitory effect of Ang II-$O_2^{\bullet-}$ on these channels (Li and Schultz 2006). It is well known that NO plays an important role in the tonic control of CB chemosensitivity (Nurse 2010). Both nNOS and endothelial NOS (eNOS) are present in the CB, and NO is inhibitory to CB activity (Sun et al. 1999). Because O_2 is essential for biosynthesis of NO, normal basal production of NO acts as an amplifier of O_2 to suppress CB chemoreceptor discharge in normoxic conditions (Sun et al. 1999). However, basal NO production and nNOS/eNOS expression within the CB are depressed in CHF (Ding et al. 2008; Li et al. 2010; Schultz and Li 2007). Thus, the tonic inhibitory effect of NO on the activity of CB chemoreceptors and its enhancement of I_{Kv} in glomus cells, demonstrated in the CB of normal rabbits, is virtually absent in CHF rabbits (Li et al. 2004).

The opposing effects of Ang II and Ang-(1–7) on CB chemo-sensitivity may provide a mechanism by which CB function can be finely tuned by the balance of factors influencing Ang production and metabolism in the CB. Under normal conditions, Ang-(1–7) may act to maintain NOS expression and NO restraint of CB chemoreceptor activity. When CB Ang II level is elevated, such as in CHF, its excitatory influence on CB chemoreceptor activity overrides Ang-(1–7) and may even participate in the downregulation of MasR-NO effects seen in the CHF state. This balance is likely to be controlled by the relative expression of ACE and ACE2 activities in the CB. Consistent with this notion, ACE2 and MasR are suppressed, whereas ACE and AT_1R are upregulated in the CB in CHF rabbits (Schultz 2011). Ultimately these mechanisms control redox balance in the CB such that oxidative stress enhances CB sensitivity in CHF, whereas inhibitory nitrosative influences on CB sensitivity are downregulated.

52.3 What Are the Functional Implications of Altered Redox Balance in the CB in CHF?

Recent studies have demonstrated that both copper/zinc superoxide dismutase (CuZnSOD) and manganese SOD (MnSOD), endogenous scavengers of $O_2^{\bullet-}$, are suppressed in glomus cells of the CB in CHF (Ding et al. 2009, 2010b). These changes serve to enhance the Ang II-$O_2^{\bullet-}$ activation of CB chemoreceptor activity. Gene transfer to restore expression of either of the SOD isoforms in the CB reverses the enhanced CB chemoreceptor and chemoreflex activation, and suppression of glomus cell I_{Kv} observed in CHF (Ding et al. 2009, 2010b), implicating a dual role of both cytosolic and mitochondrial $O_2^{\bullet-}$ in the Ang II-NOX signaling cascade within CB glomus cells.

As discussed earlier, the marked down regulation of endogenous ACE2 and nNOS and NO in the CB also plays a major role in the enhanced CB chemoreceptor activity in CHF rabbits (Li et al. 2004, 2005, 2010). Adenoviral gene transfer of nNOS to the CB in CHF rabbits reverses the enhanced CB chemoreceptor activity seen in the CHF state (Li et al. 2005). Furthermore, gene transfer of nNOS to the CB in CHF rabbits reduces resting RSNA. Preliminary studies suggest that gene transfer of ACE2 to the CB also exerts similar beneficial effects.

These results provide important evidence of a measurable contribution of enhanced CB chemoreceptor input to elevated sympathetic outflow in CHF. Importantly, they also demonstrate that normalization of oxidative and nitrosative redox pathways in the CB can reverse its excitatory influence on SNA in CHF and likely reduce breathing instability as well.

52.4 Does Blood Flow Play a Role in CB Function in CHF?

The factors responsible for upregulation of local ACE and Ang II-AT_1R-$O_2^{\bullet-}$ signaling and downregulation of ACE2 and Ang-(1–7)-MasR-NO in CB glomus cells during CHF are not clear. Adaptations to hypoxia are not likely to play a major role. CB sensitivity progressively increases during the development of CHF while arterial PO_2 remains stable. Intermittent hypoxia can result from Cheyne-Stokes breathing and central apneas induced by CHF, but breathing instability in animal models of CHF appears only after alterations in CB function are evident (Marcus and Schultz 2011). This raises the question whether impaired blood flow secondary to cardiac failure may play a role. Although acute changes in systemic hemodynamics are known to have little effect on chemoreceptor activity, chronic changes in blood flow could impact the CB via alterations in endothelial function and signaling pathways.

Recent evidence indicates that a chronic reduction in CB blood flow mimics the changes in CB afferent function seen in CHF (Ding et al. 2010a). A sustained reduction in CB blood flow over 3 weeks, imposed with adjustable cuff occludes on the carotid arteries in rabbits, enhances CB chemoreceptor discharge and its reflex hypoxic ventilatory and sympathetic responses. The reduction in blood flow and functional CB effects are similar to changes observed in CHF rabbits over a similar time course. The chronic nature of the flow effect is evident by the fact that chemoreflex and afferent responses are unaffected by only one day of carotid occlusion. Immuohistochemical examination of the CB after chronic carotid occlusion revealed an increase in AT_1R and decrease in nNOS expression (Ding et al. 2010a).

These results support the concept that a chronic reduction in blood flow to the carotid body precipitates altered CB function in CHF. However, the signaling link between reduced CB flow and altered AT_1R and NOS function in the CB remains to be identified. The transcriptional regulation of ACE and eNOS expression (and possibly ACE2 expression) in vascular endothelial cells is regulated by shear stress (Dekker et al. 2005; Miyakawa et al. 2004). This association raises the question whether an endothelial response to changes in blood flow in the CB plays an important role in regulating Ang and NO effects on CB chemoreceptor function in CHF.

52.5 Is KLF2 a Transcriptional Mediator of Altered CB Function in CHF?

Kruppel-like factor 2 (KLF2) is a mechano-activated transcription factor that is important in the transduction of shear-stress effects on endothelial function (Dekker et al. 2005; Miyakawa et al. 2004). KLF2 represses transcription of angiotensin converting enzyme (ACE) and induces transcription of endothelial nitric oxide synthase (eNOS) in endothelial cells (Dekker et al. 2005) as well as activating

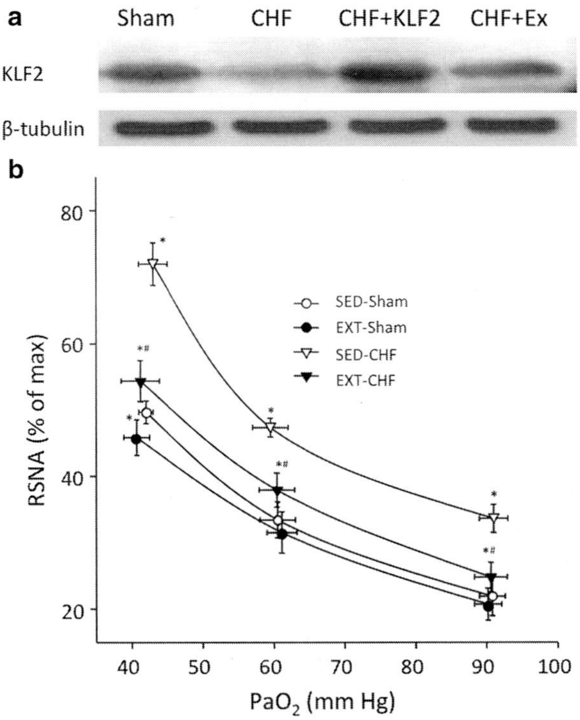

Fig. 52.1 **a** Representative immunoblots of KLF2 protein levels in the CBs from sham and CHF rabbits, CHF rabbit 4 days after adenoviral gene transfer of KLF2 to the CB (1 x 10^8 pfu), and CHF rabbit after 4 weeks of regular exercise (Ex: 15–18 m/min, 30 min/day, 6 days/week). **b** The CB RSNA-PaO2 relationship in SED-sham, EXT-sham, SED-CHF, and EXT-CHF rabbits. SED; sedentary; EXT: regular exercise as in A. Values are means ± SE; $n = 8$ in each group. $^*P < 0.05$ vs. SED-sham, $\#P < 0.05$ vs. SED-CHF (Adapted from Li et al. 2008)

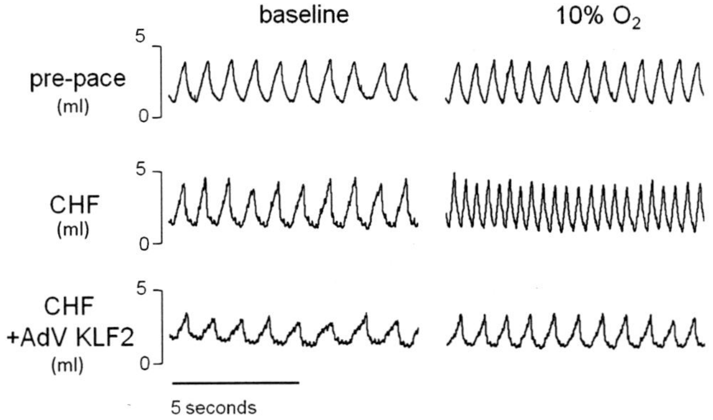

Fig. 52.2 KLF2 overexpression in the carotid bodies attenuates enhanced ventilatory chemoreflex sensitivity in CHF rabbits. Representative tracings of the ventilatory responses to isocapnic hypoxia (10% FIO_2) in the same rabbit prior to cardiac pacing, after cardiac pacing to induce CHF, and in CHF 4 days after adenoviral transfer of KLF2 (1 x 10^8 pfu) to the carotid bodies. KLF2 reversed the enhanced carotid body chemoreflex sensitivity induced by CHF

expression of Nrf2 and repressing NFkB that regulate expression of anti-oxidant/oxidant mediators (Boon and Horrevoets 2009). The impact of chronic reduced blood flow on CB function and its relation to altered Ang and NO metabolism in the CB suggest that KLF2 may be an important transcriptional mediator of changes that occur in the CB function in CHF.

Preliminary results indicate that KLF2 expression is decreased in the CB in CHF (Fig. 52.1a) (Marcus and Schultz 2011). Adenoviral gene transfer of KLF2 to the CB normalizes the hypoxic ventilatory response and decreases the incidence of periodic breathing in CHF (Fig. 52.2). Of note, the hypoxic ventilatory responses in CHF animals are complete normalized, even suppressed below normal levels, with CB KLF2 overexpression. This complete normalization of CB function in CHF is in contrast to that seen by overexpression of SOD or nNOS in the CB, which individually attenuated but did not completely normalize CB function in CHF animals. These findings suggest that KLF2 plays a fundamental role in the alteration in CB function that occurs in CHF. Further studies will be needed to confirm whether these functional effects of KLF2 are linked to normalization of oxidative and nitrosative redox pathways in the CB as discussed above.

52.6 What Is the Clinical Impact of Enhanced CB Function in CHF?

Can normalization of CB functional have a beneficial effect in CHF, and how might this be achieved? These are important translational questions that have yet to be adequately addressed. Undoubtedly, high CB chemosensitivity is an independent predictor of early death from CHF (Ponikowski et al. 2001; Giannoni et al. 2009) Enhanced CB chemosensitivity not only contributes to sympathetic activation and breathing instability, but it also contributes to exercise intolerance, impaired baroreflex function, and reduced heart rate variability: conditions that are maladaptive in the CHF state. Thus, it is within reason to suggest that normalizing CB function would be efficacious in CHF.

Transient hyperoxic inhibition of the CB reduces sympathetic activity and breathing instability and improves exercise tolerance in CHF patients, but has limited efficacy in an ambulatory setting. Pharmacological approaches such as opiates or dopamine can reduce CB chemoreflex drive in CHF patients but have other contraindicative effects (Van De Borne and Somers 1999). ACE inhibitors and AT1R blockers (ARBs) are routinely used to treat CHF with variable success (Morrissey et al. 2011), but there has been no effort to relate the clinical efficacy of these drugs to effects on chemoreflex function. Further, the potential adverse implications of CB inhibition in patients with marked hypoxemia or tenuous cardio-respiratory status must be realized.

Exercise conditioning has come to light as an effective therapeutic modality in CHF patients. Moderate regular exercise has significant beneficial effects in CHF (Downing and Balady 2011) that include enhancement of endothelial function and normalization of sympatho-vagal balance. The effects of regular exercise on CB chemoreflex function in CHF, therefore, are of obvious relevance.

Regular exercise is effective in normalizing CB chemoreflex function in CHF rabbits (Fig. 52.1b) via enhancing NO and reducing Ang II- $O_2^{\bullet-}$ redox influences on the CB (Li et al. 2008). These changes may be due to the effects of exercise on endothelial function in the CB. Preliminary evidence also suggests that exercise attenuates the suppression KLF2 expression the CB in CHF animals (Fig. 52.1a). Although these studies illustrate the beneficial effect of exercise on CB function in CHF, further studies are needed to assess its long-term impact on cardiac function and mortality.

52.7 Summary

The causes of enhanced CB function in CHF are multifactorial (Fig. 52.3). CB afferent activity is elevated under both normoxic and hypoxic conditions as a result of a shift in redox balance in the CB from inhibitory nitrosative pathways to excitatory oxidative pathways. Altered angiotensin metabolism and its signaling pathways in the CB play an important role in this shift. These alterations are mediated by upregulation of ACE, Ang II, AT1R and NOX2 and downregulation of ACE2, Ang 1–7, e–and nNOS, and NO in the CB. Human and animal studies indicate that the elevated CB activity in CHF contributes to sympathetic activation, breathing instability, exercise intolerance, impaired baroreflex function, and impaired cardiac sympatho-vagal balance. A precipitating factor involved in altering CB function in CHF may be related to reduced blood flow and its hindrance of normal

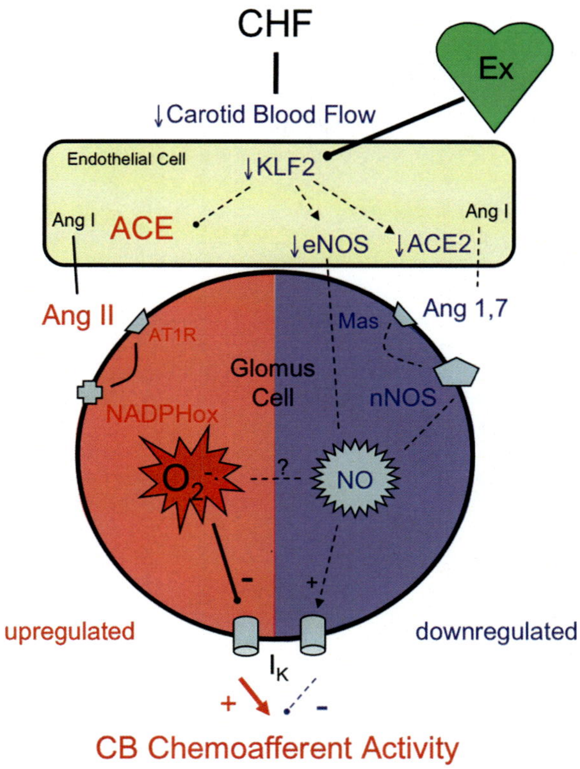

Fig. 52.3 Illustrative model of cellular pathways known to contribute to the enhanced CB chemoafferent sensitivity in CHF. Details and abbreviations as discussed in the text. Ex:regular exercise

endothelial function and signaling in the CB. Recent studies suggest that downregulation of the endothelial transcription factor KLF2 in response to a reduction in blood flow may mediate these changes in CB function. The beneficial effect of exercise to normalize CB function in CHF may be related to improving endothelial function and counteracting these phenomena in the CB. The translational impact of these effects deserves further study. Future inroads in assessing the therapeutic value of normalizing CB function in CHF patients will rely upon developing more refined and clinically safe techniques to selectively target the CB.

References

Boon RA, Horrevoets AJ (2009) Key transcriptional regulators of the vasoprotective effects of shear stress. Hamostaseologie 29(1):39–40, 41–33. doi:09010039 [pii]

Dekker RJ, van Thienen JV, Rohlena J, de Jager SC, Elderkamp YW, Seppen J, de Vries CJ, Biessen EA, van Berkel TJ, Pannekoek H, Horrevoets AJ (2005) Endothelial KLF2 links local arterial shear stress levels to the expression of vascular tone-regulating genes. Am J Pathol 167(2):609–618. doi:S0002-9440(10)63002-7 [pii] 10.1016/S0002-9440(10)63002-7

Ding Y, Li YL, Schultz HD (2008) Downregulation of carbon monoxide as well as nitric oxide contributes to peripheral chemoreflex hypersensitivity in heart failure rabbits. J Appl Physiol 105(1):14–23. doi:01345.2007 [pii] 10.1152/japplphysiol.01345.2007

Ding Y, Li YL, Zimmerman MC, Davisson RL, Schultz HD (2009) Role of CuZn superoxide dismutase on carotid body function in heart failure rabbits. Cardiovasc Res 81(4):678–685. doi:cvn350 [pii] 10.1093/cvr/cvn350

Ding Y, Li YL, Schultz HD (2010a) Role of blood flow in carotid body chemoreflex function in heart failure. J Physiol. doi:jphysiol.2010.200584 [pii] 10.1113/jphysiol.2010.200584

Ding Y, Li YL, Zimmerman MC, Schultz HD (2010b) Elevated mitochondrial superoxide contributes to enhanced chemoreflex in heart failure rabbits. Am J Physiol Regul Integr Comp Physiol 298(2):R303–R311. doi:00629.2009 [pii] 10.1152/ajpregu.00629.2009

Downing J, Balady GJ (2011) The role of exercise training in heart failure. J Am Coll Cardiol 58(6):561–569. doi:S0735-1097(11)01780-3 [pii] 10.1016/j.jacc.2011.04.020

Esler M (2010) The 2009 Carl Ludwig lecture: pathophysiology of the human sympathetic nervous system in cardiovascular diseases: the transition from mechanisms to medical management. J Appl Physiol 108(2):227–237. doi:00832.2009 [pii] 10.1152/japplphysiol.00832.2009

Feng Y, Xia H, Cai Y, Halabi CM, Becker LK, Santos RA, Speth RC, Sigmund CD, Lazartigues E (2010) Brain-selective overexpression of human angiotensin-converting enzyme type 2 attenuates neurogenic hypertension. Circ Res 106(2):373–382. doi:CIRCRESAHA.109.208645 [pii] 10.1161/CIRCRESAHA.109.208645

Giannoni A, Emdin M, Bramanti F, Iudice G, Francis DP, Barsotti A, Piepoli M, Passino C (2009) Combined increased chemosensitivity to hypoxia and hypercapnia as a prognosticator in heart failure. J Am Coll Cardiol 53(21):1975–1980. doi:S0735-1097(09)00745-1 [pii] 10.1016/j.jacc.2009.02.030

Li YL, Schultz HD (2006) Enhanced sensitivity of Kv channels to hypoxia in the rabbit carotid body in heart failure: role of angiotensin II. J Physiol 575(Pt 1):215–227. doi:jphysiol.2006.110700 [pii] 10.1113/jphysiol.2006.110700

Li YL, Sun SY, Overholt JL, Prabhakar NR, Rozanski GJ, Zucker IH, Schultz HD (2004) Attenuated outward potassium currents in carotid body glomus cells of heart failure rabbit: involvement of nitric oxide. J Physiol 555(Pt 1): 219–229. doi:10.1113/jphysiol.2003.057422 jphysiol.2003.057422 [pii]

Li YL, Li YF, Liu D, Cornish KG, Patel KP, Zucker IH, Channon KM, Schultz HD (2005) Gene transfer of neuronal nitric oxide synthase to carotid body reverses enhanced chemoreceptor function in heart failure rabbits. Circ Res 97(3):260–267. doi:01.RES.0000175722.21555.55 [pii] 10.1161/01.RES.0000175722.21555.55

Li YL, Xia XH, Zheng H, Gao L, Li YF, Liu D, Patel KP, Wang W, Schultz HD (2006) Angiotensin II enhances carotid body chemoreflex control of sympathetic outflow in chronic heart failure rabbits. Cardiovasc Res 71(1):129–138. doi:S0008-6363(06)00144-1 [pii] 10.1016/j.cardiores.2006.03.017

Li YL, Gao L, Zucker IH, Schultz HD (2007) NADPH oxidase-derived superoxide anion mediates angiotensin II-enhanced carotid body chemoreceptor sensitivity in heart failure rabbits. Cardiovasc Res 75(3):546–554. doi:S0008-6363(07)00161-7 [pii] 10.1016/j.cardiores.2007.04.006

Li YL, Ding Y, Agnew C, Schultz HD (2008) Exercise training improves peripheral chemoreflex function in heart failure rabbits. J Appl Physiol 105(3):782–790. doi:90533.2008 [pii] 10.1152/japplphysiol.90533.2008

Li YL, Zheng H, Ding Y, Schultz HD (2010) Expression of neuronal nitric oxide synthase in rabbit carotid body glomus cells regulates large-conductance Ca2+-activated potassium currents. J Neurophysiol 103(6):3027–3033. doi:jn.01138.2009 [pii] 10.1152/jn.01138.2009

Marcus NJ, Schultz HD (2011) Role of carotid body chemoreflex function in the development of cheyne-stokes respiration during progression of congestive heart failure. FASEB J 25:841.7

Miyakawa AA, de Lourdes JM, Krieger JE (2004) Identification of two novel shear stress responsive elements in rat angiotensin I converting enzyme promoter. Physiol Genomics 17(2):107–113. doi:10.1152/physiolgenomics.00169.200300169.2003 [pii]

Morrissey RP, Czer L, Shah PK (2011) Chronic heart failure: current evidence,challenges to therapy, and future directions. Am J Cardiovasc Drugs 11(3):153–71 doi:10.2165/11592090-000000000-00000 PMID: 21619379

Nurse CA (2010) Neurotransmitter and neuromodulatory mechanisms at peripheral arterial chemoreceptors. Exp Physiol 95(6):657–667. doi:expphysiol.2009.049312 [pii] 10.1113/expphysiol.2009.049312

Peng YJ, Yuan G, Jacono FJ, Kumar GK, Prabhakar NR (2006) 5-HT evokes sensory long-term facilitation of rodent carotid body via activation of NADPH oxidase. J Physiol 576(Pt 1):289–295. doi:jphysiol.2006.116020 [pii] 10.1113/jphysiol.2006.116020

Peng YJ, Raghuraman G, Khan SA, Kumar GK, Prabhakar NR (2011) Angiotensin-II evokes sensory long-term facilitation of the carotid body via NADPH oxidase. J Appl Physiol. doi:japplphysiol.00022.2011 [pii] 10.1152/japplphysiol.00022.2011

Ponikowski P, Banasiak W (2001) Chemosensitivity in chronic heart failure. Heart Fail Monit 1(4):126–131

Ponikowski P, Chua TP, Anker SD, Francis DP, Doehner W, Banasiak W, Poole-Wilson PA, Piepoli MF, Coats AJ (2001) Peripheral chemoreceptor hypersensitivity: an ominous sign in patients with chronic heart failure. Circulation 104(5):544–549

Schmidt H, Francis DP, Rauchhaus M, Werdan K, Piepoli MF (2005) Chemo- and ergoreflexes in health, disease and ageing. Int J Cardiol 98(3):369–378. doi:S0167527304000932 [pii] 10.1016/j.ijcard.2004.01.002

Schultz HD (2011) Angiotensin and carotid body chemoreception in heart failure. Curr Opin Pharmacol 11(2):144–149. doi:S1471-4892(10)00181-5 [pii] 10.1016/j.coph.2010.12.004

Schultz HD, Li YL (2007) Carotid body function in heart failure. Respir Physiol Neurobiol 157(1):171–185. doi:S1569-9048(07)00055-9 [pii] 10.1016/j.resp. 2007.02.011

Schultz HD, Li YL, Ding Y (2007) Arterial chemoreceptors and sympathetic nerve activity: implications for hypertension and heart failure. Hypertension 50(1):6–13. doi:HYPERTENSIONAHA.106.076083 [pii] 10.1161/HYPERTENSIONAHA.106.076083

Sun SY, Wang W, Zucker IH, Schultz HD (1999) Enhanced activity of carotid body chemoreceptors in rabbits with heart failure: role of nitric oxide. J Appl Physiol 86(4):1273–1282

Van De Borne P, Somers VK (1999) Dopamine and congestive heart failure: pharmacology, clinical use, and precautions. Congest Heart Fail 5(5):216–221

Yang RF, Yin JX, Li YL, Zimmerman MC, Schultz HD (2010) Angiotensin-(1–7) increases neuronal potassium current via a nitric oxide-dependent mechanism. Am J Physiol Cell Physiol. doi:ajpcell.00369.2010 [pii] 10.1152/ajpcell.00369.201

Concluding Remarks

Chris Peers

As is the tradition of the International Society for Arterial Chemoreception, it is my pleasure to begin these concluding remarks by thanking our host for the XVIIIth ISAC meeting, Professor Colin Nurse, for such an enjoyable and memorable meeting. In his role as President of the Society, Professor Nurse led a local organizing committee – Nikol Piskuric, Cathy Vollmer, Karen Haines, Simon Livermore, Shaima Salman, Min Zhang, Dianne Carment, Wendy Read and Paula Nurse – who worked tirelessly during the meeting, and for many months leading up to it, to ensure that all who attended had an enjoyable and engaging experience both scientifically and socially. The attendees were treated to an excellent programme of science, as well as a number of wonderful social events, all of which ensured the meeting will remain memorable for many years to come. The organizers can all rest assured that their efforts were completely successful and greatly appreciated by all.

This volume of Advances in Experimental Biology and Medicine contains 46 short articles together with eight more extensive review papers which collectively summarise the work presented at the Hamilton meeting and also the present state of understanding of the general field of oxygen sensing. In all, the meeting comprised six keynote / plenary lectures, over 40 shorter oral presentations and a similar number of posters. The posters were available for viewing throughout the meeting, and collectively covered a wide range of topics associated with the subject matter covered by each of the symposia. Accounts of most of these are found in this volume.

The meeting commenced on Sunday, 10th July with a Plenary lecture from Professor Gregg Semenza which was presented in the delightful Art Gallery of Hamilton. Professor Semenza captivated his audience with a presentation entitled "Regulation of Oxygen Homeostasis by Hypoxia-Inducible Factors". The talk covered an impressive range of fields of study, including developmental biology, cellular metabolism, acute oxygen sensing and proliferation, in each of which HIFs have been found to play vital roles. The presentation extended into mechanisms underlying major diseases such as cardiovascular disease and cancer. In all of these fields Professor Semenza's research group has made important contributions and it was a pleasure for all of those present to hear such an authoritative account of this work.

The rest of the meeting was held at the McMaster University Campus, from Monday 11th–Friday 15th July. The meeting was divided into five symposia, each of which commenced with a keynote lecture. The first symposium was chaired by Nurse and Gauda and was entitled "*Comparative and*

C. Peers (✉)
Division of Cardiovascular and Neuronal Remodelling, LIGHT, Faculty of Medicine and Health,
University of Leeds, Clarendon Way, Leeds LS2 9JT, UK
e-mail: c.s.peers@leeds.ac.uk

C.A. Nurse et al. (eds.), *Arterial Chemoreception: From Molecules to Systems*,
Advances in Experimental Medicine and Biology 758, DOI 10.1007/978-94-007-4584-1,

Developmental Aspects of Acute O_2 Sensing". It commenced with a Keynote lecture from Naweed Syed, who described intermittent hypoxia-induced plasticity of rhythmic central pattern generators in the pond snail *Lymnea stagnalis* (a useful model system, since its central nervous system is well defined). This presentation was followed by shorter talks from Nurse and Jonz who provided insight into the contrasts and similarities of O_2 sensing during development in water breathers and air breathers and highlighted the fact that fundamental similarities in O_2 sensing can be found across diverse species and environments. This was extended by the presentation of Buck and colleagues, who described the role of GABA in suppressing neuronal activity in the anoxia-tolerant turtle brain.

The symposium continued with a focus on developmental aspects of O_2 sensing. Shirahata and colleagues informed about further differences in protein expression profiles between different strains of mice, an issue which is important to bear in mind as transgenic models pertinent to O_2 sensing become increasingly available. Fournier and Kinkhead described how stress in fetal life can adversely affect respiratory stability and promote apnea in the newborn, whilst Bairam and co-workers used transgenic mice lacking progesterone receptors to demonstrate their involvement in developing chemoreflexes; both studies are clearly of immediate clinical significance.

Symposium 2, *"O_2 sensing in mammals"*, was chaired by Prabhakar and Peers and began with a Keynote lecture from Nino Ramirez focussed on the effects of hypoxia on the central respiratory network of the venterolateral medulla. The profound central effects of acute (sustained or intermittent) hypoxia were described, occurring in part via suppression of synaptic inhibition, whereas chronic intermittent hypoxia led to permanent circuit remodelling; as in earlier presentations the plasticity of the CNS in the face of hypoxia was most impressive. Prabhakar then presented the first of six short talks, concentrating on the contrasting effects of acute hypoxia in mice genetically engineered to express suppressed levels of either HIF1α or HIF2α, and how this is associated with altered expression of enzymes associated with cellular redox status. Following this, Evans and colleagues presented two talks discussing the role of AMP-activated protein kinase (AMPK) in acute O_2 sensing by the carotid body. This is a 'work in progress', but evidence is steadily accumulating to suggest that AMPK may well be necessary and sufficient for hypoxic chemotransduction in the carotid body. It is of interest that, to date, no evidence has emerged to counter Evans' hypothesis which, as these talks indicated, has been examined at multiple levels, from ion channel regulation in isolated type I cells, to intact organ and ventilatory responses.

Ruangkittisakul and Ballanyi returned the focus back to the central nervous system, and how bursting activity of inspiratory preBotzinger complex neurones was silenced by cyanide when extracellular K^+ levels were low (physiological). This did not involve a rise of intracellular Ca^{2+} levels. However, intriguingly, a rise of intracellular Ca^{2+} was noted in adjacent astrocytes, suggesting they might have a potential role to play in central oxygen sensing, in addition to CO_2 / pH sensing. Responses to low pO_2 stimuli will be of great future interest.

Symposium 2 concluded with two contrasting presentations concerning the role of hydrogen sulphide (H_2S) in acute O_2 sensing by the carotid body. In the first, Prabhakar and co-workers demonstrated that mice lacking cystathionine-γ-lyase (CSE), an enzyme required for H_2S formation, had very poor carotid body and ventilatory responses to acute hypoxia, and that hypoxia stimulated H_2S formation, probably due to suppression of heme oxygenase-2 activity which in normoxia inhibits CSE. Pharmacological inhibition of H_2S formation also suppressed the secretory responses of isolated type I cells to hypoxia. In stark contrast, Buckler indicated that, in young rat type I cells, such pharmacological inhibition of H_2S formation was without effect on the rises of intracellular Ca^{2+} levels triggered by hypoxia. Buckler described evidence to suggest H_2S essentially mimics the effects of hypoxia by inhibiting mitochondrial respiration. These findings are difficult to reconcile with Prabhakar's observations in $CSE^{-/-}$ mice and this led to much excited debate at the symposium. Further reports of the importance of H_2S to acute O_2 sensing are eagerly awaited.

On the second full day of the meeting, Symposium 3 (chaired by Kumar and Buckler), entitled *"Polymodal Chemosensing: Central and Peripheral Mechanisms"* began with a Keynote Lecture from

Patrice Guyenet which was an excellent, authoritative guide through central respiratory chemoreceptors (particularly the retrotrapezoid nucleus) and their integration with peripheral chemoreceptors. Whilst many questions in this field remain, the audience was particularly impressed with the advances presented despite the significant technical difficulties associated with such research. Kumar then focussed on polymodal sensing by the carotid body, shedding new light on its role as a glucose sensor (an issue which has proved to be of much interest since the last ISAC meeting). He showed that prior exposure to hypoxia (or indeed hyperoxia) reduced the delay in the response to glucose removal. How this "priming" of glucose sensitivity arises awaits further investigation. The following presentation from Joyner's group extended this theme by providing evidence in healthy humans that the carotid bodies play an important role in the regulation of blood pressure in the face of hypoglycaemia. Zapata and colleagues then reminded us of the complex integration of the carotid body into overall control of ventilation: they employed the Dejours' test of replacing air with pure O_2 to determine the previous influence of chemoreceptors under normoxic conditions, and demonstrated that the hyperoxic ventilatory depression was proportional to the degree of prior hypoxia. However, responses did not lead to expected changes in depth and timing of breaths for reasons which are unknown but emphasise the complexity of breathing control.

Jonz's group then returned the audience to the sensory properties of fish chemoreceptors, discussed in the first symposium. This time, CO_2 sensing was described in zebrafish gill neuroepithelial cells. Whilst parallels could be drawn with hypercapnic sensing in mammalian carotid body cells, differences were also noted. In particular, rises of intracellular Ca^{2+} levels evoked by hypercapnic acidosis in these cells was not fully dependent on Ca^{2+} influx, suggesting a significant degree of Ca^{2+} release from internal stores. The symposium progressed with presentations from the two groups within the Society who focus on mammalian pulmonary neuroepithelial bodies (NEBs). Cutz and colleagues presented work which added to their substantial body of evidence that NEBs act as airway O_2 sensors, by demonstrating that they are also sensitive to hypercapnia – NEB cells released the neurotransmitter 5-HT in response to elevated CO_2 in a manner which depended on carbonic anhydrase activity. In contrast to this study, Adriaensen and co-workers, who questioned the O_2– sensing properties of NEBs at the previous ISAC meeting in Valladolid, used hyperosmotic solutions as a challenge to NEB cells in order to ascribe mechanosensory properties to these enigmatic cells: hyperosmotic solutions evoked ATP release from NEB cells, triggered by Ca^{2+} influx which appeared to arise due to activation of the canonical transient receptor potential channel, TRPC5. There is clearly still contention within the NEB field which will provide further scientific intrigue and debate at future meetings.

The diversity of the effects of erythropoietin (Epo) on central and peripheral chemoreception and ventilation was described by Soliz, who used transgenic mice over-expressing Epo either centrally or both centrally and peripherally to demonstrate that hypoxic responses at both sites could be modified by Epo. Furthermore, Epo also appears to work at both sites to suppress hypercapnic responses, indicating a more diverse influential role of this hormone. Gauda and colleagues then presented work to indicate that peripheral chemoreceptor drive influenced central control of anoxia-induced gasping and suggested this was influenced by altered expression of brain derived neurotrophic factor and / or altered glutamate receptor expression. Zachar and Jonz concluded the symposium by returning our attention to gill neuroepithelial bodies, but this time in the anoxia-tolerant goldfish. These studies were preliminary but worthy of the audience's consideration, and it was of interest to hear that at least some of the properties of mammalian chemoreceptor cells could be found in this interesting model. Future work will hopefully develop this interesting project further.

Shirahata and Bairam co-chaired Symposium 4 *"Neurotransmitters and Neuromodulators in Chemoreception"*, on the penultimate day of the meeting. Our host, Professor Nurse, commenced by introducing the plenary speaker, Ellis Cooper who discussed the regulation by ROS of nicotinic acetylcholine receptors (nAChRs) in sympathetic neurones. This presentation was a remarkable tour-de-force, taking the audience from the molecular basis of ROS-mediated regulation of nAChRs to its relevance in ganglionic transmission and hence autonomic function in the intact animal.

The importance of such regulation in ROS-associated diseases such as diabetes was emphasised, much detail of which was directly relevant to ongoing chemoreceptor research. Shirahata then discussed the diversity of expression of both nAChR subunits and muscarinic receptors amongst different species commonly employed in chemoreceptor research. Again, Dr Shirahata indicated that expression also varied between different strains of mice, an issue which must be taken into account in future studies employing transgenic models. Bairam then discussed excellent data to indicate that nAChRs (along with those for adenosine, but not purinergic receptors) played an important role in the postnatal maturation of carotid body function and its influence on development of the hypoxic ventilatory response.

Salman and colleagues provided insight into the molecular basis of nicotine-induced blunting of hypoxic secretory responses, employing immortalised neonatal chromaffin cells. Data indicate that nicotine up-regulated HIF2α which in turn controls the expression of a subunit of a key K^+ channel (the K_{ATP} channel) whose activity leads to blunting of the hypoxic response. These findings may be of tremendous importance in understanding the link between maternal smoking and sudden infant death syndrome. Fitzgerald returned the audience to the topic of H_2S and its role in chemotransduction, suggesting that it may act via opening of K_{ATP} channels in glomus cells of the cat carotid body. If true, this would be of great interest since these channels are not thought to be present in glomus cells of other species (e.g. rat). Conde presented new neurochemical results from the long-standing collaboration between Lisbon and Valladolid, in which ATP and adenosine levels were monitored during carotid body development. Evidence suggested differential release of each transmitter as a function of hypoxic stimulus intensity, the role of adenosine being disproportionally greater under mild hypoxia. Caroll and colleagues continued the reports of purinergic influences in the carotid body, indicating that ATP can inhibit the hypoxic response via P2Y12 receptors, and that this inhibition is more prominent in neonates. This latter finding suggests that the reduced inhibitory effects of ATP with age may contribute to postnatal maturation processes. Monteiro then presented further findings from the Lisbon/Valladolid collaboration in which they investigated the content, synthesis and acute hypoxia-evoked release of dopamine from rat carotid bodies, all of which were increased by chronic hypoxia and by dietary caffeine, the latter finding suggesting a sensitizing effect of adenosine at the carotid body, yet an opposing central effect on ventilation. Obeso continued the neurochemical focus on the carotid body, describing the presence and turnover of 5-HT under normoxic and hypoxic conditions, and suggesting that it may be a modulator of catecholamine release from type I cells.

Wilson and co-workers then described a novel role for thermosensitive TRPV1 channels to enhance the activity of the carotid sinus nerve when stimulated with an inhibitor of TASK K^+ channels, anandamide. TRPV1 was located in petrosal neurones, but not glomus cells, implicating a postsynaptic role for these receptors, perhaps as mediators of increased carotid sinus nerve during exercise, when body temperature and anandamide levels are elevated. Di Giulio and colleagues then provided some rare insights into the expression of proteins key to the overall oxygen-sensing properties of the carotid body in young and aged human samples. Access to such samples is an unusual opportunity, and these workers showed variable age-related changes in neuroglobin, HIF-1α, iNOS and VEGF which, when the functional roles of these proteins in development and ageing are understood using animal models, will provide important translational information. Nunes and collaborators next demonstrated that soluble adenylate cyclase is present and active in type I cells of the carotid body. Its role, however, appears less clear since, whilst it contributes significantly to cellular cAMP levels, it is unlikely to play an important role in hypercapnic chemoreception.

The symposium was rounded off with a presentation from O'Conner, O'Halloran and Jones who examined the paraganglia of the superior laryngeal nerve (SLN) as a chemoreceptor structure with cardiorespiratory influence. They demonstrated that SLN stimulation evoked hindlimb vasodilation, but that this was not mediated by adrenaline. The talk served in part to remind the audience that non-carotid body chemoreceptors, which are relatively under-studied, exert significant physiological influence on the cardiovascular system.

On the final day of the meeting, an innovative Symposium 5 *"Translational impact of O_2 sensing"*, co-chaired by Gonzalez and Cutz, commenced with a Keynote lecture from Jerome Dempsey who described in detail the interdependence of central and peripheral chemoreceptors and how this influenced sleep apnea. The audience were treated to an excellent and clear explanation of some complicated concepts and a comprehensive review of the involvement of peripheral chemoreceptors in the progression of apneas and periodicity in breathing. Following this lecture, six successive presentations were heard concerning various aspects of the effects of intermittent hypoxia (IH), a topic of major clinical importance and clearly of developing interest within the Society. Martin began by discussing the clinical implications of IH episodes in preterm infants, which can be life-threatening, and emphasised the differential influence of peripheral chemoreceptor input during maturation and in lung disease. Nanduri et al. then described effects of neonatal IH on breathing and blood pressure in rats, emphasizing the involvement of ROS and of DNA methylation, an epigenetic phenomenon which caused altered expression of antioxidant enzyme genes. Gonzalez and colleagues continued the theme by describing the chemoreceptor and sympathetic autonomic influence in IH-induced development of hypertension. An enhanced carotid chemosensitivity was clearly influential in this regard, and Iturriaga and co-workers provided evidence in the following presentation that this enhancement could arise from a more pronounced effect of acute hypoxia to inhibit TASK-like currents in type I cells following IH. Edge, Bradford and O'Halloran then demonstrated that chronic IH increased the frequency of occurrence of central apneas via a ROS-dependent mechanism, speculating that this may arise due to oxidative damage of medullary rhythm generating neurones. Of more immediate importance is the suggestion that antioxidants may be beneficial in sleep apnea. Del Rio et al. then suggested that inflammation may also be involved in the cardiorespiratory consequences of IH, since they showed that ibuprofen could reverse some of these deleterious effects.

Schultz then discussed the increased carotid body sensitivity to hypoxia observed in congestive heart failure. He proposed that this adaptive response, dependent in part on oxidative stress with numerous other striking neurochemical changes, was attributable to reduced blood flow through the carotid body which could act at least in part by altering tonic carotid body endothelial shear stress. Strohl et al. provided a genetic analysis of two strains of mice, only one of which suffers spontaneous apneas following recovery from acute hypoxic exposures. Their analysis identified a region on chromosome 1 which confers post hypoxic spontaneous apneas. Pokorski and colleagues provided the final presentation of the meeting, describing the peripheral effects of mangiferin (an antioxidant found in mangos which can also chelate iron). A single bolus injection of this compound led to significant depression of the peak ventilatory reponse to acute hypoxia (presumably via a carotid body-targeted effect) suggesting possible involvement of oxidative signalling in oxygen sensing.

The ISAC business meeting concluded the academic aspect of the meeting. As is tradition, the location of the next meeting was discussed and it was agreed to be held in Yorkshire, UK in the summer of 2014. As co-host alongside Dr Prem Kumar, I am pleased to invite all present and future members to join this Society meeting. We hope to maintain the high standards of science and hospitality which make these meetings both important and highly pleasurable.

Index

C.A. Nurse et al. (eds.), *Arterial Chemoreception: From Molecules to Systems*, Advances in Experimental Medicine and Biology 758, DOI 10.1007/978-94-007-4584-1,

Printed by Publishers Graphics Canada Inc
CAMO20121026.19.32.18